MATHEMATICAL ANALYSIS FOR ECONOMISTS

BY

R. G. D. ALLEN

The general science of mathematics is concerned with the investigation of patterns of connectedness, in abstraction from the particular relata and the particular modes of connection.

ALFRED NORTH WHITEHEAD, *Adventures of Ideas*

To connect elements in laws according to some logical or mathematical pattern is the ultimate ideal of science.

MORRIS R. COHEN, *Reason and Nature*

MACMILLAN AND CO., LIMITED
ST. MARTIN'S STREET, LONDON
1938

PRINTED IN GREAT BRITAIN

FOREWORD

THIS book, which is based on a series of lectures given at the London School of Economics annually since 1931, aims at providing a course of pure mathematics developed in the directions most useful to students of economics. At each stage the mathematical methods described are used in the elucidation of problems of economic theory. Illustrative examples are added to all chapters and it is hoped that the reader, in solving them, will become familiar with the mathematical tools and with their applications to concrete economic problems. The method of treatment rules out any attempt at a systematic development of mathematical economic theory but the essentials of such a theory are to be found either in the text or in the examples.

I hope that the book will be useful to readers of different types. The earlier chapters are intended primarily for the student with no mathematical equipment other than that obtained, possibly many years ago, from a matriculation course. Such a student may need to accustom himself to the application of the elementary methods before proceeding to the more powerful processes described in the later chapters. The more advanced reader may use the early sections for purposes of revision and pass on quickly to the later work. The experienced mathematical economist may find the book as a whole of service for reference and discover new points in some of the chapters.

I have received helpful advice and criticism from many mathematicians and economists. I am particularly indebted to Professor A. L. Bowley and to Dr. J. Marschak and the book includes numerous modifications made as a result of their suggestions on reading the original manuscript. I am also indebted to Mr. G. J. Nash who has read the proofs and has detected a number of slips in my construction of the examples.

R. G. D. ALLEN

THE LONDON SCHOOL OF ECONOMICS
October, 1937

CONTENTS

A SHORT BIBLIOGRAPHY

It is essential to keep in mind, at all times, what mathematical analysis is about and the way in which it has developed over the centuries. Not only beginners, but also accomplished mathematicians, can profit by reading such short introductions as

Whitehead: *An Introduction to Mathematics* (Home University Library, 1911).

Brodetsky: *The Meaning of Mathematics* (Benn's Sixpenny Library, 1929).

Rice: *The Nature of Mathematics* (An Outline of Modern Knowledge, 1931);

and longer accounts such as

Dantzig: *Number* (1930).

Forsyth: *Mathematics in Life and Thought* (1929).

Hogben: *Mathematics for the Million* (1936).

The history of mathematics has been the subject of many books: good short treatments are

Rouse Ball: *A Short Account of the History of Mathematics* (6th Ed. 1915).

Sullivan: *The History of Mathematics in Europe* (1925).

The reader wishing to revise his knowledge of " matriculation " algebra and geometry and to obtain a " practical " introduction to the calculus can consult

Durell: *Elementary Geometry* (1919).

Durell, Palmer and Wright: *Elementary Algebra* (1925) (including an introduction to the calculus).

Thompson: *Calculus made Easy* (2nd Ed., 1919).

Irving Fisher: *A Brief Introduction to the Infinitesimal Calculus* (3rd Ed., 1909).

General text-books on mathematical analysis vary a good deal in their scope and in their treatment. The following selection includes works ranging from the elementary to the advanced in content and from the practical to the rigidly theoretical in method. They are to be regarded as complementary rather than competitive commodities.

Hardy : *A Course of Pure Mathematics* (3rd Ed., 1921).

Griffin : *An Introduction to Mathematical Analysis* (no date).

Osgood : *Introduction to the Calculus* (1922) ; and *Advanced Calculus* (1925).

de la Vallée Poussin : *Cours d'Analyse Infinitésimale* (5th Ed.), Vol. I (1923), Vol. II (1925).

Courant : *Differential and Integral Calculus* (English Ed.), Vol. I (1934), Vol. II (1936).

Pavate and Bhagwat : *The Elements of Calculus* (2nd Ed., 1932).

THE USE OF GREEK LETTERS IN MATHEMATICAL ANALYSIS

THE English alphabet provides insufficient material for the notation of mathematical analysis and greater range and flexibility are acquired by using letters from the Greek alphabet. There are no fixed rules governing the mathematical use of Greek letters but the following table indicates a practice which is fairly general. It should be noticed that the English letters given are those which correspond to the Greek letters in mathematical, not literal, usage.

Greek		English	General usage
α	alpha	a	
β	bēta	b	constants
γ	gamma	c	
κ	kappa	k	
λ	lambda	l	constants, parameters
μ	mu	m	
ν	nu	n	
ξ	xi	x	
η	ēta	y	variables
ζ	zēta	z	
π	pi	p	special constants or variables (e.g. π as the constant ratio of the circumference to the diameter of a circle and ρ as a rate of interest)
ρ	rho	r	
σ	sigma	s	
τ	tau	t	
ϕ	phi	f	
Φ	cap. phi	F	functional operators
ψ	psi	g	
Ψ	cap. psi	G	
δ	delta	d	operators indicating increments in variables
Δ	cap. delta	D	
Σ	cap. sigma	S	summation sign
ϵ	epsilon	–	a small positive constant
θ	thēta	–	a positive fraction

In trigonometry, α, β, γ can denote constant angles and θ, ϕ, ψ variable angles.

CHAPTER I

NUMBERS AND VARIABLES

1.1 Introduction.

IT is conventional to divide mathematics into two separate studies, geometry and analysis. In geometry, the study of space and of spatial relationships, we investigate the nature and properties of curves, surfaces and other configurations of points in space. Mathematical analysis, including arithmetic and algebra, deals with numbers, the relationships between numbers and the operations performed on these relationships. The distinction between the two studies lies in the difference between the basic " raw materials ", the spatial points of geometry and the numbers of analysis.

As the mathematical technique is developed, however, we find that the distinction between geometry and analysis becomes less clear and less relevant. The *intended applications* remain very different but the *methods* are seen to be abstract and essentially similar in nature. Mathematics involves the definition of symbols of various kinds and describes the operations to be performed, in definite and consistent ways, upon the symbols. The distinction between geometrical and analytical symbolism is more or less arbitrary. Further, we shall see that it is a simple matter to devise a method of connecting points of space and numbers, geometry and analysis. That such a connection is possible is seen by the use of graphical methods in algebra and by the fact that elementary trigonometry is an application of algebraic methods to a study of spatial configurations.

The mathematical technique is abstract and must be developed apart from its applications. It is also essentially logical in character. Elementary geometry, for example, is largely an exercise in formal logic, the deduction of the consequences of consistently framed assumptions. The development of mathematical methods, however, soon necessitates the introduction of a concept foreign to formal logic, the concept of "infinity", the infinitely large and the indefinitely continuous. It can be said, speaking broadly, that the methods of mathematics form a branch of formal logic extended in a particular direction to include the infinite as well as the finite. The only requirements of mathematics, additional to those of logical reasoning, are connected with the introduction of the infinite.

The popular belief that logical and mathematical methods are in conflict has, therefore, no foundation. Mathematics, as a symbolised and extended form of logic, can only be regarded as an alternative to logic in a special sense. Some problems are adequately treated by relatively simple logical reasoning and the introduction of mathematical symbolism is then only destructive of clarity. But formal logic, with its limited range, is clearly insufficient for the elucidation of a vast number of problems and this provides the case for the development and application of mathematics.

Though an abstract development, mathematics is not just a fascinating game of chess played in n-dimensional space with pieces of fantastic design and according to rules laid down in an arbitrary way. The methods of mathematics are designed primarily for actual or prospective application in the field of science, in the interpretation of phenomena as observed, abstracted and classified. The nature of the application must not, however, be exaggerated. The patterns of scientific phenomena, the laws obeyed and the uniformities displayed, are not provided by mathematics. They are assumed into the system and mathematics can only express and interpret them, help us to deduce their consequences, or to forecast what will happen if they hold, and tell us where to look for verification or contradiction of our hypotheses.

The methods of mathematics apply as soon as spatial or numerical attributes are associated with our phenomena, as soon as objects can be located by points in space and events described by properties capable of indication or measurement by numbers. The main object of the following development is the description of certain mathematical methods. At most points, however, the nature and field of application of the various methods will be noticed and discussed, the applicability of mathematical methods in economic theory receiving special consideration.

In concluding these introductory remarks, we can note that the fundamental bases of the mathematical technique have received much attention in recent years. It is indeed a fascinating set of questions that are posed in what can be described, for want of a better phrase, as mathematical philosophy. The nature of some of these questions can be observed at various stages in the following treatment, but the temptation to pursue them must be resisted. The further we probe into the fundamentals of mathematics, the more shaky does the whole familiar structure appear. This need not worry us unduly ; the crazy system works as no other system can.

1.2 Numbers of various types.

Numbers are usually taken as self-evident or undefinable entities which do not require examination. It is easily seen, however, that the number system is far from simple and that numbers of very different kinds are freely used in arithmetic and algebra. It is convenient, therefore, to spend a little time sorting out the various types of numbers and showing how they have developed side by side with, and by reason of, the growth of mathematical analysis from the simplest notions of arithmetic. Some light is then thrown on the nature and applicability of mathematical analysis in its present highly developed form.

We all begin, in arithmetic, with some idea of the *integers,* the natural or whole numbers, as concepts intimately connected with the process of counting or enumeration. It is not generally recognised that counting, and hence the integers, involves two quite distinct notions, the ordinal and the cardinal notions. The fundamental property of integers is that they can be written down in succession without end:

$$1, 2, 3, 4, 5, \ldots .$$

The integers can thus be used to indicate any order or sequence and this is one of the objects of counting. We have here the *ordinal* aspect of counting and of the integers. But the integers do more than indicate order, for we can speak of four men, of four hats and of four sticks without any notion of order being necessarily involved. We simply mean that we have a collection of men, a collection of hats and a collection of sticks, and that the collections have something in common expressed by the number four. There is, in fact, a complete correspondence between the members of the three collections; each man, for example, may be wearing a hat and carrying a stick. This is the *cardinal* aspect of counting and of the integers, an aspect which enables us to say how many men, hats or sticks there are and which will later provide us with the basis of measurement.

Developing the arithmetic of integers, we define the processes of addition and multiplication, the sum and product of two or more integers being themselves integers. Our first difficulty arises when we consider the converse process to multiplication. The division of one integer by another does not, in the vast majority of cases, produce an integer. We can only say, for example, that "three into seven won't go exactly" or that we "put down two and carry one". In order to introduce uniformity, the number system is extended by defining the *fraction* as a new type of number. The wider set of numbers, integers and fractions, is termed the system of *rational numbers*. Like the integers, the rational numbers can be written down in a limitless sequence of ascending magnitude, the order being indicated most clearly in the decimal method of writing the numbers.* The ordered rational numbers, however, display a property not shown by the ordered integers. The sequence of rationals is not only of limitless extent but also of limitless "density". As many fractions as we like can be written down to lie between any two numbers of the order, between $\frac{1}{2}$ and 1 for example. The processes of addition, multiplication and division can be extended to apply to the whole system of rational numbers.

Even with the rational numbers, however, it is found that the processes of arithmetic still lack uniformity. Consider the process of extracting the square root of a given number. The square roots of a few numbers, such as 36, 169 and $6\frac{1}{4}$, are found at once as rational numbers. But, if rational numbers are alone admitted, we can only say that the square roots of most numbers, e.g. 2, 3 and 5, do not exist and we have again reached the unsatisfactory situation in which we have to own that a simple arithmetic process breaks down. For uniformity, therefore, a further type of number is introduced, the *irrational number*. Irrational numbers include the square

* A certain amount of duplication must be eliminated by writing $\frac{2}{4}, \frac{3}{6}, \frac{4}{8}, \ldots$, for example, in the simplest equivalent form $\frac{1}{2}$.

roots of numbers which are not "perfect squares" and also the solutions of many equations and such numbers as those denoted by π and e which play extremely important parts in mathematical analysis. The definition and nature of irrational numbers will not be discussed here but it can be assumed that they fit into their appropriate places in the order of numbers according to magnitude.* Further, with some difficulty, we find it possible to extend the arithmetical processes to apply to irrational as well as to rational numbers.

The next extension of the number system is useful in arithmetic but only becomes vital when algebra is developed. We have not yet referred to subtraction, the process converse to addition. When a smaller (rational or irrational) number is taken from a larger one, the result is an ordinary number of the same kind. This is not true when the numbers are reversed. In arithmetic it is desirable, and in algebra essential, that the difference between *any* two numbers should be a number of the same system. This uniformity can only be achieved by "doubling" the number system so far described by the distinction between *positive and negative numbers* and by adding a new number *zero* indicating "nothing". The difference between any two numbers is now a number (positive, negative or zero) and subtraction is a uniform process. By the adoption of further conventions, all arithmetic processes can be made to apply to positive and negative numbers and (with the exception of division) to the number zero.

An important practical fact can be noticed in passing. The last extension of the number system makes possible a great simplification of algebraic work. In algebra we deal frequently with equations such as the following:

$$x^2+2x+1=0\,;\quad x^2=2x+3\,;\quad 3x=2x^2+1\,;\quad x^2=5.$$

We recognise these as examples of the same "form" of equation and (with the aid of zero and negative numbers) we can include all of them in one uniform expression. We write the equations:

$$x^2+2x+1=0\,;\quad x^2+(-2)x+(-3)=0\,;\quad 2x^2+(-3)x+1=0\,;$$
$$x^2+(0)x+(-5)=0.$$

* See 4.1 (footnote) below.

In the symbolism of algebra all equations of these kinds can be included in the general quadratic form :

$$ax^2 + bx + c = 0$$

where a, b and c stand for some definite numbers (positive, negative or zero). The use of symbolic letters, such as the a, b and c in the quadratic equation above, will be described more fully at a later stage.

1.3 The real number system.

The number system as now extended consists of the neutral number zero and of all positive and negative numbers, rational or irrational. This is the system of *real numbers*, as commonly used in algebra and mathematical analysis. The rational and irrational numbers can be arranged in an order of limitless extent and density. The same is true of the real numbers except that the order is of limitless extent in two directions instead of one. All negative numbers appear before the positive numbers in the order of magnitude (with zero separating them) and we can proceed through larger and larger negative numbers as well as through larger and larger positive numbers. The real number system has also another property, i.e. the property that there are no " gaps " left in the order of the numbers. There are " gaps " in the order of the rationals but these are completely filled by the insertion of irrational numbers.*

The familiar rules of arithmetic and algebra apply, therefore, to a system of real numbers with the properties :

(1) The numbers can be arranged in a definite *order*, the order of magnitude, of greater and less.

(2) The number order is of limitless extent in both directions, i.e. the numbers form a *doubly infinite system.*

(3) The number order is of limitless density and without " gaps ", i.e. the numbers form a perfectly *continuous* system.

The development of the real number system from the basic idea of a sequence of integers marks the end of a certain line of evolution and provides a convenient stopping place. One guiding principle stands out in what we have said. The introduction of each new number type was designed to impose uniformity where uniformity

* This is a statement we cannot justify here. It is based on the work of Dedekind and Cantor ; see Hardy, *Pure Mathematics* (3rd Ed., 1921), pp. 1-31.

was previously lacking and, at the same time, to preserve all the essential rules of arithmetic and algebra. This is a most important principle which will be found at work throughout the development of mathematical analysis, the constant striving after generality, uniformity and simplicity. Herein lies much of the fascination of the study of mathematics.

Even with the real number system, however, uniformity is not completely attained and, sooner or later, analysis reaches a stage where new extensions of the number system are required. The process of division, for example, still fails in one case since a real number divided by zero does not give a real number. We must turn a more or less blind eye to this deficiency though it will be partly side-tracked later when the idea of a " limit " is introduced. A more important lack of uniformity occurs when the solution of quadratic equations is considered in algebra (see 2.8 below). If real numbers are alone admitted, some quadratic equations have two solutions (distinct or coincident) and others have no solution at all. It is clearly desirable, from the point of view of uniformity, to be able to say that all quadratic equations have two solutions. This uniformity is, in fact, achieved by introducing what are termed " complex " or " imaginary " numbers, a step which opens up entirely new fields of mathematical analysis. But we content ourselves here with the real number system which is sufficient for most of the mathematical methods with which we are concerned.

Finally, the fact that the real number system is of limitless extent and density makes it only too easy to speak of infinitely large numbers and of numbers whose differences are infinitely small. Right at the beginning of our development, therefore, we find ourselves approaching the difficult country of the " infinite ", of the infinitely large and the indefinitely small and continuous. A more determined expedition into this dangerous country must be left until a somewhat later stage ; it is sufficient here to remark on the need for such an expedition.

1.4 Continuous and discontinuous variables.

The generalisation of arithmetic into algebra and of algebra into modern analysis is largely dependent on the extension of the symbolism. The number system remains the basic element but its uses are made more flexible by the symbolism. The first and most obvious distinction between algebra and analysis on the one hand and arithmetic on the other lies in the use of "variable" numbers, denoted by symbolic letters, instead of particular numbers.

A *variable number* is any number, an unspecified number, from a certain given set of real numbers and it is always symbolised by a letter such as x, y or t. Particular numbers for which the variable can stand are called the *values* of the variable, and the whole set of possible values makes up the *range* of the variable. The use of a variable number necessarily implies a range of variation, a set of numbers from which the values of the variable can be selected at will. Many variables have a range including all real numbers and, in this case, no explicit reference to it need be made. Other variables have more restricted ranges, e.g. the range of positive numbers, of numbers between zero and unity or of integers only. Here the range should be clearly indicated in the definition of the variable.

One particular case merits separate notice. If a variable x can take, as its values, all real numbers lying between two given numbers a and b, then its range is called the *interval* (a, b). We write $a<x<b$ or $a \leqslant x \leqslant b$ according as the values a and b themselves are excluded or included. For example, $0<x<1$ or $0 \leqslant x \leqslant 1$ indicate that the variable x has a range consisting of the interval of positive proper fractions, the numbers zero and unity being excluded in one case and included in the other.

The notion of an interval provides a concept which later proves of great convenience. We often make statements about the values of a variable x "near" to, or in the "neighbourhood" of, a particular value a. The statements are made in this vague way but their meaning can be made precise and should be borne in mind. A *neighbourhood* of the value $x=a$ is defined as an interval of values of x having the value a as its middle point. In symbols, if k denotes a given positive number, a neighbourhood of $x=a$ is defined by the interval $(a-k)<x<(a+k)$. Here, $(x-a)$ is numerically less than k and the total "length" of the interval is $2k$. Though k is, in general, any positive number we care to fix, it is usually convenient to take k as small, i.e. to take a small neighbourhood of $x=a$.

An important distinction can now be made. A variable is *continuous* if its range is either the whole set of real numbers or any interval of the set. The adjective " continuous " here is of exactly the same connotation as the same adjective applied to the number system itself; the values of a continuous variable can be ordered so that they are indefinitely dense and without gaps. It is often convenient, for example, to think of a continuous variable as taking values successively in increasing order of magnitude and we describe this by saying that " the variable increases continuously in value over such and such an interval ". Notice, however, that any changes in the value of a variable are essentially timeless; the values of the variable make up a range of variation which must be considered as a whole, though it may be convenient to pick out or arrange the values in any definite way we like.

A *discontinuous* variable, on the other hand, has a range which is neither the whole set of real numbers nor any interval of the set. The values of such a variable cannot be arranged in order of magnitude without gaps. Any set of numbers of a particular type, e.g. the set of integers or of multiples of $\frac{1}{2}$, provides a range of variation which is discontinuous. For example, if the price of a commodity is x pence quoted only in halfpennies, then x is a discontinuous variable taking values which are positive multiples of $\frac{1}{2}$.

1.5 Quantities and their measurement.

It is convenient, at this stage, to consider the use of our number system in the interpretation of scientific phenomena. One use springs to the mind at once, the use in the enumeration of discrete physical objects or events. In all such uses, whether we are counting (e.g.) men, road accidents or sums of money, the only numbers required are the integers themselves. The " unit " of counting is here the number one corresponding to a single object or occurrence. For convenience, however, larger " units " can be adopted; we can count men or accidents in hundreds or thousands instead of in ones, sums of money in pounds, shillings and pence instead of in the smallest coin (e.g. halfpennies). The result of enumeration then appears as a fraction or a decimal instead of as a simple integer. Thus we can say that the number of road accidents is 53·7 thousands instead of specifying the number as the integer 53,700. Or we can say that a sum of money is £1 11s. 4½d. or £$1\frac{273}{480}$ instead of specifying that it contains 753 halfpennies. Hence, even in the use of numbers in enumeration, the range of the numbers employed can be quite extensive and the question of the " unit of measurement " arises as it does in more complicated cases.

Enumeration is not sufficient for the description of scientific phenomena. We derive, from our observations, what can only be called abstracted properties—shapes, colours, temperatures, pitches of notes, lengths, masses, time-intervals, and so on—and it is essential to connect these properties with numbers if possible. The two aspects of number, the ordinal and the cardinal, now become important. The ordinal aspect of number applies at once if the property considered is capable of arrangement in some order, and this is the essential prerequisite for any association of numbers with the property. Some of the properties abstracted from scientific phenomena (e.g. temperatures and lengths) can be presented in order according to a criterion of greater or less, higher or lower or whatever it may be; other properties (e.g. shapes and colours) cannot be ordered in this way.* Hence we arrive at our first distinction. Properties not capable of arrangement in order can be termed *qualities* and numbers are not directly applicable to them. Properties to which a natural order is attached can be termed *magnitudes* and numbers can be associated with them at once. The number attached to a magnitude simply serves to *indicate* its position in its appropriate order and the association is loose and by no means unique. The association is, however, of considerable importance in the scientific field, the Centigrade and Fahrenheit readings of temperature providing a good example, and more will be said about it later on.

A second distinction can be drawn amongst magnitudes according as the cardinal aspect of number can or cannot be applied in addition to the ordinal aspect. If a magnitude displays the *additive* property that any two of its values can be taken and added together to give a third value of the same magnitude, then numbers can be used in their cardinal aspect and the processes of arithmetic apply. Magnitudes with the additive property can be called directly measurable *quantities* and we shall show how numbers can measure, and not only

* It may be possible, in the construction of physical theories, to analyse a quality such as colour and to connect it indirectly with numbers (e.g. by means of "wave lengths"). This merely indicates that our distinctions are not hard and fast—as, of course, they cannot be.

order, quantities. Lengths, masses, time-intervals and electric charges are examples of quantities; temperatures and pitches of notes are instances of magnitudes which are not quantities.

In order to fix ideas, let us consider the quantity length; what follows applies, with suitable modifications, to other quantities such as mass or time-interval. The additive property of length makes it possible to construct a *scale* for measuring length. A definite length is chosen as standard and a number of such lengths are placed end to end to make the scale. This is possible since we know what is meant by equal lengths and since we can add lengths together to get other lengths. The standard length, definable in any convenient way, will be assumed here to be that length known as an inch. Any given length can be compared now with the scale of inches and defined as covering so many scale intervals, so many inches. All that is required is the process of counting and we get lengths of 2 inches, 10 inches, 143 inches, and so on. In other words, any given length is specified or measured by a number, such as 2, 10 or 143, when compared with the standard inch scale.

A difficulty now presents itself. In almost all cases, the scale reading of any given length must be taken to the nearest number of inches. As a result, two lengths may have the same measure though direct comparison shows them to be unequal; or the sum of two or more lengths may give a scale reading which is not the sum of the separate scale readings. The fundamental ordered and additive properties of length appear to have been lost in the process of measurement, a state of affairs which cannot be tolerated. The measurement of length must be *made* to preserve the fundamental properties.

The first step is to admit fractional as well as integral numbers in our measure of length. We can subdivide the scale unit by taking (e.g.) tenths or hundredths of an inch and then compare the given length with the subdivided scale. Alternatively, we can apply a given length repeatedly to the fixed inch scale until the multiple length coincides with a scale reading. Thus, the length of $2\frac{3}{4}$ or 2·75 inches is such that it covers 275 of the hundredth inch units, or such that a hundred such lengths together cover 275 of the inch units.

In practice, this process will suffice since there must be some physical limit to the comparisons that can be made between a given length and any actual inch scale. But the measurement of lengths by means of rational numbers fails in our theoretical requirements of length. Pythagoras' well-known theorem, for example, requires that the diagonal of a square of one inch side should be represented by $\sqrt{2}$ inches, and $\sqrt{2}$ is not a rational number. We must, therefore, conceive of lengths measured by irrational numbers, even though actual measurement with a scale of inches does not support us.

What has happened now is that we have associated the concept of the infinitely small and continuous with the observable property of length and, like the abstract number system itself, *we assume that length is continuously variable.* In other words, a new and abstract concept of length is introduced; we assume that there is a "true" length which is continuous but incapable of actual measurement. This assumption of continuity, which must be recognised as something outside the sphere of actual observation, is imposed upon length for purposes of mathematical convenience, in order that the measure of length should display the additive property and obey the rules of arithmetic exactly as we desire. The assumption of continuity into the phenomena we study is of frequent occurrence and the present instance is only the first of many. The only difference between this and some later instances is that we have grown used to the assumption in this case, but not in others, so that we think it "reasonable" here but not there.

In the description of scientific phenomena, therefore, there appear certain directly measurable quantities. Only three of these quantities are required in mechanics, length for the description of space, mass for the description of weight or inertia, and time-interval for the description of temporal changes. Other fundamental quantities must be added as the range of physics is widened. In electricity, for example, we need the new fundamental quantity of electric charge to describe the phenomena then included. Having selected a standard scale unit, we can express each quantity as a variable number. The recognised units of theoretical physics for measuring length, mass and time are the centimetre, gramme and second respectively. Other units are possible and are often used in more mundane descriptions; the units of a foot and a pound can replace, for example, the standard centimetre and gramme units.

Finally, the property of continuity is imposed upon the measure of each of the fundamental quantities. When we speak of a length of x centimetres, a mass of y grammes or a time-interval of t seconds, the numbers x, y and t are assumed to be continuous variables subject to the rules of arithmetic and algebra. This seems reasonable enough, but the remarks above serve to indicate, though inadequately, the serious abstractions and assumptions that lie behind bald statements about the measurement of fundamental quantities.

1.6 Units of measurement.

The measurement of a quantity, as defined above, is only unique if the standard scale unit is given. The choice of the standard unit of each quantity is open and can be made in a large number of ways. As we have noted, the scale unit of length adopted in theoretical work is almost invariably the centimetre. But inches, feet, yards, miles, millimetres, kilometres and rods or poles provide other recognised examples of length units. It appears, therefore, that there is a large element of arbitrariness in the measure of any quantity and the question arises whether there is any real distinction between quantities, with measures arbitrarily dependent on the choice of scale unit, and magnitudes in general, indicated by numbers without any pretence at uniqueness. It will be shown that the distinction is real, that the arbitrary element in the measure of a quantity is much more limited and easily dealt with than in the indication of non-measurable magnitudes.

There exists a very simple rule connecting the measures of a given quantity on two different scales. Let the second scale unit be measured and found to contain λ of the first scale units. If a quantity has measures x and y in the first and second scale units respectively, then we have $x=\lambda y$, a relation which holds for any one quantity whatever of the kind considered.

Hence, the numerical measure of a given quantity changes when the unit of measurement is changed, but the change in the measure is quite simple and follows the familiar rule of proportions. The proportion is fixed by λ, the measure of one unit in terms of the other. For example, a length can be expressed either as x feet or as $12x$ inches, whatever the value of x. The length measure changes according to the fixed proportionality rule independently of the length we happen to take. The connection between units may be expressed in fractional, or even irrational, form. One inch is measured as 2·54... on the centimetre scale and a length of x inches is thus $(2{\cdot}54\ldots)x$ centimetres.

The measures of a series of quantities of the same kind on one scale are exactly proportional to the measures on a second scale. For example, the same set of lengths can be measured by

$$6,\quad 12,\quad 18,\quad 24,\quad 30,\ \ldots \text{ inches};$$

$$\tfrac{1}{2},\quad 1,\quad \tfrac{3}{2},\quad 2,\quad \tfrac{5}{2},\ \ldots \text{ feet};$$

or

$$\tfrac{1}{6},\quad \tfrac{1}{3},\quad \tfrac{1}{2},\quad \tfrac{2}{3},\quad \tfrac{5}{6},\ \ldots \text{ yards}.$$

The limitation on alternative measures can be expressed by saying that *the ratio of the measures of two quantities of the same kind is independent of the units chosen.* The ratio of the measures of the first two lengths instanced above is always $\frac{1}{2}$; the first length is half the second, a statement independent of scales.

The arbitrary element in the indication of a magnitude which is not a directly measurable quantity is subject to no such limitation. A set of magnitudes of the same (non-measurable) kind arranged in ascending order can be indicated by any rising set of numbers whatever:

$$1,\quad 2,\quad 3,\quad 4,\quad 5,\ \ldots;$$

$$1,\quad 3,\quad 5,\quad 7,\quad 9,\ \ldots;$$

$$1,\quad 4,\quad 9,\quad 16,\quad 25,\ \ldots, \text{ and so on.}$$

There is no connection between the various numbers that can indicate any *one* magnitude; the question of changing units does not arise.

1.7 Derived quantities.

From the limited number of fundamental quantities necessary in the description of scientific phenomena can be obtained a large number of secondary or derived quantities. The two most frequent ways of defining derived quantities are by the multiplication and division of fundamental quantities of the same or different kinds. It is sufficient to quote, as examples of derived quantities defined as products, the cases of superficial *area* and of cubic *volume*. Area is obtained as the product of two, and volume as the product of three, length measures. Probably more important, and certainly more frequent, are the derived quantities obtained by dividing one fundamental quantity by another. The quotient of distance travelled by an object divided by the time of transit gives the derived quantity known as *speed*. If the mass of a body is divided by its volume, we obtain the derived quantity of *density*. The quantity of electric charge passing a given point in a wire divided by the time it takes to pass provides the derived quantity of *electric current*. Further, to quote a simple example from the economic sphere, a sum of money divided by the number of units of a commodity bought with the sum gives the *price* of the commodity.

From their definition it follows that derived quantities can be measured, but only in terms of two or more standard units. A scale unit for each fundamental quantity used in defining a derived quantity must be chosen before the latter can be expressed in numerical terms. This is made clear by the way in which derived quantities are commonly specified ; speeds are indicated by measures such as 30 miles per hour or 25 feet per second, densities by so many pounds per cubic inch and prices by so many pence per pound. The measures of derived quantities, like those of the fundamental quantities, are numbers which are taken as continuously variable.

It will be seen later that derived quantities include two distinct types, the " average " type and the " marginal " type. Average speed illustrates the first type ; if a train travels 45 miles in an hour and a half, then we say that its speed, on the average, is 30 miles per hour. The second type is illustrated by velocity or instantaneous speed ; if the velocity of a train is 30 miles per hour at any moment, this implies that the distance travelled in a very short time-interval from the given moment, divided by the time-interval, is approximately equivalent to 30 miles in an hour. The price of a commodity, as usually defined, is an average price. The corresponding marginal concept can be defined, however, and given the name " marginal revenue ", a concept of considerable importance in economic theory. Both corresponding types of derived quantities are measured in the same way and in terms of the same pair of scale units.

The various scales necessary for the measure of a derived quantity are at choice and, if any one unit is changed, then the numerical measure of the derived quantity is changed. But the change is governed by a rule of proportions similar to that already described.

Suppose, for example, that a derived quantity is obtained by the division of one quantity A by another quantity B. The units of measurement are changed so that one unit of the second scale of A equals λ_a units of the first scale, and one unit of the second scale of B equals λ_b units of the first scale. A given value of the derived quantity is measured by x on the first scales and by y on the second scales. Then we have

$$x = \frac{\lambda_a}{\lambda_b} y.$$

For example, the average speed of a train is 30 miles per hour. Now, 30 miles in an hour implies 30×5280 feet in an hour, and so $\frac{30 \times 5280}{3600}$ feet in a second. The speed is thus 44 feet per second. In general, a speed of x feet per second corresponds to y miles per hour if

$$x = \tfrac{5280}{3600} y = \tfrac{22}{15} y.$$

Again, if the average price of sugar is 3d. per pound, what is the price in shillings per cwt.? A cwt. or 112 pounds costs 336d. or 28 shillings. The required price is 28s. per cwt. In general, a price of xs. per cwt. corresponds to yd. per pound if

$$x = \tfrac{112}{12} y = \tfrac{28}{3} y.$$

So, one measure of a derived quantity is a constant multiple of an alternative measure obtained by changing the units of measurement. When the derived quantity is the quotient of two quantities, as often happens, the multiple is simply the ratio of the scale multiples of the separate quantities.

1.8 The location of points in space.

Our main purpose is to develop the methods of mathematical analysis. Any introduction of geometrical or spatial considerations, therefore, will be subsidiary to, and illustrative of, the analytical development. It will not be necessary to enlarge upon the fundamental geometrical problems concerned with space as an abstract conception and defined in any consistent way we please. We avoid difficulties by using only the space of our senses, taken as possessing the properties of being infinite, continuous and metrical. Primary sensations give us a conception of space and spatial configurations and tell us that we can assume that distance between points of space is measurable (in definite units) and continuously variable. Further, space can be regarded as of limitless extent in one, two or three dimensions as the case may be.

The geometrical considerations given here can, therefore, be illustrated by drawing actual diagrams in the plane of the paper (two dimensions) or by sketching plane representations of spatial models (three dimensions). It is possible, by giving up visual representations, to extend our geometry to spaces of more than three dimensions with the same properties as those given above. But this would require a precise re-definition of all the geometrical concepts which we are taking as self-evident, a task which it is scarcely appropriate to undertake here. Occasionally, however, we shall proceed by analogy and refer to n-dimensional "distances", "flats" and "hypersurfaces" as the analogues of three-dimensional distances, planes and surfaces. It must be remembered that such geometrical terms are introduced only as illustrative of the analytical methods.

The spatial properties we have assumed here are clearly analogous to the properties of the real number system. This analogy makes it possible to locate points in our metrical space by means of numbers, the method being derived from the way in which we measure distances in space and depending on the number of dimensions of the space.

Space of one dimension consists of a continuous and indefinitely extended straight line which we can denote by L. A practical method of fixing the position of a point on a line is to state the distance, in definite units, of the point from a given base point: the position of a train on a railway track, for example, can be defined in this way. We can easily express this method in a precise form suitable for the line representing our one-dimensional space. A convenient point O is selected on L and a definite scale of measurement, e.g. one unit $= \frac{1}{2}$ inch, is taken. Taking for convenience of wording the line L as "horizontal", a convention is adopted whereby all distances measured from O to the right are positive and all distances to the left negative. Then each point on the line L is located without ambiguity by a number, the number of units (e.g. $\frac{1}{2}$ inches) of the distance of the point from O. The number can be positive (if the point is to the right of O) or negative (if the point is to the left of O). So, the point located by the number $+3$ is 3 units of distance to the right of O and the point located by $-\frac{1}{2}$ is a half unit of distance to the left of O. Since the number system and the points on L are continuous, each point on L is represented by one definite number and each number corresponds to one definite point on L. The method depends entirely on the fixing of O and of a definite scale; if a different base and scale are chosen, an entirely different correspondence, still unique within itself, is set up.

The next step is to locate points in two dimensions, i.e. in a plane which can be visually represented by the plane of the paper. Experience again provides a practical indication of the appropriate method. A position in a flat country is conveniently located, as on an Ordnance map, by giving the number of miles the position is north or south of a given base point, and the number of miles it is east or west of the same point. The essential fact is that two numbers or measurements are required. To make the method precise, fix two straight lines L_1 and L_2 in the plane which cut at right angles in a point O. In the visual representation, and for convenience of wording, the first line L_1 can be taken as horizontal and the second line L_2 as vertical, the positive direction of L_1 being from left to right and of L_2 upwards. A definite unit is chosen which serves for measuring distances along both lines. From a point P in the plane, perpendiculars PM and PN are drawn to the lines L_1 and L_2. The point P is then located without ambiguity by two numbers, the first being the number of units in the distance OM and the second the number of units in the distance ON. In Fig. 1, P is located by the number pair $(4, 3)$, Q by $(3, 4)$, R by $(\frac{7}{2}, -2)$ and S by $(-4, -\frac{3}{2})$.

Of the two numbers which locate the position of a point P, the first is the distance of P from O parallel to L_1, i.e. to the right or left of L_2; if this number is positive, P is to the right of L_2, and if negative, P is to the left of L_2. The second number is the distance of P from O parallel to L_2, i.e. the height of P above or below L_1; P is above L_1 if the number is positive and below L_1 if the number is negative. The lines L_1 and L_2 divide the whole plane into four " quadrants ". In the N.E. quadrant both numbers are positive, in the N.W. quadrant the first number is negative and the second positive, in the S.W. quadrant both numbers are negative, and in the S.E. quadrant the first number is positive and the second negative. The order of writing the numbers is important; an interchange of the numbers (if unequal) alters the position of the point (see P and Q of Fig. 1). From the continuity of the number system and of the plane, each point of the plane is represented by a definite pair of numbers and each pair of numbers by a definite point of the plane. Relative to the fixing of L_1 and L_2 and of the scale of measurement, the correspondence between number pairs and points is quite unambiguous.

Suppose a table is given showing corresponding values of two variable quantities. Taking squared paper and ruling two perpendicular lines along which values of the two quantities are respectively measured, each pair of values can be plotted, by the above method, as a point on the squared paper. The result is a *graph*, showing a

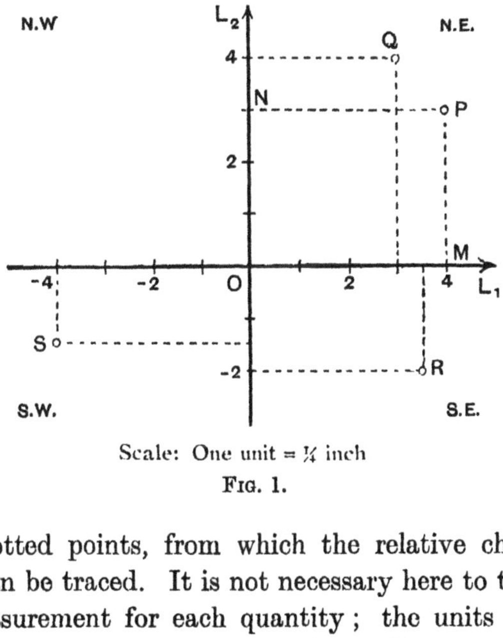

Scale: One unit = ¼ inch

FIG. 1.

group of plotted points, from which the relative changes in the quantities can be traced. It is not necessary here to take the same scale of measurement for each quantity; the units in which the quantities are measured are more or less arbitrary and each scale can be chosen for convenience of plotting. Examples I provide practical instances of graphical representations of this kind.

The location of points in space of three dimensions can be carried out in an exactly similar way. Three straight lines L_1, L_2 and L_3 are fixed to intersect at right angles in a point O. With definite positive directions along the lines and a definite scale of measurement for all distances, the position of a point P is located in space as follows. A rectangular " box " is constructed with three adjacent sides OM,

ON and OL along the lines L_1, L_2 and L_3 respectively, and with P as the vertex opposite to O. The position of P is then located by the set of three numbers which represent respectively the number of units in the distances OM, ON and OL, due regard being paid to sign. Remarks similar to those given in the case of two dimensions apply to this three-dimensional method of locating points.

1.9 Variable points and their co-ordinates.

The importance of the methods of locating points in space is seen when they are applied, with suitable modifications of terminology and notation, to connect variable points and variable numbers. A variable point in space is a point which can take any position from a given range of possible positions. As a point varies in position, so the numbers which serve to locate the point vary in value. From this simple fact, we can construct a most powerful theoretical tool applying to both analysis and geometry.

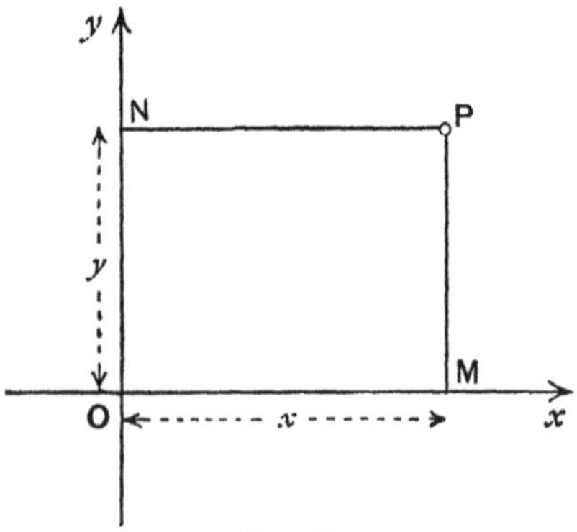

FIG. 2.

We can start with the two-dimensional case and, since we are considering variable points in general, we can give up any attempt at scale drawings and rely upon geometrical reasoning illustrated by simple diagrams. Denote two given perpendicular lines by Ox and Oy, drawn with positive directions as shown in Fig. 2 and with the line Ox taken horizontally and the line Oy vertically. The point O, in which the lines intersect, is called the *origin* of co-ordinates and the lines Ox and Oy are called the *co-ordinate axes* or the axes of reference. Any point P is taken in the plane Oxy and perpendiculars PM and PN are drawn to the axes. With a definite scale of measurement for all distances, let OM contain x units and ON y units of distance. The point P is fixed by the number pair (x, y), where the numbers must be written in this definite order. The point P is said to have co-ordinates (x, y), the first being the x-co-ordinate and the second the y-co-ordinate. Alternatively, the co-ordinate along the horizontal axis (Ox here) is called the *abscissa* and the co-ordinate along the vertical axis is called the *ordinate* of P.

Once the axes and the scale of measurement are fixed, all points are located uniquely by their co-ordinates. As the point P varies in the plane, so the co-ordinates become variable numbers with certain definite ranges of variation. If P varies over the whole plane, then the variables x and y each have the whole set of real numbers as their range of variation. If P varies in a more restricted way, then the variables x and y have ranges which may be continuous or discontinuous and related in all kinds of ways. To a variable point in two dimensions, therefore, there corresponds a pair of variable numbers ; much use of this will be made later.

There are several ways of interpreting the co-ordinates of a point. The point (x, y) can be regarded as distant x units perpendicularly from the y-axis and y units perpendicularly from the x-axis. In other words, the point is a distance x units to the right or left of the vertical base line and a height y units above or below the horizontal base line. From another point of view, the point (x, y) is reached by going a distance x units from O along Ox and then a distance y units parallel to Oy, or conversely.

The co-ordinates of particular points should be noticed. The origin O is itself a point in the plane and its co-ordinates are seen to be $(0, 0)$. Any point on Ox must have zero for its y-co-ordinate and it appears as $(x, 0)$, where the (positive or negative) distance of the point from O is x units. In effect, therefore, a point on the line which is taken as the x-axis need have only one co-ordinate, and a variable point on this line is represented by one variable number x. Similar remarks apply to a point with co-ordinates $(0, y)$ lying on the y-axis. Finally, points in the various quadrants of the plane have co-ordinates whose signs are determined by the rules already given (1.8 above).

The extension of the representation of points by co-ordinates to the case of three-dimensional space presents no essential difficulty. Three perpendicular co-ordinate axes Ox, Oy and Oz must now be fixed, intersecting in the origin O and with positive directions as shown in Fig. 3. For convenience, the plane Oxy can be taken as horizontal, the axis Oz then being vertical. Dropping perpendiculars from any point P in space to the planes Oxy, Oyz and Ozx (called the co-ordinate planes) and completing the rectangular "box" with sides OM, ON and OL along the axes, let OM contain x units, ON y units and OL z units of distance. Then the position of the point P is fixed by three co-ordinates (x, y, z) written in this definite order. For fixed co-ordinate axes and a fixed scale of distance measurement, the variation of a point in space is described by three variable numbers, the variable co-ordinates of the point. Two of the variables, x and y, determine points in the horizontal plane Oxy while the third variable z determines the height of the variable point in space above or below the corresponding point in the horizontal plane.

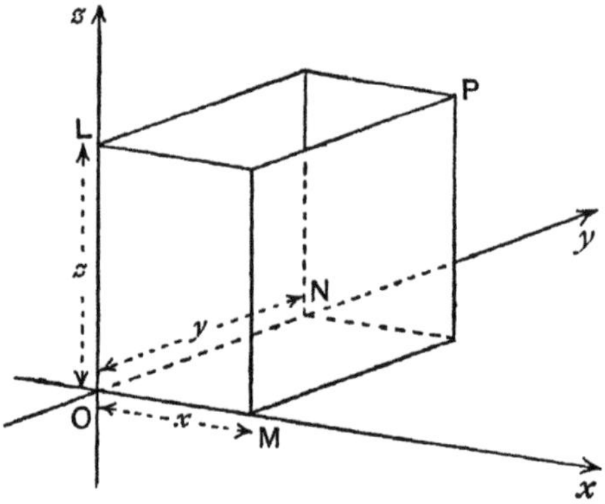

FIG. 3.

There is no need, of course, to adopt only the point of view taken so far in which variable points are represented by variable co-ordinates. Since the connection between points and co-ordinates is quite unambiguous (relative to fixed axes and a fixed scale of measurement), we can equally well start with variable numbers and represent them by a variable point in space. For example, given two variables x and y related in any definite way, we can fix two axes Ox and Oy in a plane together with a scale of measurement and so obtain a corresponding variable point P (x, y) in two-dimensional space. All this is independent of the symbols adopted for the variables concerned; to a pair of variables denoted by p and q, there corresponds a variable point (p, q) in a plane referred to two perpendicular axes Op and Oq.

The perfect "two-way" connection we have described between variable points and variable numbers clearly provides the link we require between geometry and analysis. This link, as we shall see, enables us to translate an analytical problem into a geometrical or diagrammatic one, and conversely. We have, in fact, established one of the most useful methods of the mathematical technique.

EXAMPLES I

The measurement of quantities *

1. Explain how a scale of measurement for mass, considered as a fundamental quantity, can be constructed. How is a given mass compared with the scale? Must continuity be imposed upon the measure of mass? Consider the other fundamental quantity, time-interval, on the same lines.

2. An isosceles triangle has two equal sides of one inch enclosing an angle of variable size. Show how the length measure of the third side can give a scale for measuring angles of less than two right angles, thus deriving an angular measure from length measure. What is the measure, on this system, of angles of 60° and 90°?

3. Assuming that lengths can be measured along the circumference of a circle, obtain an alternative length measure of angles, showing that angles up to four right angles can now be measured. How can this be extended to angles of any size? What is the relation of this length measure to the familiar measure in degrees?

4. Illustrate the change of measure of a quantity by expressing 321 inches in feet; 1·63 yards in inches; $25\frac{1}{3}$ cwts. in pounds; 4897 pounds in tons and 0·285 hours in seconds.

5. Given the connection between a foot and a centimetre, find an approximate measure of a centimetre in inches. Express 5 inches in centimetres, $3\frac{1}{2}$ yards in metres and 1 mile in kilometres.

6. From the measure of a pound in grammes, find the approximate measure of a gramme in ounces. Express 2 ounces in grammes and 12 pounds in kilogrammes.

7. Express 5072 square inches in square yards and 0·038 cubic feet in cubic inches. One acre is 4840 square yards; how many acres are there in a square mile?

8. Express a speed of 20 feet per second in miles per hour. What is the measure of a speed of 60 miles per hour in feet per second and in metres per second?

9. A uniform body weighs 6 cwts. and has a volume of 5 cubic feet. Find the density in pounds per cubic foot and in ounces per cubic inch.

10. A cube of uniform substance has a side of 2 inches and weighs 3 pounds. Find the density in pounds per cubic inch and in grammes per cubic centimetre.

* In these examples, take 1 foot = 30·48 centimetres and 1 pound = 453·6 grammes.

11. For a particle of mass m and velocity v, we define

$$\text{Momentum} = m \,.\, v \quad \text{and} \quad \text{Kinetic Energy} = \tfrac{1}{2} m \,.\, v^2.$$

How are the measures of momentum and kinetic energy expressed in terms of length, mass and time units?

12. A body of mass 2 pounds is moving at a uniform speed of 30 feet per second. Find its momentum and kinetic energy in pound-feet-second units and also in centimetre-gramme-second units.

13. Illustrate the fact that the derivation of secondary quantities can be carried beyond the first stage by expressing acceleration (the change of velocity over time) in distance and time units. The acceleration due to gravity is approximately 32 in feet-second units; what is its measure in centimetres and seconds?

14. A retail tobacconist buys a brand of tobacco at £5 9s. 10d. per 10 pound bag wholesale. Adding 15% for his expenses and profit, what should be the retail price per pound (to the nearest penny) and per ounce (to the nearest halfpenny)?

15. The number of acres of land used in wheat production, the amount of wheat obtained in bushels and the total cost of production are all known. Explain the derived nature of the concepts of average product per acre and of average cost per bushel of wheat.

16. From the data given in the table:

Land in	Area (acres)	Wheat production (bushels)	Total cost (£'s)
Great Britain -	12	467	97
Canada - -	185	2236	325
France - -	57	1304	233

find the average product per acre and the average cost per bushel of wheat in each case. How can these derived figures be used to compare wheat production in the different regions named?

17. Assuming that a bushel of British wheat, produced under the conditions of the previous example, is sufficient for 60 pounds of bread, calculate the cost of production of the wheat content of a two pound loaf. The wheat producer makes a profit of $7\frac{1}{2}\%$ and the loaf sells at 4d.; what proportion of the selling price goes in baking and distribution?

18. A small chocolate firm employs 15 hands at an average wage of 45s. per week; overheads and materials cost £41 5s. per week and the output of the firm is 34 tons of chocolate per year. Find the average product in pounds per hand per week and the average cost of a $\frac{1}{2}$ pound block of chocolate in pence.

19. Show that the rate of simple interest is a number expressed only in time units, varying inversely with time. How is the rate of interest per year reduced to a rate of interest per week? What is the amount of £100 at $r\%$ simple interest per year at the end of n years?

20. A man pawns his watch for 18s. and is charged $\frac{1}{2}$d. per week by way of interest. Calculate the rate of (simple) interest per week that is implied by this charge. If 12% per year is taken as a standard rate of interest for this type of loan, is the pawnbroker extortionate in this case?

Graphical methods

21. Selecting suitable rectangular axes in a plane, plot the points with co-ordinates $(4, -1)$; $(-\frac{3}{2}, 3)$; $(2, \frac{1}{2})$; $(2{\cdot}3, -1{\cdot}7)$ and $(-2{\cdot}8, 0)$.

22. Show that the three points whose co-ordinates referred to rectangular axes in a plane are A (5, 7), B (9, 3) and C (−2, −4) form an isosceles triangle ABC. Which are the equal sides? By direct measurement, determine whether the other side is longer or shorter.

23. By plotting a graph, show that the points with co-ordinates $(-3, -1)$; $(6, 2)$; $(1, -3)$; $(-3, 5)$; $(5, -1)$ all lie on a certain circle. Locate the centre and measure the radius of the circle.

24. Four points O, P, Q and R are marked in a plane country. OR is a line in a W.E. direction and R is 12 miles from O. The position of P is 3 miles E. and 4 miles N. of O, and that of Q is 4 miles W. and 5 miles N. of R. What axes and scales are suitable for locating the points graphically? Find the co-ordinates of the points in the system you choose, plot the points on graph paper and measure the distances OP, PQ and QR. Estimate the length of the shortest route from O to R calling at P and Q and check your result by using Pythagoras' Theorem.

25. Illustrate the arbitrary element in the location of points in a plane by rectangular co-ordinates by considering the following.

Three points A, B and C lie in a plane in which lines L_1 and L_2 (L_1 horizontal and L_2 vertical) are fixed. A is 3 inches above L_1 and 2 inches to the left of L_2; B is 6 inches above L_1 and 10 inches to the right of L_2; C is 1 inch below L_1 and 8 inches to the right of L_2. Find the co-ordinates of the points referred to L_1 and L_2 as axes when the scale unit is equal to (*a*) one inch, (*b*) one-tenth of an inch, and (*c*) one foot.

A point O' is now taken 2 inches below L_1 and 1 inch to the right of L_2. Straight lines L_1' and L_2' are drawn through O' parallel to L_1 and L_2 respectively. Find the co-ordinates of the three points on each of the above scales when L_1' and L_2' are axes.

26. Three mutually perpendicular axes L_1, L_2 and L_3 are fixed to intersect in a point O in space (L_3 being vertical). Find the co-ordinates of the point A which lies $3\frac{1}{2}$ inches from O on L_2; of the point B in the plane of L_1 and L_2 which is $5\frac{1}{2}$ inches from O parallel to L_1 and 2 inches parallel to L_2; and of the points C and D which are 4 inches above and below B respectively. Draw a rough diagram to illustrate.

27. Choosing convenient scales, plot a graph of the pairs of values of the variables x and y given below:

x	1	2	3	4	5	6	7	8	9	10
y	108	81	63	50	40	32	26	21	$16\frac{1}{2}$	13

Draw a freehand curve through the plotted points. Is there any evidence of a "law" connecting changes in x and y? Find an approximate value of y when $x = 3{\cdot}6$.

28. The following pairs of values of x and y are given:

x	−9	−7	−5	−3	−2	0	1	3	4	6	8	9
y	10·0	9·4	7·7	5·0	3·4	0	−1·7	−5·0	−6·4	−8·7	−9·8	−10·0

Plot the corresponding points on graph paper, draw a smooth curve through the points and find the values of y when x is −6 and 5. What can be deduced about the variation of y when x varies?

29. Show, by a graphical method, how the pressure on a piston of a steam engine varies at different positions of the stroke when the following facts are given:

Position of piston (inches from beginning of stroke) - -	0	1	4	6	8	11	15	19	20
Pressure (000 lbs.) - - -	36·0	36·5	36·5	33·5	25·5	17·5	13·7	9·0	1·85

30. The following table (data from Schultz: *Statistical Laws of Demand and Supply*) gives p, the yearly average N.Y. wholesale price of sugar, and q, the yearly U.S.A. consumption of sugar, for the period from 1900 to 1914.

Year	p (cents per pound)	q (Mn. short tons)	Year	p (cents per pound)	q (Mn. short tons)
1900	5·32	2·49	1908	4·96	3·57
1901	5·05	2·66	1909	4·77	3·65
1902	4·46	2·87	1910	4·97	3·75
1903	4·64	2·86	1911	5·34	3·75
1904	4·77	3·10	1912	5·04	3·93
1905	5·26	2·95	1913	4·28	4·19
1906	4·51	3·21	1914	4·68	4·21
1907	4·65	3·35			

Plot a graph, one point for each year, to show the related variations of p and q. (Such a graph is called a "scatter diagram".) Is there any evidence that a rise in consumption accompanies a fall in price?

31. Points in a plane can be located by means of co-ordinates with reference to axes which are not perpendicular (oblique co-ordinates). If lines L_1 and L_2, intersecting in O at any angle, are taken as axes, the co-ordinates of a point P are (x, y), where P is reached from O by going a distance x units along L_1 and then a distance y units parallel to L_2. Illustrate with a diagram.

32. The axes are at an angle of 45°; plot the points whose co-ordinates are given by the following pairs of numbers:

$$(2, 2); \quad (0, 3); \quad (-4, 0) \quad \text{and} \quad (-3, -2).$$

(Scale: one unit = one inch.)

33. A parallelogram has a horizontal side OP of 4 inches and a side OR of 3 inches at an angle of 60° to OP. Draw the parallelogram on graph paper and find the co-ordinates of the corners and of the point of intersection of the diagonals when the axes are taken as (*a*) OP and OR, and (*b*) horizontal and vertical lines at O.

34. An entirely different method of locating points in space is by means of polar co-ordinates. A point O and a direction Ox (taken horizontally) are fixed in a plane. The polar co-ordinates of a point P are defined as (r, θ), where the distance of P from O is r units and where OP makes an angle of θ°, in the counterclockwise direction, with Ox. Illustrate with a diagram and by plotting a graph showing the points whose polar co-ordinates are

$$(3, 60^\circ)\ ; \quad (1, 300^\circ)\ ; \quad (\tfrac{3}{2}, 0^\circ) \quad \text{and} \quad (\tfrac{5}{2}, 135^\circ).$$

35. Referred to rectangular axes, four points have co-ordinates

$$(1, 1)\ ; \quad (-2, -2)\ ; \quad (3, 0)\ ; \quad (0, -2).$$

Fixing the point O and the direction Ox, find the polar co-ordinates of the points. Illustrate by plotting the points on graph paper.

36. An observer, standing at a position O in a plane country, locates four landmarks A, B, C and D. A is 10 miles away in a N.N.E. direction ; B is due S. and at a distance of 7 miles ; C is W.N.W. and 8 miles away and D is distant 15 miles in a S.E. direction. Plot a graph showing the four landmarks and find the polar co-ordinates of the plotted points referred to O and the E. direction through O. (Scale : 5 miles = one inch.)

CHAPTER II

FUNCTIONS AND THEIR DIAGRAMMATIC REPRESENTATION

2.1 Definition and examples of functions.

In mathematical analysis we are concerned with variable numbers. The important thing about variables is not the way in which each varies by itself, but the way in which different variables are related one to another. Mathematical analysis, in short, is the study of relationships between variable numbers. A simple technical term is used to describe and symbolise a relationship between variables, the term " function ".

The notion of a function is an abstract and general one but essentially very simple. In order to emphasise the simplicity of the notion, we can conveniently introduce at once a few examples of functions as *applied* in scientific work and everyday life. In applying mathematical analysis to actual phenomena, the physical and other quantities of the phenomena are expressed by variable numbers. Any observed or assumed relation between the quantities then corresponds to a functional relation between the variables. The following examples make this clear.

If a lead pellet is dropped from a height, we know that the distance it has fallen (in feet or some other unit) depends on the time-interval (in seconds or some similar unit) since the moment of dropping. There is a relation between distance and time and the distance fallen is a " function " of the time-interval. Similarly, if a gas is subjected to pressure at a constant temperature, its volume varies with, i.e. is a " function " of, the pressure. To take a more everyday example, the amount paid in inland postage on a parcel (in some monetary units) depends on, and is a " function " of, the weight of the parcel (in pounds or a similar unit). Finally, in economics, the amount of a certain commodity demanded by a given market is taken as connected in some way with the market price of the commodity, i.e. demand is a "function" of price.

In each of these examples, there are two variable quantities represented, in specific units, by variable numbers. The quantities, and the associated variables, do not change independently of each other; there is a connection between corresponding values, a dependence of one quantity upon the other. The idea of a function, therefore, involves the concepts of a *relation* between the values of two variables and the *dependence* of one variable on the other.

Formal definitions of two technical terms can now be given. There is an *implicit function* or functional relation between two variables x and y, with given ranges of variation, if the values that x and y can take are not independent of each other but connected in some definite way. If the value of x is known, then the value or values that y can take are fixed and not arbitrary, and similarly conversely. An implicit function is thus a mutual relation between two variables and either variable "determines" the other. The variable y is an *explicit function* of the variable x if the value or values of y depend in some definite way upon the value which is allotted arbitrarily to x. In this case, it is the variable x which "determines" the variable y. In the same way, x may be given as an explicit function of y.

It follows at once that a given implicit function between x and y fixes two explicit functions—y as an explicit function of x and x as an explicit function of y. These two explicit functions are said to be *inverse* to each other. Conversely, each given explicit function must arise from some implicit function and has its corresponding inverse function. The difference between implicit and explicit functions is mainly one of point of view or emphasis. If the relation between x and y is regarded as mutual, then the term implicit function is appropriate; the variables are on an equal footing. If the relation is regarded from a definite angle, e.g. y as depending on x, then we use the term explicit function. Here, the variables are arbitrarily separated and y is called the *dependent* and x the *independent* variable. It is to be noticed, however, that a function is not a causal relation even from this latter point of view. There is no discrimination between the variables except for convenience, and one variable does not "cause" the other. Causal relations occur only between the quantities of actual phenomena and, when such a relation is *interpreted* by a function, it merely happens that one view of the function is dominant and the other neglected. Every mathematical function can be regarded from either point of view; every function has its inverse.

The concept of a function is an extremely wide one. The following examples serve to show this and will, also, help us to classify functions in a convenient way for detailed consideration.

Ex. 1. The variables x and y take any numerical values which are such that the value of y is double that of x. In symbols

$$2x - y = 0$$

is the expression of this implicit function. The two explicit and inverse functions are

$$y = 2x \quad \text{and} \quad x = \tfrac{1}{2}y.$$

The range of variation of x and y consists of all real numbers.

Ex. 2. The variable y depends on the variable x, which can take any numerical value whatever, according to the algebraic processes indicated by the symbolic expression

$$y = x^2 + 3x - 2.$$

Here we have y as an explicit function of x and the range of variation of x consists of all real numbers. The implicit function from which it arises is expressible

$$x^2 + 3x - y - 2 = 0.$$

The inverse function is more difficult to derive, but it is seen that, to each given value of y, there corresponds either a pair of values of x or no value of x at all. Solving the above quadratic equation for a given value of y, we find (see 2.8 below) that

$$x = \tfrac{1}{2}(-3 \pm \sqrt{4y + 17}).$$

This gives x as an explicit function of y in which the range of values of y, for a real value of x, must be limited so that negative values numerically greater than $4\tfrac{1}{4}$ are excluded.

Ex. 3. The variables x and y take any values whose product is 3. In symbols, the implicit function is

$$xy = 3$$

and the two explicit and inverse functions are

$$y = \frac{3}{x} \quad \text{and} \quad x = \frac{3}{y}.$$

Again, the range of x or y consists of all real numbers.

Ex. 4. The variable y is defined as the result of the process

$$y = 100(1 \cdot 05)^x$$

where the range of values of x is taken as consisting of positive integers only. The inverse function can be expressed only in terms of logarithms (see 10.2). This is an example of a function of a discontinuous variable and it can be noticed that it expresses the amount £y that results when £100 is left for x years at 5% interest compounded yearly.

Ex. 5. The variables x and y are related by the equation

$$x^2 + y^2 = 16,$$

an implicit function between x and y. The explicit functions are

$$y = \pm\sqrt{16 - x^2} \quad \text{and} \quad x = \pm\sqrt{16 - y^2}.$$

The range of values of x for a real value of y consists of all numbers not numerically greater than 4, and similarly for the range of y. Except when x or y is ± 4, there are two values of y corresponding to each given value of x, and conversely.

Ex. 6. An implicit function between x and y is defined by

$$x^3 + y^3 - 3xy = 0.$$

The two explicit functions here cannot be expressed in any direct algebraic way. This does not mean that the functions do not exist. Given any value of x, a "cubic" equation determines the corresponding values of y and we know (see 2.8 below) that there is either one value or three values of y, according to the value originally allotted to x. The dependence of y on x is definite and so is that of x on y.

Ex. 7. Corresponding values of x and y are defined by the table:

$0 < x \leqslant 3$	$3 < x \leqslant 4$	$4 < x \leqslant 5$	$5 < x \leqslant 6$	$6 < x \leqslant 7$	$7 < x \leqslant 8$	$8 < x \leqslant 15$
$y = 6$	$y = 7$	$y = 8$	$y = 9$	$y = 10$	$y = 11$	$y = 12$

Here, y is a function of x and the range of x consists of all positive numbers not greater than 15. On the other hand, when x is considered as a function of y, the range of the latter is discontinuous and consists only of the positive integers between 6 and 12. This function expresses the inland postage, y pence, as dependent on the weight of a parcel, x pounds (see *P.O. Guide*, 1936). A function of this nature is called a *step function.*

Ex. 8. The variable y is defined as the number of marks obtained by the candidate whose official number was x at the June Matriculation Examination of London University, 1936. Here, a table is necessary showing the numbers of all candidates with the number of marks obtained entered against each case. The range of x consists of those positive integers allotted as numbers to candidates who sat the examination, and the range of y consists of those positive integers corresponding to total marks obtained by at least one candidate. Both variables have discontinuous ranges.

2.2 The graphs of functions.

We have shown (1.8 above) how points in a plane can be located by means of numbers. Direct use can be made of this method in the graphing of functions. The practical process is described in text-books on algebra and we need give here only a general account and a few examples of the process.

From the definition of a given function, a table of corresponding values of the two variables x and y can be constructed and made as detailed as we wish. It is usually a matter of giving definite values to one of the variables and solving the formula of the function for the corresponding value or values of the other variable in each case. If the range of variation is continuous, there is no limit to the number of entries that can be inserted.

Co-ordinate axes Ox and Oy are ruled on a sheet of graph paper (with, e.g., Ox horizontal and Oy vertical) and a definite scale of measurement is selected. Each pair of values from the table obtained from the function is plotted as a point on the graph paper. As more and more points are plotted, it is found, in the case of any ordinary function, that a freehand curve can be drawn through them. Either the collection of plotted points, or the freehand curve, is called a graph of the function. The following examples fully illustrate the process.

Ex. 1. $y=x^2+3x-2$.

The following table of values of x and y is obtained from the formula of the function :

x	-4	-3	-2	$-1\frac{1}{2}$	-1	0	1	2
y	2	-2	-4	$-4\frac{1}{4}$	-4	-2	2	8

Plotting the corresponding points, it is seen that a U-shaped curve can be drawn through them (Fig. 4). Between the points where x equals -2 and -1, more detail is required than elsewhere and one intermediate point has been plotted. It is found that the lowest point of the curve is at or very near to the point $(-1\frac{1}{2}, -4\frac{1}{4})$. The graph represents a curve known as the parabola (see 3.4 below).

Ex. 2. $xy=3$.

The graph of the function can be plotted from the table :

x	-4	-3	-2	-1	0	1	2	3	4
y	$-\frac{3}{4}$	-1	$-\frac{3}{2}$	-3	?	3	$\frac{3}{2}$	1	$\frac{3}{4}$

It is seen (Fig. 5) that a smooth curve can be drawn through the set of plotted points, except that there is some doubt about its course when x is small. When x is actually zero, the formula of the function provides

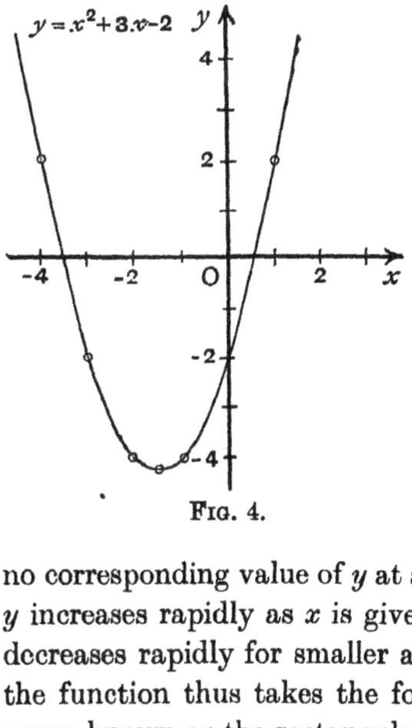

FIG. 4.

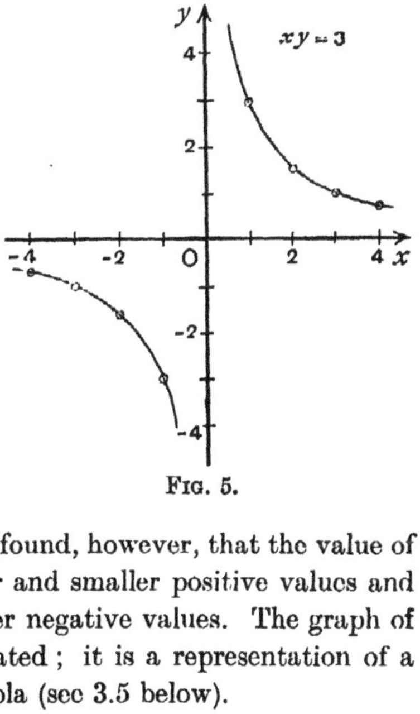

FIG. 5.

no corresponding value of y at all. It is found, however, that the value of y increases rapidly as x is given smaller and smaller positive values and decreases rapidly for smaller and smaller negative values. The graph of the function thus takes the form indicated ; it is a representation of a curve known as the rectangular hyperbola (see 3.5 below).

Ex. 3. $x^2+y^2=16$.

The following table is derived for values of x from -4 to $+4$:

x	-4	-3	-2	-1	0	1	2	3	4
y	0	$2{\cdot}6$ $-2{\cdot}6$	$3{\cdot}5$ $-3{\cdot}5$	$3{\cdot}9$ $-3{\cdot}9$	4 -4	$3{\cdot}9$ $-3{\cdot}9$	$3{\cdot}5$ $-3{\cdot}5$	$2{\cdot}6$ $-2{\cdot}6$	0

the values of y being taken to one decimal place. Outside the range shown, there exist no values of y at all. At $x=\pm 4$ there is a single value of y while at other values of x there are two values of y. The graph of the function takes the form shown (Fig. 6); it is recognised as a circle of radius 4 units with centre at the origin of co-ordinates.

Ex. 4. $x^3 + y^3 - 3xy = 0$.

The table of corresponding values of x and y, from which a graph of the function can be drawn, now appears:

x	−4·9	−3·9	−2·8	−1·7	0	0·7	1·3	1·5	1·9	2·9	3·9
y	3·9	2·9	1·9	0·9	0	1·3 0·15 −1·5	1·6 0·7 −2·25	1·5 0·9 −2·4	−2·8	−3·9	−4·9

Here, in order to obtain the values of y corresponding to any given value of x, it is necessary to solve a cubic equation. The values above

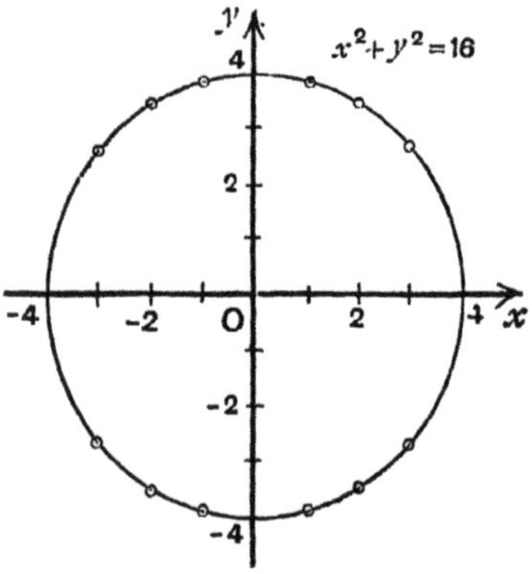

Fig. 6.

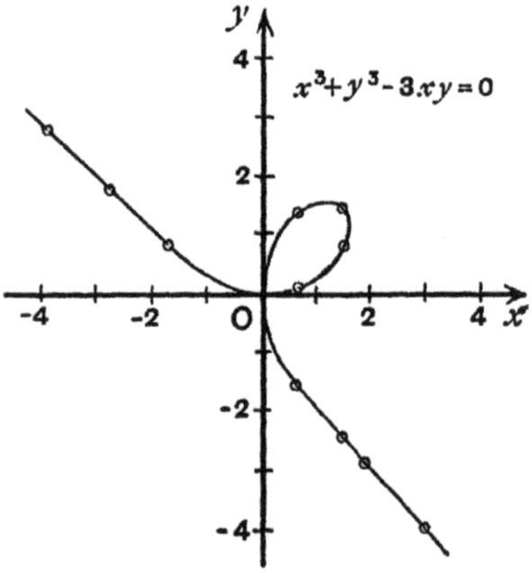

Fig. 7.

are correct to one decimal place and are obtained by a graphical method described later (see 2.8 below). The set of points plotted from the table can be joined by a smooth curve (Fig. 7) consisting of a "loop" and two "tails". The graph represents a curve, known as the Folium of Descartes, of a more unusual type than those obtained in the previous examples.

Ex. 5. The step function defined by the table of 2.1 (Ex. 7) can be plotted in a similar way. No smooth curve can be drawn through the plotted points which can, in fact, be connected only by a set of seven disconnected lines parallel to the x-axis (Fig. 8). This is a function which will later be described as discontinuous.

A graph is clearly of great service in the consideration of the properties of particular functions : it provides, at least, a visual representation of the related variation of the two variables. For example, the varying height of the curve shown on the graph (above or below the horizontal axis Ox) indicates the way in which y varies as x is changed in a definite way. The graph can also be used to replace numerical calculations, provided that approximate results suffice. In plotting the graph, only certain convenient (e.g. integral) values of the variables are used and, once this is done, corresponding values of the variables at intermediate positions can be read off approximately from the curve shown. This method of graphical "interpolation" is used in the solution of equations (see 2.8 below) and in many other connections.*

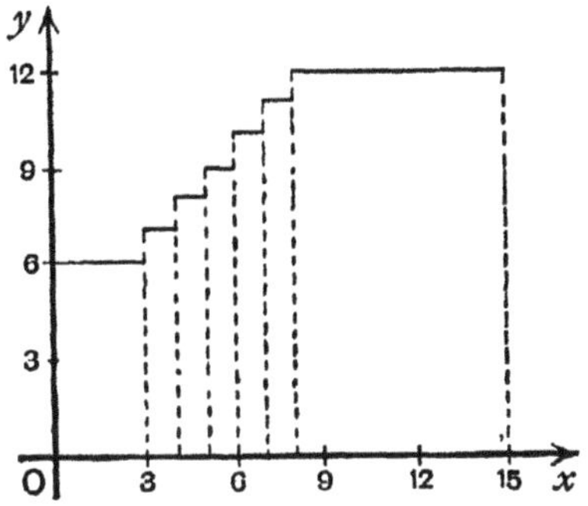

FIG. 8.

The method of graphing we have described illustrates the relation between two variables connected by a given mathematical function. But the same method can be used when the variables represent physical or other quantities given by observed phenomena and when we know that no " perfect " functional relation can exist between them, either in the nature of things or because some suspected relation is disturbed by random deviations or " errors ". Here we obtain what can be called a " statistical ", rather than a functional, graph.

Two cases can be profitably distinguished. One of the observed quantities x may be ordered in space or time and so that a single value of the other observed quantity y corresponds to each value of x. We take x along the horizontal and y along the vertical axis and the graph shows a series of points proceeding from left to right, varying height illustrating the variation of the observed y as the observed x varies according to its ordered progression. The graph

* For example, the graph of $y=\sqrt{x}$ can be plotted by giving x values which are perfect squares. The values of square roots such as $\sqrt{2}$ or $\sqrt{3}$ can then be obtained approximately as the heights of the graph corresponding to (e.g.) $x=2$ or 3. Alternatively, the graph of $y=x^2$ can be plotted, using integral values of x, and the approximate values of (e.g.) $\sqrt{2}$ and $\sqrt{3}$ can then be read off the points on the axis Ox corresponding to heights of 2 and 3 on the graph. See Hogben, *Mathematics for the Million* (1936), p. 415.

may show, for example, the varying observed volume of a gas at constant temperature as the observed pressure increases over time. From this graph, we could see whether, apart from errors and random deviations, a definite functional relation between volume and pressure can be reasonably assumed. Again, a graph can be used to show the value of imports into the United Kingdom over a period of years, or the variation in some other statistical time series (see 10.4 below). In these cases, no functional relation (e.g. between imports and time) can exist and the graph is mainly descriptive.

On the other hand, neither of the two observed quantities may possess a natural spatial or temporal order. We may have, for example, a series of observed values of the market price (p) of a commodity and the corresponding market consumption (q). Such a correspondence can be obtained by taking a succession of yearly quotations of p and q (see Examples I, 30). Each pair of values of p and q can be plotted as a single point referred to two axes Op and Oq and the whole set of observations is then represented by a graphical " cluster " of points. Such a graph is called a " scatter diagram " and its uses, at least for purposes of illustration, are evident. It may be suspected, for instance, that there exists a definite relation between p and q, a relation which is disturbed by errors and other deviations in the actual observations. The correctness of this supposition can be examined and, if a relation is found to exist, its nature can be discovered.

2.3 Functions and curves.

The graphical method of representing functions can be extended to establish a general connection between functions and curves. If we are given a functional relation between two variables x and y of continuous variation, then there is no limit to the number of entries that can be inserted in the table of corresponding values of the variables and, hence, no limit to the number of points that can be plotted in the plane Oxy. The function gives rise, therefore, to an indefinitely large number of points in the plane and only a few of them are shown in any actual graph. Just as the variables are related in an ordered way by the function, so the corresponding points must display a definite characteristic, i.e. must make up a *locus* or *curve* in the plane Oxy. Hence, to each given function relating variables x and y, there corresponds a set of points comprising a curve in the plane Oxy; the analytical property defined by the function is reflected in the geometrical property common to all points on the curve.

Conversely, a curve, as a geometrical concept, is a collection of points in a plane with a common characteristic. Fixing co-ordinate axes and attaching co-ordinates to each point of the plane, the property defining the curve can be translated into an analytical relation between x and y satisfied by all points (x, y) on the curve, a relation which is called the *equation* of the curve. Consider, for example, a circle with radius 4 units and centre at a fixed point O of a plane, i.e. the curve consisting of all points distant 4 units from O. Fix co-ordinate axes to pass through O, let P with co-ordinates (x, y) be any point on the circle and drop perpendiculars PM and PN to the axes (Fig. 9). From Pythagoras' Theorem

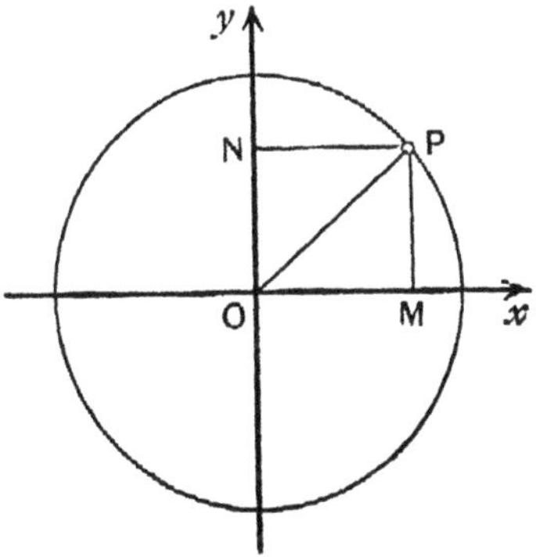

FIG. 9.

$$OP^2 = OM^2 + MP^2 = OM^2 + ON^2 = x^2 + y^2,$$

since $OM = x$ units and $ON = y$ units. But OP must equal 4 units for all positions of P on the circle. Hence,

$$x^2 + y^2 = 16$$

is the relation satisfied by all points (x, y) on the given circle, i.e. the equation of the circle (cf. 2.2, Ex. 3, above).

The correspondence between the functions of analysis and the curves of geometry is thus a perfect one. To each function relating variables x and y there corresponds a definite curve in the plane Oxy, and conversely. It is important to remember, however, that the uniqueness of the connection depends entirely on the fixing of co-ordinate axes in the plane. A given curve has different equations when different axes are selected.

It is now clear that a graph of a given function is simply an actual representation on squared paper of the curve which corresponds to the function. The function is an abstract concept relating two variables x and y; the curve is an equally abstract concept relating points in two-dimensional space. If the variables are continuous, there is an indefinitely large number of pairs of variable values and of points on the curve. A graph, on the other hand, consists of a finite collection of points selected arbitrarily from the indefinitely numerous set possible and designed to indicate clearly the shape of the curve. But the curve itself remains an abstract concept, of which the graph is only a visual, and more or less imperfect, representation.

Finally, notice that a curve, as defined above, need not be "smooth" in the sense that it can be shown graphically as a continuous line without breaks or angles. Nor need the curve be described in any temporal way by a moving point. It is often convenient to think of a point moving along a curve, but this is only a device to facilitate exposition. A curve, like a function, is defined and considered as a whole, as a collection of points in a plane united by a common characteristic.

2.4 Classification of functions.

We can now proceed to distinguish functions of different kinds according to the nature of their definitions or of their symbolic expressions. It is necessary, for this purpose, to introduce a number of technical but very useful terms.

A function is *analytical* if it can be expressed in symbols by means of a single formula connecting the variables, a formula which is the "general law" behind the function. From the formula, as we have seen, values of one variable, corresponding to given values of the other, can be derived by carrying out the algebraic or other operations indicated. The first six examples of 2.1 give analytical functions; the other two functions are not analytical.

The next distinction is an extremely important one. Here, we regard functions as explicit, taking *either* y as a function of x *or* x as a function of y. Explicit functions can be divided into two classes, those which are single-valued and those which are multi-valued. A function such that *one and only one* value of y corresponds to each given value of x is said to define y as a *single-valued function* of x. A function for which this is not true defines y as a *multi-valued function* of x, and we have double-valued, triple-valued and higher valued functions. The inverse function, x as a function of y, can be considered in the same way.

The functions of the first four examples of 2.1, as well as those of the last two examples, all give y as a single-valued function of x. In the fifth example, y is a double-valued function, and in the sixth example y is a triple-valued function of x. On the other hand, only in the first, third and fourth examples is x a single-valued function of y. This introduces a very important point ; *the inverse of a single-valued function is not necessarily itself a single-valued function.* This is clear from the inspection of even such a simple function as that of the second example of 2.1.

In graphical terms, if y is a single-valued function of x, then the curve corresponding to the function is cut by any line parallel to Oy in only a single point. But, if y is multi-valued, then the curve can be cut by such lines in two or more points. In the same way, x is a single-valued (or multi-valued) function of y if the curve is cut by lines parallel to Ox in only a single point (or in several points). There is no necessary connection between these two ways of finding the intersections of a line and the curve ; a given curve can be cut by a line parallel to one axis in only a single point while lines parallel to the other axis cut it in two or more points. The single-valued property does not necessarily apply to both the inverse explicit functions.

One reason for the importance of single-valued functions is clear. When y is a single-valued function of x, we can take the function as described by values of x which increase steadily in value ; the independent variable can then be completely ignored and attention concentrated upon the variation in y. The value of y may be found to increase, to decrease, to have alternating increasing and decreasing " stretches " or to vary in a more or less erratic way. Graphically, we take x along the horizontal axis and describe the curve from left to right. The varying height of the curve above or below the horizontal shows the variation of y, and we can see when the curve rises (y increasing) and when it falls (y decreasing).

The last remarks lead to the definition of a particularly useful sub-class of single-valued functions. If y is a single-valued function of a continuous variable x, and if y increases in value as x increases, then y is called an *increasing function* of x. Similarly, if the value of y decreases as x increases, we have a *decreasing function* of x. The class of increasing and decreasing functions, taken together, comprise what are called *monotonic functions*. The graph of a monotonic function of x, the axis Ox being taken as horizontal, either rises or falls without interruption from left to right, rising for an increasing function and falling for a decreasing function.

Subject to minor qualifications concerning the " continuity " of the function (4.6 below), a significant property of a monotonic function can be established. *The inverse of a single-valued and monotonic function is also single-valued and monotonic in the same sense as the original function.* For example, if y is a single-valued increasing function of x, it follows that only one value of x can correspond to each value of y and that this value must increase as y increases. This is quite clear graphically, since the curve representing y as a single-valued increasing function of x (axis Ox horizontal) rises throughout from left to right and so the distance of the curve from the vertical axis increases as we proceed upwards. It is thus only in the case of monotonic functions that we can always say that both inverse functions are single-valued.

The first and third examples of 2.1 show functions which are monotonic, the first being an increasing function and the other a decreasing function. In each case, what is true of y as a function of x is also true of x as a function of y. The second example of 2.1 provides a case of a single-valued function which is not monotonic. As x increases from large negative values up to $x=-\frac{3}{2}$, the value of y decreases, whereas y increases as x is increased beyond $x=-\frac{3}{2}$. The graph falls and then rises. The inverse function, x as a function of y, is not single-valued, but double-valued, here.

Returning to the case of multi-valued functions, it is sometimes possible to divide such a function into two or more " branches ", each branch being a single-valued function. The corresponding curve is also divided into two or more distinct sections, each cut by lines parallel to the appropriate axis in only a single point. The function of the fifth example of 2.1 is a case in point. The explicit expression, giving y as a double-valued function of x, is

$$y=\pm\sqrt{16-x^2}$$

which can be divided into two single-valued functions,

$$y = +\sqrt{16-x^2} \quad \text{and} \quad y = -\sqrt{16-x^2}.$$

The corresponding curve, a circle, is likewise divided into two single-valued sections, one above the axis Ox and the other below. In the same way, the second example of 2.1 gives x as a double-valued function of y divisible into the two branches

$$x = \tfrac{1}{2}(\sqrt{4y+17} - 3) \quad \text{and} \quad x = -\tfrac{1}{2}(\sqrt{4y+17} + 3).$$

The corresponding curve, a parabola, is divided into two sections by a line parallel to Oy at $x = -\frac{3}{2}$, each section being cut in only a single point by lines parallel to Ox (see Fig. 4). The first branch of this function is an increasing function of y, the corresponding section of the curve rising from left to right ; the second branch is a decreasing function of y and this section of the curve falls from left to right. Such divisions of multi-valued functions into single-valued branches are clearly very useful.

Finally, it can be noticed that some functions can be described as *symmetrical*. Symmetry can be defined in various ways, of which the following is perhaps the most important. An implicit function relating variables x and y is said to be symmetrical in x and y if an interchange of these variables leaves the function unaltered in form. This implies that y as an explicit function of x is of exactly the same form as x as an explicit function of y. The function $xy = 3$ is symmetrical on this definition and the inverse functions, $y = \frac{3}{x}$ and $x = \frac{3}{y}$, are seen to be of identical form. Graphically, such a symmetrical function is represented by a curve symmetrical about lines bisecting the angles between the axes. The part of the curve on one side of the line is the reflection in the line of the part of the curve on the other side. The graph of $xy = 3$ illustrates this fact (see Fig. 5).

2.5 Function types.

So far we have considered only a number of particular functions such as appear in the simplest operations of algebra. Our next task is to group functions into types and to devise a symbolic notation to include all functions of one type in a single formula. The grouping of functions into types proceeds by means of the notion of analytical " form ". This notion, which must be " sensed " rather than defined, is essential to the extension of the processes of algebra into the more powerful methods of analysis, and its nature is indicated most easily by giving concrete examples.

The explicit function $y=2x$ is clearly but one example of a wide range of functions of the same " form ". Other examples are

$$y=-3x\,; \quad y=5x+2\,; \quad y=x-3 \quad \text{and} \quad y=1-2x.$$

All such functions can be included in the single formula

$$y=ax+b,$$

where a and b denote any definite numbers, positive, negative or zero. The function type, represented by this " portmanteau " formula, is described as an explicit *linear function* of x.* In the same way, we can define x as a linear function of y. Further, putting the linear function into implicit form, we can write

$$ax+by+c=0,$$

where a, b and c stand for any three numbers whatever. The implicit linear function gives rise to two explicit linear functions :

$$y=\left(-\frac{a}{b}\right)x+\left(-\frac{c}{b}\right) \quad \text{and} \quad x=\left(-\frac{b}{a}\right)y+\left(-\frac{c}{a}\right),$$

the coefficients of which are expressed in terms of the three coefficients of the original implicit function.†

As a second example, we see that the explicit function

$$y=x^2+3x-2$$

is one example of the function type, called the explicit *quadratic function*, which can be symbolised

$$y=ax^2+bx+c,$$

where a, b and c denote any definite numbers. In the same way, we can have x as an explicit quadratic function of y. Putting either of these functions into implicit form, it appears as a " polynomial " expression, containing no powers or products of the variables of

* The term " linear " is used since, as we see later, such a function is represented graphically by a straight line.

† The cases where a or b equals zero are a little troublesome (see 3.3 below) but need not delay us here. Further, it may seem odd that there are three coefficients in the implicit form and only two in the explicit form. This is, however, only a matter of convenience (see 3.7 below).

higher degree than the second, equated to zero. Other implicit functions are also of this form. The function type $xy=a$, of which $xy=3$ is one case, and the function type $ax^2+by^2=c$, of which $x^2+y^2=16$ is an instance, are examples. We derive, therefore, a wider function type, symbolised by the formula

$$ax^2+2hxy+by^2+2gx+2fy+c=0,$$

which can be called the *implicit quadratic function* relating the variables x and y. This includes, for certain values of the coefficients a, b, c, f, g and h, the explicit quadratic function and also many others. The grouping of functions into types is not necessarily unique ; there are several overlapping types possible according to the particular point of view adopted.

The number of different function types that can be distinguished is indefinitely large. Amongst " algebraic " functions, we can go on to separate " cubic " functions of explicit and implicit form, " quartic " functions, and so on. There are also numerous functions of forms we have not yet introduced and some of these types will be considered at a later stage. The exponential, logarithmic and trigonometric functions are notable instances.

The grouping of functions into types, largely a matter of convenience, is characterised by the use of symbolic letters other than those standing for the variables themselves. The familiar algebraic device of representing unspecified numbers by letters has been significantly extended and, before proceeding, it is essential to get some idea of this extension. The symbolic letters, such as a, b and c in the above formulae, must be " variable " since they stand for unspecified numbers, but there must be some difference between them and the original variables x and y. This difference, on which the whole grouping of functions into types depends, must be made clear.

To fix ideas, consider the quadratic function type which we can write $y=ax^2+bx+c$. Two points of view can be adopted. From the first point of view, we investigate the relation between the variables x and y and the numbers a, b and c must be taken as fixed, as having the same numerical values whatever they may be. From the other point of view, the function is treated as a whole and the type includes a whole set of particular functions united by the common property of being " quadratic ". The formula, in fact, represents *any* function of quadratic form. Here, x and y cease to be of importance and the whole function is made to vary by changing the values of a, b and c. These latter numbers are *fixed* within one function of the quadratic type but *variable* from one quadratic function to another.

Hence, the characteristic of letters such as a, b and c is that they lead double lives; they are "variable constants". To suggest this double use, we describe such symbolic letters as *parametric constants* and one of two shorter terms can then be used according to the point of view. When attention is directed to the relation between the variables of the function, the symbols denote fixed numbers and can be termed *constants*. When the important thing is the variation of the function as a whole, the variables of the function fading out of the picture, we describe the symbols as *parameters*.*

We have tried to show how the functional notation is made more general and flexible by the use of parametric constants. The quadratic function type, for example, becomes a particular function once numerical values are allotted to the parameters. But, taking a more general view, we can attempt to derive properties, not of one quadratic function, but of all quadratic functions. Or we may know that some quadratic function satisfies a given condition and then proceed to determine the values of the parameters for which the condition holds. This, for example, is the idea behind the fitting of a "parabolic trend" to a statistical time-series.

A simple algebraic instance can be quoted. It is required to find the height of an open box of square base of side 4 inches so that the surface area is 48 square inches. If the height is x inches, the surface area (y square inches) is a function of x:

$$y=16(x+1).$$

Putting $y=48$, the value of x is found to be 2 inches. This problem can be generalised and solved for all cases by the use of parameters. If y square inches is the surface area of a box of square base of given side a inches and of variable height x inches, then

$$y=a^2+4ax \quad \text{and} \quad x=\frac{1}{4a}(y-a^2).$$

* The term parameter comes from παραμετρεω which means "to compare" or "to measure one thing by another".

This is a functional relation between x and y. If the surface area is known to be b square inches, then

$$x=\frac{b-a^2}{4a} \text{ inches.}$$

We have solved all problems of this type, and the solution of any particular problem is derived by allotting the parameters a and b the appropriate numerical values.

Finally, the use of parameters is carried over into the mathematical sciences, as is seen, for example, by the parametric rôle of market prices in economic theory. A firm (under conditions of " pure competition ") must take as given the market prices that happen to exist and we can assume its output is known for each given price system. In combining the output decisions of all firms, we treat the price system as variable and attempt to discover a set of prices consistent with equilibrium (however the latter is defined). Prices are parametric constants, constant in the examination of the decision of a single firm, parameters when we combine the separate decisions of a whole group of firms.

2.6 The symbolic representation of functions of any form.

A further generalisation of the functional notation now takes us outside the limited field of algebra, to which we have been confined so far, into the wider territories of modern analysis. The idea involved is simple. Just as a symbol is introduced in algebra to stand for *any* unspecified number, so now a symbol is introduced to stand for *any* function. In algebra, we deal with the properties of functions of particular types ; with the new notation, mathematical analysis goes far beyond this and introduces more powerful processes, such as those of the calculus, to deal with functions of any kind whatever. The development of mathematics, here as elsewhere, is dependent on the development of its notation.

An implicit function relating two variables x and y can be represented, whatever its form may be, by the notation

$$f(x, y)=0.$$

This is certainly appropriate for any analytical function, the symbolic $f(x, y)$ standing for some expression involving x and y. To obtain, from the general notation, a particular function, we need only specify the exact form of $f(x, y)$; for example, if we take $f(x, y)$ as $2x-y$, x^2+y^2-16 or x^3+y^3-3xy, we obtain a particular function $y=2x$, $x^2+y^2=16$ or $x^3+y^3-3xy=0$. The notation can also be applied, without confusion, to include even non-analytical functions.

Again, if y is an explicit function of x of any form whatever, we can denote it by the notation

$$y=f(x).$$

This notation is clearly applicable to the case of a function which is single-valued and analytical.* In such a function, y is defined as equal to the value of a certain expression involving x, and $f(x)$ can be taken as a convenient way of representing this expression in x, no matter what its form may be. To derive a particular function, we supply a definite form for $f(x)$; if we take (e.g.) $f(x)$ as $2x$, x^2+3x-2 or $\frac{3}{x}$, we obtain three well-known single-valued functions. The notation can also be extended to apply to functions which are not single-valued or even analytical; in its broadest use, $y=f(x)$ signifies only that we are taking y as *some* function of x. Notice that, in this notation, we need not refer to the variable y at all; we can say that $f(x)$ is a function of x. For example, we often describe (x^2+3x-2) as a function of x. This is a matter of convenience only; a function always relates the values of two variables whether we care to suppress reference to one of them or not.

Since the importance of a good notation, both general and flexible, is evident, we can insert here a number of remarks on the symbolism of mathematical analysis. Letters are used to denote both variables and parameters, and, in order to distinguish one from the other, it is usual to reserve the later letters of the alphabet for variables and the earlier and middle letters for parametric constants. The letters x, y, z, u, v and t are most frequently used for variables, and, if these do not suffice, the Greek letters ξ, η and ζ are called into service. When parametric constants are regarded primarily as constants,

* An alternative, and in some ways a superior, notation is occasionally used. Instead of writing $y=f(x)$ to denote the single-valued function, we write $y=y(x)$. This economises letters and enables us to write the dependent variable by the single letter y when the dependence on x is not stressed, and by the same letter with x in brackets when the dependence on x becomes important. For an actual use of the notation in economics, see Frisch, *New Methods of Measuring Marginal Utility* (1932).

the letters $a, b, c, \ldots$, or the corresponding Greek $\alpha, \beta, \gamma, \ldots$, are conventionally used. When the parametric property is the more important, it is often useful to take k, l, m, n, p, q and r, or $\kappa, \lambda, \mu, \nu$ and ρ, to denote the parameters. These remarks serve only as general guides; the context should be sufficient, in any particular case, to make clear the nature of the symbols used.

The letters denoting functions are of different nature from the variable or parametric letters; they do not denote anything to which numerical values can be allotted but stand for the complex notion of the form of a function. It is usual to reserve the letters f and g, and the Greek ϕ and ψ, together with the corresponding capitals, for the denotation of functions. It is also possible, for greater variety, to add suffixes to the functional letter, $f_1, f_2, f_3, \ldots$ and $\phi_1, \phi_2, \phi_3, \ldots$ being examples.

It is, of course, essential to denote different variables, parameters or functions by different letters. For example, if we wish to say that any implicit function gives rise to two explicit and inverse functions, we must symbolise somewhat as follows. From the given implicit function $f(x, y)=0$, we derive the two functions $y=\phi(x)$ and $x=\psi(y)$ which are inverse to each other.

Functions refer to *operations* performed on variables, and it is thus necessary to denote a function, not only with its operational or functional letter, but also with an indication of the variable or variables to which the operation applies. The notation $f(x)$ for an explicit function makes this clear, the letter f being the functional letter and the letter x in brackets denoting the variable to which the operation applies. Other operations, such as those of derivation and integration, will appear later and will be denoted by further symbolic devices of the same kind.

One more notational device remains for consideration, i.e. the method of introducing *particular* values of the variables into a general function. Suppose $y=f(x)$ is some explicit function in which we give, to the variable x, definite numerical values such as $x=0$, $x=\frac{1}{2}$, $x=1$. The corresponding values of y are then written $f(0)$, $f(\frac{1}{2})$, $f(1)$. Further, we can allot values to x which are fixed but not specified. Such values of x can be denoted by $a, b, c, \ldots$, or by the letter x with suffixes or primes added:

$$x_1, x_2, x_3, \ldots \quad \text{or} \quad x', x'', x''', \ldots .$$

The value of y when x is given the fixed value (e.g.) a or x_1 is written $f(a)$ or $f(x_1)$. These fixed, but unspecified, values of the independent variable are, of course, instances of parametric constants. To illustrate with a particular function, if

$$f(x)=x^2+3x-2,$$

then $$f(0)=(0)^2+3(0)-2=-2\ ;\ f(\tfrac{1}{2})=(\tfrac{1}{2})^2+3(\tfrac{1}{2})-2=-\tfrac{1}{4},$$
$$f(1)=(1)^2+3(1)-2=2\ ;\ \text{and so on.}$$

Further, $$f(a)=a^2+3a-2.$$

Again, if $f(x, y)=0$ is an implicit function, then $f(a, y)=0$ indicates an equation to be solved for the values of y corresponding to $x=a$, and $f(x, b)=0$ an equation giving the values of x corresponding to $y=b$. Specific numbers can, of course, be substituted for a or b.

As an example in the use of the functional notation, consider the following operations of which much use will be made later. If a and $(a+h)$ are two particular values of a variable x, where a and h are any positive or negative values, then $f(a)$ and $f(a+h)$ are the corresponding values of a function $f(x)$. The change (or increment) in the value of the function corresponding to the given change in x is

$$f(a+h)-f(a).$$

The change in the function *per unit change in* x is then

$$\frac{f(a+h)-f(a)}{h}$$

and this is defined as the *average rate of change* of the function when x changes from a to $(a+h)$. The actual changes shown in x and $f(x)$ can be positive or negative according as they are increases or decreases.

2.7. The diagrammatic method.

We have seen that a definite curve referred to selected axes in a plane corresponds to each particular function we care to define, and conversely. Further, when we are dealing with definite functions and curves, the *graphical method*, in which the graph of a function or curve is plotted on squared paper, is of great service. But it is easily seen that the connection between functions and curves remains of the greatest importance when we pass from particular functions to function types. It is reasonable to suppose, and it is in fact true, that the curves corresponding to a set of functions of the same type form a class of the same general nature. To a function type there corresponds a curve class, and properties common to all functions of the type are paralleled by properties common to all curves of the class. This correspondence is investigated in the branch of mathematics known as *analytical geometry*, where we consider what function types correspond to particular curve classes and what analytical properties of the functions reflect various geometrical properties of the curves. Some account of analytical geometry will be given in the following chapter.

The function-curve correspondence would appear to be useless, however, when we come to "arbitrary" functions of unspecified form. All that can correspond to an arbitrary function is an arbitrary curve, and neither graphical methods nor analytical geometry avail. But it always happens, in our analysis, that functions, though otherwise arbitrary, are limited by certain general properties. It may be, for example, that a function is limited by the conditions that it is single-valued, positive and decreasing, or that it decreases up to a certain point and then increases. The curve which corresponds to the function, though still largely arbitrary in position, now displays sufficient properties to make the drawing of an illustrative diagram worth while. In the first of the above instances, the diagram would show a curve lying completely above the horizontal axis and falling steadily from left to right.

It is here that the *diagrammatic method* of illustrating analytical arguments becomes important. The least service diagrams can perform is to illustrate and check our analytical development. It is possible, in some cases where the properties of our functions are very definite, to make more positive use of diagrams by employing them in conjunction with simple geometrical arguments. But we must be careful to avoid mistaking casual properties of the curves in our diagrams for essential general ones, and to prevent ourselves accepting apparently self-evident results without formal proof.

It is to be noticed that one curve serves to represent a given function, whether the latter is expressed in implicit form or in either of the inverse explicit forms. We need examine the shape of only one curve to illustrate properties of a functional relation between x and y, of y as a function of x and of x as a function of y. To see, for example, how many values of y correspond to a given value of x, and conversely, we look for the number of points of intersection of the curve and lines parallel to the axes. The use of the diagrammatic method on these lines is illustrated in 2.4 above.

A minor difficulty in the use of diagrams concerns the fact that we can fix the co-ordinate axes in two different ways, even when we adopt the established convention of drawing one axis horizontal. A given curve may look quite different when one variable is measured along the horizontal axis than it does when the other variable is so measured. Since it is important to be able to recognise the same curve however the diagram is drawn, we can indicate here a way in which a curve shifts when the axes are transposed. Fig. 10 shows

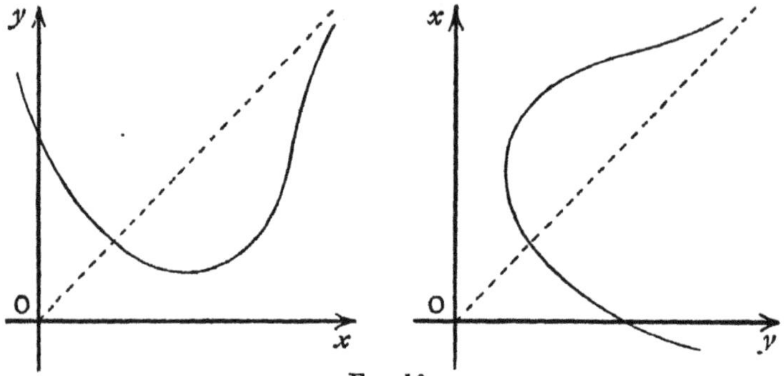

FIG. 10.

the curve corresponding to a given function relating x and y, firstly when Ox is horizontal and secondly when Oy is horizontal. The two positions of the curve are such that one position is the "reflection" of the other in the bisector of the angle between the axes, i.e. in the broken line shown in the figure. Hence, if we start with the curve in the first position and rotate the plane containing it through 180° about the bisector indicated, the new position taken up by the curve is the position with axes interchanged.

2.8 The solution of equations in one variable.

As an exercise in the use of analytical methods supported by graphs or diagrams, we can consider here the important question of the solution of equations, to which many mathematical problems reduce. An equation in a single variable x can be written, in general, as $f(x)=0$, where $f(x)$ denotes some given expression or function of x. Our problem, in solving the equation, is to find the value or values of x which make the value of $f(x)$ equal to zero.

The simplest type of equation in one variable occurs when $f(x)$ is a *polynomial*, i.e. the algebraic sum of terms which involve only various powers of x, and such equations can be classed according to their *degree*, as determined by the highest power of x involved. There are linear equations (of the first degree), quadratic equations (of the second degree), cubic equations and so on. A solution of such an equation, a value of x which makes the polynomial zero, is usually called a *root* of the equation. The process of finding roots of polynomial equations, at least in simple cases, is fully described in text-books on algebra.

The general linear equation is $ax+b=0$ and this has one root, $x=-\dfrac{b}{a}$, in all cases. The general quadratic equation is

$$ax^2+bx+c=0,$$

of which the solution is known to be

$$x=\frac{-b\pm\sqrt{b^2-4ac}}{2a}.$$

It follows that there are two real roots if $b^2>4ac$, and no real root if $b^2<4ac$. If $b^2=4ac$, then the two real roots coincide and there is only one value of x satisfying the equation, which reduces, in this case, to a perfect square equated to zero. The formula provides the roots indicated in the following examples of quadratic equations :

$$2x^2-x-1=0, \quad x=-\tfrac{1}{2} \text{ and } x=1,$$
$$x^2-3x+1=0, \quad x=\tfrac{1}{2}(3+\sqrt{5}) \text{ and } x=\tfrac{1}{2}(3-\sqrt{5}),$$
$$x^2+x+1=0, \quad \text{no real roots.}$$

In the first example, the roots can also be found by a simple process of factorising the quadratic expression.

Cubic and higher degree equations present much more difficulty. It is occasionally possible to solve such an equation by factorisation, as in the case of the cubic equation

$$x^3-2x^2-2x+1=0.$$

By trial and error we see that $x=-1$ satisfies the equation. So one root is -1 and $(x+1)$ must be a factor of the cubic expression. Taking out this factor, the equation becomes

$$(x+1)(x^2-3x+1)=0.$$

The other roots of the cubic are thus those of the quadratic

$$x^2-3x+1=0.$$

The complete solution of the cubic then gives three roots,

$$x = -1, \quad x = \tfrac{1}{2}(3 + \sqrt{5}) \quad \text{and} \quad x = \tfrac{1}{2}(3 - \sqrt{5}).$$

In general, however, analytical methods of solving cubic and higher degree equations are difficult to devise and will not be considered here. Instead we can indicate a graphical method of solving, not only a polynomial, but also any given equation. By its nature, the method can only produce *approximate* results, the approximation being closer the more accurately the graphs are drawn. The following examples suffice to illustrate the method.

Ex. 1. $x^3 - 2x^2 - 2x + 1 = 0$.

The graph of the function $y = x^3 - 2x^2 - 2x + 1$ is constructed as shown in Fig. 11, where (for convenience) different scales for x and y are adopted.

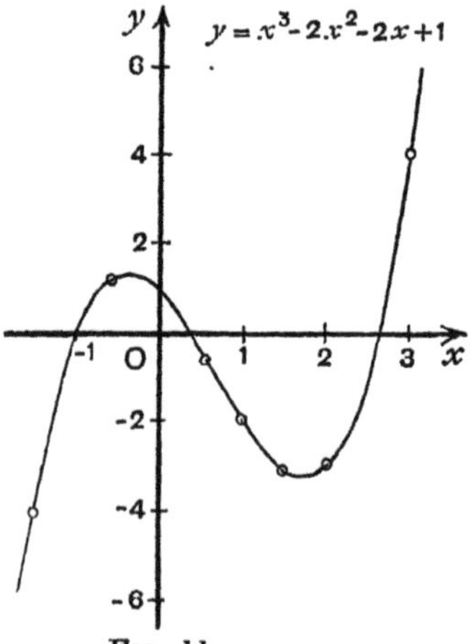

FIG. 11.

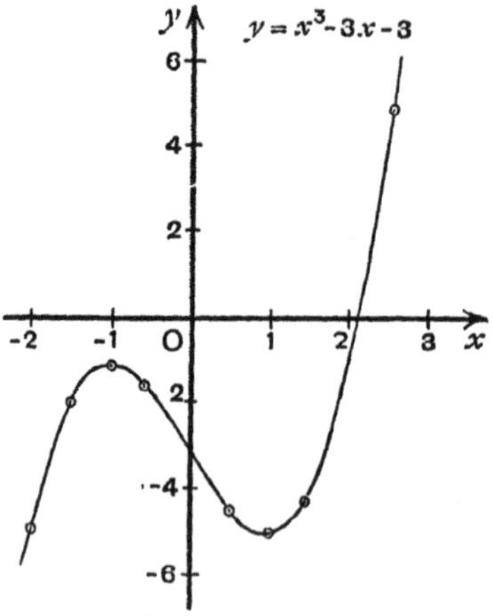

FIG. 12.

The roots of the equation are those values of x which make y zero, and so must be obtained from the points *where the curve of the graph cuts Ox*. In this case, it is seen that the plotted curve cuts Ox in three points and the approximate values of the three roots are read off as

$$x = -1, \quad x = 0{\cdot}4 \quad \text{and} \quad x = 2{\cdot}6.$$

These are the values, to one decimal place, of the exact roots already found.

Ex. 2. $x^3 - 3x - 3 = 0$.

Fig. 12 shows the graph of the function $y = x^3 - 3x - 3$, and the curve plotted is seen to cut Ox in one point only where the value of x is a little greater than 2. The cubic equation has thus only one root which is approximately equal to 2.1.

Ex. 3. A graphical method slightly different from that of the previous examples can also be used. Taking the cubic equation of the last example, we can write it in the form

$$x^3=3x+3.$$

Fig. 13 shows the graphs of the two functions $y=x^3$ and $y=3x+3$, plotted on the same graph paper and using the same scales. The first function gives a well-known curve and the second a straight line. At any point where the curves intersect, the values of y are equal, i.e. the corresponding value of x satisfies the given cubic equation. Fig. 13 shows that there is only one point of intersection and that the only root of the equation is approximately 2·1 (as before).

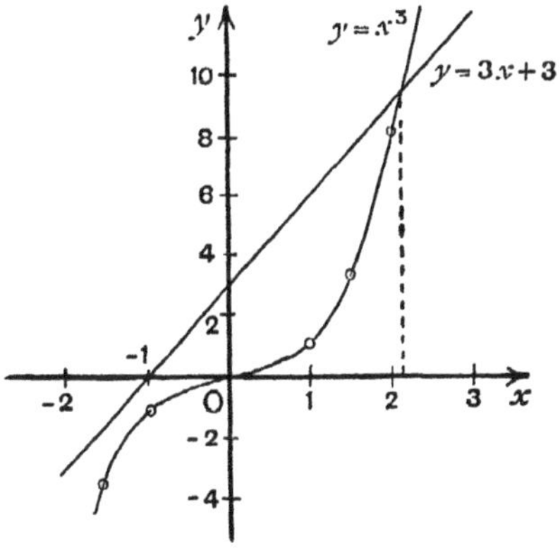

FIG. 13.

Hence, to solve any given equation $f(x)=0$ approximately, we need only plot the graph of $y=f(x)$ and determine where it cuts the axis Ox. Or, we can split the equation so that it appears in the form $\phi(x)=\psi(x)$, and then find the x-co-ordinates of the points where the graphs of $y=\phi(x)$ and $y=\psi(x)$ intersect. In either of these ways two things are determined, the *number* of the roots of the equation and the *approximate value* of each root.*

Notice that the first piece of information is often of use quite apart from the second. If this is all we require, the graphical method can be generalised into a diagrammatic method of determining the number of roots possessed by an equation of given type. For example, any cubic equation can be written in the form

$$x^3=ax^2+bx+c,$$

where a, b and c are constants. The curve $y=x^3$ has been plotted in Fig. 13. The quadratic function $y=ax^2+bx+c$ ($a\neq 0$) is represented by a U-shaped curve (known as the parabola, see 3.4 below), which

* Various numerical methods of improving upon the approximate roots obtained graphically have been devised. The best known is that of Horner; see Burnside and Panton, *The Theory of Equations*, Vol. I (8th Ed., 1918), pp. 225 *et seq.*

cuts the first curve either in one point or in three points.* The cubic equation, therefore, has sometimes one real root and sometimes three real roots according to the values of the coefficients.

2.9 Simultaneous equations in two variables.

From a single equation in one variable, we pass now to the next simplest case where two equations are given connecting two variables x and y. In general, the equations can be written

$$f_1(x, y)=0 \quad \text{and} \quad f_2(x, y)=0,$$

where f_1 and f_2 denote two given functional expressions. Such simultaneous equations are said to be solved when we have determined a pair, or a number of pairs, of values of x and y satisfying both equations. The simplest analytical device of solving simultaneous equations is to obtain, from one equation, an expression for one variable y in terms of the other variable x, and to substitute this expression into the second equation. The result is an equation in x only, which can be solved in the ways already indicated. The following examples illustrate the method.

Ex. 1. $x+y-3=0$ and $x-3y+1=0$.

From the first equation, we obtain $y=3-x$, and, on substituting in the second equation, we find that $x=2$. The corresponding value of $y=3-x=1$. The simultaneous equations have the unique solution $x=2$ and $y=1$.

The result can be generalised. The general pair of simultaneous linear equations

$$a_1x+b_1y+c_1=0 \quad \text{and} \quad a_2x+b_2y+c_2=0$$

has a unique solution:

$$x=\frac{b_1c_2-b_2c_1}{a_1b_2-a_2b_1} \quad \text{and} \quad y=\frac{c_1a_2-c_2a_1}{a_1b_2-a_2b_1}.$$

Ex. 2. $2x+y-1=0$ and $x^3+y^3-3xy=0$.

The first equation is linear and gives $y=1-2x$. Substituting into the second equation and simplifying, we obtain

$$7x^3-18x^2+9x-1=0.$$

This cubic equation is found, by graphical methods, to have three roots which are approximately equal to

$$x=0{\cdot}2, \quad x=0{\cdot}45 \quad \text{and} \quad x=1{\cdot}8.$$

* A similar result holds when $a=0$. Here the second curve is a straight line representing $y=bx+c$, which again cuts the first curve in one or in three points. (Cf. Fig. 13.)

The corresponding values of y, from $y=1-2x$, are

$$y=0{\cdot}6, \quad y=0{\cdot}1 \quad \text{and} \quad y=-2{\cdot}6.$$

There are, therefore, three solutions of the given simultaneous equations.

A direct graphical method can be used to give the approximate solution of any specified pair of simultaneous equations, $f_1(x, y)=0$ and $f_2(x, y)=0$. We plot the graph of each of these implicit functions, using the same pair of axes Ox and Oy and the same scales in each case. At any point of intersection of the curves of the graph, the co-ordinates give values of x and y satisfying both the equations, i.e. give a solution of the simultaneous equations. The graph thus tells us how many solutions there are and gives us approximate values of x and y for each solution.

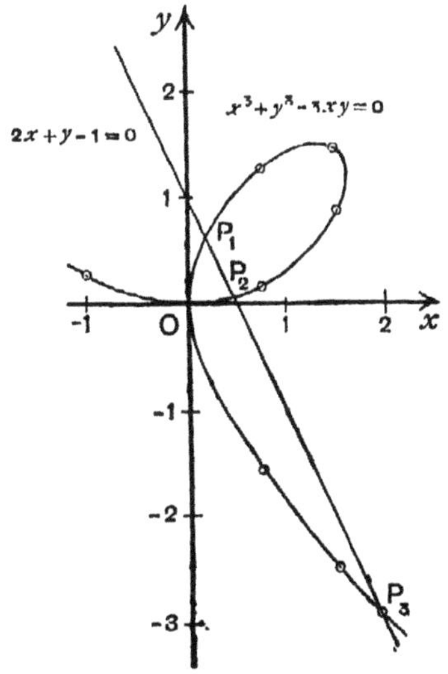

FIG. 14.

To illustrate the graphical method, the equations of Ex. 2 above can be used. Fig. 14 shows the graphs of the implicit functions

$$2x+y-1=0 \text{ and } x^3+y^3-3xy=0.$$

The first curve is a straight line and the second a looped curve which has been plotted already (in 2.2 above). There are three points of intersection, P_1, P_2 and P_3, which provide, by reading off the co-ordinates, the three pairs of values of x and y given above.

If the simultaneous equations are not given in specified form, then the graphical method breaks down. But a diagram can still be drawn, in many cases, to indicate the number of solutions, but not their numerical values. For example, the solution of any linear equation $ax+by+c=0$ and the equation $x^3+y^3-3xy=0$ can be indicated by the points of intersection of some straight line (representing $ax+by+c=0$, as shown in 3.3 below) and the looped curve of Fig. 14. There is either a single solution, or three solutions, of the simultaneous equations according to the values of a, b and c and the position of the straight line.

It appears that, in general, two equations in two variables x and y

provide a "determinate" solution, i.e. there is only a definite number of pairs of values of x and y satisfying the equations. This is a question which will be taken up at a later stage (see 11.6 below). We can, however, consider here one objection to the method of solving equations as applied (e.g.) in economic theory. It is said that a "circular" argument is involved when we write two equations in two variables x and y; the first equation gives y as a function of x, the second x as a function of y and these are inconsistent. This is an argument that has no justification.

If no relation connects the variables x and y, then each can take any value independently of the other and, if they have continuous ranges, we have a "doubly infinite" set of values of the pair (x, y). If one relation is known to exist between x and y, then the number of possible pairs (x, y) is much restricted. There is still an indefinitely large number of such pairs but the range is only "singly infinite". Finally, if another relation is known and added to the first, the possibilities are still further restricted and, except in odd cases, there is only one or a finite number of possible pairs (x, y). In this case, x and y are said to be determined. In diagrammatic terms, a pair of values of x and y is represented by a point P in a plane. If no relation between x and y is given, P can move at will over the plane. If one relation exists between x and y, then P can move only in a restricted way, i.e. along the curve representing the relation. If two relations between x and y are given, the possible positions of P are confined to the points of intersection of two curves and they are, in general, only of finite number.

To take a concrete "applied" example: if the output x of an industry is known to be related to the price p of the commodity by a demand relation (giving the amounts the market would take at various prices), then the possible values of x and p are limited but still indefinitely large in number. But, if we also know that the output is related to the price by a supply relation (giving the amounts the industry would offer at various prices), the values of x and p are much more limited and there is, in general, only one or a few pairs of values possible. Here we say that output and price are determined. The two relations of demand and supply are independent of each other, and neither alone determines both output and price. There is no circular argument.

EXAMPLES II

Functions and graphs

1. A railway ticket is in the form of a rectangle $ABCD$ with sides $AB=2$ inches and $BC=3$ inches. The corner D is clipped off by cutting along the line EF, where E lies on CD and F on AD so that $CE=\frac{1}{2}AF$. Denoting CE by x inches, find the area of the clipped ticket as a function of x and represent it graphically.

2. A variable point P is taken on a semicircle drawn on the diameter $AB=4$ inches. PN is perpendicular to AB and N is x inches along AB from A. Find the length of PN in terms of x. Deduce the area of the triangle APN as a function of x, and plot a graph of the function for $0 \leqslant x \leqslant 4$.

3. A beam has a rectangular cross-section of sides x and $(x+2)$ feet and its length is y feet. Find y as a function of x if the volume of the beam is 100 cubic feet, and represent graphically.

4. The appended table shows a simple way of obtaining values of the quadratic function $y=2x^2-2x+1$. Extend the table to give the values of the function for integral and half-integral values of x from -3 to $+3$, and plot a graph of the function. Find an explicit expression for x as a function of y and deduce that two values of x correspond to each value of $y>\frac{1}{2}$. Examine the cases where $y \leqslant \frac{1}{2}$.

x	1	2
$2x^2$	2	8
$-2x$	-2	-4
$+1$	1	1
y	1	5

5. Plot a graph of the function $y=5x-2x^2$. Show that y has a greatest but no least value. Locate the greatest value as accurately as you can from the graph. Between what values of x is y positive?

6. Graph $y=2x-1+\dfrac{1}{x}$ for positive values of x, and show how the graph can be obtained by the addition of those of $y=2x-1$ and $y=\dfrac{1}{x}$. Establish that y is always positive.

7. Plot a graph of $y=\dfrac{2x}{x^2+1}$ for positive values of x, and show how the graph for negative values of x can be deduced. What are the greatest and least values of y? Explain what happens to the value of y as x is increased indefinitely.

8. A single man earns £x a year. His taxable income, obtained by deducting one-fifth of the total income plus his personal allowance of £100, is taxed at 4s. 6d. in the £, except that the first £175 is taxed at only half the full rate. If the tax is equivalent to y shillings in the £ of original income, show graphically the variation of y for incomes between £100 and £1000.

New regulations are introduced whereby the first £135 of taxable income is taxed at one-third of the standard rate and the rest at the full rate. Plot a new graph (on similar lines) and compare the new values of y with the old values.

9. Find the function inverse to $y=\dfrac{2x+1}{x-1}$ and show that it is single-valued. Represent graphically, and give some account of the behaviour of the graph in the neighbourhood of $x=1$ and of $y=2$.

10. A chord of a circle of radius 3 inches has length x inches and is distant y inches from the centre. Find a relation between x and y. Express y as an explicit function of x and x as an explicit function of y. Are these functions single-valued or not?

11. Obtain an explicit expression for the function inverse to $y = x + \frac{1}{x}$ and show that it is not single-valued.

12. By selecting a sufficiently large number of values of x and finding the corresponding values of y, indicate that the function $y = 3x^3 + 3x^2 + x - 1$ is monotonic. Does y increase or decrease as x increases? Illustrate graphically.

13. Show that $x^2 + bx + c$ can be written in the form $(x + \frac{1}{2}b)^2 - \frac{1}{4}(b^2 - 4c)$, and deduce that $y = x^2 + bx + c$ has a least value when $x = -\frac{1}{2}b$ but no greatest value. Hence indicate the general shape of the curve $y = x^2 + bx + c$. What is the condition that the curve lies entirely above the axis of x? Consider the curve $y = -x^2 + bx + c$ in a similar way.

14. Combine the results of the previous example to indicate the nature of the curves represented by the general quadratic equation $y = ax^2 + bx + c$. Show that the curves are of the same type and can be divided into two groups, one group consisting of curves with a lowest but no highest point, and the other group of curves with a highest but no lowest point.

15. A ball, thrown vertically into the air with a given velocity v, reaches a height x after time t. It is known that x is the difference between two independent factors. The first is the effect of the initial velocity v and is measured by v times t. The second is the effect of gravity and is represented by a constant $\frac{1}{2}g$ times t^2. Express x as a function of t involving the two parameters v and g. Taking $g = 32$, find the height of the ball after 4 seconds if the initial velocity is 75 feet per second.

16. The volume of a rectangular box of sides x, $(x+1)$ and $(x+2)$ inches is y cubic inches. Of what type is y as a function of x? Generalise to give the volume when the smallest side is x inches and the other sides are respectively a inches and b inches longer. Arrange the function to show its type and its dependence on the parameters a and b.

17. Write down the general symbolic form for y as a cubic polynomial function of x, and show that functions such as $y = x^3$, $y = 1 - x^3$, $y = x(1 + x^2)$ and $y = (x+1)(x+2)(x+3)$ are all included.

18. How many parameters are needed to describe the general cubic functional relation between x and y? Show that the general expression includes such simple functions as

$$y = \frac{1}{x^2}, \quad y = \frac{x+1}{x(x-1)} \quad \text{and} \quad y^2 = x - \frac{1}{x}.$$

19. Show that the function type represented by $y = \dfrac{ax+b}{cx+d}$, where a, b, c and d are parameters, is a particular case of the general implicit quadratic function. Deduce that this type is such that both inverse functions are single-valued and of the same type.

20. Find the values denoted by $f(-1)$, $f(-\frac{1}{2})$, $f(0)$, $f(\frac{1}{2})$ and $f(1)$, when (a) $f(x) = 2x + 3$, (b) $f(x) = \dfrac{x-1}{2x+1}$ and (c) $f(x) = 2x^2 - 4x + 3 + \dfrac{1}{x}$. Explain why $f(-\frac{1}{2})$ cannot be defined in the second case and $f(0)$ in the last case.

21. If $f(x) = \dfrac{x^2 + 3x - 2}{x^2 + 2x + 4}$, express $f(2a)$ in terms of a.

22. Show that, for the function $f(x) = 1 - 2x^2 + 3x^4$, we have $f(-a) = f(a)$ for all values of a. Indicate that this is true of any polynomial containing only even powers of x.

Further, if $f(x) = 2x + 5x^3 - x^5$, show that $f(-a) = -f(a)$ for any value of a. For what general polynomial function of x is this relation valid?

23. Find the value of $f\left(\frac{1}{a}\right)$ in terms of a, when $f(x) = x^2 + 3x - 2$ and when $f(x) = \frac{(2x-1)(x-2)}{1-2x+x^2}$. In the latter case, show that $f\left(\frac{1}{a}\right) = f(a)$ for any value of a.

24. Find the value of $f(a+h)$ in terms of a and h when $f(x) = \frac{1}{x}$, $f(x) = x^2$, $f(x) = 1 - 2x^2$ and $f(x) = \frac{x-1}{2x+1}$. Deduce an expression for the "incrementary ratio"

$$\frac{f(a+h) - f(a)}{h}$$

in terms of a and h in each case.

25. Find y as an explicit function of t when $y = x^2 + 3x - 2$ where $x = 1 + t$ and when $y = 1 - x^2$ where $x = \frac{t-1}{2t+1}$.

26. Obtain a relation between x and y when it is given that $x = t^2$ and $y = 2t$. Plot a graph of the relation by giving various values to t.

27. The variables x and y are both given functions of t and, by eliminating t, an implicit function relating x and y is found. Obtain the relation in the following cases:

$$(a)\ x = \frac{2t}{1+t^2},\ y = \frac{1-t^2}{1+t^2}; \quad (b)\ x = t(t+2),\ y = t(t-1).$$

28. A wooden letter tray is in the form of an open rectangular box, of which the base sides are respectively twice and three times the depth. Express the volume of the tray and the area of wood used as functions of the variable depth of the tray. Deduce a relation between the volume and area of wood. What is the volume if 4 square feet of wood are used in the construction of the tray?

The solution of equations

29. By writing factors, find the roots of each of the equations $x^2 - 3x + 2 = 0$, $2x^2 + 5x + 2 = 0$, $6x^2 + x - 2 = 0$, $x^3 + 6x^2 + 11x + 6 = 0$ and $2x^3 - 5x^2 + x + 2 = 0$. In the first three cases, check your results by means of the formula for the general solution of a quadratic equation.

30. From the general formula, solve the quadratic equations $x^2 + 2x - 3 = 0$, $x^2 + 2x + 3 = 0$ and $3x^2 - 7x - 3 = 0$.

31. A beam has a rectangular cross-section of sides x and $(x+a)$ feet and a volume of b cubic feet. If the length of the beam is c feet, show that algebra indicates that there are two possible values of x, one of which does not "fit" the problem. Find an expression for the solution in terms of the parameters a, b and c.

32. Show that each of the cubic equations $x^3 + 5x^2 + 4x - 4 = 0$, $x^3 + 1 = 0$, and $2x^3 + x^2 - 9x - 2 = 0$ has one integral root. Hence, complete the solution of each equation.

33. By graphical methods, find the number and the approximate values of the roots of each of the equations:

$$(a)\ x^3+x^2-2x-1=0,\quad (b)\ x^3-2x-5=0,\quad (c)\ x^3-3x^2-1=0.$$

34. Verify the approximate solutions of the previous example by plotting graphs of

$$(a)\ y=\frac{1}{x}\text{ and }y=x^2+x-2,\quad (b)\ y=x^3\text{ and }y=2x+5,\quad (c)\ y=x^2\text{ and }y=\frac{1}{x-3}.$$

35. Show graphically that the equation $x^3-7x+7=0$ has two positive roots almost equal in value, and evaluate them approximately.

36. Find the approximate values of the roots of $x^4-12x+7=0$ and of the negative root of $x^4-12x^2+12x-3=0$.

37. By considering the forms of the general curves representing

$$y=a_1x^3+a_2x^2+a_3x+a_4 \quad\text{and}\quad y=\frac{a_5}{x}$$

where the coefficients are constants, show that the general quartic equation must have either four, or two, or no real roots.

38. By algebraic methods, solve each of the following pairs of simultaneous equations: $3x+6y=11$ and $6x-30y+41=0$; $2x-y+3=0$ and $xy=2$; $x-2y+1=0$ and $x^2+y^2=9$.

39. Find the solution of the general pair of linear equations

$$a_1x+b_1y+c_1=0 \quad\text{and}\quad a_2x+b_2y+c_2=0.$$

Show that the solution fails only when $\frac{a_1}{a_2}=\frac{b_1}{b_2}$. Illustrate this case of failure by attempting to solve the equations $2x-y+3=0$ and $4x-2y+5=0$. What is the graphical reason for the case of failure?

40. How many solutions are to be expected in each of the types of simultaneous equations: (*a*) $xy=1$ and $ax+by+c=0$, (*b*) $xy=1$ and $y=ax^2+bx+c$? Find, by graphical methods, the approximate solutions of

$$xy=1 \quad\text{and}\quad y=5x-2x^2.$$

41. The prices per bushel of wheat and rye are p_1 and p_2 respectively. The market demand for wheat is given by $x_1=4-10p_1+7p_2$ and for rye by $x_2=3+7p_1-5p_2$. The supply of wheat is related to the prices by the relation $x_1=7+p_1-p_2$ and the supply of rye by the relation $x_2=-27-p_1+2p_2$. Find the pair of prices which equate demand and supply both for wheat and for rye.

42. In the problem of the previous example, a tax of t_1 per bushel is imposed on wheat producers and a tax of t_2 per bushel on rye producers. Find the new prices for the equation of demand and supply (substituting p_1-t_1 for p_1 and p_2-t_2 for p_2 in the supply relations above). Show that the wheat price increases by an amount $\frac{1}{13}(9t_2-t_1)$ and the rye price by an amount $\frac{1}{13}(14t_2-3t_1)$. Deduce that

(*a*) a tax on wheat alone reduces both prices;

(*b*) a tax on rye alone increases both prices, the increase in the rye price being greater than the tax.

(See Hotelling: *Edgeworth's Taxation Paradox*, Journal of Political Economy, 1932, especially pp. 602-3.)

CHAPTER III

ELEMENTARY ANALYTICAL GEOMETRY

3.1 Introduction.

WE propose, in the present chapter, to follow up a line of development already indicated (2.7 above) and examine, in some detail, the relation between function types and curve classes. Our investigations belong to what is termed "analytical geometry" and our method is to take certain well-known classes of curves in turn, determining what type of function corresponds to each class. We have then an analytical method of treating the geometrical properties of the curves and a diagrammatic method of illustrating analytical properties of the functions.

Three simple, but essential, formulae must first be established. Two fixed points have co-ordinates $P(x_1, y_1)$ and $Q(x_2, y_2)$ referred to rectangular axes fixed in a plane Oxy. Then:

(1) The *distance between P and Q* is $\sqrt{(x_2-x_1)^2+(y_2-y_1)^2}$.

(2) The *mid-point between P and Q* has co-ordinates

$$\left(\frac{x_1+x_2}{2}, \frac{y_1+y_2}{2}\right).$$

(3) The *point which divides PQ in the ratio p : q* has co-ordinates

$$\left(\frac{qx_1+px_2}{p+q}, \frac{qy_1+py_2}{p+q}\right).$$

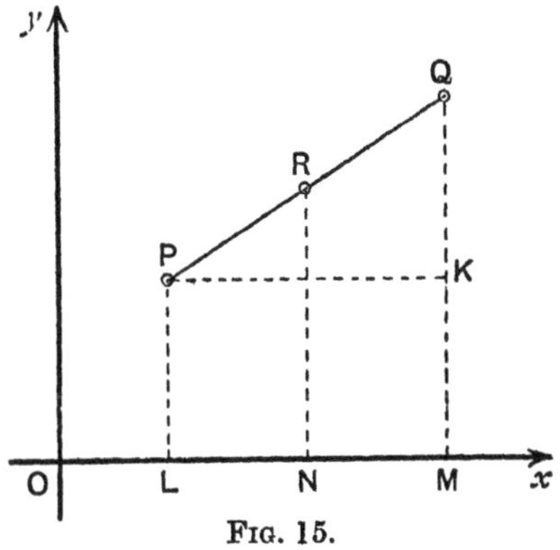

FIG. 15.

The proofs of these formulae, involving quite simple geometrical notions, make use of the notation of Fig. 15 and apply no matter what positions the fixed points P and Q take up in the plane.

(1) Pythagoras' Theorem, applied to the right-angled triangle PQN, gives

$$PQ^2 = PK^2 + KQ^2 = (x_2 - x_1)^2 + (y_2 - y_1)^2,$$

since $$PK = LM = OM - OL = x_2 - x_1,$$

and $$KQ = MQ - MK = MQ - LP = y_2 - y_1.$$

Hence, $$PQ = \sqrt{(x_2 - x_1)^2 + (y_2 - y_1)^2}.$$ Q.E.D.

(2) If R is the mid-point of PQ, then a well-known property of parallel lines tells us that N is the mid-point of LM, i.e. $LN = NM$. If the co-ordinates of R are (x, y), then

$$LN = ON - OL = x - x_1 \quad \text{and} \quad NM = OM - ON = x_2 - x.$$

So $$x - x_1 = x_2 - x,$$

i.e. $$2x = x_1 + x_2.$$

Hence, $$x = \frac{x_1 + x_2}{2}, \quad \text{and similarly} \quad y = \frac{y_1 + y_2}{2}.$$ Q.E.D.

(3) If R divides PQ in the given ratio $p : q$, then

$$\frac{LN}{NM} = \frac{PR}{RQ} = \frac{p}{q},$$

and, if the co-ordinates of R are (x, y) so that

$$LN = x - x_1 \quad \text{and} \quad NM = x_2 - x,$$

then $$\frac{x - x_1}{x_2 - x} = \frac{p}{q}.$$

Simplifying, we obtain $$x = \frac{qx_1 + px_2}{p + q},$$

and similarly $$y = \frac{qy_1 + py_2}{p + q}.$$ Q.E.D.

Formula (2) is a particular case of formula (3) with $p = q = 1$. It is interesting to notice the connection between these results and the notion of an *average*. The mid-point between P and Q is a kind of simple average point and each of its co-ordinates is the simple average of the corresponding co-ordinates of P and Q. In the same way, the point R dividing PQ in the ratio $p : q$ is in the nature of a weighted average point between P and Q. Each of its co-ordinates is the weighted average of the corresponding co-ordinates of P and Q, P's co-ordinates being weighted with q and Q's with p. In fact, the larger is q relative to p, the nearer is R to P, and the larger is the weight of P's co-ordinates in those of R.

The co-ordinates of the origin O are $(0, 0)$. If P (x, y) is any fixed point, the following are particular cases of the above results :

(1) The distance $OP = \sqrt{x^2 + y^2}$.

(2) The mid-point of OP has co-ordinates $(\frac{1}{2}x, \frac{1}{2}y)$.

(3) The point dividing OP in the ratio $p : q$ has co-ordinates

$$\left(\frac{px}{p+q}, \frac{py}{p+q}\right).$$

3.2 The gradient of a straight line.

The simplest curve class is the class of straight lines.* A straight line is fixed if two points on it are specified. We can, therefore, speak of the straight line PQ, P and Q representing two fixed points on the line, but we must always remember that the straight line extends indefinitely in both directions.

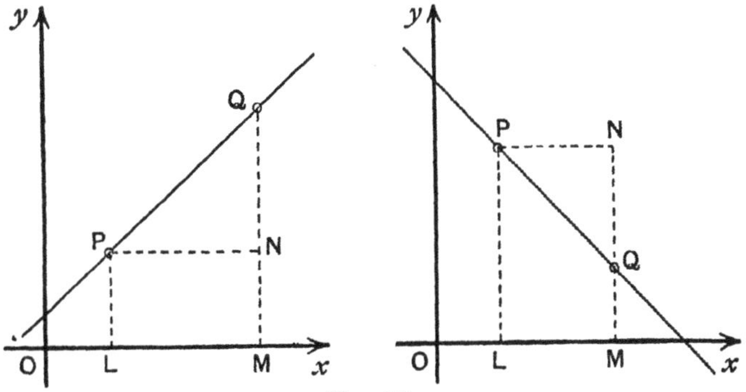

FIG. 16.

A most important property of a given straight line is its *direction* relative to a pair of fixed axes Ox and Oy, defined by means of the angle α (e.g. in degrees) that the line makes with the positive direction of the horizontal axis Ox. If the angle α is acute, the line slopes upwards from left to right; if α is obtuse, the line slopes downwards from left to right. The angular measure of the direction of a line, however, is not metrical. It is, therefore, inconvenient for most analytical purposes and a metrical indicator of direction is needed. Such an indicator is readily provided by making precise

* A straight line is a locus of points in a plane and thus a curve. It can, however, be regarded as the "limiting" case of a curve which does not "curve" at all.

our everyday notion of a "gradient". We say, for example, that a railway track has a gradient of 1 in 10 if it rises one yard vertically for every ten yards horizontally. With the notation of Fig. 16, the straight line PQ "rises" a distance NQ over a "horizontal" distance PN. The ratio $\frac{NQ}{PN}$ measures the amount of the rise per unit horizontal distance and can be called the gradient of the line. Hence:

DEFINITION: The gradient of the straight line PQ referred to the axis Ox is the ratio of NQ to PN.

Here P and Q are any two points on the straight line and N is the point where the parallel to Ox through P cuts the parallel to Oy through Q.*

It is clearly essential that the gradient should depend *only* on the direction of the straight line and not on the positions of the points P and Q selected to express it. This point needs investigation. Let P' and Q' be any other pair of points on the given line. Then, from Fig. 17, PQN and $P'Q'N'$ are similar triangles, and a well-known geometrical property of such triangles gives

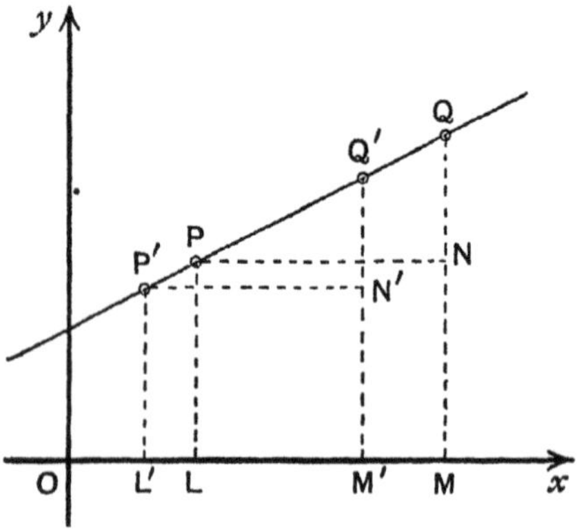

FIG. 17.

$$\frac{NQ}{PN} = \frac{N'Q'}{P'N'}.$$

The ratio is thus the same for all selections of P and Q. Exactly the same result follows if P' and Q' are any points on a line parallel to PQ. Hence, the gradient of a line depends only on its direction and all parallel lines have the same gradient.

It must be stressed that the lengths NQ and PN of the gradient ratio must be given signs according to the usual conventions. It is easily seen that the gradient of a line sloping upwards from left to

* The gradient is sometimes referred to as the "slope" of the line. If α is the angle the line makes with the axis Ox, then the gradient is the trigonometric tangent of α, i.e. gradient $= \tan \alpha$.

right is positive and that the gradient of a downward sloping line is negative. In the first diagram of Fig. 16, where the line slopes upwards, NQ and PN are both positive and the gradient is positive. In the second diagram, where the line slopes downwards, NQ is negative and PN positive, i.e. the gradient is negative.

Finally, we can show how the magnitude of the gradient indicates the steepness of the line. Since all parallel lines have the same gradient, we can draw lines through O for convenience and Fig. 18 shows a number of such lines sloping upwards with increasing steepness. The points P_1, P_2, P_3, ... are taken on the lines with the same abscissa OM. Then

$$\frac{MP_1}{OM} < \frac{MP_2}{OM} < \frac{MP_3}{OM} < \dots,$$

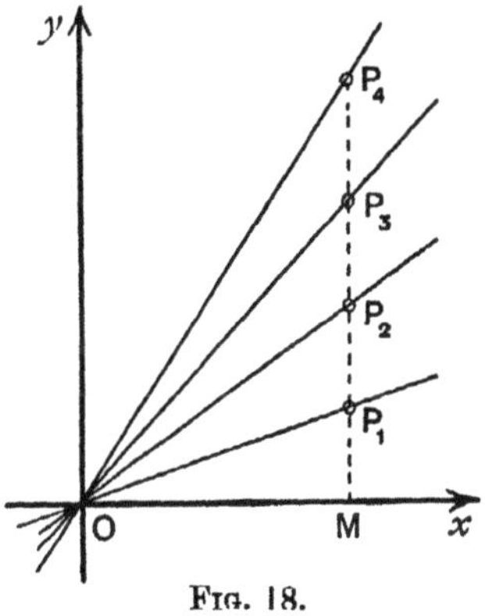

FIG. 18.

i.e. the gradient of the line increases as the line becomes steeper. In the same way, if a line slopes downwards, its gradient is negative but increases numerically as the line becomes steeper. The following results are thus established:

(1) The gradient of a line is a metrical indicator of its direction, all parallel lines having the same gradient.

(2) A line sloping upwards from left to right has a positive gradient, and a downward sloping line a negative gradient.

(3) The steeper the line, the larger is the numerical value of its gradient.

Two limiting cases arise when a given line is parallel to one or other of the axes of reference. In Fig. 18, as the line becomes less steep and tends to coincide with Ox, the gradient decreases and tends to zero. Again, as the line becomes more steep and tends to coincide with Oy, the gradient increases indefinitely.* Hence, *the gradient of a line parallel to Ox is zero and the gradient of a line parallel to Oy is indefinitely large.*

The angle between two given straight lines is independent of the

* In the terminology of the following chapter, the gradient "tends to infinity", and the gradient of a line parallel to Oy is "infinite".

positions of the lines as long as the gradients are fixed. In particular, the conditions that two lines are parallel or perpendicular involve only the gradients of the lines. If m_1 and m_2 are the gradients, then *the lines are parallel if* $m_1 = m_2$ and *perpendicular if* $m_1 \times m_2 = -1$. The first of these results needs no further proof; we have seen that parallel lines have the same gradient. The proof of the second result proceeds:

In Fig. 19, P and Q are two points with the same abscissa OM on two perpendicular lines drawn, for convenience, through the origin. The triangles OMP and QMO are similar and, since MP and MQ are necessarily lengths of opposite sign, the ratio of MP to OM equals minus the ratio of OM to MQ. Hence,

$$\frac{MP}{OM} \times \frac{MQ}{OM} = -1 \quad \text{or} \quad m_1 \times m_2 = -1.$$

The condition can also be written

$$m_1 = -\frac{1}{m_2} \quad \text{or} \quad m_2 = -\frac{1}{m_1},$$

i.e. perpendicular lines are such that the gradient of one is minus the reciprocal of the gradient of the other.

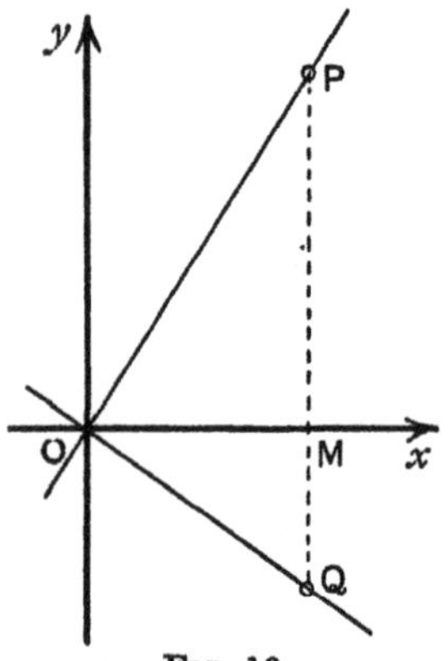

FIG. 19.

3.3 The equation of a straight line.

A straight line is fixed in position if two things, e.g. two points on the line or one point on the line and the gradient, are known about it. Our problem now is to find the equation of the line referred to some selected axes of reference, i.e. the relation between the co-ordinates of a variable point on the line.

A line passes through the fixed point P with co-ordinates (x_1, y_1) and its gradient is known to be m. Then, from Fig. 16,

$$m = \frac{NQ}{PN} = \frac{MQ - LP}{OM - OL} = \frac{y - y_1}{x - x_1},$$

for any position of a variable point Q (x, y) on the line. Hence,

The equation of the straight line with gradient m passing through the fixed point (x_1, y_1) is

$$y - y_1 = m(x - x_1). \quad \text{.........................(1)}$$

Next, suppose that the line with equation (1) also passes through a second fixed point with co-ordinates (x_2, y_2). These co-ordinates must, therefore, satisfy the equation, i.e. equation (1) holds when x_2 is substituted for x and y_2 for y:

$$y_2 - y_1 = m(x_2 - x_1) \quad \text{or} \quad m = \frac{y_2 - y_1}{x_2 - x_1}.$$

Hence,

The gradient of the straight line passing through the two fixed points (x_1, y_1) and (x_2, y_2) is $(y_2 - y_1)/(x_2 - x_1)$, and the equation of the line is

$$y - y_1 = \frac{y_2 - y_1}{x_2 - x_1}(x - x_1). \qquad \text{.........................(2)}$$

The equation (2) can also be shown to be algebraically equivalent to the alternative form

$$y - y_2 = \frac{y_2 - y_1}{x_2 - x_1}(x - x_2).$$

Attention is drawn to the values m, x_1 and x_2 appearing in the equation (1), and to the values x_1, x_2, y_1 and y_2 in the equation (2). For any given line, these values must be fixed but, by varying the values, different lines are obtained. The values are, in fact, parameters and the equation (1) or (2) is a function type representing the whole class of straight lines. Various sub-classes of the complete class of lines are obtained by varying the parameters in defined ways. For example, if the point (x_1, y_1) is kept fixed and the parameter m varied, the equation (1) represents a set of lines of varying gradient all passing through a fixed point, a set technically known as a " pencil " of lines. Again, if m is fixed and the point (x_1, y_1) varied, we obtain a set of parallel lines with a given gradient m.

The equation of the straight line, in form (1) or (2), is seen to be of the " linear " type, i.e. x and y appear only to the first degree. The converse is also true and any linear relation between x and y represents a straight line. The relation

$$ax + by + c = 0 \qquad \text{.........................(3)}$$

can be written*

$$y = -\frac{a}{b}\left(x + \frac{c}{a}\right).$$

* This arrangement of the relation holds, provided that $b \neq 0$. When $b = 0$ we have a straight line parallel to Oy (see below).

On comparison with the equation (1), we see that (3) must represent a straight line of gradient $-\frac{a}{b}$ and passing through $\left(-\frac{c}{a}, 0\right)$. So,

The equation $ax+by+c=0$ always represents a straight line and the equation of any line can be written in this form.

There is, therefore, a perfect correspondence between the class of straight lines and the linear functional relation.

We have shown, incidentally, how the gradient of a straight line can be determined from the equation of the line. The gradient of the line with equation (3) is given by $-\frac{a}{b}$, i.e. the gradient is minus the ratio of the coefficients of x and y when the equation is written with both terms on the same side. It follows that

$$a_1x+b_1y+c_1=0 \quad \text{and} \quad a_2x+b_2y+c_2=0$$

represent parallel lines if $-\frac{a_1}{b_1}=-\frac{a_2}{b_2}$, i.e. if

$$a_1b_2-a_2b_1=0$$

and perpendicular lines if $\left(-\frac{a_1}{b_1}\right)\left(-\frac{a_2}{b_2}\right)=-1$, i.e. if

$$a_1a_2+b_1b_2=0.$$

For completeness, we must examine the cases where a line is parallel to one or other of the axes. It is easily seen that *the equation of a line parallel to Ox is of the form $y=$constant, and the equation of a line parallel to Oy is of the form $x=$constant.* This follows since the y-co-ordinate of any point on a line parallel to Ox has the same value as the y-co-ordinate of any other point on the line, and similarly for the x-co-ordinates when the line is parallel to Oy. The results can also be derived from the equation (1). A line parallel to Ox has gradient m equal to zero and so its equation is

$$y-y_1=0 \quad \text{or} \quad y=y_1=\text{constant}.$$

A line parallel to Oy has an indefinitely large gradient. The equation (1) can be written in the form

$$x-x_1=\frac{1}{m}(y-y_1)$$

and, when the value of m is increased indefinitely, the equation tends to assume the form

$$x - x_1 = 0 \quad \text{or} \quad x = x_1 = \text{constant},$$

which is the equation of a line parallel to Oy. Notice, also, that the equation (3) represents a line parallel to Ox when $a=0$ and a line parallel to Oy when $b=0$. The axes themselves are straight lines with equations $y=0$ (the x-axis) and $x=0$ (the y-axis).

Finally, the equation of a straight line passing through the origin with gradient m has equation $y=mx$. So, the equation (3) represents a line through the origin if $c=0$, when it appears in the form

$$ax + by = 0.$$

3.4 The parabola.

A parabola is a curve defined as the locus of a point which is equidistant from a given point S and a given line d in a plane. The point S is called the *focus*, the line d is called the *directrix* and the

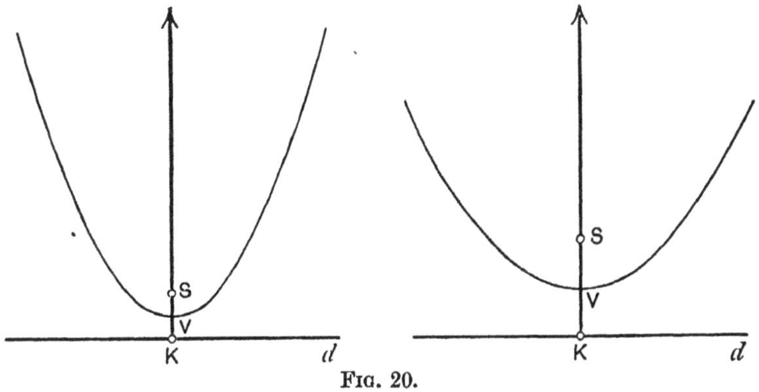

FIG. 20.

line KS perpendicular from S to d is called the *axis* of the parabola. The distance of S from $d\,(SK)$ is denoted by 2α, where α is some given constant. The general form of the parabola is shown in Fig. 20. The curve is symmetrical about its axis KS and must pass through the mid-point of KS, called the *vertex* V of the parabola.

A whole set of different parabolas are obtained by varying the positions of S and d; the general shapes of all parabolas are similar but the particular position and shape of a parabola depend on the location of S and d. The most important thing distinguishing one parabola from another is the distance of S from d, i.e. the value of the parameter α. If α is small, then S is close to d and the parabola is elongated as shown in the first curve of Fig. 20. If α is large, then S is distant from d and the parabola is flat as shown in the second curve of Fig. 20.

The equation of a given parabola can be obtained when the axes of reference are selected. We propose, here, to select the axes in such a way that the axis of the parabola is vertical, i.e. parallel to Oy. The position of the origin is still at choice and it is found that the simplest result is obtained when the origin is taken at the vertex of the parabola. In this case, the equation of the parabola is derived as follows.

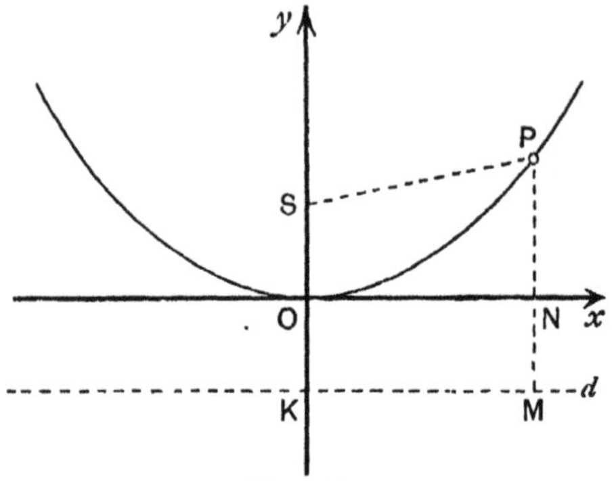

FIG. 21.

Take axes of reference as shown in Fig. 21 and suppose that the focus S is *above* the vertex of the parabola. The point S has co-ordinates $(0, \alpha)$ and the line d is parallel to Ox and a distance α below it. If P is any point (x, y) on the parabola, then the definition of the curve gives $SP = MP$. But

$$SP = \sqrt{x^2 + (y - \alpha)^2} \quad \text{and} \quad MP = NP + MN = y + \alpha.$$

So,
$$x^2 + (y - \alpha)^2 = (y + \alpha)^2$$

and this, on reduction, gives the equation of the curve in the form

$$y = \frac{1}{4\alpha} x^2.$$

On the other hand, if S is *below* the vertex, the co-ordinates of S are $(0, -\alpha)$ and d is parallel to Ox and a distance α above it. An argument exactly similar to that given above then shows that the equation of the parabola is

$$y = -\frac{1}{4\alpha} x^2.$$

The following result is thus established:

The equation of a parabola with axis vertical and vertex at the origin of co-ordinates is $y = ax^2$ where a is some constant.

The *sign* of the constant a indicates the position of the parabola. If a is positive, the axis of the parabola points upwards (as in Fig. 20); if a is negative, the axis of the parabola points downwards. The *numerical value* of a is the reciprocal of twice the distance of S from d. If a is large, then S is near d and the parabola is elongated; if a is small, S is distant from d and the parabola is flat.

Suppose, now, that the axes are selected so that the vertex of the parabola has co-ordinates (ξ, η), the axis of the parabola being parallel to Oy as before. Then, if S lies above the vertex, the co-ordinates of S are $(\xi, \eta+\alpha)$ and the directrix d is parallel to Ox and a distance $(\eta-\alpha)$ above it. If P is any point (x, y) on the parabola, the condition that P is equidistant from S and d gives

$$\sqrt{(x-\xi)^2+(y-\eta-\alpha)^2}=y-(\eta-\alpha).$$

On squaring each side of this equation and expanding, we obtain

$$(x-\xi)^2+(y-\eta)^2-2\alpha(y-\eta)+\alpha^2=(y-\eta)^2+2\alpha(y-\eta)+\alpha^2,$$

i.e.
$$(y-\eta)=\frac{1}{4\alpha}(x-\xi)^2.$$

Similarly, if S lies below the vertex, the equation is

$$(y-\eta)=-\frac{1}{4\alpha}(x-\xi)^2.$$

So,

The equation of a parabola with axis vertical and vertex at the point (ξ, η) is $(y-\eta)=a(x-\xi)^2$.

The constant a is to be interpreted, in sign and magnitude, as before. The only difference between this general form of the equation of a vertical parabola and the previous simpler form lies in the substitution of $(x-\xi)$ for x and $(y-\eta)$ for y. When $\xi=\eta=0$, the vertex is at the origin and the general case reduces to the special case.

The most important feature of the general equation of a vertical parabola is that it gives y as a quadratic function of x. The converse of this is also true. The general quadratic function

$$y=ax^2+bx+c$$

can be written (by completing the square in x) in the form

$$y=a\left(x+\frac{b}{2a}\right)^2-\frac{b^2-4ac}{4a},$$

i.e
$$\left(y+\frac{b^2-4ac}{4a}\right)=a\left(x+\frac{b}{2a}\right)^2.$$

Comparing this with the equation of the parabola above, we deduce :

The equation $y=ax^2+bx+c$, for any values of the constants a, b and $c(a\neq 0)$, represents a parabola with axis vertical.

The constant a is to be interpreted exactly as before and the vertex of the parabola is at the point with co-ordinates

$$\left(-\frac{b}{2a},\ -\frac{b^2-4ac}{4a}\right).$$

There is, therefore, a correspondence between the quadratic function type and the class of parabolas with axis vertical.*

3.5 The rectangular hyperbola.

A rectangular hyperbola is a curve which can be defined as the locus of a point the product of whose distances from two fixed perpendicular lines is a positive constant α^2. The fixed lines are called the *asymptotes* and their point of intersection the *centre* of the rectangular hyperbola. A convention is required regulating the signs to be allotted to the distances of a point on the curve from its asymptotes. Here, we draw one asymptote horizontally and the other vertically, taking distances above the first asymptote and to the right of the second as positive. The general form of the curve is shown in Fig. 22. There is one portion of the curve in the positive (or N.E.) quadrant of the plane, and, as a point P moves along this portion to the right, the distance PM from the horizontal asymptote

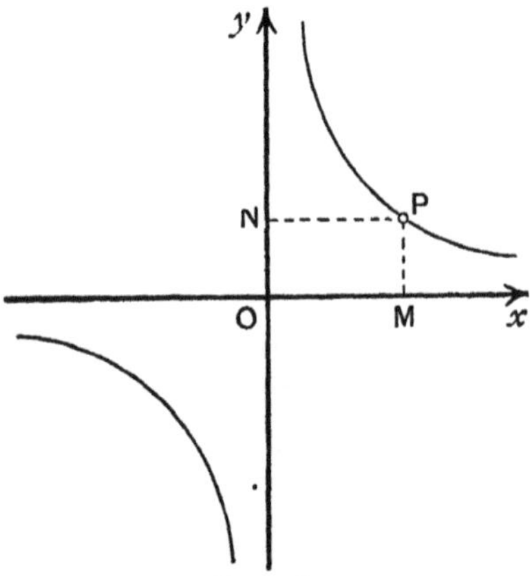

FIG. 22.

* The quadratic function and its graphical representation have been considered in the previous chapter (e.g. Ex. 1 of 2.2). From our point of view, parabolas are important because they represent such a simple and useful function type as the quadratic.

decreases in the same proportion as the distance PN from the vertical asymptote increases. The area of the rectangle $ONPM$ remains constant (α^2), and the curve approaches but never cuts the asymptotes. There is a second portion of the curve, exactly similar in shape, in the S.W. quadrant.

A whole set of different rectangular hyperbolas is obtained by varying the positions of the asymptotes and the value of the constant α^2. The general shape of all curves is the same but the particular shape of one curve, when the asymptotes are fixed, is determined by the value of α^2. Fig. 23 (where, for convenience, only the portions of the curves in the positive quadrant are drawn) shows a set of different rectangular hyperbolas corresponding to the values of α^2 indicated.

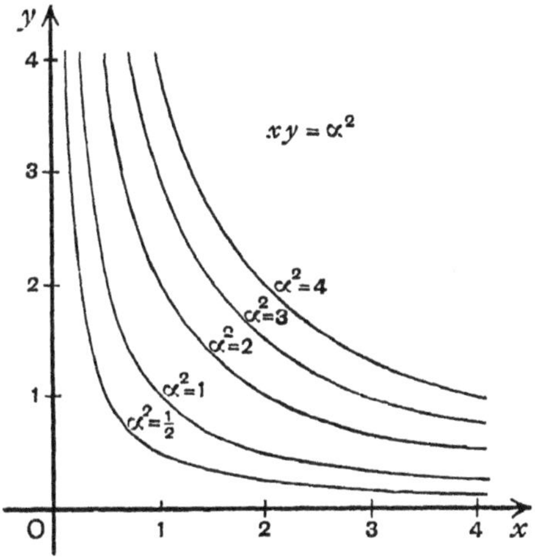

FIG. 23.

In obtaining the equation of a rectangular hyperbola here, we take the co-ordinate axes parallel to the asymptotes of the curve, only the position of the origin being at choice. The simplest case arises when the origin is taken at the centre of the curve, i.e. when the axes are the asymptotes. If P is any point (x, y) on the rectangular hyperbola in this case, then in Fig. 22

$$NP \times MP = \alpha^2.$$

But, $\quad NP = OM = x \quad$ and $\quad MP = ON = y.$

So, $\quad xy = \alpha^2.$

The equation of a rectangular hyperbola referred to its asymptotes as axes is $xy = \alpha^2$, where α^2 is a constant.

The equation can be written $y = \frac{\alpha^2}{x}$ or $x = \frac{\alpha^2}{y}$, showing that y is a single-valued decreasing function of x, and conversely.

In the general case, where the asymptotes are parallel to the axes and the centre of the curve is at the point (ξ, η), the distances of any point (x, y) on the rectangular hyperbola from the asymptotes are seen to be $(x-\xi)$ and $(y-\eta)$ respectively. The equation of the curve is then provided by the condition that

$$(x-\xi)(y-\eta)=\alpha^2.$$

The equation of a rectangular hyperbola with asymptotes parallel to the axes and centre at the point (ξ, η) is $(x-\xi)(y-\eta)=\alpha^2$.

This general form is only different from the previous special case in that $(x-\xi)$ replaces x and $(y-\eta)$ replaces y. When $\xi=\eta=0$, the centre is at the origin and the special form is obtained.

The general equation of the rectangular hyperbola in the form just obtained is found to give y as the ratio of two linear expressions in x, y being a single-valued decreasing function of x. Similar remarks apply to x as a function of y. The converse is also true. The function type

$$y=\frac{a_1x+b_1}{a_2x+b_2}$$

can be written in the form

$$(a_2x+b_2)y-(a_1x+b_1)=0,$$

i.e.

$$xy-\frac{a_1}{a_2}x+\frac{b_2}{a_2}y-\frac{b_1}{a_2}=0.$$

On factorising, this reduces to the form

$$\left(x+\frac{b_2}{a_2}\right)\left(y-\frac{a_1}{a_2}\right)=\frac{a_2b_1-a_1b_2}{a_2^2}.$$

Comparing this with the above equation of the rectangular hyperbola, we deduce :

The equation $y=\dfrac{a_1x+b_1}{a_2x+b_2}$, for any values of the constants $(a_2\neq 0)$, represents a rectangular hyperbola with asymptotes parallel to the co-ordinate axes.

The centre of the curve is at the point $\left(-\dfrac{b_2}{a_2}, \dfrac{a_1}{a_2}\right)$. The class of rectangular hyperbolas, with asymptotes parallel to the axes, is in correspondence with a simple algebraic function type.*

3.6 The circle.

A circle is the locus of a point which moves at a constant distance from a fixed point. The fixed point is the *centre* and the constant distance the *radius* of the circle. The circle is a curve discussed in all elementary geometries and needs no description here. It can be stressed, however, that there is a whole class of different circles obtained by varying the position of the centre and the length of the radius.

A circle is symmetrical about any diameter. The choice of the origin, therefore, is more important than the direction of the axes in obtaining a simple equation of a given circle. If the origin is taken at the centre of the circle, the equation of the circle is found (by the method described in 2.3) to be

$$x^2+y^2=a^2,$$

where a is the radius. This is the simplest equation possible for the circle. Similarly, if the centre has co-ordinates (ξ, η) referred to the selected axes, the equation of the circle is

$$(x-\xi)^2+(y-\eta)^2=a^2.$$

This is the general equation of a circle.

If the squares in this latter equation are expanded, the equation of a circle is seen to be a quadratic relation between x and y of such form that the coefficients of x^2 and y^2 are equal and the term in xy absent. The converse of this result is also true. By completing the squares in both x and y, the quadratic relation

$$x^2+y^2+ax+by+c=0$$

can be written in the form

$$(x+\tfrac{1}{2}a)^2+(y+\tfrac{1}{2}b)^2=\tfrac{1}{4}(a^2+b^2-4c).$$

* Our expression of this result is not quite complete. The function type only represents a rectangular hyperbola of the form considered if $(a_2b_1-a_1b_2)$ is positive. If that quantity is negative, the curve represented by the function type is of the form $(x-\xi)(y-\eta)=-\alpha^2$. This is a rectangular hyperbola lying in opposite quadrants of the plane to that shown in Fig. 22.

The function type represented by a rectangular hyperbola (i.e. the ratio of linear expressions) will be considered particularly in the following chapter (see Ex. 4 of 4.2).

The equation thus represents a circle with centre at the point with co-ordinates $(-\frac{1}{2}a, -\frac{1}{2}b)$ and with radius $\frac{1}{2}\sqrt{a^2+b^2-4c}$.

This statement, however, needs some qualification. It is only when $c<\frac{1}{4}(a^2+b^2)$ that the radius has a definite and positive value. If $c=\frac{1}{4}(a^2+b^2)$, the radius is zero and the circle reduces to a point. If $c>\frac{1}{4}(a^2+b^2)$, the square of the radius is a negative quantity, i.e. no real radius or circle exists at all. In this last case, the locus represented by the equation contains no points whatever. Hence,

If $c<\frac{1}{4}(a^2+b^2)$, the equation $x^2+y^2+ax+by+c=0$ represents a definite circle with centre at the point $(-\frac{1}{2}a, -\frac{1}{2}b)$ and with radius $\frac{1}{2}\sqrt{a^2+b^2-4c}$.

3.7 Curve classes and curve systems.

We can now pass to a number of more general considerations in the field of analytical geometry. The curves we have discussed belong to particular curve classes, the class of straight lines, of parabolas, of rectangular hyperbolas or of circles. In each case, an equation has been derived to represent the whole curve class and the equation is a particular function type involving certain parameters. Analytical geometry, in fact, is simply the study of curve classes in relation to the corresponding function types, and, if a definite curve of a certain class is given, we need only substitute the appropriate values of the parameters in the corresponding equation or function type. Our results can be summarised :

Curve class	General equation
Straight lines - - - - - -	$ax+by+c=0$
Parabolas (axis vertical) - - - -	$y=ax^2+bx+c$
Rectangular hyperbolas (asymptotes parallel to the axes) - - - - - -	$y=\dfrac{a_1x+b_1}{a_2x+b_2}$
Circles - - - - - - -	$x^2+y^2+ax+by+c=0$

A number of useful comparisons can be made. The general equation of a straight line or rectangular hyperbola gives y as a single-valued monotonic function of x, and conversely. The general equation of a parabola gives y as a single-valued (but not monotonic) function of x while x is a double-valued function of y. The general equation of a circle gives y as a double-valued function of x, and conversely. These results correspond to obvious geometrical properties of the curves themselves. For example, any parabola is cut by a line parallel to its axis in only a single point, whereas it is cut by lines perpendicular to its axis either in no point or in two points. The number of parameters in the equations of the different curve classes is significant. The general equation of a straight line contains two parameters, the ratios of a, b and c.* This corresponds to the fact that two things (e.g. two points on the line) are needed to define the position of the line. On the other hand, the general equations of the parabola (axis vertical), the rectangular hyperbola (axis parallel to the asymptotes) and the circle are function types containing three parameters. Three things are required to fix the position of one of these curves; a definite vertical parabola can be drawn, for example, to pass through three points known to lie on the curve.

Special sub-classes of a complete curve class can also be considered. In a sub-class, the number of parameters in the corresponding function type or equation is less than the number in the complete class, and the most useful sub-classes are those defined by a single parameter. A sub-class of curves of this latter kind is described as a *system of curves*. We can define, therefore, various systems of straight lines, parabolas, rectangular hyperbolas or circles, each system corresponding to a functional equation with one parameter. The following examples are instances of some of the many ways in which systems of curves can be specified.

Ex. 1. The two most important systems of straight lines have been referred to already (3.3 above). They are the system (or pencil) of straight lines through a given point and the system of parallel straight lines with a given gradient.

* In the linear equation $ax+by+c=0$, both a and b cannot be zero. If $b\neq 0$, we can write $y=\alpha x+\beta$ where $\alpha=-\frac{a}{b}$ and $\beta=-\frac{c}{b}$; if $a\neq 0$, we can write $x=\alpha' y+\beta'$ where $\alpha'=-\frac{b}{a}$ and $\beta'=-\frac{c}{a}$. There are only two independent parameters, either α and β or α' and β'. The same reasoning shows that the equation of the rectangular hyperbola $y=\frac{a_1x+b_1}{a_2x+b_2}$ contains only three independent parameters, the ratios of a_1, b_1, a_2 and b_2. We have only to divide numerator and denominator by a_2 (which cannot be zero) to see this.

Ex. 2. The system of parabolas with axis vertical and vertex at the origin is defined by the equation $y=ax^2$ with one parameter a. A less obvious system of parabolas is given by the equation

$$y+k=\frac{1}{a^2}\{x-h(a-1)\}^2 \quad 0\leqslant x\leqslant a(h-\sqrt{k})-h$$

where h and k are fixed positive numbers ($h>\sqrt{k}$) and a is a positive parameter. Any parabola of the system has vertex at a point $\{h(a-1), -k\}$ lying on a fixed line parallel to Ox, and its focus is a distance $\frac{1}{2}a^2$ from the directrix. As the parameter a increases, the vertex moves to the right and the parabola becomes flatter. A parabola of the system cuts Ox where

$$y=0 \text{ and } x=h(a-1)\pm a\sqrt{k}.$$

It cuts Ox, therefore, at only one point, at the end of the range of values of x considered, i.e. at the point where $x=a(h-\sqrt{k})-h$. The curve also cuts Oy at one point where $x=0$ and $y=h^2\left(1-\frac{1}{a}\right)^2-k$.

As a increases, the parabola cuts the axes at points farther and farther from the origin. Notice that the range of x is so chosen, in each case, that only the part of the parabola in the positive quadrant is taken. The system of parabolas is illustrated in Fig. 24, where certain curves of the system obtained when $h=10$ and $k=4$ are drawn.

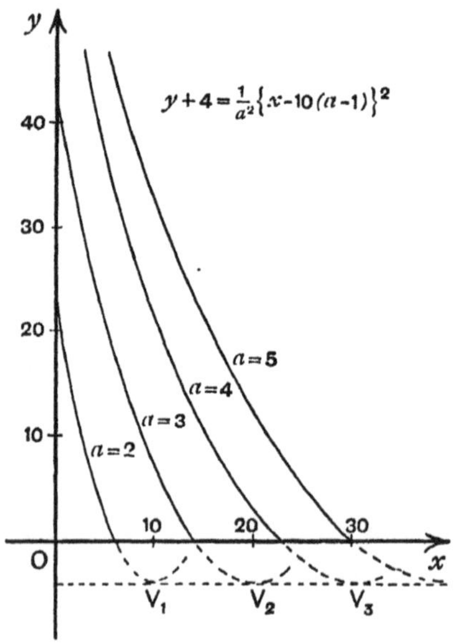

Fig. 24.

Ex. 3. The system of rectangular hyperbolas with asymptotes as axes is defined by the equation $xy=\alpha^2$ with one parameter α. This system is illustrated in Fig. 23.

A rather different system of rectangular hyperbolas is defined by

$$y=\frac{a}{x+h}-k, \quad 0\leqslant x\leqslant\frac{a}{k}-h,$$

where h and k are fixed positive numbers and a is a parameter. Here

$$(x+h)(y+k)=a,$$

i.e. the rectangular hyperbolas have fixed asymptotes parallel to the axes and with centre at $(-h, -k)$. The range $0 \leqslant x \leqslant \frac{a}{k} - h$ limits each curve of the system to the positive quadrant. Certain curves of the system in the case $h=2$, $k=1$ are shown in Fig. 25.

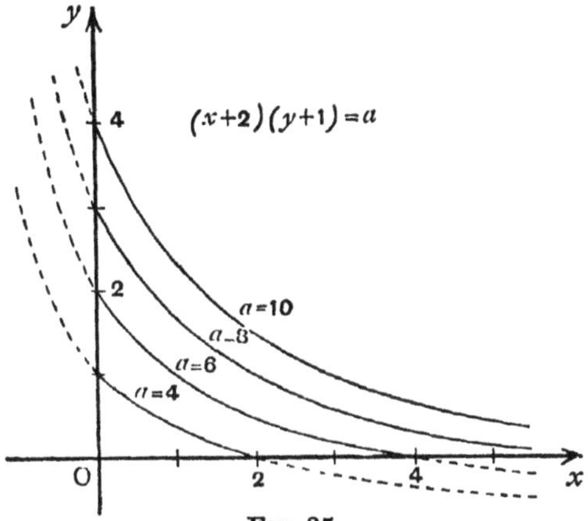

FIG. 25.

Ex. 4. The equation $x^2+y^2=a^2$ represents a system of concentric circles, a being a parameter. A different system of circles is defined by the equation

$$x+y+\sqrt{2xy}=a, \quad 0 \leqslant x \leqslant a,$$

where a is again the parameter. (The square root is assumed positive.) The equation can be written

$$\sqrt{2xy}=a-x-y,$$

and, on squaring and collecting terms, we obtain

$$(x-a)^2+(y-a)^2=a^2.$$

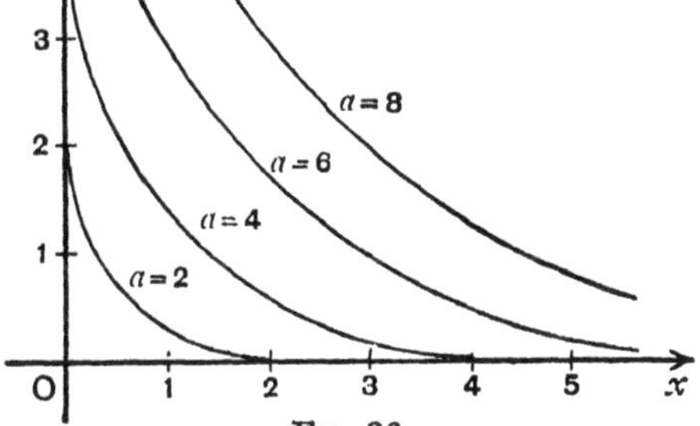

FIG. 26.

The circle of the system with parameter a thus has centre (a, a) and radius a; as the parameter a increases the centre of the circle moves away from O along a line bisecting the angle between the axes and the radius of the circle increases. From the equation, it follows that $0 \leqslant x+y \leqslant a$ and, since x is limited to the range $0 \leqslant x \leqslant a$, both x and y are positive and less than a. This limits each circle of the system to one quarter of its circumference, as shown in Fig. 26.

Any system of curves defined by a single parameter a can be represented by the equation

$$f(x, y; a)=0,$$

where, for convenience, the parameter is included in the functional expression. An important kind of curve system arises when the curves do not intersect each other, one and only one curve of the system passing through each point of the relevant part of the plane. Here, each pair of values of x and y is associated with only *one* value of the parameter a, i.e. the value defined by the curve of the system passing through the point (x, y). The parameter a can, therefore, be separated off explicitly and the equation of the curve system can be written in the form

$$f(x, y)=a.$$

Each of the curve systems instanced above is of this type.* In particular, Figs. 24, 25 and 26 show non-intersecting curve systems in the positive quadrant of the plane and their equations can be written

$$\frac{x+h}{h-\sqrt{y+k}}=a, \quad (x+h)(y+k)=a \quad \text{and} \quad x+y+\sqrt{2xy}=a$$

respectively, i.e. in the form $f(x, y)=a$. Non-intersecting curve systems are of particular interest in economic theory (see 5.7 below).

3.8 An economic problem in analytical geometry.

The consumers of a certain commodity are distributed over a geographical area which can be considered as a two-dimensional plane. The commodity is produced by two firms situated respectively at the points A_1 and A_2 in the plane. The problem is to determine how the whole area is divided into two parts by a curve, the consumers of one part obtaining their supplies of the commodity from the firm at A_1 and the consumers in the other part buying from the firm at A_2.†

The solution of the problem depends on what is assumed about the price " at works " charged by each firm, and about the nature of the transport costs from factory to consumer. It is assumed, here,

* If the curve systems of Figs. 24 and 26 were not confined to the positive quadrant in the way shown, they would cease to be of the non-intersecting type.

† The solution of this problem, as given here, is based on the work of Prof. Schneider. See *Bemerkungen zu einer Theorie der Raumwirtschaft*, Econometrica, 1935.

that the factory price p per unit of the commodity is the same for both firms. Further, it is taken that the transport cost varies directly with the distance of the consumer from the factory concerned, the distance being measured in a straight line. The transport cost per mile is, however, different for the two firms, being t_1 per unit of the commodity for the firm at A_1, and t_2 for the firm at A_2. To fix ideas, we can take $t_2>t_1$, so that it is more expensive, other things being equal, to buy from the firm at A_2.

If a consumer is s_1 miles from A_1 and s_2 miles from A_2, then he pays $(p+t_1s_1)$ per unit of the commodity if he buys from A_1 and $(p+t_2s_2)$ if his supply comes from A_2. Assuming that consumers buy from the firm for which total price is cheaper, the boundary between the two areas required is defined by the points in the plane where the total prices for the two supplies are equal:

$$p+t_1s_1=p+t_2s_2 \quad \text{or} \quad t_1s_1=t_2s_2.$$

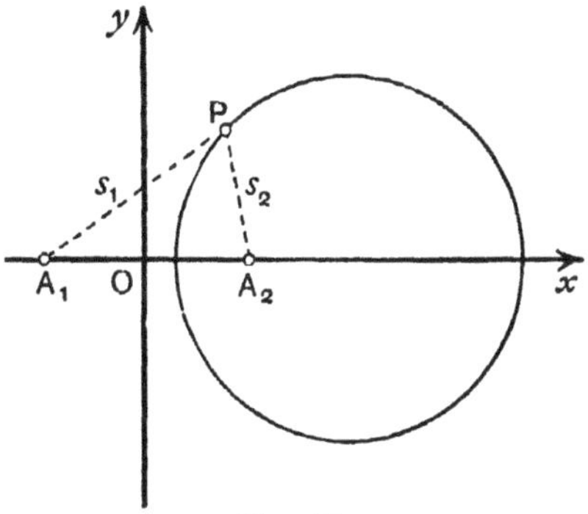

FIG. 27.

Choose axes of reference as shown in Fig. 27, where the origin O is taken as the mid-point of A_1A_2. If the firms are $2a$ miles apart, A_1 is the point $(-a, 0)$ and A_2 is the point $(a, 0)$. If P is any point (x, y) on the boundary, then

$$s_1=\sqrt{(x+a)^2+y^2} \quad \text{and} \quad s_2=\sqrt{(x-a)^2+y^2}$$

and
$$t_1\sqrt{(x+a)^2+y^2}=t_2\sqrt{(x-a)^2+y^2},$$

i.e.
$$t_1^2(x^2+y^2+2ax+a^2)=t_2^2(x^2+y^2-2ax+a^2).$$

Collecting terms and dividing through by $(t_2^2-t_1^2)$, we have

$$x^2+y^2-2a\frac{t_2^2+t_1^2}{t_2^2-t_1^2}x+a^2=0,$$

as the equation of the boundary referred to our selected axes. Write

$$b=\frac{t_2^2+t_1^2}{t_2^2-t_1^2}a>a.$$

The boundary is then seen to be a circle with centre at $(b, 0)$ and with radius $\sqrt{b^2-a^2}$. The circle has centre beyond A_2 on the line A_1A_2, and encloses the point A_2 as shown in the figure. On the assumptions we have made, therefore, the firm at A_2 supplies all consumers within this definite circle, and the firm at A_1 supplies all consumers outside the circle. Many other problems of this kind can be devised by varying the assumptions; the method of solution is similar in all cases.

EXAMPLES III

The straight line

1. Find the distance between the points (1, 2) and (– 2, 1) and the co-ordinates of the mid-point between them. Verify your results by drawing an accurate graph.

2. Obtain an expression for the distance between the points $(at_1^2, 2at_1)$ and $(at_2^2, 2at_2)$, where a, t_1 and t_2 are constants.

3. Show that the points (1, 1), (– 3, – 1) and (– 4, 1) form a right-angled triangle.

4. Show that the point (2, – 1) forms an isosceles triangle, and the point $(\sqrt{3}, 1-\sqrt{3})$ an equilateral triangle, with the pair of points (1, 2) and (– 1, 0).

5. A quadrilateral is formed by the points A (5, 2), B (20, 10), C (3, 8) and D (– 12, 0). Write down the equations of the sides and show that the quadrilateral is a parallelogram. Show that the diagonals AC and BD have the same mid-point, i.e. bisect each other.

6. By considering an isosceles right-angled triangle, show that the gradient of a line at 45° to the horizontal is unity.

7. An equilateral triangle has a side of 2 inches; show that the perpendicular from a vertex to the opposite side is of length $\sqrt{3}$ inches. Deduce that the gradients of lines making angles 30° and 60° with the horizontal are $\frac{1}{\sqrt{3}}$ and $\sqrt{3}$ respectively.

8. Find the equations of the lines passing through the point (2, 1) (*a*) with a gradient of 2, (*b*) making an angle of 60° with Ox.

9. Show that the line joining the points (a, b) and (ka, kb) passes through the origin whatever the values of a, b and k.

10. Find the equation of the line joining the points (4, 3) and (– 2, – 1) and deduce that the point (1, 1) is collinear with these two points.

11. Show that the line joining the points $(-1, -\frac{1}{2})$ and (2, 1) passes through the origin, which is one point of trisection.

12. Find values of a, b and c so that the line $ax+by+c=0$ passes through the points $(1, \frac{1}{2})$ and (– 1, 2). Verify your result by writing down the equation of the line joining these points.

13. Find the gradient of the line $2x-3y+1=0$ and the co-ordinates of the points where it cuts the axes. Hence, describe the simplest method of plotting the line on squared paper.

14. By finding the condition that the point (x, y) is equidistant from the points (3, 1) and (– 1, 2), obtain the equation of the line which is the perpendicular bisector of the line joining these points.

15. Write down the gradients of the lines $x - y + 1 = 0$, $x + y + 7 = 0$ and $2x - 3y + 1 = 0$, and show that they form a right-angled triangle. Plot a graph to illustrate this fact.

16. Obtain the equations of the lines which pass through the point (2, 2) and are respectively parallel and perpendicular to the line $2x + y - 3 = 0$. Verify graphically.

17. Find the co-ordinates of the vertices of the triangle formed by the lines $x = 1$, $x - 3y - 1 = 0$ and $x + y - 5 = 0$. Show that the line joining the point (2, 1) to any vertex is perpendicular to the opposite side. What is the geometrical meaning of this result?

18. Show that the lines $3x + y - 4 = 0$, $x - 2y + 1 = 0$ and $x + 5y - 6 = 0$ all pass through the same point. Show that the equation of the third line can be written in the form $(3x + y - 4) - \lambda(x - 2y + 1) = 0$ for some value of λ. Why should we expect this?

Curves and curve systems

19. Use the definition of the curve to find the equation of the parabola with focus at the point (2, 1) and directrix $y = -\frac{1}{2}$.

20. Plot a graph of the curve $y = x^2$. Mark the position of the focus and directrix of this parabola and verify that points on the curve are equidistant from focus and directrix.

21. Locate the focus and the directrix of each of the parabolas

$$y = x^2 + 3x - 2, \quad y = 2x^2 - 2x + 1 \quad \text{and} \quad y = 5x - 2x^2.$$

Compare the parabolas as regards size and position.

22. By considering the parabolas which represent the function, show that $y = ax^2 + bx + c$ has a greatest value if a is negative and a least value if a is positive. At what value of x do these greatest and least values occur? (Cf. Examples II, 14.)

23. Write down the equation of the rectangular hyperbola with asymptotes parallel to the axes, with centre at $(-1, 2)$ and with $\alpha^2 = 3$.

24. Find the centre and the asymptotes of the rectangular hyperbola $xy - 2x - y - 1 = 0$. (Cf. Examples II, 9.)

25. If a point moves so that the difference of its distances from the fixed points (a, a) and $(-a, -a)$ is always $2a$, show that the curve described is the rectangular hyperbola $xy = \frac{1}{2}a^2$.

26. If the two fixed points of the previous example are $(\sqrt{2}a, 0)$ and $(-\sqrt{2}a, 0)$, show that the equation of the rectangular hyperbola is $x^2 - y^2 = a^2$. Plot a graph of the curve when $a = 1$ and verify that it is the same curve as in the previous example with different axes of reference. Where are the asymptotes now?

27. Write down the equation of the circle with centre at the point $(2, \frac{3}{2})$ and with radius $\frac{5}{2}$. Where does the circle cut the axes?

28. Find the co-ordinates of the centre and the radius of the circle whose equation is $x^2 + y^2 - 3x = 0$.

29. Show that $(x + 1)(x - 2) + y^2 - 1 = 0$ is the equation of a circle with the fixed points $(-1, 1)$ and $(2, -1)$ at the ends of a diameter.

30. From the results of Examples II, 38, deduce the length of the chord cut off the line $2x - y + 3 = 0$ by the rectangular hyperbola $xy = 2$, and the length of the chord cut off the line $x - 2y + 1 = 0$ by the circle $x^2 + y^2 = 9$.

31. Re-examine the problems of Examples II, 40, in the light of the results of the present chapter.

32. Draw a diagram to illustrate the system of curves $y = \sqrt{a^2 - x^2}$, where $0 \leqslant x \leqslant a$ and where the positive square root is taken. What are the curves ?

33. If h and k are fixed positive numbers and a a positive parameter, show that $y = k - \frac{1}{a^2}(x + h)^2$, where $0 \leqslant x \leqslant a\sqrt{k} - h$, represents a system of parabolas, only the parts in the positive quadrant being taken. Draw a graph of certain curves of the system when $h = 5$ and $k = 25$, and verify that each curve is concave to the origin. Compare with the system of parabolas of Fig. 24, where each curve is convex to the origin.

34. Show that $(x - h)(y - k) = a$, where $0 \leqslant x \leqslant h - \frac{a}{k}$, and where h and k are fixed positive numbers and a a positive parameter, represents a system of rectangular hyperbolas confined to the positive quadrant. Taking $h = 2$ and $k = 1$, plot a graph showing the four curves of the system corresponding to $a = \frac{1}{4}, \frac{1}{2}, 1$ and $\frac{3}{2}$. Compare with the curve system of Fig. 25.

35. Consider the curve system $y = \sqrt{(a - x)^3}$ where a is a positive parameter and $0 \leqslant x \leqslant a$. Draw a diagram showing the curves for $a = 1$, 2, 3 and 4, verifying that the curves are non-intersecting and convex to the origin.

36. In the problem of 3.8, two firms are 10 miles apart, the price of their product at works is the same for each and transport cost per mile per unit of the commodity is three times as high for one firm as for the other. Show that the former firm supplies an area within a circle of radius $3\frac{3}{4}$ miles. Draw a graph to illustrate the distribution of the market.

37. In the problem of 3.8, the two firms are $2c$ miles apart, the prices at works are £p_1 for one firm and £p_2 for the other and transport costs are £t per mile per unit of the commodity for each. Show that the curve separating the areas supplied by the firms has equation (referred to the same axes as in 3.8) $\frac{x^2}{a^2} - \frac{y^2}{b^2} = 1$, where $a = \frac{p_2 - p_1}{2t}$ and $b^2 = c^2 - a^2$. Plot a graph to illustrate the distribution when $p_1 = £32$, $p_2 = £40$, $t = £1$ and $2c = 10$ miles.

CHAPTER IV

LIMITS AND CONTINUITY OF FUNCTIONS

4.1 The fundamental notion of a limit.

We are now in a position to begin the promised discussion of the concept of the "infinite" so essential for the full appreciation both of the power and of the limitations of mathematical analysis. We must first go back to the number system itself and elaborate the associated notions of order, continuation and limit.*

Real numbers are capable of arrangement in order of magnitude and we can select, from this order, a particular set of numbers according to any rule we care to specify. For example,

$$1,\quad 2,\quad 3,\quad 4,\quad 5,\quad 6,\quad \ldots,$$

$$\tfrac{1}{2},\quad \tfrac{2}{3},\quad \tfrac{3}{4},\quad \tfrac{4}{5},\quad \tfrac{5}{6},\quad \tfrac{6}{7},\quad \ldots,$$

are two instances of *increasing sequences* of real numbers and it is clearly possible to quote many other examples. All such sequences display the property that the numbers of the sequence can be written down one after the other, by the rule of selection, without coming to an end. This is the property of indefinite or endless continuation.

Increasing sequences can be divided into two classes. In one class, of which the first sequence above is a member, the numbers of the sequence get larger and larger without limit as we proceed through the order. Given any number we like to name, no matter how large, we can always find a number of the sequence which exceeds it. The numbers of an increasing sequence of this class are said to *tend to infinity* and here we have the idea of the infinitely large, the endless continuation of an order of larger and larger numbers.

* Much of the detail of the present chapter is, perhaps, unsuitable at a first reading. In any case, only general ideas concerning limits and continuity need be obtained at first and, after the development of the differential calculus in subsequent chapters, the reader can return profitably to a further study of the present chapter in conjunction with the examples set at the end.

In the other class, of which the second sequence above is a member, the numbers increase as we proceed through the order but not without limit. On the contrary, the particular example above is such that the numbers steadily approach nearer and nearer to the value 1, without ever attaining this value. We can get a number as close as we like to 1 by going far enough through the order. In this case, the numbers of the sequence are said to *tend to* 1, and 1 is the *limit*, or limiting value, of the sequence. Each increasing sequence of this second class tends to some definite number as a limit and, since the interval between any member of the sequence and the limit is subdivided ever more and more finely by later members, we have the basis of the idea of the indefinitely small or continuous, an idea again dependent on endless continuation.*

In the same way, we can select *decreasing sequences* of real numbers, i.e. sequences in which the order of magnitude of the numbers is reversed. Two examples of decreasing sequences are

$$1, \quad 0, \quad -1, \quad -2, \quad -3, \quad -4, \quad \ldots,$$
$$2 \quad \tfrac{3}{2}, \quad \tfrac{4}{3}, \quad \tfrac{5}{4}, \quad \tfrac{6}{5}, \quad \tfrac{7}{6}, \quad \ldots.$$

Again, there are two classes, one class illustrated by the first and the other by the second example. In a sequence of the first class, the numbers grow indefinitely larger numerically, but through negative values instead of positive values. Such a sequence is said to *tend to minus infinity* and we have the idea of the infinitely large and negative, of " minus infinity " as the counterpart to " infinity ". In the second class of sequence, the numbers tend to a limiting value, decreasing to the limit instead of increasing to it. In the second sequence above, the numbers decrease and approach the value 1, i.e. the sequence tends to 1.

* It may be noticed that we have given no justification for the statement that there are only two classes of increasing sequences, those tending to infinity and those tending to a finite limit. We have not excluded the possibility of other cases of increasing sequences. It is, however, a fundamental property of the real number system that cases other than the two here mentioned do not exist. We must assume the result without proof here. In passing, we can remark that the same property is not true of the system of *rational* numbers only. An increasing sequence of rationals may tend neither to infinity nor to a rational number as limit. This is, in fact, the way in which *irrational* numbers can be defined—as the limits of increasing sequences of rationals not tending to infinity or to a rational limit. For a discussion of these fundamental points, based on the work of Dedekind and Cantor, see Hardy, *Pure Mathematics* (3rd Ed., 1921), pp. 1-31.

Finally, sequences of real numbers can be written down in which the order does not proceed according to magnitude at all, sequences which are not increasing or decreasing. Any of the three possibilities already described may obtain for a general sequence of this kind, i.e. the sequence may tend to infinity, to minus infinity or to a numerical limit. For example, the sequence

$$\tfrac{1}{2},\ \tfrac{3}{2},\ \tfrac{3}{4},\ \tfrac{5}{4},\ \tfrac{5}{6},\ \tfrac{7}{6},\ \ldots$$

(obtained by "crossing" two previously quoted sequences) is neither increasing nor decreasing, but tends to the limit 1. An additional complication is here introduced. The sequence may be such that none of the three possibilities holds; the numbers of the sequence may "oscillate" in value without displaying any of the three regular progressions discussed above. An example of such a sequence is

$$1,\ \tfrac{3}{2},\ 3,\ \tfrac{5}{4},\ 5,\ \tfrac{7}{6},\ \ldots .$$

Here the sequence is said to *oscillate* without tending to a limit.

A given sequence of numbers, therefore, can behave in one of four ways. The sequence may

(1) tend to infinity,

or (2) tend to minus infinity,

or (3) tend to a numerical limit,

or (4) oscillate without tending to a limit,

as we proceed through the order of the sequence.

All concepts of the infinite in mathematics, whether of the infinitely large or of the indefinitely small or continuous, are based on the idea of an endless sequence of numbers. In general terms, the infinitely large corresponds to a process of multiplication without end, the indefinitely small to a process of division without end. The indefinitely large and small are thus closely related (as multiplication is to division) and they are merely two aspects of the same essential notion of order and continuation. Further, the notions of the infinite are implicit in the real number system itself. The system is of infinite extent since the order of the numbers can be continued indefinitely; it is continuous or indefinitely divisible since we can continue indefinitely to insert numbers between any two given numbers. In the present chapter, the ideas of a limit, of the infinite, and of continuity will be developed. But the development must not be allowed to hide completely the fundamental ideas we have attempted to set out briefly above.

4.2 Examples of the limit of a function.

Our next consideration is the extension of the concept of a limit, as described above, to apply to functions. The most convenient way of introducing the variety of cases of a limit of a function is by considering a number of actual examples. A careful study of the examples below is recommended; they will be found to cover most of the possible cases of limits. We can then proceed to more precise definitions.

Ex. 1. $y = 1 - \frac{1}{x}$.

Giving x in turn the sequence of positive integral values and the sequence of negative integral values, the corresponding sequences of values of y can be obtained from the function as shown below:

x	1	2	3	4	5	6	7	...
y	0	$\frac{1}{2}$	$\frac{2}{3}$	$\frac{3}{4}$	$\frac{4}{5}$	$\frac{5}{6}$	$\frac{6}{7}$	...

x	-1	-2	-3	-4	-5	-6	-7	...
y	2	$\frac{3}{2}$	$\frac{4}{3}$	$\frac{5}{4}$	$\frac{6}{5}$	$\frac{7}{6}$	$\frac{8}{7}$	...

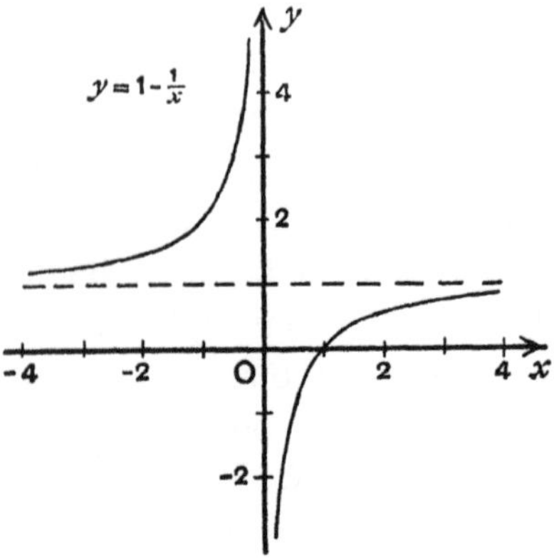

FIG. 28.

The curve representing the function, a rectangular hyperbola, can be graphed from these tables (Fig. 28).

The tables show four sequences of numbers which have been met before. The two sequences of the first table are associated in such a way that, as the x-sequence increases indefinitely through integral values, the y-sequence increases and approaches the limit 1. The variable x is actually continuous but it is clear that the insertion of further values into the x-sequence makes no essential difference; the y-sequence always increases to the limit 1. This idea of a correspondence between x- and

y-sequences is expressed by saying that *the function* $y=1-\frac{1}{x}$ *tends to* 1 *as* x *tends to infinity.* The graph of the function illustrates this fact. As x increases through positive values, the curve shown in the graph rises from left to right and approaches the horizontal line drawn at unit distance above the axis Ox.

In the same way, the second table provides a further pair of associated sequences, the x-sequence tending to minus infinity and the y-sequence decreasing to the limit 1. This enables us to say that *the function* $y=1-\frac{1}{x}$ *tends to* 1 *as* x *tends to minus infinity.* The graph again illustrates the tendency to the limit, the curve falling from right to left and approaching the same horizontal line as before as x decreases through negative values.

Ex. 2. $y=x^2+3x-2$.

The following tables of corresponding values are obtained:

x	1	2	3	4	5	...
y	2	8	16	26	38	...

x	-1	-2	-3	-4	-5	-6	...
y	-4	-4	-2	2	8	16	...

It follows that, when x is given an increasing sequence of values tending to infinity, the corresponding sequence of values of y also tends to infinity. A similar result holds when x is given a decreasing sequence of values tending to minus infinity. Hence, we say that *the function* $y=x^2+3x-2$ *tends to infinity as* x *tends to infinity or to minus infinity.* Fig. 4 illustrates these facts, the curve shown (a parabola) rising indefinitely to the right and left.

In the same way, the function $y=x^3-3x-3$ tends to infinity as x tends to infinity. But as x is allotted a decreasing sequence of values tending to minus infinity, the corresponding values of y are seen to form a similar sequence. Hence, *the function* $y=x^3-3x-3$ *tends to minus infinity as* x *tends to minus infinity.* Fig. 12 illustrates this limiting tendency, the curve shown falling indefinitely to the left.

Ex. 3. $y=\frac{3}{x^2}$.

The following tables of corresponding values are constructed:

x	± 1	± 2	± 3	± 4	± 5	...
y	3	$\frac{3}{4}$	$\frac{1}{3}$	$\frac{3}{16}$	$\frac{3}{25}$	...

x	± 1	$\pm\frac{1}{2}$	$\pm\frac{1}{3}$	$\pm\frac{1}{4}$	$\pm\frac{1}{5}$	...
y	3	12	27	48	75	...

and the graph shown in Fig. 29 is obtained.

In this case, it is immaterial whether positive or negative values of x are taken, the value of y being independent of the sign of x. As x is given an indefinitely increasing sequence of values, the corresponding sequence of values of y gets smaller and smaller. Hence, *the function* $y=\frac{3}{x^2}$ *tends to zero as* x *tends to infinity or minus infinity.* The value of y is not defined at all when x is zero. But, as x is given a decreasing sequence of values tending to zero, the corresponding sequence of values of y is seen to increase indefinitely. Hence, *the function* $y=\frac{3}{x^2}$ *tends to infinity as* x *tends to zero.* The graph of the function illustrates these limits. The curve falls and approaches the axis Ox as we move to the right or to the left indefinitely, and it rises indefinitely and approaches the axis Oy as x decreases through smaller and smaller values to zero.

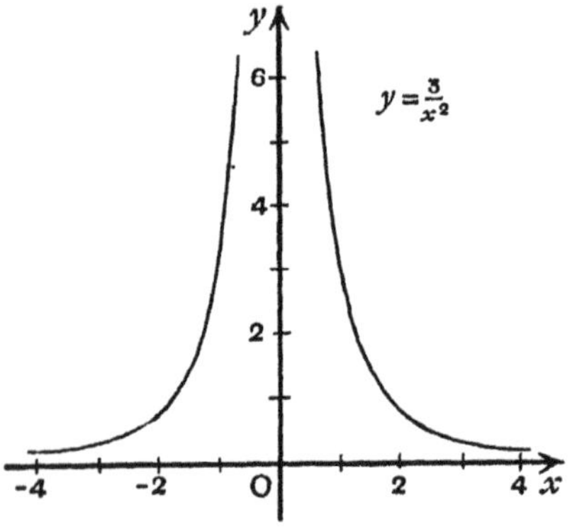

FIG. 29.

The function $y=\frac{3}{x}$, shown graphically in Fig. 5, behaves in a similar way, except in one important particular. As x decreases in numerical value, the numerical value of y increases indefinitely, but y is positive when x is positive and negative when x is negative. Hence, *the function* $y=\frac{3}{x}$ *tends to infinity as* x *tends to zero through positive values, and tends to minus infinity as* x *tends to zero through negative values.*

Ex. 4. $y=\frac{2x+1}{x-1}$.

The graph of this function is shown in Fig. 33 below, the curve represented being a rectangular hyperbola. It is only necessary to construct suitable tables of corresponding values of x and y, particular attention being paid to the approach of x to the value 1 (where y is not defined). It follows that *the function* y *tends to* 2 *as* x *tends to infinity or to minus infinity.* Further, though not defined when $x=1$, *the function* y *tends to infinity as* x *tends to* 1 (*through values greater than* 1) *and to minus infinity as* x *tends to* 1 (*through values less than* 1).

Ex. 5. $y=\frac{(1+x)^2-1}{x}$.

In this case, no value of y corresponds to zero value of x since, if we substitute $x=0$, the expression for y takes the meaningless form of zero divided by zero. The value of y is, however, obtained for any non-zero value of x and the following tables can be constructed :

x	1	$\frac{1}{2}$	$\frac{1}{3}$	$\frac{1}{4}$	$\frac{1}{5}$	...
y	3	$2\frac{1}{2}$	$2\frac{1}{3}$	$2\frac{1}{4}$	$2\frac{1}{5}$	...

x	-1	$-\frac{1}{2}$	$-\frac{1}{3}$	$-\frac{1}{4}$	$-\frac{1}{5}$	...
y	1	$1\frac{1}{2}$	$1\frac{2}{3}$	$1\frac{3}{4}$	$1\frac{4}{5}$	...

From these tables, and from the corresponding graph of the function shown in Fig. 30, it is evident that the value of y approaches the limiting value 2 as x is given a sequence of smaller and smaller values, whether positive or negative. Though x can never assume the value zero nor y attain the value 2, we can say that *the function* $y=\frac{1}{x}\{(1+x)^2-1\}$ *tends to* 2 *as* x *tends to zero.*

Ex. 6. Consider the step-function of Ex. 7 in 2.1 above. At the value $x=3$, the corresponding value of $y=6$. Further, as x approaches the value 3 *through values smaller than* 3, the corresponding value of y remains 6. But, as x approaches 3 *through values greater than* 3 (e.g. as x decreases from 4 to 3), the corresponding value of y is fixed at 7. Hence, we must say that, for this step-function, y *tends to the value* 7 *as* x *tends to* 3 (*through values greater than* 3) *and tends to the value* 6 *as* x *tends to* 3 (*through values less than* 3). The nature of the graph of the function (Fig. 8) near the point where $x=3$ illustrates this unusual limiting case.

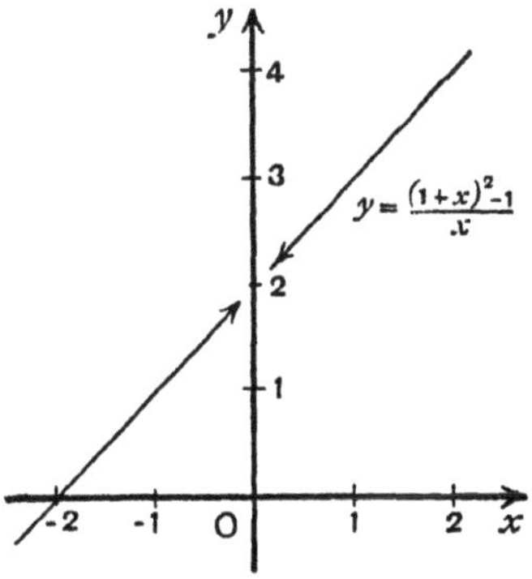

Fig. 30.

4.3 Definition of the limit of a single-valued function.

The examples of the previous section show that the notion of a limit of a function is to be derived from that of the limit of a sequence of numbers. If y is a given single-valued function of x, a sequence of values can be allotted to x according to some prescribed rule and the limit of the function is then dependent on the behaviour of the corresponding sequence of values of y. The limit is essentially a relative concept, the value approached by the sequence of y *as the* *independent variable x is made to change in some given way.* We have, therefore, to fill in the two blanks in the stock phrase

"y tends to ——, as x tends to ——."

The second blank is at our choice; the first blank is then determined from the behaviour of the function.

The following definitions include all the cases of the limit of a function $y=f(x)$ which is explicit and single-valued.*

(1) *The limits of $f(x)$ as $x \to \pm\infty$.*

For the limit of the function as x tends to infinity, we observe the behaviour of the sequence of values of the function corresponding to a selected sequence of indefinitely increasing (positive) values of x. There are four possibilities:

(*a*) The values of $f(x)$ may be positive and becoming larger and larger as x increases. The value of $f(x)$ can be made to exceed any given number, no matter how large, simply by allotting a sufficiently large value to x. Here, we say that "$f(x)$ tends to infinity as x tends to infinity", a phrase which can be symbolised

$$f(x) \to \infty \quad \text{as} \quad x \to \infty \quad \text{or} \quad \operatorname*{Lim}_{x\to\infty} f(x) = \infty.$$

In this case, the curve $y=f(x)$ rises indefinitely to the right (the axis Ox being horizontal) as shown in curve (*a*) of Fig. 31.

(*b*) The value of $f(x)$ may be negative and numerically larger and larger as x increases. This case is similar to the first and we say that "$f(x)$ tends to minus infinity as x tends to infinity", the symbolic expression for which is

$$f(x) \to -\infty \quad \text{as} \quad x \to \infty \quad \text{or} \quad \operatorname*{Lim}_{x\to\infty} f(x) = -\infty.$$

The curve $y=f(x)$ now falls indefinitely to the right (see curve (*b*) of Fig. 31).

(*c*) The values of $f(x)$ may form a sequence approaching a definite value λ as x increases. The value of $f(x)$ can be made to differ from λ by as little as we please simply by allotting to x a sufficiently large value. Here we say that "$f(x)$ tends to the limit λ as x tends to infinity" and we write

$$f(x) \to \lambda \quad \text{as} \quad x \to \infty \quad \text{or} \quad \operatorname*{Lim}_{x\to\infty} f(x) = \lambda.$$

* The definitions given here, though clear enough in a general way, are not sufficiently rigorous from a purely analytical point of view. For more precise definitions, see Hardy, *op. cit.*, pp. 162-8. See also Examples IV, 19 and 20.

The curve $y=f(x)$ now approaches nearer and nearer (from one side or the other) the fixed horizontal line $y=\lambda$ as we move indefinitely far to the right (see curve (*c*) of Fig. 31).

(*d*) The values of $f(x)$ may behave in none of the previous three ways, oscillating without settling down to any "trend" as x increases. In this case, the function $f(x)$ has no limiting value as x tends to infinity and the curve $y=f(x)$ continues its finite oscillations as we proceed indefinitely to the right (see curve (*d*) of Fig. 31).

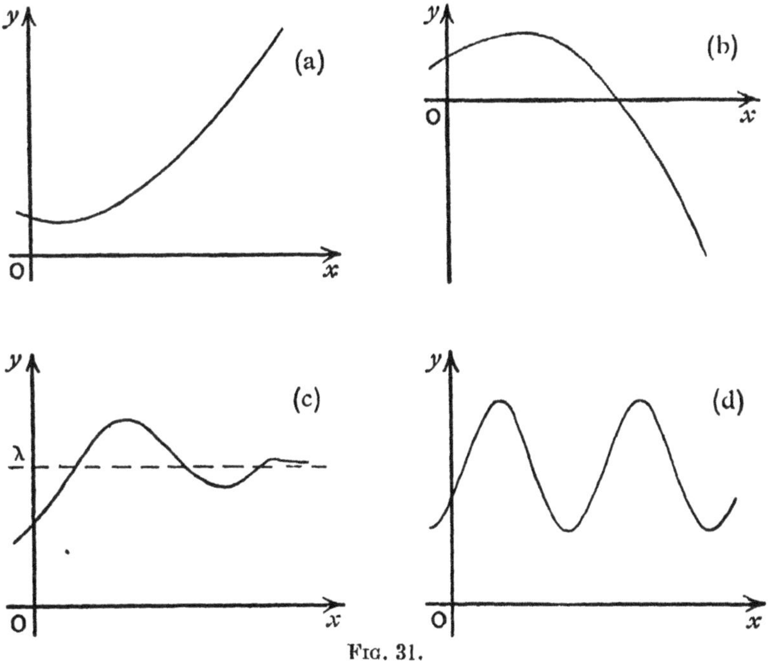

FIG. 31.

The following examples illustrate this class of limits :

$$1-\frac{1}{x}\to 1 \quad \text{as} \quad x\to\infty; \qquad \frac{2x+1}{x-1}\to 2 \quad \text{as} \quad x\to\infty;$$

$$x^2+3x-2\to\infty \quad \text{as} \quad x\to\infty; \qquad x^3-3x-3\to\infty \quad \text{as} \quad x\to\infty;$$

$$\frac{3}{x^2}\to 0 \quad \text{as} \quad x\to\infty; \qquad \frac{3}{x}\to 0 \quad \text{as} \quad x\to\infty.$$

The limit of $f(x)$ as x tends to minus infinity is defined and denoted in similar ways. For example :

$$1-\frac{1}{x}\to 1 \quad \text{as} \quad x\to-\infty; \quad x^3-3x-3\to-\infty \quad \text{as} \quad x\to-\infty.$$

(2) *The limits of $f(x)$ as $x \to a$.*

The problem of definition here is to trace the behaviour of the values of $f(x)$ as the variable x is allotted a sequence of values approaching nearer and nearer to the given value $x=a$. In no case does x ever actually attain the value $x=a$ in the limiting process.

Suppose, first, that the x-sequence approaches $x=a$ only through values *greater* than a. The corresponding sequence of values of $f(x)$ is subject to the same four possibilities as before, i.e. the function $f(x)$ can tend to infinity, to minus infinity, to a finite limit λ or to no limit at all as x approaches a through values greater than a. The symbolic notation here is similar to that adopted above, the phrase " x tends to a through values greater than a " being used and written " $x \to a+$ ". For example,

$$f(x) \to \lambda \quad \text{as} \quad x \to a+ \quad \text{or} \quad \operatorname*{Lim}_{x \to a+} f(x) = \lambda$$

indicates the case where $f(x)$ tends to the limit λ.

Similar remarks apply when the x-sequence is taken through values *less* than a. Here, x tends to a through values less than a (written $x \to a-$) and we have, for example,

$$f(x) \to \lambda \quad \text{as} \quad x \to a- \quad \text{or} \quad \operatorname*{Lim}_{x \to a-} f(x) = \lambda.$$

Finally, if it is found that $f(x)$ tends to the same limit as $x \to a+$ and as $x \to a-$, we say that $f(x)$ tends to this limit as x tends to a (written $x \to a$). So, for example,

$$f(x) \to \lambda \quad \text{as} \quad x \to a \quad \text{or} \quad \operatorname*{Lim}_{x \to a} f(x) = \lambda$$

implies that $f(x)$ tends to the limit λ as x tends to a *through values both greater and less than a*. This is the most important case in this class of limits. If the limits as $x \to a+$ and as $x \to a-$ are different, it is not possible to combine them in one notation. The diagrammatic significance of limits as x tends to a is shown in Fig 32, which exhibit the curve $y=f(x)$ in the neighbourhood of the point $x=a$. In the cases of curves (a) and (b), $f(x)$ has a single limit as $x \to a$, the limit being finite in one case and infinite in the other. In the cases of curves (c) and (d), the limits as $x \to a+$ and as $x \to a-$ are different, both being finite in one case and both being infinite in the other.

The following examples illustrate the possibilities :

$\frac{3}{x^2} \to \infty$ as $x \to 0$; $\quad \frac{(1+x)^2 - 1}{x} \to 2$ as $x \to 0$;

$\frac{3}{x} \to \infty$ as $x \to 0+$; $\quad \frac{2x+1}{x-1} \to \infty$ as $x \to 1+$;

$\to -\infty$ as $x \to 0-$; $\quad \to -\infty$ as $x \to 1-$.

The step-function of Ex. 7 of 2.1 above gives

$$y \to 7 \text{ as } x \to 3+ \text{ and } y \to 6 \text{ as } x \to 3-.$$

All the cases of Fig. 32 are illustrated by these examples.

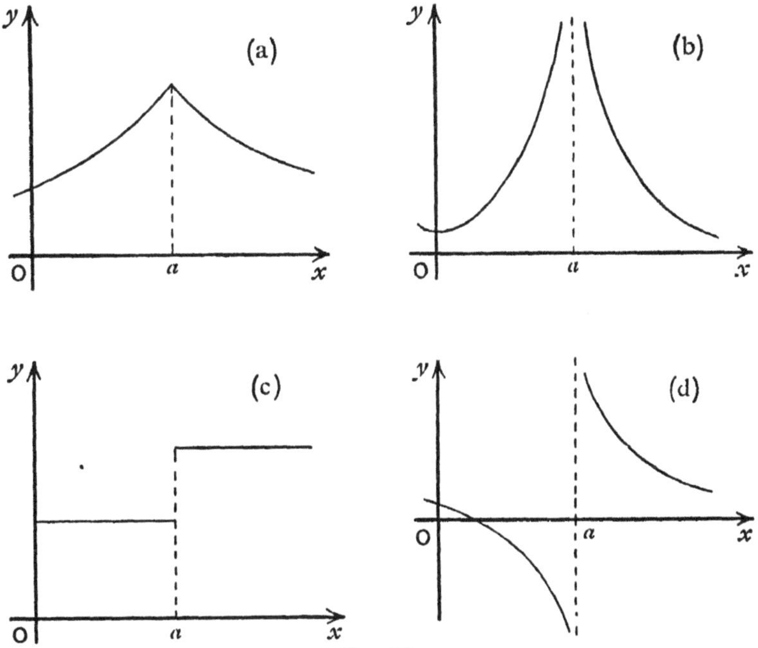

FIG. 32.

4.4 Limiting and approximate values.

It is important to point out and emphasise the connection between limiting and approximate values. If a sequence of numbers tends to a finite limit, then, by definition, the numbers of the sequence approach nearer and nearer to the limiting value without ever attaining it. It follows that any number of the sequence is an *approximate* value of the limit, and conversely. Further, the approximation becomes closer the farther advanced in the sequence is the number taken. Thus, no number of the sequence

$$\tfrac{1}{2},\ \tfrac{2}{3},\ \tfrac{3}{4},\ \tfrac{4}{5},\ \tfrac{5}{6},\ \tfrac{6}{7},\ \ldots$$

is ever equal to the limit 1. But 1 is an approximate value of any number of the sequence, and the farther we proceed through the sequence the closer is the approximation.

Similar remarks apply when we come to the case of a limit of a function. To fix ideas, suppose the function $f(x)$ tends to a finite limit λ as x tends to infinity, i.e. the value of $f(x)$ approaches, but never attains, the value λ as x increases indefinitely. It follows that λ is an approximate value of the function $f(x)$ when x is large and the approximation becomes closer as we allot larger values to x. For example, since we know that

$$\frac{2x+1}{x-1} \to 2 \quad \text{as} \quad x \to \infty,$$

it follows that this expression is approximately equal to 2 when the value of x is large. And the larger the value of x taken, the better is the approximation. To check this statement, put $x=100$:

$$\frac{2x+1}{x-1} = \frac{201}{99} = 2{\cdot}0303\ldots,$$

which is within 0·04 of the limiting value 2.

The limit of a function is thus an abstract notion in the sense that the function never attains its limiting value. On the other hand, this abstract notion is precise and does correspond to a mathematically significant property of the function. One of the main uses of the limiting concept is derived from the definition; the limiting value serves as an approximate value of the function for large values of x (if the limit is obtained as x tends to infinity) or for values of x near a (if the limit is obtained as x tends to a). In short, the precise notion of a limit covers the vaguer notions of approximate values.

In conclusion, it is hardly necessary to stress the fact that the symbol " ∞ " which appears in the symbolic expression of certain limits does not stand for a number " infinity ". The symbol means nothing by itself and must never be detached from its context. The phrase " x tends to infinity ", or its symbolic expression " $x \to \infty$ ", is to be taken as a whole; it is simply a convenient way of expressing the fact that the variable x is being allotted a sequence of larger and larger *finite* values and that the corresponding variation in the value of the function is under observation. Similarly, when we use the phrase "$f(x)$ tends to infinity", we must remember that the values of $f(x)$, however large they may be, are always finite values. The development here given of infinite limits involves only finite numbers together with the idea of processes which can be applied to finite numbers repeatedly and without end.

4.5 Some properties of limits.

Two single-valued functions $\phi(x)$ and $\psi(x)$ are given and it is found that $\phi(x) \to \lambda$ and $\psi(x) \to \mu$ as $x \to \infty$, where λ and μ are finite values. The following results can then be established:

$$\text{(1)} \quad \phi(x) + \psi(x) \to \lambda + \mu \quad \text{as} \quad x \to \infty.$$

$$\text{(2)} \quad \phi(x) - \psi(x) \to \lambda - \mu \quad \text{as} \quad x \to \infty.$$

$$\text{(3)} \quad \phi(x) \,.\, \psi(x) \to \lambda\mu \quad \text{as} \quad x \to \infty.$$

$$\text{(4)} \quad \frac{\phi(x)}{\psi(x)} \to \frac{\lambda}{\mu} \quad \text{as} \quad x \to \infty \quad (\mu \neq 0).$$

From our definitions of limits, it appears that these results are too obvious to need formal proof. Since the value of $\phi(z)$ approaches nearer and nearer to the value λ, and $\psi(x)$ nearer and nearer to the value μ, the value of (e.g.) the sum of these two functions must approach nearer and nearer to the value $(\lambda + \mu)$ as x increases indefinitely. Hence, $(\lambda + \mu)$ must be the limit of $\{\phi(x) + \psi(x)\}$ as x tends to infinity. Exactly equivalent results hold for limits as x tends to minus infinity or to a finite value a. Further, the results extend, in an obvious way, for cases of sums, differences and products or quotients involving more than two functions.*

We are now able to evaluate the limit of a complicated function, making use only of the limits of the simpler functions included in its expression and the general results set out above. The method is illustrated by the following examples:

Ex. 1. The constant value 2 has, of course, the limit 2 as $x \to \infty$. The function $\frac{3}{x} \to 0$ as $x \to \infty$. Hence, by result (1),

$$2 + \frac{3}{x} \to 2 \quad \text{as} \quad x \to \infty.$$

* With more strict definitions of limits (see 4.3, footnote), the results here given are rigidly proved only with some difficulty. See Hardy, *op. cit.*, pp. 125-30.

Ex. 2. We know that $\frac{2x+1}{x-1} \to 2$ and $1-\frac{1}{x} \to 1$ as $x \to \infty$. But

$$\frac{x^2+3x-1}{x(x-1)} = \frac{2x+1}{x-1} - \frac{x-1}{x} = \frac{2x+1}{x-1} - \left(1-\frac{1}{x}\right).$$

From result (2), it follows that

$$\frac{x^2+3x-1}{x(x-1)} \to 1 \quad \text{as} \quad x \to \infty.$$

Ex. 3. Since $\frac{2x+1}{x-1} \to 2$ and $\frac{1}{x} \to 0$ as $x \to \infty$, the result (3) shows that

$$\frac{2x+1}{x(x-1)} \to 0 \quad \text{as} \quad x \to \infty.$$

It is to be noticed that the results do not necessarily hold when the limit of either of the functions $\phi(x)$ and $\psi(x)$ is infinite. If, for example, both functions tend to infinity as x tends to infinity, then we can deduce nothing about the limit of the quotient of the functions. Numerator and denominator both increase indefinitely as x increases and the quotient can vary in diverse ways according to the forms of the two functions. If $\phi(x) = 2\psi(x)$ the quotient has the limit 2, while if $\phi(x) = \frac{1}{2}\psi(x)$ the quotient has the limit $\frac{1}{2}$. And there are many other possibilities. Care must be taken, therefore, in the use of the above results to avoid reaching fallacious conclusions about limits of functions made up of simpler functions with infinite limits.

4.6 The continuity of functions.

The definition of a function is based on the correspondence between values of two variables; to each value of one variable there corresponds a definite set of values of the other variable. The variables themselves may be continuous (in the sense that the number system is continuous) or they may be discontinuous. But the definition of the function is quite independent of any idea of continuity. The substitution of a value of one variable in the function, to give the corresponding values of the second variable, is quite independent of the substitution of other values.

The step we now propose to take is to extend the idea of continuity to qualify functions, to distinguish between functions which are continuous and functions which are discontinuous. The distinction can be expressed, in a rough preliminary way, by saying that a function is only continuous if the related variables *vary continuously together*. The variables must be separately capable of continuous variation. But this is not enough; the continuous variation of the two variables must be synchronised.

The discussion is again restricted to the case of a single-valued function of a continuous variable: $y=f(x)$. The general idea of a continuous function applies to the function as a whole and involves the continuous variation of y as x varies continuously. When we attempt to formulate a precise mathematical definition, however, we find it necessary to define first the concept of continuity *at a point*, i.e. at one value of x. Point continuity is mathematically more significant than general continuity.

The first requisite for the continuity of the function at the point $x=a$ is that $f(x)$ should be defined both for the value $x=a$ and for all values of x near $x=a$. There are now several possibilities concerning the variation of $f(x)$ as x approaches $x=a$ in relation to the value $f(a)$ assumed at $x=a$. Firstly, $f(x)$ may tend to no limiting value or to two different limiting values as x tends to a from one side and from the other. Secondly, $f(x)$ may tend to a definite limiting value as x tends to a from either side but this limit is different from $f(a)$. Thirdly, $f(x)$ may tend to the value $f(a)$ as x tends to a from either side. It is only in the third case that the function is described as continuous at $x=a$.

In diagrammatic terms, the function $y=f(x)$ is represented by a curve in the plane Oxy which is cut by lines parallel to Oy in single points. Two conditions are needed if the curve is to be continuous at the point where $x=a$. Firstly, there must be a definite point P on the curve at $x=a$; there can be no gap in the curve. Secondly, as we move along the curve towards the value $x=a$, we must arrive at this point P whether we approach from one side or the other; there can be no jump in the curve as we pass through P. A curve is continuous at a point $x=a$, therefore, if there is neither a gap nor a jump in the curve at the point. In Fig. 32, the curve (a) is the only one which is continuous at $x=a$. Curves (b) and (d) are discontinuous because there is a gap in the curve at the point, and the curve (c) because there is a jump in the curve.

The formal definition of continuity of a function at a point is:

DEFINITION: The function $y=f(x)$ is continuous at $x=a$ if

(1) $f(a)$ exists and is finite;

(2) $f(x)$ exists and is finite at all values of x near $x=a$ in such a way that $\lim_{x \to a} f(x)=f(a)$.

The definition is equivalent to the identity of limit and value of $f(x)$ at the point, and it is essential to see the necessity of the two conditions—that $f(x)$ assumes a definite value $f(a)$ at the point and that $f(x)$ tends to the same value as x approaches a from either side. The conditions correspond to the diagrammatic conditions that the curve $y=f(x)$ should have neither gap nor jump at the point in question.

It is now a simple matter to extend the definition to include continuity of a function in general. A function is continuous over a range of values of x if it is continuous at every point of the range. On the other hand, a function is not continuous if it exhibits at least one point of discontinuity in the relevant range.

4.7 Illustrations of continuity and discontinuity of functions.

If a function is continuous over a range, it is continuous at every point of the range and can be represented by a curve which has no gaps and no jumps in the range concerned. This implies, roughly speaking, that the curve can be drawn graphically on squared paper without taking pencil point from paper. It is important, however, to stress the fact that a continuous curve is not necessarily a "smooth" curve (using the everyday meaning of the term "smooth"). Smoothness is, in fact, more limited than continuity.* In Fig. 33, the curves (*a*) and (*b*) are both continuous over the whole range. The first of these, the parabola $y=\frac{1}{2}(x^2+3x-2)$, is smooth as well as continuous. The second is the curve drawn from the equation $y=1+\sqrt[3]{(x-1)^2}$ and it is continuous since it can be drawn without taking pencil from paper. But it is not smooth in the usual sense since it has a "sharp point" at the point where $x=1$.

If a function is discontinuous in a range, it must have at least one point of discontinuity where one or other of the conditions for con-

* Smoothness of a curve is a concept connected with the "derivative" of the function (see 6.5 below).

tinuity fails to obtain. In diagrammatic terms, the curve representing the function has at least one gap or jump somewhere in the range concerned. The most important example of a discontinuity arising from the non-existence of some point on the curve, i.e. by reason of a gap in the curve, is that of an *infinite discontinuity*. The value of the function $y=f(x)$ at such a point $x=a$ is infinite in the sense that $f(x)\to\pm\infty$ as $x\to a$. The curve " goes off to infinity " as we approach the point from either side. This kind of discontinuity is of frequent appearance even when we consider quite ordinary algebraic functions.

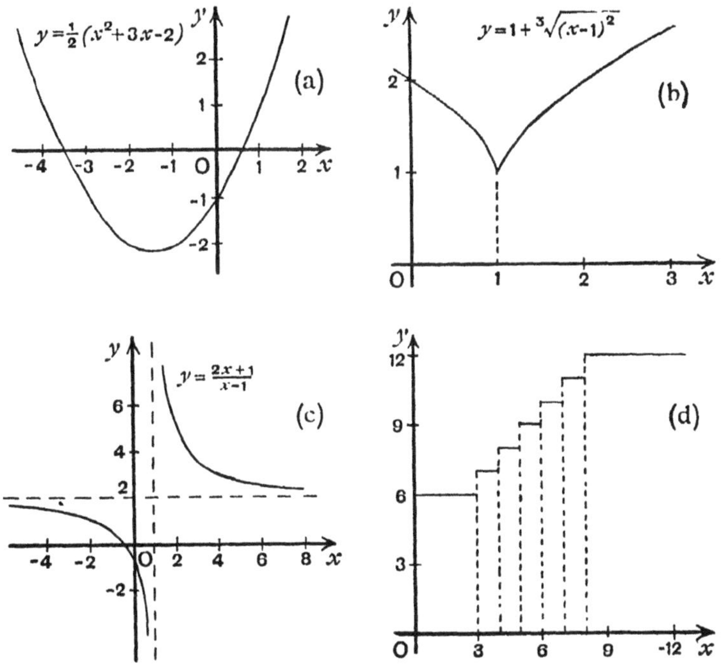

Fig. 33.

The function which represents a rectangular hyperbola with asymptotes parallel to the axes (i.e. the function which is the ratio of two linear expressions in x) always includes one infinite discontinuity, at the point where we have the vertical asymptote of the curve. The curve (c) of Fig. 33 shows a rectangular hyperbola with an infinite discontinuity at $x=1$. Infinite discontinuities may be troublesome but they can be allowed for, in mathematical analysis, by the device of infinite limits.

The case where the discontinuity arises because of a jump in the curve is very different. The curve (*d*) of Fig. 33, corresponding to the step-function of Ex. 7 of 2.1 above, displays six points of discontinuity of the "jump" kind. Such a discontinuous function is clearly of little service in mathematical analysis, and it cannot be represented by a simple analytical formula of the kind we usually consider in mathematical work. But, as we shall see in the following chapter, it is the kind of function that applies most closely to many problems of economics and other sciences. We are here faced with the difficulty that the functions most suitable for the description of economic phenomena are often those which are most inconvenient from the point of view of mathematical analysis. Some way out of this difficulty of the discontinuous element in economics must be found if the full power of mathematical analysis is to be applied.

4.8 Multi-valued functions.

Limits and continuity have been discussed above only in relation to single-valued functions. It is possible, however, to extend the definitions to apply to functions which are multi-valued and given in implicit form. Since no essentially new ideas are involved, it is not necessary here to consider the extended definitions in detail. A brief general account is all that is required.

An implicit relation between x and y usually gives y as a multi-valued function of x. The ideas of limits and continuity, as applied to such a function, necessitate the separate treatment of each of the several values of y that correspond to any given value of x. If the function is easily divisible into a number of distinct "branches" each of which is single-valued (see 2.4 above), there is little difficulty. The limit of each "branch" of the function, as x tends to infinity or to a definite value a, is defined as before and so the limits of the whole function are described. Further, each "branch" can be considered separately from the point of view of continuity. All "branches" may be continuous or one or more "branches" may show points of discontinuity.

In diagrammatic terms, taking the axis Ox as horizontal, the curve representing y as a multi-valued function of x shows several "sections" in vertical line with each other. Each "section" corresponds to one single-valued "branch" of the function. All sections may be continuous and fit on to each other in such a way that the whole curve can be drawn without taking pencil from paper. On the other hand, points of discontinuity (whether of the infinite or "jump" kind) may appear on one or more sections of the curve.

To take an actual example, the relation $x^2-y^2=9$ defines y as a double-valued function of x with the single-valued branches

$$y=\sqrt{x^2-9} \quad \text{and} \quad y=-\sqrt{x^2-9},$$

where we can consider x as taking values numerically greater than 3. Each branch is found to be continuous at all points. For the first branch, $y\rightarrow\infty$ as $x\rightarrow\pm\infty$. For the second branch, $y\rightarrow-\infty$ as $x\rightarrow\pm\infty$. Fig. 34 shows the graph of the function and illustrates the fact that it is a continuous function with the specified limiting tendencies. The curve is an example of a rectangular hyperbola but the asymptotes of the curve (shown by broken lines in Fig. 34) are not parallel to the axes of co-ordinates. The double-valued function $x^2+y^2=16$, corresponding to the circle of Fig. 6, can be shown to be continuous in the same way.

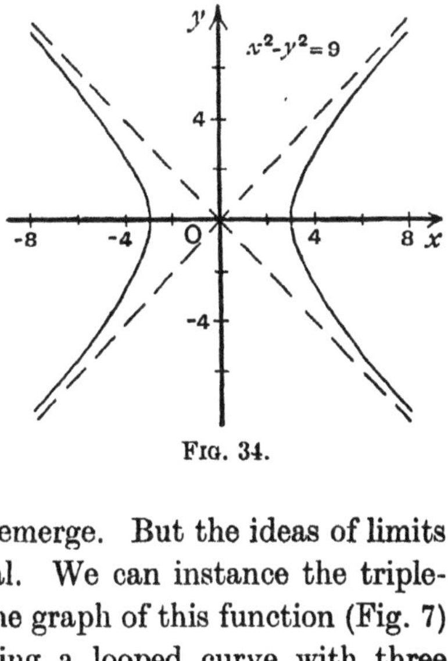

FIG. 34.

When a multi-valued function cannot be divided easily into distinct "branches", some difficulties emerge. But the ideas of limits and continuity still apply in general. We can instance the triple-valued function $x^3+y^3-3xy=0$. The graph of this function (Fig. 7) indicates that it is continuous, being a looped curve with three sections fitting continuously on to each other.

EXAMPLES IV

Limits of functions

1. By giving x an arbitrary sequence of increasing positive values, find the limit, as $x\rightarrow\infty$, of each of the following functions:

$$\frac{1}{2x+1}; \quad \frac{1-x}{1+x}; \quad \frac{1-x^2}{1+2x}; \quad \frac{2x^2+x+1}{6x^2+x-2}; \quad \frac{1+2x+3x^2}{1+2x-5x^2}.$$

Illustrate graphically. Are the same limits obtained as $x\rightarrow-\infty$?

2. What are the limits, as $x \to \infty$, of the functions:

$$(a)\ \frac{x+1}{3x-1};\quad (b)\ \frac{2x+1}{x-1};\quad (c)\ \frac{7x^2+x-2}{3x^2-4x+1};\quad (d)\ \frac{2x^2+3x+1}{3x^2-4x+1}?$$

Noticing that (c) is the sum of (a) and (b), and that (d) is the product of (a) and (b), verify the general results that the limit of a sum or product is the sum or product of the separate limits.

3. Explain why the limit, as $x \to \infty$, of $\frac{x+1}{3x-1}$ cannot be obtained by multiplying the limits of $(x+1)$ and $\frac{1}{3x-1}$.

4. Show that
$$\frac{x-1}{x+2} = \frac{x^2-2}{x+2} + \frac{1+x-x^2}{x+2}.$$

Why cannot the limit, as $x \to \infty$, of $\frac{x-1}{x+2}$ be deduced as the sum of the limits of the other two expressions?

5. Show that $\frac{3x+1}{4x-3} \to \frac{3}{4}$ as $x \to \infty$. What percentage error is involved by taking $\frac{3}{4}$ as the value of this expression in x when $x=10$ and when $x=100$? How large must x be for the expression to differ from $\frac{3}{4}$ by less than ·005?

6. Establish the general result that, as $x \to \infty$,
$$\frac{a_1x+b_1}{a_2x+b_2} \to \frac{a_1}{a_2}$$
by writing numerator and denominator as $x\left(a_1+\frac{b_1}{x}\right)$ and $x\left(a_2+\frac{b_2}{x}\right)$ respectively and by using the general results for the limits of sums, products and quotients. Check the validity of the general result by reference to the particular cases given in previous examples.

7. Find an expression for the difference between $\frac{a_1x+b_1}{a_2x+b_2}$ and $\frac{a_1}{a_2}$ and show that it tends to zero as $x \to \infty$. Hence deduce the general result of the previous example.

8. By methods similar to those of the previous two examples, show that
$$\frac{a_1x^2+b_1x+c_1}{a_2x^2+b_2x+c_2} \to \frac{a_1}{a_2} \quad \text{as} \quad x \to \infty.$$
What modifications are necessary when a_1 or a_2 is zero? Check this result by reference to the particular cases of the first two examples above.

9. If $y=ax^2+bx+c$, show that $y \to \pm\infty$ as $x \to \infty$. How does the sign of a affect the result?

10. If x is positive and $y=2x-3+\frac{1}{x}$, show that $y \to \infty$ as $x \to 0$ and as $x \to \infty$. Verify that these results can be obtained by adding the limits of $(2x-3)$ and $\frac{1}{x}$ evaluated separately.

11. A relation between x and y is defined by $x=t^2$, $y=2t$. Show that $x \to \infty$ as $y \to \infty$, and conversely. Can an exact meaning be attached to the statement that x tends to infinity "faster" than y?

12. Show that $\frac{1}{x}\{(1+x)^3-1\}\to 3$ as $x\to 0$. Generalise by finding the limit as $x\to 0$ of the expression $\frac{1}{x}\{(a+x)^3-a^3\}$.

13. Show that $\frac{1}{x}\left(\frac{1}{1+x}-1\right)\to -1$ as $x\to 0$, and generalise by finding the limit of $\frac{1}{x}\left(\frac{1}{a+x}-\frac{1}{a}\right)$ as $x\to 0$.

14. What error is involved by taking -1 as the value of $\frac{1}{x}\left\{\frac{1}{1+x}-1\right\}$ when $x=0{\cdot}1$ and when $x=0{\cdot}01$?

15. Verify that $\frac{1}{x}\{1-\sqrt{1-x}\}$ is equal to the reciprocal of $\{1+\sqrt{1-x}\}$ as long as $x\neq 0$. Deduce that, as $x\to 0$, $\frac{1}{x}\{1-\sqrt{1-x}\}\to\frac{1}{2}$. By a similar method, show that $\frac{1}{x}\{\sqrt{1+x}-\sqrt{1-x}\}\to 1$ as $x\to 0$.

16. A lead shot is allowed to fall in a vacuum. It is known that the shot falls $16x^2$ feet in x seconds. Find an expression for the average number of feet fallen per second between the ath and the $(a+x)$th second. Find the limit of the expression as $x\to 0$ and give a meaning to this limit.

17. P is the point with abscissa a, and Q the point with abscissa $(a+x)$ on the curve $y=x^3$. The line through P parallel to Ox cuts the line through Q parallel to Oy in the point L. The ratio of LQ to PL is the gradient of the chord PQ. Find an expression for this gradient in terms of a and x, evaluate its limit as $x\to 0$ and explain the meaning of the limit.

(The results of this and the previous example illustrate problems of fundamental importance in the development of the following chapters.)

18. A right-angled triangle has two equal sides of 1 inch. One of these sides is drawn horizontally and divided into $(n+1)$ equal portions. On each of the portions after the first a rectangle is formed with height equal to the vertical distance from the left-hand end of the portion to the hypotenuse of the triangle. Find an expression for the sum of the rectangle areas and evaluate the limit of the sum as $n\to\infty$. What is the meaning of the limiting value?

19. Show that a more rigid definition of the idea of a limit of a function is provided by the following:

The function $f(x)$ tends to the limit λ as x tends to a if, for any selected small quantity ϵ, we can find a small interval $a-\delta<x<a+\delta$ such that $\{f(x)-\lambda\}$ is numerically less than ϵ for any x of the interval.

20. In the same way, examine the more rigid definition:

The function $f(x)\to\infty$ as $x\to\infty$ if, for any quantity b no matter how large, we can find a quantity a so that $f(x)>b$ for all values of $x>a$.

Continuity of functions

21. Examine the function $y=\frac{1-x}{1+x}$ from the point of view of continuity and illustrate graphically.

22. From the function $xy+2x+y-1=0$, find the limit of y as $x\to -1$, and the limit of x as $y\to 1$. What restriction must be added to the statement that y is a continuous function of x, and conversely?

23. In the general relation type $axy+bx+cy+d=0$, show that y is a single-valued function of x which is continuous except for an infinite discontinuity at one point. Show also that x is a similar function of y. Relate the result to the properties of the rectangular hyperbola.

24. For what values of x does $y=\frac{x+1}{(x+2)(x+3)}$ tend to infinity? Indicate the form of the graph of the function and describe its discontinuities.

25. What are the discontinuities of the function

$$y=\frac{2x^2+3x+1}{3x^2-4x+1}?$$

Illustrate graphically.

26. Show that the function $y=\frac{a_1x^2+b_1x+c_1}{a_2x^2+b_2x+c_2}$ never tends to infinity if $b_2^2<4a_2c_2$, and tends to infinity for two values of x if $b_2^2>4a_2c_2$. Examine also the case where $b_2^2=4a_2c_2$. Interpret in terms of the continuity of the function.

27. Show that $y=\frac{x^2-1}{x-1}$ is continuous except at $x=1$. What is the nature of the discontinuity at this point? Show that the function becomes completely continuous if the value $y=2$ at $x=1$ is inserted.

28. Describe the discontinuities of the step-function, illustrating with the function defined by the table:

$0 \leqslant x<4$	$4 \leqslant x<6$	$6 \leqslant x<10$
$y=0$	$y=5$	$y=10$

29. The function $f(x)$ is defined as taking the value x when x is positive and the value $(-x)$ when x is negative. Deduce that the function is continuous but has a "sharp point" where $x=0$.

30. If a is a fixed positive number and if $n \to \infty$ through integral values, show that $a^n \to 0$ if $a<1$ and that $a^n \to \infty$ if $a>1$. What can be said of the case $a=1$? Deduce the limit of $\frac{1}{1+a^n}$ as $n \to \infty$ for various fixed values of a.

31. The function $f(x)$ is defined as taking its value from the limit of $\frac{1}{1+x^n}$ as n tends to infinity. Describe the variation of the function for positive values of x and illustrate graphically. Show that the function is discontinuous at $x=1$. What is the nature of the discontinuity?

32. Draw the graph of the function $4x^2+9y^2=36$ and show that it represents a continuous oval curve.

33. The relation $(1-x)^3-xy^2=0$ defines y as a double-valued function of x. Obtain the two single-valued "branches" and draw the graph of the function, showing that the curve is continuous with a "sharp point" at (1, 0). What are the limits of y as $x \to 0$?

34. Show that the curve representing the double-valued function

$$y^2=(x-1)^2(x-2)$$

can be taken as continuous, except that there is a point (1, 0) which must be regarded as a part of the curve but completely separate from all other parts of the curve.

CHAPTER V

FUNCTIONS AND DIAGRAMS IN ECONOMIC THEORY

5.1 Introduction.

ECONOMICS is an analytical study, concerned with the relations that exist, or can be assumed to exist, between quantities which are numerically measurable. As instances of variable quantities in economics, we need refer only to prices, interest rates, incomes, costs of production, amounts of goods bought or sold on a market and amounts of factors of production employed by a firm or industry. Some of these quantities can be measured in physical or " natural " units, others only in money or " value " units. The relevant point here is that they are measurable in some units. There are, however, other concepts of considerable importance in economics, concepts which are ordered magnitudes rather than measurable quantities and which can only be represented by " indicators " or " index-numbers ". All concepts of " utility " or " satisfaction " are of this nature, as are notions of the general level of prices or production. But, for the moment, we content ourselves with the fact that economic measurable quantities exist and leave the " ordinal " concepts for later consideration.

It follows that there can be no doubt that mathematical methods are *possible* in economics and that economic relations are expressible by means of mathematical functions. Whether it is *helpful* to introduce the mathematical technique is another question and it is not proposed to devote any space here to a discussion of this purely methodological matter. It is sufficient that mathematical analysis is applicable and can be called into service when convenient.

An important point is that the relations of economics, and the functions which express them, are usually of unspecified or unknown form. It is not often that we can say that an economic function is (e.g.) of linear or quadratic form, though it is sometimes convenient to assume that it can be approximately represented in one of these ways. Even in the most general case, however, the economic conditions of the problem impose certain limitations on the form of the functions. By considering the *economic* nature of our problem, we are usually able to say that the function concerned has the *mathematical* property (e.g.) of being single-valued and decreasing, or of being represented by a U-shaped curve. And this is enough for the profitable use of mathematical analysis.

Analytical methods, then, are directly applicable to economic problems. But diagrammatic methods are also of great service. Any economic problem capable of symbolic representation can, in general, be illustrated with the aid of diagrams. There is a curve corresponding to each function we use to interpret a relation between two economic variables ; we have demand or cost curves as well as demand or cost functions. Diagrams displaying the properties of such curves and relating one curve to another are extremely useful. At least they throw into prominence just those points we are attempting to make in our main argument.

We can now pass to a consideration of some of the functions and curves commonly employed in economic analysis, stressing their general forms and giving an indication of the more limited forms to be assumed in "normal" cases. Each problem must be considered on its own merits and allowed to dictate what is or is not to be regarded as "normal". Mathematics is the servant of the sciences, not the master. The present chapter is divided into two main sections. In the first section, we are concerned with those economic functions which are expressible directly in explicit and single-valued form, with demand and cost functions in particular. In the second section, we pass to more complicated cases where the economic functions employed appear in implicit, and not necessarily single-valued, form. Here we are concerned with function types and curve systems (see 3.7 above) as applied in the interpretation of "indifference maps".

5.2 Demand functions and curves.

Our first problem is to express, in symbolic and diagrammatic form, the conditions of demand on a consumers' goods market. We limit our discussion to the case where there is *pure competition* amongst consumers, where (to use Professor Frisch's term) each consumer acts only as a "quantity adjuster". This case occurs when no individual consumer has any direct influence on market prices (which are the same for all consumers), each consumer being able to choose only the amounts of the various goods he will take at the ruling prices. In other cases (e.g. when consumers have some direct influence on prices), the construction of the demand relations, on the lines described below, automatically breaks down.

Consider the amount of a definite good X demanded by a market consisting of a definite group of consumers. In order to obtain a simple representation of the demand for X, we must assume that

(1) the *number* of consumers in the group,
(2) the *tastes* or preferences of each individual consumer for all the goods on the market,
(3) the *income* of each individual consumer, and
(4) the *prices* of all goods other than X itself

are fixed and known. The amount of X each consumer will take can then be considered as uniquely dependent on the price of X ruling on the market. By addition, it follows that the total amount of X demanded by the market depends uniquely on the market price of X. The demand for X can only change if its market price varies.

This expression of market conditions can be translated at once into symbolic form. Let p denote, in definite units, the market price of X, and let x denote, in definite units, the amount of X demanded by the market. Then x is a single-valued function of p, which can be written, in general, in the symbolic form:

$$x = \phi(p) \quad \text{the } \textit{demand function} \text{ for } X.$$

The variables x and p take only positive values.

The demand function can be represented as a *demand curve* in a plane referred to axes Op and Ox along which prices and demands are respectively measured. Since we have taken price as the independent variable, we should, according to the mathematical convention, draw the axis Op horizontally. Since the days of Marshall, however, the economic convention has been otherwise and the axis Op has been taken vertical. We shall here adopt the economic convention and all the demand curves we use are referred to a vertical price axis.

It must be emphasised, at the very outset, that the demand relation is a completely static one and does not refer to changes in demand over time. The demand function and curve assume a fixed situation in which a set of *alternative and hypothetical* prices are given. These prices cannot be taken as those actually ruling on a market at successive points of time.

The demand function and curve can only be specified if the four factors given above are known. The *form* of the function and the *shape* of the curve depend on these factors; if changes occur in any of the factors, then the whole form of the demand function and curve changes. We can describe this by saying that a change in any one of the factors causes a *shift* in the position of the demand curve, the new demand curve being different, in shape and location, from the old in a way dependent on the nature of the change assumed in the factor. A shift in demand, corresponding to a change in the demand situation as a whole, must be carefully distinguished from a move along a given demand curve. The former implies a change of data and may easily take place over time: the demand curve can shift over time. The latter implies a hypothetical and non-temporal change in price in a given situation. Another way of putting the point is to say that the demand function $x = \phi(p)$ has form dependent on a number of *parameters*, e.g. the number of consumers, their incomes and tastes and the prices of other goods. The parameters are fixed for one demand function; they assume different values only for a shift of demand, for a change in the whole demand relation.

To take an illustrative case, suppose the market demand curve for sugar has been fixed. The population now increases, the consumers' tastes change in favour of sugar, the prices of tea and coffee fall and the income distribution changes so that the poor have more and the rich less to spend. We can say that the demand for sugar, at any one price whatever, is greater now than it was before. The whole demand curve for sugar has shifted to the right.

We must now examine the general nature of the demand function and curve in the light of the economic conditions of demand. The first point that arises here concerns the continuity of the variables x and p and the demand function itself and it is worth while treating this point at some length since our remarks apply, with little modification, to other functions introduced later. It would appear that the demand for (e.g.) sugar can only be defined for a limited number of isolated prices, for (say) prices at halfpenny intervals from 1d. to 6d. per lb. Given the relevant data, the demand " curve " for sugar then consists of an isolated set of points in the plane Oxp. The adoption of such a discontinuous representation is clearly inconvenient for a theoretical and mathematical development of the demand problem. Can price be taken, for theoretical purposes, as a continuous variable? There is certainly nothing *logically* inconsistent in the notion of a continuously variable price. Further, an examination of actual practice shows that prices are less discontinuous than they appear at first sight. Sugar may be quoted at $2\frac{1}{2}$d. or 3d. per lb. But it is possible to insert intermediate prices by quoting sugar at two lbs. for $5\frac{1}{2}$d.; at three lbs. for 8d. or $8\frac{1}{2}$d.; at five lbs. for 1s. 1d., 1s. $1\frac{1}{2}$d., 1s. 2d. or 1s. $2\frac{1}{2}$d.; and so on. Hence, unless there are strong reasons to the contrary, it can be safely assumed that the price in a demand relation is a continuous variable taking all positive values. The demand is then uniquely determined for each price of this continuous range.

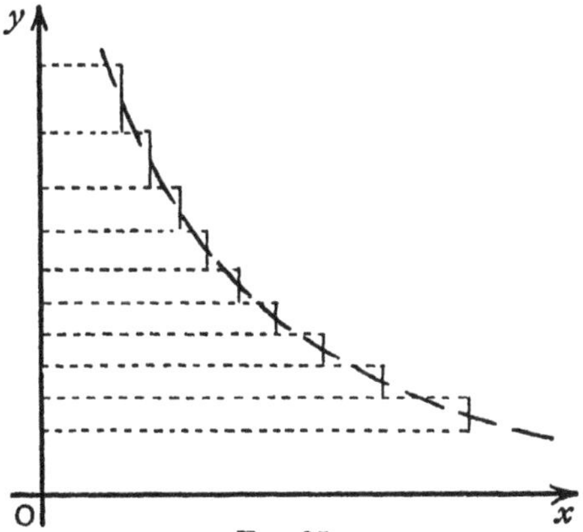

FIG. 35.

But, even when p varies continuously, it does not necessarily follow that x varies continuously with p, i.e. that the demand function is continuous. On the contrary, it is reasonable to assume that demand varies by a series of " jumps " as price increases continuously. Demand may be insensitive to small price changes and " jumps " only when price has increased by a definite amount. In this case, the demand function has " jump " discontinuities at certain prices and is a step-function of the kind already considered. The demand curve takes the form illustrated in Fig. 35.

It is clearly convenient, for theoretical purposes, to assume that the demand function and curve are continuous, that demand varies continuously with price. There is, at least, nothing that violates our logical sense in this assumption. And it is reasonable to suppose that any actual demand conditions, for a large market at least, can be " fitted " sufficiently well by a continuous demand curve for the latter to be assumed. Fig. 35 illustrates a hypothetical fitting of a continuous demand curve. It is not proposed, however, to consider here the " fit " of a continuous demand curve to actual data, even if such data can be obtained. We need only notice the convenience of the continuous demand function for theoretical purposes and point out the limitations of this continuity assumption. We proceed, therefore, on the assumption that demand is a continuous function of a continuously variable price. The assumption is made for purposes of mathematical convenience ; it can be given up, if necessary, at the cost of considerable complications in the theory.

5.3 Particular demand functions and curves.

We come now to an assumption of very different nature from that of continuity. It is assumed that the larger is the price the smaller is the market demand for the good concerned. The validity of this assumption must be examined, in the light of actual conditions, in each particular case. In some cases of demand, as Marshall has pointed out following Giffen, it is certainly not valid. But it is surely appropriate to all usual or " normal " cases of demand. Our main development can thus proceed on this assumption of falling demand, and the theory can be modified where necessary to meet any exceptional cases that arise.

The assumption implies that x decreases as p increases, i.e. that the demand function is monotonic decreasing. It follows that the inverse function is also monotonic decreasing. We can take either demand as dependent on price, or conversely :

$$x=\phi(p) \quad \text{and} \quad p=\psi(x).$$

Here, both functions are single-valued, continuous and monotonic decreasing. In diagrammatic terms, the demand curve falls continuously from left to right and it has the same general appearance whether we take Op horizontally or (as we do here) vertically.

Further restrictions on the form of the demand relation can now be added according to the particular requirements of the demand problem considered. Fig. 36 shows three general demand curves in certain hypothetical and illustrative cases. The first curve D_1 is taken to represent the demand of a single family for sugar.* The price is measured in pence per lb. and the demand in lbs. per month per adult member of the family. No sugar is demanded at prices above 1s. 3d. per lb. As the price falls, the demand increases slowly at first, then rather rapidly and finally (for prices below 2d. per lb.)

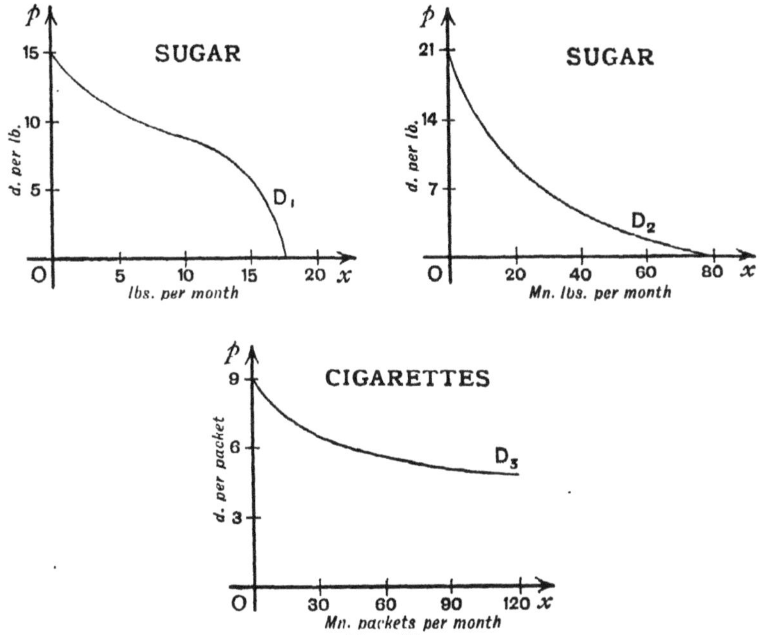

FIG. 36.

slowly again. The family never demands more than the "saturation" amount of 17 lbs. per adult per month. The demand curve cuts both axes and is first convex and then concave to the price axis. The shape of such a family demand curve can vary, of course, over a wide range for different families and different goods. The form D_1 is given merely as a fairly typical example.

The demand curve D_2 is illustrative of the demand of a large market for sugar, taken as typical of a good with no closely related substitutes. The market demand becomes zero at 1s. 9d. per lb., no family making any purchases at this or any higher price. The demand then increases at an ever increasing rate as the price falls

* An "individual" demand curve of this kind can be regarded as the limiting case of a "market" demand curve when only one consumer is taken. A demand curve, in fact, can be defined for the whole group of consumers or for any particular group of them we care to specify.

until approximately 80 Mn. lbs. per month are consumed at very low prices. The curve cuts both axes and is convex to the price axis at all points. As an approximate representation of such a demand, we could take a demand curve approaching the axes "asymptotically" in the manner of the rectangular hyperbola (see 3.5 above).

Sugar is a commodity which is well defined and clearly demarcated from its nearest substitute (say saccharine or jam). But most commercial commodities are such that there is, in each case, a number of closely allied commodities.* The variation can be obtained in many ways, e.g. by change of quality or by trade-marks. The demand curve D_3 is drawn to illustrate the demand for Player's cigarettes, taken as typical of such a commodity. Assuming that the price of competing cigarettes is 6d. per packet of ten (a fixed price by our assumptions), the demand for Player's cigarettes is very small at 9d. per packet and increases rapidly as the price falls, becoming very large when the price is 5d. per packet or less. The demand curve cuts the price axis and, since a small price fall results in a large increase in demand, it is nearly parallel to the demand axis for most of its course. The curve must cut the demand axis at some point, but only when the demand is very large compared with that at the price of (say) 6d. per packet.

It is often convenient to represent a demand law by a definite type of function, at least as an approximation for a certain range of prices. The demand curve is then of definite class, e.g. a straight line or a parabola. The following function types can be given as examples of suitable demand laws:

(1) $x=\frac{a-p}{b}, \quad p=a-bx.$ (2) $x=\frac{a}{p+c}-b, \quad p=\frac{a}{x+b}-c.$

(3) $x=\frac{a-p^2}{b}, \quad p=\sqrt{a-bx}.$ (4) $x=\frac{a-\sqrt{p}}{b}, \quad p=(a-bx)^2.$

(5) $x=\sqrt{\frac{a-p}{b}}, \quad p=a-bx^2.$ (6) $x=bp^{-a}+c, \quad p=\left(\frac{b}{x-c}\right)^{\frac{1}{a}}.$

(7) $x=ae^{-bp}, \quad p=\frac{1}{b}\log\frac{a}{x}.$ (8) $x=p^a e^{-b(p+c)}.$

Each of these demand laws satisfies the "normal" monotonic de-

* I.e. commodities of the kind considered in the economic theory of "monopolistic" or "imperfect" competition.

creasing condition.* The variables x and p take only positive values and a, b and c are positive constants in each case. If actual numerical demand data are available, it may be possible to fit actual values to the constants and to draw a definite graph of the demand curve.

The demand curve of the first law is a downward sloping straight line cutting the price axis at the point A where $p=a$ and with gradient (referred to the price axis) numerically equal to b. If the values of a and b are changed, then the demand line shifts in position. If a alone varies, the line is transposed parallel to itself. If b alone varies, the line is rotated about the point A on the price axis. If both a and b vary, the line shifts by transposition and rotation. The second demand law is $(x+b)(p+c)=a$. The demand curve is now a rectangular hyperbola with asymptotes parallel to the axes and with centre at the point where $x=-b$ and $p=-c$. The curve is restricted, of course, to that part of the rectangular hyperbola in the positive quadrant. The demand curve shifts in position when the values of a, b and c are varied. Variation in the values of b and c alone transposes the curve without change of shape. If b decreases, for example, then the curve is shifted bodily to the right. Variation in the value of a, however, changes the shape of the demand curve. Fig. 37 shows actual cases of linear and hyperbolic demand curves drawn from the demand laws:

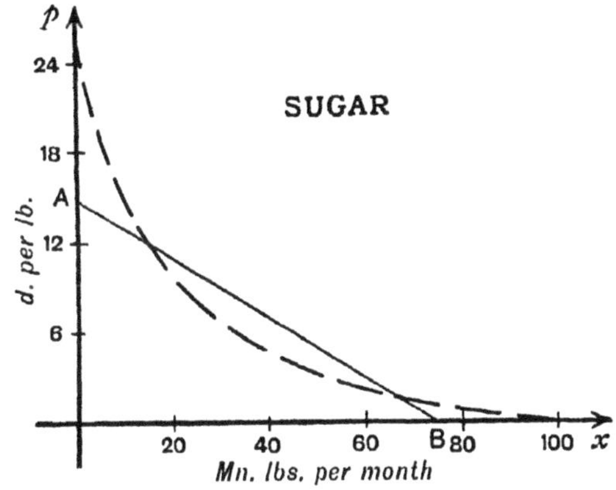

FIG. 37.

$$x=75-5p \qquad \text{or} \quad p=15-\tfrac{1}{5}x,$$

and

$$x=\frac{600}{p+5}-20 \quad \text{or} \quad p=\frac{600}{x+20}-5.$$

In each case, x is the market demand for sugar, in Mn. lbs. per month, when the price of sugar is p pence per lb.

* The last two cases involve exponential and logarithmic expressions (see 9.8 below). In case (4), x must be limited to the interval $0 \leqslant x \leqslant \dfrac{a}{b}$.

An important instance of the use of parametric constants is thus obtained. If a demand relation is known to be of one of the general types quoted above, then the position of the demand curve is fixed only when the parameters are given definite values. The values a, b and c are constants under given demand conditions. But a shift in demand results when the parameters are changed in value, the nature of the shift depending on the particular change adopted. Demand laws such as those given above suffice, not only for the description of some given demand situation, but also for the shifting of the demand situation over time.

5.4 Total revenue functions and curves.

The demand for a good X is represented by the two inverse functions, $x=\phi(p)$ and $p=\psi(x)$, of the demand law. When the demand is x and the price p, the product $R=xp$ is called the *total revenue* obtainable from this demand and price. It represents the total money revenue of the producers supplying the demand and the total money outlay of the consumers composing the demand. Now

$$R=p\phi(p)=x\psi(x),$$

i.e. R can be expressed *either* as a function of the price *or* as a function of the output (=demand). The latter is the more convenient expression in most cases (e.g. in relating demand to cost of production) and the function $R=x\psi(x)$ is called the *total revenue function* of the given demand law. This function, like the demand function itself, assumes pure competition amongst consumers. The function is represented by the *total revenue curve* referred to a horizontal axis Ox and a vertical axis OR.* The height of the total revenue curve thus measures the total revenue obtainable from the output indicated.

The total revenue function, being obtained from a continuous demand function, must itself be continuous. But it is not possible, at this stage, to indicate the " normal " form of the total revenue function consequent upon the assumption of a monotonic decreasing demand function. It will be seen later that the form of the total revenue function depends on the concept of elasticity of demand (see 10.7 below). We can simply notice here that R increases or decreases

* The fact that Ox is the obvious axis to take as horizontal for the total revenue curve is one of the reasons for taking Ox as horizontal for the demand curve itself.

with increasing output according as the demand is (in a very general sense) elastic or inelastic. By an elastic demand, for example, we mean a demand which increases in greater proportion than the corresponding decrease in price. In this case, the revenue obtained from the demand must increase as the demand increases.

If the form of the demand law is given, however, it is a simple matter to deduce the form of the total revenue function and curve. For example, in the case of the linear demand law $p=a-bx$, we have

$$R=ax-bx^2.$$

This expression can be written in the form

$$R-\frac{a^2}{4b}=-b\left(x-\frac{a}{2b}\right)^2.$$

The total revenue curve is thus a parabola with axis vertical and pointing downwards (see 3.4 above). The highest point of the curve occurs where $x=\frac{a}{2b}$. Total revenue increases as output increases at first, reaches a maximum value $\frac{a^2}{4b}$ at the output $x=\frac{a}{2b}$ and then decreases as output increases further. A total revenue curve of this kind, obtained from a linear demand law, is shown in Fig. 44.

The total revenue curve contains no reference to the price of the good. But the price can be read off the curve quite easily. If P is the point on the curve at output $x=OM$, then

$$p=\frac{R}{x}=\frac{MP}{OM}=\text{gradient of } OP.$$

Price can be regarded, in fact, as *average revenue*, i.e. as the revenue per unit of the output concerned. As such, it is measured by the gradient of the line joining the origin to the appropriate point on the total revenue curve. Since price decreases as demand increases, the line OP becomes less and less steep as we move to the right along the total revenue curve.

5.5 Cost functions and curves.

The following assumptions are needed to obtain a simple expression of the problem of cost of production. A given firm produces a single uniform good X with the aid of certain factors of production. Some of the factors are employed in fixed amounts irrespective of the output of the firm, and it is taken that the expenditure on these factors is known and fixed. The remaining factors are variable and the conditions of their supply are assumed to be known. For example, the factors may be obtainable at fixed market prices. Further, the technical conditions under which production is carried out are taken as known and fixed. Finally, the firm is assumed to adjust the employment of the variable factors so that a given output x of the good X is obtained at the smallest possible total cost Π. Then Π is determined when x is given and assumes different values as x is changed, i.e. Π depends uniquely on x:

$$\Pi = F(x) \quad \text{the } \textit{total cost function.}$$

The variables x and Π are necessarily restricted to positive values. The function is shown by a *total cost curve* referred to a horizontal axis Ox and a vertical axis $O\Pi$ drawn in a plane. The varying height of the curve shows the changing total cost of an increasing output.

The total cost function and curve are static constructions relating to various alternative and hypothetical outputs produced under given conditions. They do not describe the variation of cost and output over time. The form of the function F depends on such conditions as the technique of production and the conditions of supply of the variable factors of production. The function and curve are fixed only for given conditions and are completely changed in form if any of the conditions are varied. For example, if the technique of production is improved by an invention or if the wages of the labour employed fall, then the cost function is changed so that the cost of any given output is less than before and the total cost curve shifts to a new position below the old curve.

It should be noticed that the total cost function and curve are "minimal" concepts. By varying the employment of the variable factors, a given output can be produced at various total costs of which only the smallest is used to define the total cost function. In diagrammatic terms, the ordinate of a point on the total cost curve represents the least cost of producing the output concerned and the same output could be produced at larger costs shown by points vertically above the point taken on the curve. The positive quadrant of the plane $Ox\Pi$ is thus divided into two areas by the curve; points in the area above the curve correspond to total costs at which the outputs concerned could be produced while points in the area below the curve represent total costs too small for production of the outputs. The total cost curve is the lower boundary of the area representing possible cost situations. But, since we usually need consider only the lowest cost of any output, we are interested in the cost curve rather than in the area above it.

For purposes of mathematical convenience, the single-valued cost function can be taken as continuous. Further, under "normal" conditions with at least one factor of production fixed, it is assumed that there are fixed "overhead" costs, incurred even when no output is produced, and that variable costs increase as output increases.* The total cost curve, therefore, cuts the vertical axis at a point above the origin and rises continuously from left to right. Fig. 38 shows a

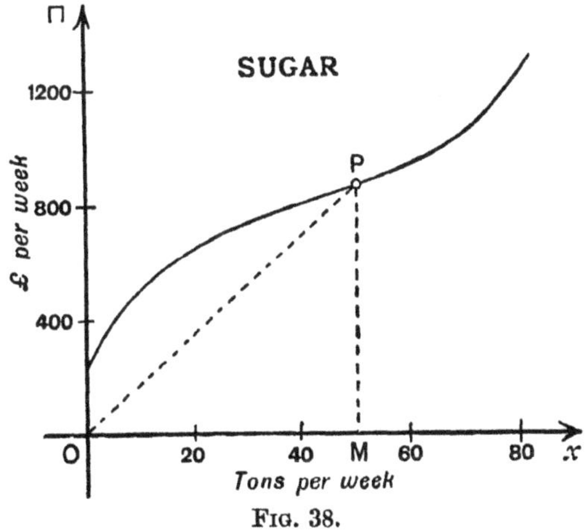

FIG. 38.

total cost curve of "normal" form appropriate to a hypothetical case of a sugar refinery; overhead costs are £250 per week and total cost increases with output, reaching a value of £1250 per week when the weekly output of sugar is 80 tons.

It is sometimes convenient, at least for approximate representations of cost over a limited range of outputs, to take the total cost

* If we consider an industry instead of a firm and take all the factors of production as variable, then overhead costs are zero. Or, we may consider only the total variable costs of a firm and neglect overhead costs. In either case, the normal form of the total cost curve is similar to that given here except that it now starts from the origin.

function of some particular form. The following are types of cost functions appropriate to the "normal" case:

(1) $\Pi = ax + b$. (2) $\Pi = ax^2 + bx + c$.

(3) $\Pi = \sqrt{ax+b} + c$. (4) $\Pi = ax^3 - bx^2 + cx + d$.

(5) $\Pi = ax\dfrac{x+b}{x+c} + d$. (6) $\Pi = ax^2\dfrac{x+b}{x+c} + d$.

(7) $\Pi = ae^{bx}$. (8) $\Pi = x^a e^{bx+c} + d$.

The parameters a, b, c and d are positive in each case. In case (2), the total cost curve is a parabola with axis pointing vertically upwards and with vertex to the left of the origin at $x = -\dfrac{b}{2a}$. Since the curve is limited to the positive quadrant, it consists of an arc of the parabola rising continuously from left to right. The curve graphed in Fig. 46 below represents the cost curve

$$\Pi = \tfrac{1}{10}x^2 + 5x + 200,$$

where £Π per week is the total cost of an output of x tons of sugar per week in a sugar refinery.

The parameters in cost functions of types such as those above enable us to allow for shifts in cost. A cost curve given by (2), for example, is fixed if a, b and c have given values. As the parameters are changed, so the whole cost curve shifts in position corresponding to some change in the conditions under which the firm operates. An increase in c represents a rise in overhead costs and the cost curve shifts upwards without change of form. Changes in a and b, on the other hand, correspond to modifications in variable cost and the cost curve is changed in shape as well as in position.

At any output, the ratio of total cost to output defines *average cost* or cost per unit of output. Denoting average cost by π and using Fig. 38, we have

$$\pi = \frac{\Pi}{x} = \frac{MP}{OM} = \text{gradient of } OP,$$

and average cost is read off the total cost curve as the gradient of the line joining the point on the curve to the origin. Since average cost varies with output, we have the *average cost function* $\pi = f(x)$ where $f(x) = \dfrac{\Pi}{x} = \dfrac{F(x)}{x}$, and a corresponding *average cost curve* referred to axes Ox and $O\pi$ in a plane. The form of this function is to be obtained from that of the total cost function. For example, if total cost is given by the function (2) above, then

$$\pi = ax + b + \frac{c}{x}.$$

Here $\pi \to \infty$ as $x \to 0$ and as $x \to \infty$, i.e. the average cost curve must fall as output increases at small outputs and rise at large outputs. In the particular case of the sugar refinery already quoted, the average cost curve is of the form shown by the curve A.C. of Fig. 46. The average cost function, unlike the total cost function, is not necessarily monotonic.

5.6 Other functions and curves in economic theory.

It is appropriate, in some problems, to represent the conditions of supply of a good in a way analogous to those of demand. Suppose, for example, that a market consists of individuals who bring fixed stocks of various goods to a market-place for exchange amongst themselves. If there is pure competition amongst the individuals, the market prices (in money terms) of all goods being given, then each individual determines by how much he will increase or decrease his stock of any one good. If he wishes to increase his stock, he forms part of the *demand* for the good ; if he decreases his stock, he contributes to the *supply* of the good. Then, if all prices other than that of the good concerned are fixed, the total demand and the total supply of the good are defined, by addition over the whole market, as dependent upon the price. The demand and supply functions are exactly similar, supply being negative demand, and it can be taken that the functions are single-valued, continuous and monotonic, demand increasing and supply decreasing as price decreases.

A different representation of demand and supply is appropriate to simple problems of international trade.* Two countries A and B exchange two goods X and Y, country A producing only X and country B only Y. It is assumed that the conditions of production and the supply of resources are given in each country. Then country A can determine uniquely the amount of X it is willing to offer in exchange for any given amount of Y, i.e. the amount of X offered

* See Marshall, *The Pure Theory of Foreign Trade* (1879).

is a single-valued function of the amount of Y taken, a function which appears as the *demand and offer curve* of country A (when referred to axes Oxy in a plane). A similar function and curve is defined for country B, the good Y being offered and X demanded. These curves differ from those previously defined in that they are referred to axes along which amounts of two goods are measured. The prices of the goods do not appear explicitly. But, if P is any point on the demand and offer curve of the country A, then the gradient, referred to Ox, of the line OP gives the amount of Y taken in exchange per unit of X, i.e. it gives the price of X in terms of Y. A similar price can be defined, as a ratio of exchange, for the country B.

We assumed, in 5.5 above, that only a single good was produced by the given firm. A first approach to a more general problem can be made by assuming that a firm produces two goods X and Y under given technical conditions and making use of fixed supplies of certain factors of production. The total cost of production is now given and the interest lies entirely in the varying amounts of the two goods that can be produced. If a given amount x of the good X is produced, then the fixed resources of the firm can be adjusted so as to produce the largest amount y of the other good Y compatible with the given production of X. Here y is a single-valued function of x which can be taken as continuous and monotonic decreasing; the larger the amount of X required, the smaller is the amount of Y obtainable. Conversely, if x is the largest amount of X that can be produced jointly with a given amount y of Y, then x is a single-valued, continuous and decreasing function of y. The two functions must be inverse to each other, i.e. two aspects of a single relation between the amounts of X and Y produced, a relation imposed by the condition of given resources. In symbols, we can write the relation

$$F(x, y)=0,$$

giving $$y=f(x) \quad \text{and} \quad x=g(y),$$

where f and g are single-valued and decreasing functions to be interpreted in the way described. The relation, $F(x, y)=0$, can be called the *transformation function* of the firm and it serves to show the alternative productive possibilities of the given supplies of the factors. The corresponding *transformation curve* in the plane Oxy

is cut by parallels to either axis in only single points and is downward sloping to both Ox and Oy.* Further, it can be taken, in the "normal" case, that the production of Y decreases *at an increasing rate* as the production of X is increased, and conversely. The transformation curve is thus concave to the origin. Fig. 39 shows a

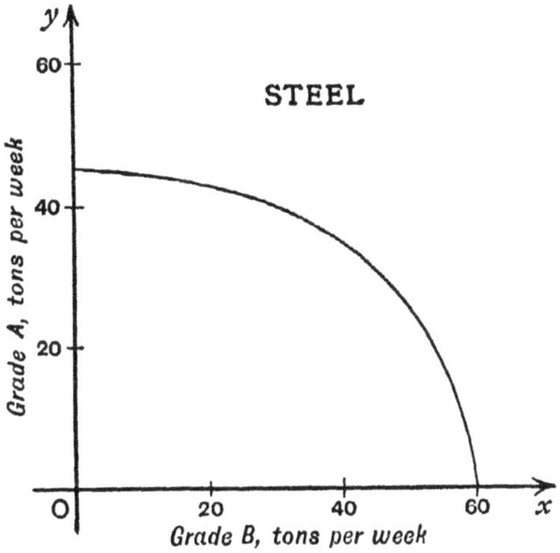

FIG. 39.

"normal" transformation curve in the hypothetical case of a firm producing two grades of steel with given supplies of labour, raw materials and equipment. The transformation curve, it should be noticed, is a "maximal" concept and forms the outer boundary of the productive possibilities. Any point within the curve corresponds to productions of X and Y possible with the given resources while productions represented by points outside the curve cannot be obtained no matter what adjustments of the given resources are made.

The analysis can be generalised to allow for the use of given resources in the production of different goods, not at the same time, but in different periods of time. In the simplest problem, it is assumed that a firm, with given technical conditions and given productive resources, can arrange production in two years in various

* The curve is sometimes called a "production indifference curve", a terminology which is not to be recommended since the curve has little in common with the indifference curves described below. The term "transformation curve" is used (e.g.) by Hayek, *Utility Analysis and Interest*, Economic Journal, 1936.

ways. It is convenient here to take *incomes* rather than physical production, i.e. to take the resources of the firm as producing, according to their allocation, various incomes this year and various incomes next year. Then, for a given income of £x this year, the firm can determine the largest income £y obtainable next year from its resources. Conversely, the largest present income (£x) can be found if next year's income (£y) is given. There is thus a transformation function relating incomes in the two years, a function very similar to the transformation function described above and subject, in the "normal" case, to similar conditions. The corresponding transformation curve can be taken as downward sloping and concave to the origin; as the income desired this year increases, the income obtainable next year decreases continuously and at an ever increasing rate.*

5.7 Indifference curves for consumers' goods.

The economic problems to which we turn now are interpreted, not in terms of a single function or curve, but with the aid of a function type or a curve system. The first of these problems is concerned with the choice of an individual consumer (e.g. a family) in respect of the whole range of consumers' goods available on a market. In order to obtain a simple representation, we take the case where only two goods X and Y appear in the consumer's budget and, to this extent, the problem is over-simplified. The basic assumption now is that the consumer distributes his expenditure on the two goods according to a definite "scale of preferences". His "tastes", on this assumption, are such that he can arrange all possible purchases of the goods in ascending order of preference and, given any two alternative sets of purchases, he can *either* tell which purchases are preferable *or* say that they are indifferent to him.

The precise expression of this assumption can be most easily given first in diagrammatic terms. Measuring purchases of the two goods along axes Ox and Oy selected in a plane, any set of purchases (so much of X and so much of Y) is represented by a point P with co-ordinates (x, y) which measure the respective amounts of X and Y purchased. If we start with a given set of purchases represented by

* This transformation curve is called, by Professor Irving Fisher, the "investment opportunity line". See Fisher, *The Theory of Interest* (1930), pp. 264 *et seq.*

a point $P_0(x_0, y_0)$, then all other possible purchases can be classed into three groups, those preferable to, those which are less preferable than, and those which are on a level of indifference with the basic purchases (x_0, y_0). The third group of purchases is represented by a set of points, $P_0, P_1, P_2, \ldots$, making up a certain curve. This is called the *indifference curve*, corresponding to the level of preference associated with the original purchases (x_0, y_0). Now take any purchases (x_0', y_0') not included in the indifference group already defined. A second group of indifferent purchases can be defined at the level of preference of (x_0', y_0') and represented by a set of points, $P_0', P_1', P_2', \ldots$, making up another indifference curve. This process can be continued until all possible sets of purchases have been included, and the final result is a whole system of indifference curves, each curve consisting of points at one level of indifference. Further, by the basic assumption, the indifference curves themselves can be arranged in ascending order of preference, i.e. according to the preferences shown by the individual for the various levels of indifference attached to the curves. The *indifference map*, consisting of the whole set of indifference curves, serves to describe completely the individual's scale of preferences for the two goods. To compare any two purchases as regards preference, we need only determine whether the corresponding points lie on " higher ", " lower " or the same indifference curves.

The basic assumption tells us only that *some* indifference map can be assumed. The continuity assumption is now added for purposes of theoretical convenience. It is assumed, firstly, that the individual can vary his purchases of each good continuously and, secondly, that the variation from one set of purchases to indifferent purchases is continuous. This assumption has, of course, no justification except as an approximation and on grounds of expediency. Each indifference curve can now be taken as continuous and the whole set of curves appear in a continuously variable order of ascending preference.

Finally, we add three assumptions limiting the form of the indifference map in the " normal " case of consumer's choice. The first is the very reasonable assumption that no indifference curve intersects itself or any other indifference curve. This means that there is only one direction of variation from a given set of purchases which leaves the consumer indifferent and that no set of purchases can be at two levels of indifference. The second assumption is that the level of preference increases as we move across the indifference map in the N.E. direction in the plane Oxy. This fixes the order of ascending preference of indifference curves and implies that the consumer deems himself " better off " whenever he increases his purchases of both goods. From this reasonable assumption, it follows that the level of preference can only remain unchanged if an increase in the purchase of one good is offset by a decrease in the purchase of the other, i.e. the indifference curves are *downward sloping*. The third assumption is that each indifference curve is *convex* to each axis of reference, implying that ever larger and larger increases in the purchase of one good are required to compensate (i.e. to preserve a given level of indifference) for a steadily decreasing purchase of the other good. It may be objected that this last assumption is not reasonable as a universal, or even usual, condition. This may be so, in which case the indifference curves must be assumed to be of other forms. But the assumption can be retained to cover " normal " cases until we find that our theory based on it fails to account for observable facts.

Since the consumer's preferences for two goods are represented by a *system* of curves, the analytical expression of the problem follows from the considerations of 3.7 above. The indifference map has the property that one and only one curve of the system passes through each point of the positive quadrant of the plane Oxy. The complete map corresponds, therefore, to a functional relation between x and y involving one parameter a which can be separated off as shown in the general expression :

$$\phi(x, y) = a.$$

Here $\phi(x, y)$ is a continuous expression in the amounts x and y of the two goods purchased by the consumer. The variables x and y, and the parameter a, take only positive values. If the value of a is fixed, we obtain the equation of one definite indifference curve of the system. Different indifference curves arise when the value of a is changed. The parameter a is thus associated with the particular level of preference which corresponds to the purchases x and y of the two goods ; it is an " indicator " of the ordinal and non-measurable concept of preference or " utility ".

The following are actual examples of functional relations suitable for the expression of a "normal" scale of preferences for two consumers' goods:

(1) $\dfrac{x+h}{h-\sqrt{y+k}}=a,$ where $0 \leqslant x \leqslant a(h-\sqrt{k})-h.$

(2) $(x+h)(y+k)=a,$ where $0 \leqslant x \leqslant \dfrac{a}{k}-h.$

(3) $x+y+\sqrt{2xy}=a,$ where $0 \leqslant x \leqslant a.$

Here a is the parameter of each preference scale and h and k are positive constants. Figs. 24, 25 and 26 show indifference maps drawn from relations of the above types and illustrates the fact that the preference scales are of "normal" form.

It must be emphasised that the whole indifference map, given by $\phi(x, y)=a$, is needed to describe the preferences of the consumer. The parameter a is an essential part of the description. To allow for changes in the consumer's tastes we must modify the form of the whole indifference map. It is the function $\phi(x, y)$ which reflects the particular tastes of the consumer; the form of ϕ is fixed for given tastes and is modified whenever tastes change in any way. These remarks can be illustrated by the relations (1) and (2) quoted above. The constants h and k have fixed values for a given indifference map but, by changing their values, we modify the whole map and allow for changes in the complex of preferences of the consumer. Hence, in representing a scale of preferences by a relation such as (1) or (2), the variation of the parameter a gives the different curves of the indifference map while the constants h and k, given in value for a given complex of tastes, are changed only when we wish to consider the results of varying tastes.*

5.8 Indifference curves for the flow of income over time.

A system of indifference curves, very similar in nature to that for

* It is important to be clear on the use of parameters here. The parameters of a demand or cost curve relate to external factors (e.g. the number of consumers or the prices of the factors of production). In a given situation, there is only a single demand or cost curve. As the situation changes, the parameters take different values and the curve shifts in position. But a given situation as regards consumers' preferences is described by a whole system of curves and a parameter is needed to take us from one curve to another. If the situation itself changes, we must add other parameters in order to change the whole system of curves. The varying tastes of a consumer are described by the shifting of a system of curves, i.e. by a system of a system of curves.

consumers' goods, can be defined to describe a rather wider economic problem. An individual consumer can be assumed to act according to a scale of preferences, not only for goods purchased in one period of time, but also for goods bought in successive periods. But the discontinuous " flow " of purchases over time can be analysed on exactly similar lines to purchases at one time. In order to obtain a convenient representation of the wider problem, severe simplifying assumptions must be made. We assume that the consumer takes into account purchases of goods made in only two successive years and that his *income* can be taken as an indicator of his preferences for the actual collection of purchases made in either year. We can then speak of the consumer's preferences for various groupings of this year's and next year's incomes, so much to spend now and so much to spend next year.

On the basic assumption of a scale of preferences for this and next year's income, the consumer's time preferences for income can be represented by a system of indifference curves in a plane referred to axes along which we measure the amounts of this year's income (x) and of next year's income (y). One income grouping (so much this year and so much next year) can be compared with another grouping by the position of the corresponding points on the indifference map. The groupings are indifferent to the consumer if the points lie on one indifference curve ; one grouping is preferred to the other if it lies on a " higher " indifference curve. The income indifference map corresponds to a function type of the form

$$\psi(x, y)=a,$$

where $\psi(x, y)$ is a given relation between x and y and where a is a positive parameter indicating the level of preference for the incomes x and y. The indifference map can be assumed to be continuous, the curves being non-intersecting, downward sloping and convex to the axes. Preference for income combinations, as indicated by a, must increase as we move across the indifference map in the N.E. direction. These limitations on the form of the indifference map are intended to apply to the " normal " case of a consumer's preferences for income over time and they are similar to the limitations imposed in the previous problem of purchases of goods made at one time.

Further properties of the indifference map for incomes can also be assumed in " normal " cases. If neither income is large, the indifference curves must be steeply inclined to Ox (the axis of present income) since some of this year's income will only be sacrificed in return for relatively large additions to next year's income. As the combined income of the two years increases, the curves tend to become less steep and, when both incomes are large, they approximate closely to straight lines inclined (negatively) at an angle of 45°. The addition to next year's income will then be little larger than the loss to this year's income for which it compensates. It is seen that the curve system of Fig. 24 is suitable for the representation of such an income indifference map. The analytical expression is

$$\frac{x+h}{h-\sqrt{y+k}}=a, \quad (0 \leqslant x \leqslant a(h-\sqrt{k})-h),$$

where a is the parameter indicating the level of preference for the incomes x and y in the two years. Fixed positive values must be given to h and k if the consumer's preferences for income are known. These values are only changed when the whole scale of income preferences is modified.

EXAMPLES V

Economic functions and curves

1. The price of a certain brand of tea is p pence per pound and it is found that the market demand is x thousand pounds per week where

p	20	24	28	30	32	36	40
x	82·5	70·8	63·1	60·7	55·0	48·9	39·8

Represent these demands graphically and show that the demand law is approximately of the linear form $x=120-2p$. Then graph the total revenue curve and find the largest revenue obtainable.

2. A cheap gramophone sells at £p and it is known that the demand, x hundred machines per year, is given by

$$x=\frac{90}{p+5}-6.$$

Plot a graph of the demand curve. At what price does the demand tend to vanish? Draw a graph of the total revenue curve and determine at what price and output the total revenue is greatest.

3. A firm selling toothpaste investigates the form of the demand for its product by fixing different prices in four successive periods. It is found that the necessary outputs are

Price (d. per tube) - -	9	12	15	18
Output (tubes per week) -	1030	900	795	715

Choosing suitable scales, plot the total revenue for these outputs graphically. Show that total revenue can be taken approximately as a linear function of output. Deduce the demand law.

4. What type of demand curve corresponds to the demand law

$$p=\frac{a}{x}-c,$$

where a and c are positive constants? Show that there is some demand no matter how large the price. Is there any limit to the extent of the demand for small prices? Show that total revenue falls steadily as output rises and compare with the case of the previous example.

5. Examine the demand curve $p=\dfrac{a}{x+b}$, where a and b are positive constants. Show that the demand increases from zero to indefinitely large amounts as the price falls. What type of curve is the total revenue curve? Show that revenue increases continuously to a limiting value and contrast this case with that of the previous example.

6. From a consideration of the total revenue curves obtained in the previous two examples, why can we say that these demand laws are not typical of ordinary demands, except as approximations for limited ranges of prices?

7. Of what type is the demand curve $p=\dfrac{a}{x+b}+c$, where a, b and c are positive constants? If $\dfrac{a}{b}$ is small, show that a relatively small fall in price increases the demand from zero to a large amount. Is this a suitable approximate demand law for (e.g.) Player's cigarettes?

8. A speculative builder of working-class tenements rents each tenement at p shillings per week. It is known that

$$p=\sqrt{225-9x},$$

where x is the number of tenements let per week. Graph this demand law. Is it the kind of law you would expect here?

9. Find x as a function of p from the demand law $p=\sqrt{a-bx}$. Show that the demand curve is an arc of a parabola with its axis parallel to Ox. Locate the vertex and indicate the shape of the curve.

10. The number (x) of persons per day using a motor coach service to Southend is related to the fare (p shillings) charged by the law

$$p=\left(3-\frac{x}{40}\right)^2.$$

Show that the demand curve is a parabola and locate its vertex. Also graph the total revenue curve, showing that revenue rises rapidly to a maximum before falling off slowly. Generalise by considering the demand law $p=(a-bx)^2$, where a and b are positive constants.

11. If x is the number (in hundreds) of business men travelling by the 8.15 a.m. train from Southend to London when the return fare is p shillings, the demand law is

$$p=4-\frac{x^2}{100}.$$

Draw the demand curve, show that it is a parabola and locate the vertex. Also draw the total revenue curve and find at what fare the revenue is greatest.

Contrast the forms and economic interpretations of the demand curves in this and the previous example.

12. At a charity performance at the local cinema, it is known that the attendance at a uniform charge of p pence will be $x = \frac{a}{p} - b$, where a and b are constants. It is found that the cinema, which has 3000 seats, is half-filled when 1s. is charged but that only one-sixth of the seats are empty at 9d. Find the constants a and b. Deduce the charge that fills the cinema and show, if all the proceeds go to the charity, that the latter benefits most in this case.

13. In the case of "constant" costs $\Pi = ax + b$, show that the average cost curve is a rectangular hyperbola, average cost falling continuously towards the value a as output increases. A builder of small bungalows has fixed "overheads" of £5000 per year and other costs are always £300 per bungalow. Graph the curve showing average cost per bungalow when a variable number x is built per year.

14. A fixed plant is used to manufacture radio sets and, if x sets are turned out per week, the total variable cost is £$(3x + \frac{1}{25}x^2)$. Show that average variable cost increases steadily with output.

15. If the overheads of the plant of the previous example are £100 per week, find the average cost in terms of the output and draw the average cost curve. What is the least value of average cost?

16. A coal retailer buys coal at a list price of 40 shillings per ton. He is allowed a discount per ton directly proportional to the monthly purchase, the discount being 1 shilling per ton when he takes 100 tons per month. His overheads are £50 per month. Obtain total and average cost as functions of his monthly purchase (x tons) and draw the corresponding curves for purchases up to 1000 tons per month.

17. The electricity works in a small town produces x thousand units per day at a total cost of £$(2\sqrt{40x - 175} + 90)$ per day. Draw the total cost curve. Express the average cost, in pence per unit, as a function of x, draw the average cost curve and show that, as output increases above a certain minimum, average cost decreases rapidly.

18. The total variable cost of a monthly output of x tons by a firm producing a valuable metal is £$(\frac{1}{10}x^3 - 3x^2 + 50x)$ and the fixed cost is £300 per month. Draw the average cost curve when cost includes (a) variable costs only, and (b) all costs. Find the output for minimum average cost in each case. Show that the output giving least average variable cost is less than that giving least average (fixed and variable) cost.

19. A firm produces an output of x tons of a certain product at a total variable cost given by

$$\Pi = ax^3 - bx^2 + cx.$$

Show that the average cost curve is a parabola, find the output for least average cost and the corresponding value of average cost.

20. A tobacco manufacturer produces x tons per day at a total cost of

$$£\frac{x(x+200)}{4(x+100)}.$$

Graph the total and average cost curves and show that average cost decreases continuously from 10s. per lb. towards a lower limit of 5s. per lb.

21. Generalise the previous example by taking the total cost as

$$\Pi = ax\frac{x+b}{x+c},$$

where a, b and c are positive constants ($b > c$). Show that total cost increases, and average cost decreases, as output increases. Of what type is the average cost curve?

22. If the tobacco manufacturer produces at a total cost of

$$£\left\{\frac{x^2(x+100)}{400(x+400)}+50\right\}$$

when his output is x tons per day, draw the graphs of the total and average cost curves. Describe the way in which total cost increases with output and find the output at which average cost is least.

Generalise to show that the total cost function

$$\Pi = ax^2\frac{x+b}{x+c}+d$$

gives total and average cost curves of "normal" form.

23. On the market of a country town, butter is brought for sale from the surrounding district weekly. It is found that the weekly supply (x lbs.) depends on the price (pd. per lb.) according to

$$x = 100\sqrt{p-12}+150.$$

From a graph of the supply curve, show that there is no supply at any price less than 1s. per lb. and that the supply increases continuously as the price increases above 1s. per lb.) Generalise by considering $x = a\sqrt{p-b}+c$, where a, b and c are positive, as a typical supply law.

24. English cloth is exchanged for German linen. The amount of cloth (y million yards per year) offered by England in exchange for a given amount (x million yards per year) of linen is given by

$$y = \frac{x(55-x)^2}{2400}.$$

The amount of linen offered by Germany for a given amount of cloth is

$$x = \frac{y(25-y)}{10}.$$

Represent these conditions diagrammatically and show that, for equilibrium, 10 million yards of cloth exchange against 15 million yards of linen.

25. With a given plant and given supplies of labour, raw material, etc., a chocolate firm can produce two "lines" in various proportions. If x thousand lbs. of one line are produced per year, then

$$y = 45 - \frac{600}{50-x} \quad (x < 50)$$

thousand lbs. per year is the output of the other line. Represent the conditions graphically. What kind of curve is this transformation curve?

26. A man derives £x from his business this year and £y next year. By alternative uses of his resources, he can vary x and y according to

$$y = 1000 - \frac{x^2}{250}.$$

Graph the transformation curve. If he wants £750 next year, how much can he take out of the business this year? Generalise by showing that $y = b - \frac{x^2}{a}$ represents a "normal" case in this type of problem if the constants a and b are such that $b > a$. What is the transformation curve?

27. If a coal mine works x men per shift, the output per shift is

$$\frac{x^2}{25}\left(3 - \frac{x}{12}\right)$$

tons of coal. Draw a graph to show the way in which output varies with the number of men and find the size of the shift for maximum output. Express the average product per man as a function of x, draw the corresponding curve and show that it is a parabola. When is average product greatest?

28. A consumer's indifference map for two goods X and Y is defined by

$$\frac{4 - \sqrt{y + 1}}{x + 4} = a.$$

Draw a graph showing the five indifference curves for the values 2, 3, 4, 5, 6 of the parameter a. Verify that they are of "normal" form.

29. The relation $(x + 1)(y + 2) = 2a$ defines a system of curves. Draw a graph showing the curves, in the positive quadrant, for $a = 2, 3, 4, 5, 6$. Are these curves suitable indifference curves for the case where an individual can derive incomes of £x and £y in two years?

30. Consider the form of the indifference map given by the relation

$$(x + h)\sqrt{y + k} = a,$$

where h and k are fixed positive constants. Draw graphically a selected number of the curves in the case $h = 2$, $k = 1$, and in the case $h = 1$, $k = 1$. Show that one case is suitable for the indifference map for two consumers' goods and the other for incomes in two years.

CHAPTER VI

DERIVATIVES AND THEIR INTERPRETATION

6.1 Introduction.

HAVING made clear the nature of some of the fundamental ideas of the mathematical technique, we find ourselves in a position to attack one of the central problems of mathematics, the problem with which we are concerned in the " differential calculus ". If y is given as a function of x, then the value of y is determined by, and changes with, the value allotted to x. It is clearly important to devise a method of comparing the changes in y with those in x from which they are derived, of measuring the *rate* at which y changes when x changes. In the differential calculus, we make precise what we mean by the rate of increase or decrease of a function, we set out a method of evaluating these rates for various functions and we systematise the problem by introducing convenient symbols, processes and rules. In the words of Professor Whitehead, the differential calculus is the " systematic consideration of the rates of increase of functions ".

In the present chapter, it is proposed first to treat the problem of defining and measuring rates of change from the purely analytical point of view. It is then found that a very important diagrammatic interpretation of the problem can be given. Finally, since the importance of any mathematical method depends on its use in interpreting scientific phenomena, we shall consider quite generally the applicability of our new concepts in a wide range of scientific studies. The development of the practical technique of dealing with our problem will be given in the following chapter.

We can begin by considering special functions of simple form. If, in the function $y=x^3$, we increase the value of x from one definite value x_1 to a larger value x_2, the value of the function increases from x_1^3 to x_2^3. The variable x is increased by amount (x_2-x_1) and the corresponding increase in the function is $(x_2^3 - x_1^3)$. The *average rate of increase* of the function per unit increase in x is then

$$\frac{x_2^3 - x_1^3}{x_2 - x_1} = x_1^2 + x_1x_2 + x_2^2,$$

an expression which depends on both x_1 and x_2. The average rate of increase is dependent on the value of x from which we start and on the amount of increase allotted to x.

Consider, now, only those increases which start from a definite value of x. Adopting a different notation, let x be the fixed original value of x and h the amount of the increase allotted. As x is increased from x to $(x+h)$, the function increases by an amount $\{(x+h)^3 - x^3\}$ and the average rate of increase is

$$\frac{(x+h)^3 - x^3}{h} = 3x^2 + 3xh + h^2,$$

which is taken as dependent on the variable increase h in x.

As the value of h becomes smaller, the average rate of increase is seen to approach the limiting value $3x^2$ which we term the (instantaneous) rate of increase of the function at the point x. Hence, by definition, the *rate of increase* of the function at the point x is

$$\lim_{h \to 0} \frac{(x+h)^3 - x^3}{h} = 3x^2.$$

The average rate of increase of the function is a perfectly definite concept for any actual increase h in x however small. On the other hand, the instantaneous rate of increase is a limiting and abstract concept, the limiting value approached by the average rate of increase as smaller and smaller increases in x are allotted. It will be found that, although scientific phenomena are expressed directly by means of average rates of increase, mathematical analysis finds it more convenient to treat the related abstract concept of the instantaneous rate of increase. The mathematically significant fact is that a function is *tending* to increase at a certain rate at a definite point, not that it *actually* increases at a certain average rate over a definite range of values of the variable.

The instantaneous rate of increase has a value for any value of x we care to select. The rate of increase of x^3 is $3x^2$ when we measure it at the point x. So, the rate of increase of x^3 at the point $x=1$ is 3, at the point $x=2$ it is 12, and so on. In the first case, the value of x^3 is tending to increase at the rate of 3 units per unit of x and similarly for the other cases. The rate of increase of x^3 at the point x changes as x changes, being always given by the expression $3x^2$. The rate of increase of a function of x is itself a function of x.

As a second example, it is seen that the function $y=\frac{1}{x}$ decreases by an amount $\left(\frac{1}{x}-\frac{1}{x+h}\right)$ as x increases by an amount h from x to $(x+h)$. The average rate of decrease of the function is thus

$$\frac{1}{h}\left(\frac{1}{x}-\frac{1}{x+h}\right)=\frac{1}{x(x+h)}.$$

The instantaneous rate of decrease of the function at the point x is

$$\operatorname*{Lim}_{h\to 0}\frac{1}{h}\left(\frac{1}{x}-\frac{1}{x+h}\right)=\frac{1}{x^2}.$$

The function $\frac{1}{x}$ decreases at various rates as x increases from different values, the rate being always given by the expression $\frac{1}{x^2}$.

The process described above can be extended and generalised to apply to any single-valued function whatever. Suppose the independent variable of the function $y=f(x)$ changes in value by an amount h from x to $(x+h)$. We can regard h as taking positive or negative values, a positive value indicating that x has increased and a negative value that x has decreased. The corresponding change in the value of the function is then of amount $\{f(x+h)-f(x)\}$ and this, again, can be positive or negative according as the value of the function has increased or decreased. The expression

$$\frac{f(x+h)-f(x)}{h},$$

then indicates the average rate of change in $f(x)$ per unit change in x. The sign of this expression is important. If the sign is positive, then $f(x)$ changes in the same sense as x itself, increasing as x increases and decreasing as x decreases. If the sign is negative, then $f(x)$ changes in the opposite sense to x, increasing when x decreases and decreasing when x increases. The important question is whether the average rate of change tends to approach any definite limiting value as h gets smaller through positive and negative values. If a limit exists, then it represents the instantaneous rate of change of $f(x)$ at the point x, i.e. the rate at which $f(x)$ is tending to increase or decrease (according to the sign of the limit) as x changes from the point x. If no limit exists, then the value of the function changes in an erratic way as smaller and smaller changes are made in the variable x, changes which cannot be described by any definite instantaneous rate of change. The whole problem of the existence and value of the rate of change of a function $y=f(x)$ thus turns on the existence and value of the limit of the "incrementary ratio" $\frac{1}{h}\{f(x+h)-f(x)\}$ as h tends to zero at a definite point x.

In the first example considered above, the rate of change of x^3 at the point x is given by $3x^2$ and, since this is positive, we know that x^3 increases as x increases. On the other hand, the rate of change of the second function $y=\frac{1}{x}$ at the point x is

$$\operatorname*{Lim}_{h\to 0}\frac{1}{h}\left(\frac{1}{x+h}-\frac{1}{x}\right)=\operatorname*{Lim}_{h\to 0}\left\{-\frac{1}{x(x+h)}\right\}=-\frac{1}{x^2}.$$

Except for the sign, this is identical with the rate of decrease we obtained previously. In fact, the negative sign of the rate indicates that it is a rate of decrease as x increases and the numerical value, i.e. $\frac{1}{x^2}$, then measures the actual rate of decrease at any point. A rate of change, having sign as well as magnitude, is a wide concept including both rates of increase and rates of decrease.

6.2 The definition of a derivative.

If $y=f(x)$ is a single-valued function of a continuous variable, and if the incrementary ratio $\frac{1}{h}\{f(x+h)-f(x)\}$ tends to a definite limit as the value of h tends to zero through positive or negative values, then we say that the function has a derivative at the point x given in value by the limit of the incrementary ratio. If the incrementary ratio has no limiting value, then the function has no derivative at the point x.

A number of different notations for the derivative of a function are in current use. The two following notations are the most common and they will be used throughout the present development. The derivative of $y=f(x)$ at the point x is written *either* as

$$\frac{dy}{dx}=\frac{d}{dx}f(x),$$

or as

$$y'=f'(x).$$

Since the function itself can be written as y or as $f(x)$, each of these notations appears in two equivalent forms. Hence:

DEFINITION: The derivative of the function $y=f(x)$ with respect to the variable x at the point x is

$$\frac{dy}{dx}=\frac{d}{dx}f(x)=y'=f'(x)=\operatorname*{Lim}_{h\to 0}\frac{f(x+h)-f(x)}{h},$$

if the limit exists.

It is convenient, at this stage, to set out a number of observations on the definition of the derivative:

(1) The concept of a derivative, from its very nature, applies only to a function of a continuous variable. Further, the definition given here applies only to functions which are single-valued and the extension of the concept to cover multi-valued functions is postponed.

(2) The incrementary ratio used in the definition is not defined when h is actually zero; in this case it assumes the meaningless form of zero divided by zero. Thus the derivative cannot be regarded as the ratio formed by putting $h=0$ in the incrementary ratio. The derivative, in fact, is the result of an *operation* performed on the function, the operation of obtaining the limit of the incrementary ratio. In the notation $\frac{dy}{dx}$ or $\frac{d}{dx}f(x)$, the $\frac{d}{dx}$ must be regarded as an operational symbol applying to the function and its use is very similar to that of the symbol f in the functional notation $f(x)$. In particular, it must be noted that the symbol $\frac{dy}{dx}$ does not imply that the derivative is a ratio of one thing (dy) to another (dx).

(3) The incrementary ratio can be expressed in an alternative form which is sometimes useful. Corresponding changes or "increments" in the variables x and y connected by the function can be denoted by Δx and Δy respectively. The incrementary ratio is then

$\frac{\Delta y}{\Delta x}$, and the derivative appears as the limit of this ratio as Δx tends to zero. Both Δx and Δy take finite values and, if a value is allotted to Δx, the corresponding value of Δy is found from the function itself.

(4) The value of the derivative must be expressed as so many units of y per unit of x. If we change the units in which either variable is measured, then we must also change the numerical value of the derivative at any point. The rule for change of units, considering the derivative as a quantity derived from x and $y = f(x)$, is given in 1.7 above. If an old unit of x equals λ new units and an old unit of y equals μ new units, then the new value of the derivative is $\frac{\mu}{\lambda}$ times the old value. The fact that the derivative value depends, according to the familiar proportionality rule, on the units of measurement of both x and y must always be remembered.

(5) It is important to be clear about the use of the value x that appears in the definition of the derivative. The derivative is obtained by finding the limit of the incrementary ratio for a *constant* value of x. The limit, once found, has a value dependent on the particular value of x selected, and different limiting values result when different constant values are given to x. Here we have another example of a parameter: x is a constant in finding the limit (when h is the only variable) but parametrically variable from one limit to another. The value of the derivative thus depends on the value fixed for x, i.e. *the derivative is itself a function of* x. There are, in fact, two points of view; the derivative may be obtained as a definite value for a given x (the derivative at a given point) or it may be taken as a function of x (the derivative at a variable point). From the latter point of view, the notation $f'(x)$ is particularly appropriate and an alternative term "derived function" can be used. From the original function $f(x)$, we derive a second function $f'(x)$ by the process of finding derivatives of $f(x)$ for all the various possible values of x.

(6) It must not be assumed that the derivative of $f(x)$ necessarily exists at any point. In fact, the derivative cannot be used unless we have determined that the incrementary ratio has a limit at the point in question. A function may have a derivative at some points

and not at others; some functions may have a derivative at a point and others not. It will be seen (6.5 below) that the non-existence of a derivative is exceptional and of very special significance.

(7) The derivative notations we have given apply to any function $y=f(x)$. The notations must be suitably modified when a particular form is given for a function. So, $\frac{d}{dx}(x^3)$ and $\frac{d}{dx}\left(\frac{1}{x}\right)$ denote the derivatives of the particular functions x^3 and $\frac{1}{x}$ respectively. The notation also needs modification when we wish to write the value of the derivative, not at a general point x, but at some definite point. The following examples illustrate the flexibility of the notation.

$$\left[\frac{d}{dx}f(x)\right]_{x=0}=f'(0) \quad \text{and} \quad \left[\frac{d}{dx}f(x)\right]_{x=\frac{1}{2}}=f'(\tfrac{1}{2})$$

denote the derivatives of $f(x)$ at the particular points $x=0$ and $x=\frac{1}{2}$ respectively. More generally, the derivative of $f(x)$ at $x=a$ is

$$\left[\frac{d}{dx}f(x)\right]_{x=a}=f'(a).$$

For a particular function, say $y=x^3$, the notations

$$\left[\frac{d}{dx}(x^3)\right]_{x=0}, \quad \left[\frac{d}{dx}(x^3)\right]_{x=\frac{1}{2}} \quad \text{and} \quad \left[\frac{d}{dx}(x^3)\right]_{x=a}$$

indicate the derivative value at $x=0$, $x=\frac{1}{2}$ and $x=a$ respectively.

(8) Finally, it is scarcely necessary to note how convenient it is to have a special notation and terminology for the derivative concept. Limits of the kind indicated by the derivative are of frequent appearance in mathematical theory and applications. The short and flexible derivative notation, in replacing the clumsy limit notation, makes it possible to deal with these limits with great ease. In particular, the process of systematising the operation of finding the limit, to be discussed in the next chapter, is facilitated. Without this labour-saving device, scarcely any progress is possible.

6.3 Examples of the evaluation of derivatives.

The considerations of 6.1 show that

$$\frac{d}{dx}(x^3)=3x^2 \quad \text{and} \quad \frac{d}{dx}\left(\frac{1}{x}\right)=-\frac{1}{x^2}.$$

The following examples are further illustrations of the way in which derivatives of particular functions are obtained from the definition:

Ex. 1. The derivative of $(2x-1)$ at any point x is obtained as the limit as h tends to zero of the ratio

$$\frac{\{2(x+h)-1\}-(2x-1)}{h}=\frac{2x+2h-1-2x+1}{h}=\frac{2h}{h}=2.$$

So $$\frac{d}{dx}(2x-1)=2.$$

The derivative, in this case, is a constant. A similar result holds in the case of the derivative of any linear function :

$$\frac{d}{dx}(ax+b)=a.$$

Ex. 2. The incrementary ratio for the function (x^2+3x-2) is

$$\frac{1}{h}[\{(x+h)^2+3(x+h)-2\}-(x^2+3x-2)]$$

$$=\frac{1}{h}[(x^2+2hx+h^2)+(3x+3h)-2-(x^2+3x-2)]$$

$$=\frac{2hx+3h+h^2}{h}$$

$$=2x+3+h$$

$$\to 2x+3 \quad \text{as} \quad h\to 0.$$

So $$\frac{d}{dx}(x^2+3x-2)=2x+3.$$

In the same way, the derivative of the general quadratic function is

$$\frac{d}{dx}(ax^2+bx+c)=2ax+b.$$

Ex. 3. For the function $\dfrac{2x+1}{x-1}$, the incrementary ratio is

$$\frac{1}{h}\left\{\frac{2(x+h)+1}{(x+h)-1}-\frac{2x+1}{x-1}\right\}=\frac{(x-1)(2x+2h+1)-(2x+1)(x+h-1)}{h(x-1)(x-1+h)}$$

$$=\frac{-3h}{h(x-1)(x-1+h)}$$

$$=\frac{-3}{(x-1)(x-1+h)}$$

$$\to -\frac{3}{(x-1)^2} \quad \text{as} \quad h\to 0.$$

So $$\frac{d}{dx}\left(\frac{2x+1}{x-1}\right)=-\frac{3}{(x-1)^2}.$$

The result, as given, holds for any value of x not equal to 1. If $x=1$, the function itself has an infinite discontinuity and the question of the existence of the derivative at this point needs further consideration (see 6.5 below).

6.4 Derivatives and approximate values.

The analytical interpretation of the derivative as the measure of the rate of change of the function is clear from 6.1 above. Further, since the rate of change is the limiting value of the average rate of change over a small interval, the derivative is also an approximate value of the average rate of change, the smaller the interval of the average rate the closer being the approximation. Hence :

If $f(x)$ has a derivative $f'(x)$ at the value x, then $f'(x)$ measures the rate of change of $f(x)$ at the point x, and is an approximate measure of the average rate of change of $f(x)$ for any small interval from the point x.

As already noted, the rate of change can be a rate of increase or a rate of decrease according to the sign of the derivative at the point. In either case, the derivative as the rate of change must be expressed as so many units of the function per unit of x. Two units of measurement are involved in the derivative and, if either unit is changed, so is the expression of the derivative.

The use of the derivative as the approximate value of the incrementary ratio or average rate of change is an important one. If Δx is any *small* increment in the value of x, and if Δy is the corresponding small increment in the value of $y=f(x)$, then the derivative $f'(x)$ is the limiting and approximate value of the quotient of Δy divided by Δx. Hence

$$\Delta y = f'(x)\Delta x \quad \text{approximately when } \Delta x \text{ is small.}$$

In other words, if we multiply a small increment in x by the rate of change $f'(x)$, we obtain an approximate value of the corresponding small increment in the function $y=f(x)$.

Again, with a slightly different notation, the value of the derivative $f'(a)$ at the point $x=a$ is an approximate value of the expression $\frac{1}{h}\{f(a+h)-f(a)\}$ for a small value of h, i.e.

$$f(a+h)=f(a)+hf'(a) \quad \text{approximately when } h \text{ is small.}$$

This result enables us to trace the effect of small " errors " or deviations in the variable x. The value of x is erroneously taken as a instead of the correct value $(a+h)$, where h is a small "error". The above result shows that the "error" in the value of the function, resulting from the insertion of a instead of $(a+h)$, can be taken approximately as $hf'(a)$. For example, the volume of a sphere of radius x inches is $\frac{4}{3}\pi x^3$ cubic inches. If the radius of a given sphere is estimated as a inches, then the estimate of the volume is $\frac{4}{3}\pi a^3$ cubic inches. The error involved in the volume estimate is approximately $4\pi a^2 h$ cubic inches, where h is the small error in the original estimate of the radius. $\left(\frac{d}{dx}(\frac{4}{3}\pi x^3) = 4\pi x^2\right)$.

It is often convenient, for a definite small range of values of x, to replace the value of a given function $f(x)$ by an approximate *linear* expression in x.* The result we have given enables us to do this. If the approximation is to serve for values of x near $x=a$, we substitute $h=x-a$ in our result and write

$$f(x) = f(a) + (x-a)f'(a)$$

approximately for small values of $(x-a)$. For example, if

$$f(x) = \frac{2x+1}{x-1} \quad \text{and} \quad f'(x) = -\frac{3}{(x-1)^2},$$

then $\qquad f(3) = \frac{7}{2} \quad \text{and} \quad f'(3) = -\frac{3}{4}.$

So
$$\frac{2x+1}{x-1} = \frac{7}{2} + (x-3)(-\tfrac{3}{4}),$$

i.e.
$$\frac{2x+1}{x-1} = \frac{23}{4} - \frac{3}{4}x$$

approximately when $(x-3)$ is small. In general, therefore, it is possible to represent any given function by an approximate linear expression, $f(x) = \alpha + \beta x$, for a small range of values of x. The constants α and β are determined in the way described.

6.5 Derivatives and tangents to curves.

One of the important problems of geometry is that of finding the tangent at any point on a given curve. The tangent describes the direction of the curve at the point and, to define it, we must introduce the idea of a limiting process. The line joining any two points P and

* See, for example, Wicksell, *Über Wert, Kapital und Rente* (1893, reprinted 1933), pp. 31-2.

Q on a curve is called a *chord* of the curve and its position varies as either P or Q is varied. Suppose that P is fixed and that Q moves along the curve and approaches P from one side or the other. It is clear that it is usually, but not always, the case that the chord PQ tends to take up a definite limiting position PT as Q approaches P. Fig. 40 illustrates a case where this happens. It is possible, however,

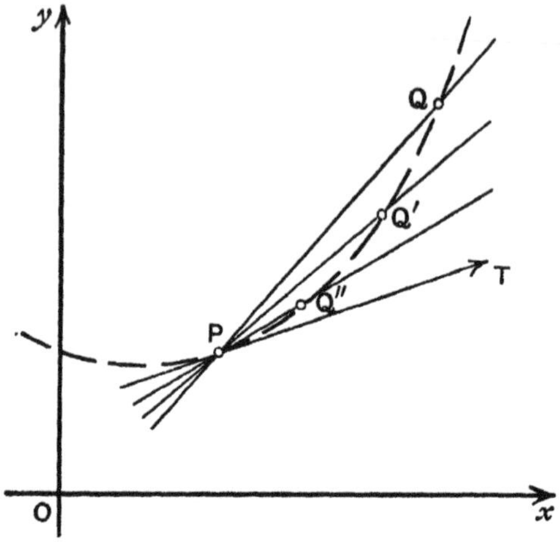

FIG. 40.

that no such limiting position PT exists at certain points, e.g. at points where the curve is discontinuous. If the limiting position PT of the chord exists, then it is called the *tangent* to the curve at P. If no limiting position exists, then the curve has no tangent at P.

The instantaneous direction of the curve at any point, i.e. the direction in which the curve is tending to rise or fall, is indicated by the tangent to the curve at the point. A convenient *measure* of the direction of the curve is thus provided by the gradient of the tangent. This gradient varies as the tangent varies from point to point on the curve and is always given as the limiting value of the gradient of any chord PQ as Q approaches P.

Our discussion is now limited to the case of a curve represented by a single-valued function $y=f(x)$, there being only one point on the curve above or below each point on the horizontal axis Ox. Let P and Q be two points on the curve with abscissae x and $(x+h)$ respectively, x being fixed and h any positive or negative amount. The gradient of the chord PQ is the ratio of the difference between the ordinates of Q and P to the difference between the abscissae of these points (see 3.3 above). So

$$\text{Gradient of } PQ = \frac{f(x+h)-f(x)}{h}.$$

The gradient of the tangent at P is the limiting value (if it exists) of this expression as Q approaches P, i.e. as $h \to 0$. Hence

$$\text{Gradient of tangent at } P = \lim_{h\to 0} \frac{f(x+h)-f(x)}{h} = f'(x).$$

The problem of finding the tangent and its gradient at any point on a curve is thus equivalent to the problem of finding the derivative of the function $y=f(x)$ which represents the curve. So

The tangent to the curve $y=f(x)$ at the point with abscissa x exists if the function has a derivative at the point x and the tangent gradient $=f'(x)$.

From 3.3 above, a straight line through a fixed point P (x_1, y_1) on the curve $y=f(x)$ and with gradient m has equation

$$y - y_1 = m(x - x_1).$$

This line is the tangent at P if the gradient m is equal to $f'(x_1)$. So

The equation of the tangent to the curve $y=f(x)$ at the point with co-ordinates (x_1, y_1) is $y - y_1 = f'(x_1)(x - x_1)$.

Since the point P lies on the curve, we have $y_1 = f(x_1)$. The tangent equation thus involves only one parameter, i.e. the abscissa x_1 of the selected point on the curve.

Taking the parabola $y=x^2$ as an example, we can show that $\frac{d}{dx}(x^2)=2x$, i.e. the tangent gradient of the parabola is $2x$ at the point with abscissa x. The tangent equation at (x_1, y_1) is

$$y - y_1 = 2x_1(x - x_1) \quad \text{where} \quad y_1 = x_1^2,$$

i.e.

$$y - 2x_1 x - x_1^2 + 2x_1^2 = 0,$$

i.e.

$$y - 2x_1 x + x_1^2 = 0.$$

For example, the point (1, 1) lies on the parabola $y=x^2$, the tangent at the point has gradient 2 and the equation of the tangent is $y - 2x + 1 = 0$. The tangent gradient and equation at any point on a given rectangular hyperbola can be found in an exactly similar way. It is only necessary to evaluate the derivative of the function which represents the curve.

The diagrammatic interpretation of the derivative as a tangent gradient is a very important one and much use of it will be made later when we come to the applications of the derivative in such problems as that of maxima and minima. In the meantime, it can be used to supplement points already discussed.

It has been seen that the process of derivation enables us to derive, from a given function $y=f(x)$, a second function $y=f'(x)$. The curves representing these two functions can be plotted with reference to the same axes and the connection between them is interesting. The ordinate of the derived curve $y=f'(x)$ at any point is equal to the tangent gradient at the corresponding point on the original curve $y=f(x)$. The original curve shows the way in which

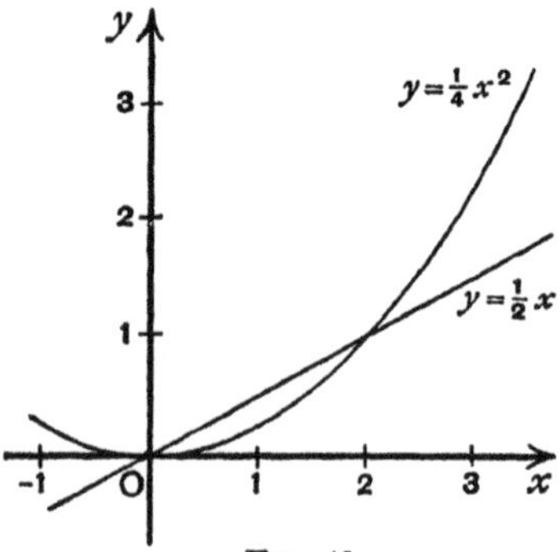

FIG. 41.

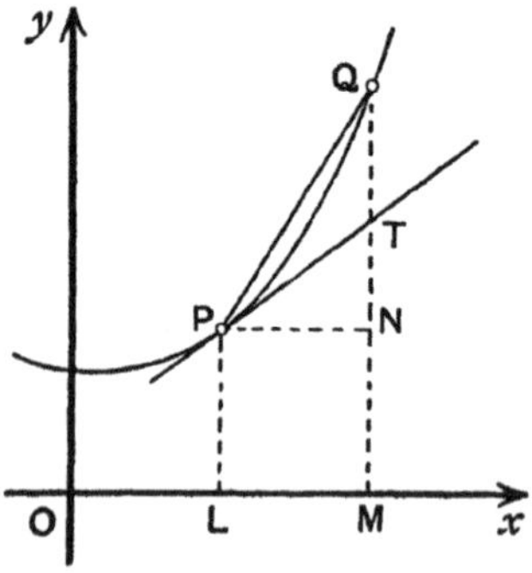

FIG. 42.

the value of the function $f(x)$ changes ; the derived curve shows the variation of the rate of change of the function. For example, the curve $y=\frac{1}{4}x^2$ and the corresponding derived curve $y=\frac{1}{2}x$ are shown in Fig. 41. The first curve shows that x^2 increases at an increasing rate as x increases. The second curve, which is a straight line, shows that the rate of increase of x^2 actually increases at a *steady* rate.

The nature of the approximation involved in writing $f'(x)\Delta x$ for the increment in the function $f(x)$ is clearly seen in Fig. 42. If P is the point (x, y) on the curve $y=f(x)$ and Q the point with abscissa $x+\Delta x$, then $\Delta x=LM=PN$. The increment in the function is $\Delta y=MQ-LP=NQ$. Now, $f'(x)=$ gradient of the tangent $PT=\dfrac{NT}{PN}$, i.e. $f'(x)\Delta x = NT$. The approximation of $f'(x)\Delta x$ to Δy is thus represented by the approximation of NT to NQ. The difference between these lengths clearly decreases as Δx decreases and Q approaches P.

It is obvious that there is some relation between the continuity of a function and curve and the existence of a derivative and tangent. It can be shown, in fact, that the existence of a derivative and tangent is a more severe restriction on the function and curve than the condition of continuity. Of the two continuous curves of Fig. 33, the first has a tangent at all points while the second has one point (a sharp point) where no tangent exists. The fact that a curve is continuous does not imply that a tangent exists at all points. A continuous curve can be drawn without taking pencil from paper; a curve which has tangents at all points is not only continuous but also "smooth" in the ordinary sense of this term. All the ordinary functions of mathematical analysis are continuous and have derivatives (except perhaps at very special points) and they are represented by continuous and smooth curves. Further, when we introduce a function such as a demand or cost function, it is usual to assume, not only that it is continuous, but also that it has a derivative, excluding the case where the corresponding curve has a sharp point.

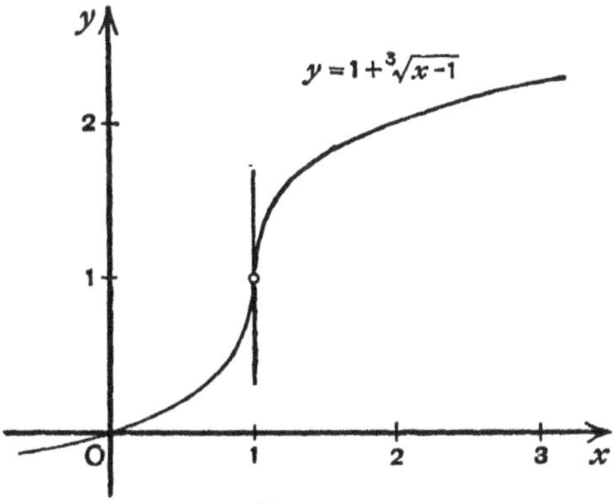

FIG. 43.

Finally, nothing has been said so far about the possibility that the incrementary ratio defining the derivative has an infinite limit. If $\frac{1}{h}\{f(x+h)-f(x)\} \to \infty$ as $h \to 0$ at a point where $f(x)$ exists, we say that the derivative $f'(x)$ is "infinite" at the point. In diagrammatic terms, there is a definite tangent to the curve $y = f(x)$ at the point P with abscissa x and the gradient of the tangent is infinite, i.e. the tangent is parallel to Oy. The case is illustrated by the curve $y = 1 + \sqrt[3]{x-1}$, shown in Fig. 43, which has a vertical tangent at the point where $x=1$. Infinite derivatives can be obtained even in cases of ordinary continuous and smooth curves. Further, if $f(x)$ is itself infinite at the point in question, we can still say, in many cases, that the derivative exists and is infinite at the point. The rectangular hyperbola of Fig. 33 is a case in point. Both the function $\frac{2x+1}{x-1}$ and its derivative $-\frac{3}{(x-1)^2}$ are infinite at $x=1$. The curve "goes off to infinity" at this point and has a vertical tangent "at infinity", i.e. the asymptote $x=1$. To deal with many infinite discontinuities, therefore, we have only to allow for the possibility of both the function and its derivative being infinite.

6.6 Second and higher order derivatives.

The derivative of a function of a variable x is also a function of x. It is possible, therefore, to apply the process of derivation all over again to the derivative and to obtain what is called the *second derivative* of the original function. The second derivative, if it exists, can be obtained for various values of x and is again a function of x. We can thus proceed to find the derivative of the second derivative, i.e. the *third derivative* of the original function. In the same way, derivatives of the fourth, fifth and higher orders can be obtained in succession. Extending the notations already adopted for the derivative, the sequence of successive derivatives of a given function $y=f(x)$ can be written

$$y, \frac{dy}{dx}, \frac{d^2y}{dx^2}, \frac{d^3y}{dx^3}, \ldots; \quad f(x), \frac{d}{dx}f(x), \frac{d^2}{dx^2}f(x), \frac{d^3}{dx^3}f(x), \ldots,$$

$$\text{or } y, y', y'', y''', \ldots; \quad f(x), f'(x), f''(x), f'''(x), \ldots.$$

The sequence of derivatives can be extended indefinitely, with the sole limitation that the derivatives at and beyond a certain order may fail to exist on account of the breakdown of the limiting process which defines them. For all ordinary mathematical functions, however, the derivatives of all orders exist.* But it is found that the first and second derivatives are sufficient for the description of the main properties of functions, and, in the applications of mathe-

* It often happens, of course, that the derivatives beyond a certain stage become zero, as in the case of the function $y=x^2$. This is not the same thing as saying the derivatives do not exist.

matical analysis, derivatives of higher order than the second are seldom needed.

The analytical interpretation of the second derivative is easily provided. The value of $f'(x)$ measures the rate of change of $f(x)$ as x increases. The value of $f''(x)$, being the derivative of $f'(x)$, thus indicates the rate of change of $f'(x)$, i.e. the rate of change of the rate of change of $f(x)$, as x increases. For example, the function x^2 increases at an increasing rate as x increases and this "acceleration" in the value of x^2 is measured by the second derivative of the function. The value of $f''(x)$, it should be noticed, must be expressed in so many units of $f'(x)$ per unit of x, i.e. in so many units of $f(x)$ per unit of x, per unit of x. As in the case of the first derivative, a change in the units of measurement necessitate a change in the value of the second derivative.

In diagrammatic terms, the value of $f''(x)$ at any point measures the tangent gradient of the derived curve $y=f'(x)$ at the point. A more useful interpretation, however, connects the second derivative with the curvature of the original curve $y=f(x)$, curvature being defined by means of the rate of change of the tangent gradient as we move along the curve. This connection will be investigated at a somewhat later stage (see 8.3 below).

6.7 The application of derivatives in the natural sciences.

In the interpretation and explanation of scientific phenomena, we attempt to throw into prominence the causal or other relations between the concrete quantities or attributes with which we deal. Mathematics is designed as a tool to aid us in this attempt. Now, a glance at the problems of the natural or social sciences shows that they are largely, but not exclusively, concerned with the question of *change*, with the way in which the variable quantities change together or in opposition. The practical method of expressing change is by means of the average rate of change of one quantity as another quantity varies. Hence, when we come to express any assumed connection between quantities by means of a functional and mathematical relation between variable numbers, we find that our main concern is with the average rates of change of these functions.

These remarks apply even if discontinuous variables and functions are used to represent the necessarily discontinuous quantities of scientific phenomena. But it is found mathematically convenient to assume, wherever possible, that the variables and functions are continuous and then to deal with instantaneous rates of change as the limiting and approximate values of average rates of change. Sometimes it is necessary to retain discontinuous variables, but the theory then becomes mathematically much more difficult and much less elegant. We are led, therefore, to the study of the derivatives of continuous functions and their use in the interpretation of the scientific phenomena under consideration.

The differential calculus is thus of almost immediate application in all scientific studies. Any relation between variable quantities is represented by a function, and the derivative of the function *measures* the instantaneous rate of growth of one quantity with respect to the other and *approximates* to the actual average rate of growth when the variation considered is small.

Examples of the use of derivatives can be provided from most scientific theories. In the science of physics, derivatives are employed to express physical movement of any kind. As a simple example, consider a train moving along a railroad track. The distance travelled by the train from a fixed point depends on the time that has elapsed, i.e. the distance travelled is a function of the time-interval.* All that can be actually observed about the train's motion is its average speed (over some time-interval, long or short) and this can be connected with the " incrementary ratio " of the distance function. But, from the observable concept of average speed, we derive the limiting concept of the *velocity* of the train at a certain time. Velocity is simply the limiting value of average speed as the time-interval of the latter is made shorter. Hence, we need the derivative of the distance function to measure the velocity of the train at any moment and to approximate to the average speed of the train over any short time-interval from the given moment.

The importance of the limiting idea involved in the derivative is seen by the way in which we have come to express physical motion. Velocities are not " natural " concepts ; they can be derived only by a process of abstraction, of finding limiting values. When we say

* For example, if x is the distance travelled in time t, then $x = at$ represents motion at a uniform speed and $x = at^2$ motion at a uniform acceleration, a being a constant in each case.

that a train is travelling at 60 m.p.h., we are expressing a rather complicated idea. We do not mean that the train will actually travel 60 miles during the following hour, one mile during the next minute or even 88 feet during the next second. The speed of the train will generally change during any time-interval however short. The statement means that the distance travelled by the train per unit of time will correspond more and more closely to 60 m.p.h., or any equivalent, as the time-interval is made shorter and shorter. In physics, therefore, we have accepted the mathematically significant limiting concept of velocity as replacing the practical concept of average speed.

The second derivative of distance travelled as a function of time is the derivative of the first derivative and so measures the rate of change of velocity over time. The rate of change of velocity is commonly described as the *acceleration* and this is again a limiting concept, the limiting rate at which velocity is changing. Acceleration is, in fact, another concept to which we have grown accustomed in the description of movement and its abstract nature has ceased to be a difficulty.

The units of measurement of both distance and time must appear in the derivatives which measure velocity and acceleration, a fact which is clear from the way in which velocities and accelerations are specified. We say, for example, that the velocity of a train is 44 feet per second (30 m.p.h.) and that its acceleration is 2 feet per second per second at a given moment. This means that the train is tending to cover distance at a rate proportional to 44 feet in one second and that its velocity is tending to increase at a rate equivalent to 2 additional feet per second in a second. In the same way, when we say that the constant acceleration of a body moving under gravity is 32 feet per second per second, we imply that the velocity of the body is tending to increase always by 32 feet per second for every second that elapses.

In other branches of mechanics or physics, the derivatives of functions are used to interpret motion, whether the motion is of liquids in hydro-mechanics, of air and gases in acoustics and the theory of sound, of electric currents in electro-mechanics or of heat in thermo-mechanics. Further, in a less "exact" science such as biology, derivatives again have their uses. For example, the changes in the proportion of dominant to recessive genes through successive generations can be interpreted, in a broad view, as the derivative of the proportion as a function of time.*

6.8 The application of derivatives in economic theory.

Finally, we come to economics as the most " exact " of the social sciences. We have seen that a relation assumed between economic variables, e.g. the relation between demand and price, can be expressed by means of a function and curve. It is usually assumed that the function and curve are continuous and smooth, i.e. the function has a derivative and the curve a tangent at all points. The derivative of the function and the tangent gradient of the curve must now be considered in some detail and their economic meaning made clear.

We are accustomed, in economics, to describe the variation of one quantity Y with respect to another quantity X by means of two concepts, an *average* concept and a *marginal* concept. The average concept expresses the variation of Y over a whole range of values of X, usually the range from zero up to a certain selected value. Thus average cost is the relation of total cost to the whole of the output concerned. Marginal concepts, on the other hand, concern the variation of Y " on the margin ", i.e. for very small variations of X from a given value. So the marginal cost at a certain level of output is the change in cost that results when output is increased by a very small unit amount from this level. It is clear that a marginal concept is only precise when it is considered in the limiting sense, as the variations in X are made smaller and smaller. It is then to be interpreted by means of the derivative of the function which relates X and Y. Average and marginal concepts are not new or peculiar to economics. In describing the motion of a train, the average concept is the average speed of the train over (e.g.) its first hour's run while the marginal concept is the velocity of the train at the end (e.g.) of the hour's run.

Some actual examples can be taken to illustrate the average and marginal concepts. We have seen that the demand for a good can be expressed, under certain conditions, by the relation $p = \psi(x)$, giving price as a continuous and decreasing function of the demand.

* See Haldane, *The Causes of Evolution* (1932), Appendix.

But the price p can be regarded as the *average revenue* obtained from the demand x, i.e. total revenue divided by the amount demanded or produced. The ordinary demand function and curve are thus equivalent to the average revenue function and curve. Total revenue $R = x\psi(x)$ is also a continuous function of demand. From the total revenue curve, we read off average revenue (or price) as the gradient of the line OP joining O to the appropriate point P on the curve.

If output is increased by a small amount Δx from a certain level x, suppose that total revenue is found to change by an amount ΔR. There is an increase or a decrease in revenue according to the sign of ΔR. The added revenue per unit of added output is then the ratio of ΔR to Δx, i.e. the "average revenue" for outputs from x to $x + \Delta x$. As the output change becomes smaller, we obtain the rate of change of revenue on the margin of the output x as the limit of $\frac{\Delta R}{\Delta x}$ as $\Delta x \to 0$. This is termed the *marginal revenue* at the output x and it is measured by the derivative of R as a function of x:

$$\text{Marginal revenue} = \frac{dR}{dx} = \frac{d}{dx}\{x\psi(x)\}.$$

Marginal revenue is thus an abstract concept only definable for continuous variations in revenue and output. But it is always approximately equal to the added revenue obtained from a small unit increase in output from the level x.* In diagrammatic terms, the marginal revenue at an output x is measured by the tangent gradient to the total revenue curve at the point with abscissa x.

Marginal revenue is itself a function of the output x, its value depending on the particular margin of output considered. We can thus draw the *marginal revenue curve* to show the variation in marginal revenue as output increases. As in the cases of the total and average revenue curves, the outputs are measured along the horizontal axis. Since marginal revenue is the tangent gradient of the total revenue curve, the form of the marginal revenue curve can be deduced at once from that of the total revenue curve.

* The term "marginal increment of revenue", which is sometimes used, is not to be confused with "marginal revenue". The former is the increment ΔR in revenue resulting from any increase Δx in output from the level x. Now $\Delta R = \frac{dR}{dx}\Delta x$ approximately when Δx is small. Hence the marginal increment of revenue for a small increase in output is approximately equal to the marginal revenue times the increment in output.

Two particular types of demand law can be taken for purposes of illustration. The linear demand law $p=a-bx$ gives

$$R=ax-bx^2 \quad \text{and} \quad \frac{dR}{dx}=a-2bx \quad \text{(see 6.3, Ex. 2).}$$

The total revenue curve is a parabola with axis pointing vertically downwards. The average and marginal revenue curves are straight lines sloping downwards and the gradient (referred to Ox) of the latter is twice that of the former. Fig. 44 shows the three curves in the particular case where the demand law for sugar is $p=15-\frac{1}{5}x$ (see 5.3 above). Here, as in all cases, it is found convenient to plot

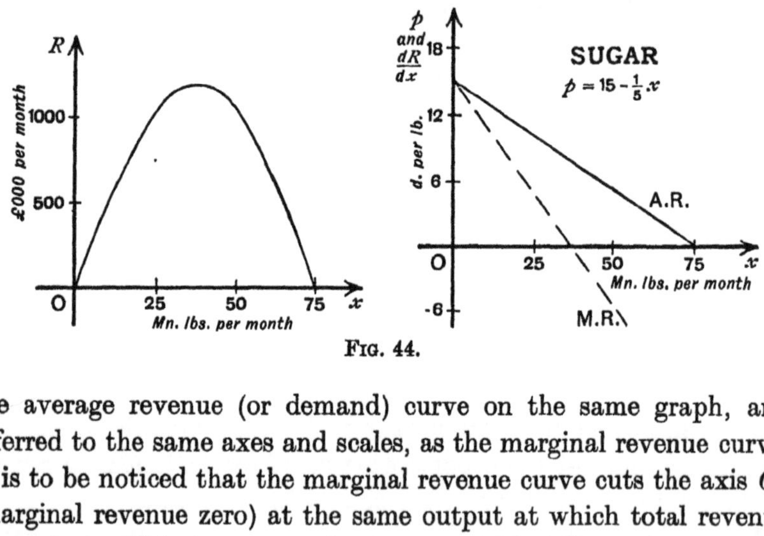

FIG. 44.

the average revenue (or demand) curve on the same graph, and referred to the same axes and scales, as the marginal revenue curve. It is to be noticed that the marginal revenue curve cuts the axis Ox (marginal revenue zero) at the same output at which total revenue is greatest. This is a general property which follows since, at this output, the tangent to the total revenue curve is horizontal with zero gradient.

From the demand law $p=\dfrac{a}{x+b}-c$, we derive

$$R=\frac{ax}{x+b}-cx \quad \text{and} \quad \frac{dR}{dx}=\frac{ab}{(x+b)^2}-c \quad \text{(see 7.4, Ex. 15).}$$

In Fig. 45 are shown the total, average and marginal revenue curves obtained in the particular case where the demand law for sugar is $p=\dfrac{600}{x+20}-5$ (see 5.3 above). Again the marginal revenue curve falls continuously, lies under the average revenue (or demand) curve and cuts Ox at the output where the total revenue is greatest.

The three concepts of total, average and marginal cost can be defined, under the conditions of the cost problem of 5.5 above, in a very similar way. The total and average cost functions and curves have been sufficiently described already, average cost being read off the total cost curve as the gradient of the line *OP* joining *O* to the point *P* with the appropriate abscissa on the total cost curve.

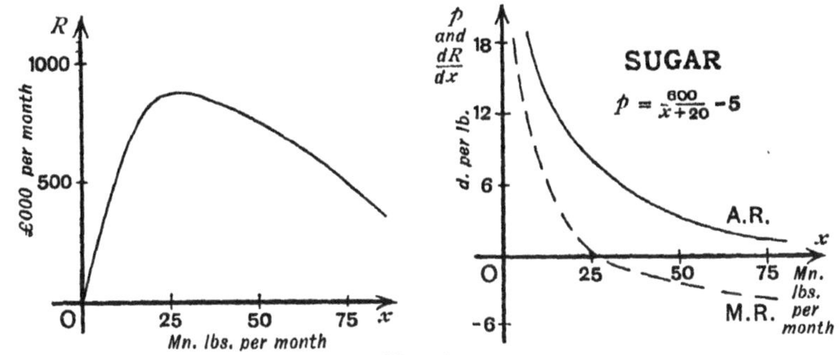

FIG. 45.

If output is increased by an amount Δx from a certain level x and if the corresponding increase in cost is $\Delta \Pi$, then the increase in cost per unit increase in output is $\dfrac{\Delta \Pi}{\Delta x}$. Marginal cost is defined as the limiting value of this ratio as Δx gets smaller, i.e. marginal cost is the derivative of the total cost function $\Pi = F(x)$. It measures the rate of increase of total cost and approximates to the cost of a small additional unit of output from the given level. Further, the marginal cost of any output is given as the tangent gradient of the total cost curve at the appropriate point. Since marginal cost varies with the output at which it is measured, we have a marginal cost function and a corresponding marginal cost curve.*

If the total cost function is $\Pi = ax^2 + bx + c$, then

$$\text{Average cost } \pi = ax + b + \frac{c}{x} \quad \text{and} \quad \text{Marginal cost } \frac{d\Pi}{dx} = 2ax + b.$$

The total cost curve is the rising portion of a parabola with axis pointing vertically upwards. The average cost curve is found to be U-shaped and the marginal cost curve is a straight line sloping

* Marginal cost is described as, or equated to, "marginal supply price" by some writers. See, for example, Bowley, *The Mathematical Groundwork of Economics* (1924), p. 34.

upwards with gradient $2a$. The three curves are shown in Fig. 46 in the particular case where $\Pi = \frac{1}{10}x^2 + 5x + 200$ is the cost relation for a sugar refinery (see 5.5 above). It is to be noticed that the marginal cost line passes through the lowest point of the average cost curve at an output OM where the tangent to the total cost curve goes through the origin. At such an output, average cost is smallest and equal to marginal cost.

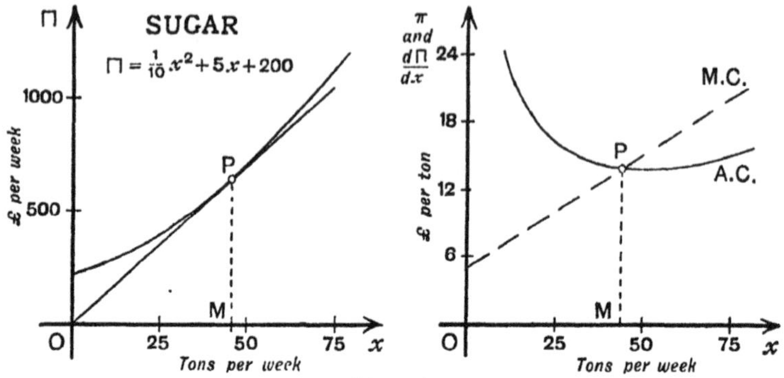

FIG. 46.

Average and marginal concepts can be usefully defined in other economic problems. The transformation problems of 5.6 above provide examples. If P is a point on the transformation curve for the production of two goods X and Y with given resources, then the co-ordinates of P represent a possible production. The gradient of OP (referred to Ox) is the average output of Y per unit output of X, the amount of Y produced as a proportion of the amount of X produced. The tangent gradient of the transformation curve at P (referred to Ox), i.e. the derivative of y with respect to x, is negative and its numerical value measures the marginal rate of substitution of Y production for X production, the rate at which the production of Y is increased when the production of X is decreased. In the same way, the tangent gradient of the transformation curve for incomes in two years is used to define what Professor Irving Fisher calls the marginal rate of return over cost.* Marginal concepts are also fundamental in the use of indifference curves for choice between two goods or two incomes. The " marginal rate of substitution " of one good for another in consumption is defined as the numerical value

* See 9.7 below and Examples VI. 31.

of the tangent gradient at the appropriate point on an indifference curve for the two goods. Again, the numerical value of the tangent gradient of an indifference curve for two incomes is used to define the "marginal rate of time preference" for present over future income in the case of the individual concerned. These concepts will be considered in detail at a later stage.

EXAMPLES VI

Evaluation and interpretation of derivatives *

1. Find, from the definition, the derivatives of x^2 and $(1+x)^2$.

2. Show that the derivative of a constant is zero and that the derivative of x is unity for any value of x. Interpret these results in terms of rates of change.

3. Use the definition of the derivative to establish the general result that the derivative of (ax^2+bx+c) is $(2ax+b)$.

4. Evaluate $\frac{d}{dx}\{x(1+x)\}$ and $\frac{d}{dx}\{x(1-x)\}$ from the definition. Show that these derivatives are respectively the sum and the difference of the derivatives of x and x^2. Does this suggest a general rule?

5. From the definition, show that $\frac{d}{dx}\left(\frac{a_1x+b_1}{a_2x+b_2}\right)=\frac{a_1b_2-a_2b_1}{(a_2x+b_2)^2}$ and hence verify the derivative of $\frac{2x+1}{x-1}$.

6. Show that the derivative of a step-function is zero except at certain points where it does not exist at all. Illustrate the statement that a function has no derivative at a point of discontinuity.

7. Find the average rate of increase of x^2 when x increases from 1 to 1·1. What percentage error is involved when the derivative of x^2 at $x=1$ is used to represent this average rate of increase?

8. Find the average rate of decrease of $\frac{2x+1}{x-1}$ as x increases from 2 to each of the values 3, 2·5, 2·1, 2·05 and 2·01. Find, from the derivative, the instantaneous rate of decrease at $x=2$ and compare with each of the average rates.

9. Evaluate the derivatives of $\frac{1}{x^2}$ and $\frac{x}{x+1}$. Show that $\frac{1}{x^2}=3-2x$ and $\frac{x}{x+1}=\frac{1}{4}(x+1)$ approximately when x is nearly equal to 1. What percentage errors are involved in the use of these approximations when $x=1·1$?

10. If x is small, show that $f(x)=f(0)+xf'(0)$ approximately. Find the derivative of $\frac{1}{1+x}$ and show that this expression can be represented approximately by $(1-x)$ when x is small. Express the difference between the expression and its approximation in terms of x.

* Further examples on the interpretation of derivatives are given in Examples VII, 24-37 below.

11. Show that $f(a+x)=f(a)+xf'(a)$ approximately when x is small. From the derivatives of x^3 and $\frac{1}{x^2}$, show that

$$(1+x)^3=1+3x \quad \text{and} \quad \frac{1}{(2x+3)^2}=\frac{1}{9}-\frac{4}{27}x$$

approximately when x is small. Check the first result by multiplying $(1+x)^3$ out as a cubic expression in x. How good is the second approximation when $x=0{\cdot}5$?

12. Use the derivative of x^2 to show that the approximate increase in the area of a circle is $2\pi ah$ square inches when the radius is increased by a small amount h from the value a. Calculate the actual increase in area when the radius increases from 10 to 10·1 inches and compare with the approximate value given by the above formula.

13. A circular ink-blot grows at the rate of 2 square inches per second. Show that the radius is increasing at the rate of $\frac{1}{\pi r}$ inches per second at the time when its length is r inches.

14. The surface area and volume of a sphere of radius r are $S=4\pi r^2$ and $V=\frac{4}{3}\pi r^3$. Find approximate expressions for the increases ΔS and ΔV when the radius is increased from r by an amount Δr. Deduce that the proportionate increase in S is approximately twice, and in V approximately three times, the proportionate increase in radius.

15. The pressure of a gas at constant temperature is p lbs. per square inch and the volume v cubic inches where $pv=10$. Find the rate of decrease of volume as pressure increases and the approximate decrease in volume when the pressure is increased from p by an amount Δp.

16. At what point on the parabola $y=x^2$ does the tangent make equal angles with the axes? Use the derivative of $(1+x)^2$ to obtain the equation of the tangent at (0, 1) on the parabola $y=x^2+2x+1$.

17. What is the equation of the tangent to the parabola $y=x^2+3x-2$ at the point with abscissa $x=a$? Show that there is one point on the parabola with tangent parallel to Ox. What is this point?

18. From the derivative of (ax^2+bx+c), show that the tangent at the point (x_1, y_1) on the parabola $y=ax^2+bx+c$ has equation

$$y+y_1=2axx_1+b(x+x_1)+2c.$$

19. Show that the tangent at the point (1, 1) on the rectangular hyperbola $xy=1$ cuts equal lengths off the axes.

20. Show that $xy_1+yx_1=2\alpha^2$ is the tangent at the point $P(x_1, y_1)$ on the rectangular hyperbola $xy=\alpha^2$. If the tangent cuts the axes in A and B and if PM and PN are perpendicular to the axes, show that M bisects OA and N bisects OB. Show also that the area of the triangle OAB is a constant independent of the position of P on the curve.

21. From the derivative of $\frac{2x+1}{x-1}$, find the equation of the tangent at the point (x_1, y_1) on the rectangular hyperbola $(x-1)(y-2)=3$. Show that the tangent gradient tends to become infinite as $x_1 \to 1$. What is the relation of this fact to the vertical asymptote of the curve?

22. Find the derivatives of ax and ax^2, where a is a constant.

A ball rolls at feet in t seconds. Show that the average speed over any time-interval and the velocity at any time are both constant.

A lead pellet falls at^2 feet in t seconds. What is the average speed between the t_1th and the t_2th seconds? Find the velocity after t seconds. How good an approximation is the velocity after 3 seconds to the average speed between times $t=3$ and $t=3{\cdot}5$ seconds?

23. Show that the acceleration of the lead pellet of the previous example is constant over time.

24. What is the derivative of $(1+2x)$? Show that the electric current in a wire is constant if $(1+2t)$ units of electricity pass a point in the wire in t seconds.

25. Draw the total revenue curve on one graph and the demand and marginal revenue curves on another graph in the case of the demand for gramophones of Example V, 2. Verify that the marginal revenue is zero at the output giving greatest total revenue.

26. If the market demand for tea is given by the law of Examples V, 1, draw the total revenue curve and read off the tangent gradient to give the marginal revenue at demands of 50, 60 and 70 thousand lbs. per week. Check these values from the expression for marginal revenue. Draw on one graph the demand and marginal revenue lines. At what price does marginal revenue vanish?

27. Evaluate the derivative of $(ax+bx^3)$. If $100x$ passengers travel on the train of Examples V, 11, find an expression for the marginal revenue derived by the railway company. Draw the marginal revenue curve, showing that it is a part of a parabola falling to the right. Find the fare at which marginal revenue is zero and verify that this fare produces the greatest total revenue.

28. A firm produces radio sets according to the conditions of Examples V, 15. Draw two graphs, one showing the total cost curve and the other the average and marginal cost curves. Verify that average and marginal cost are equal when the former assumes its least value.

29. If the cost of a coal merchant's supply of coal is of the form described in Examples V, 16, show that the marginal cost decreases steadily as the monthly purchase increases.

30. From the derivative of (ax^3+bx^2+cx), find the marginal cost of any output in the case of the firm of Examples V, 18. Show that the marginal cost curve is a parabola and locate the output for least marginal cost. Draw the average and marginal cost curves on one graph and show that the output for least average cost is greater than the output for least marginal cost.

31. A business produces an income of £x this year and £y next year, where these values can be varied according to the relation $y=1000-\frac{x^2}{250}\cdot$ Explain how $\left\{\left(-\frac{dy}{dx}\right)-1\right\}$ can be interpreted as the marginal rate of return over cost. Show that the value of this marginal rate is $\frac{x-125}{125}$ when this year's income is £x.

CHAPTER VII

THE TECHNIQUE OF DERIVATION

7.1 Introduction.

THE first object of the differential calculus is to systematise the process of finding the derivatives of functions, to make the writing down of a derivative an almost mechanical matter. Text-books on the subject necessarily devote a considerable amount of space to this stage, but it must be remembered that the systematic evaluation of derivatives is not an end in itself. We evaluate derivatives only because they are useful to us in the application of mathematical methods in the natural or social sciences. These applications would be extremely laborious were it not for the technique of the differential calculus; we should find ourselves repeating over and over again the same kind of algebraic process, the process of finding the limits of certain expressions, whenever we need the derivatives of particular functions. We have, therefore, to learn to evaluate derivatives easily before we can apply them fruitfully in economics or any other scientific study.

The systematised technique of derivation involves two steps. The first consists of the evaluation (from the definition) of the derivatives of the simplest functions, the results, which are called "standard forms", being set out in tabular form and memorised. The table of standard forms, once obtained, is taken for granted and the derivatives it contains simply quoted whenever they are required. All the functions considered in the present chapter are actually derivable from one simple function, the "power" function. Our table of standard forms, therefore, need contain only a single entry at the moment, i.e. the derivative of the simplest power function. Many other functions, such as the trigonometric, exponential and logarithmic functions, are used in mathematical analysis and some of these will be introduced at a later stage. It will then be necessary to extend the table of standard forms by evaluating the derivatives of the simplest examples of the new functions.

The second step is to frame a set of rules which serve for the derivation of more complicated functions. A given function, no matter how involved, is reduced to a combination of simpler functions the derivatives of which appear in the table of standard forms. The rules for derivation simply tell us how the derivative of a combination of this kind can be obtained in terms of the simpler standard form derivatives. The combinations may involve sums, differences, products or quotients. But, as we shall see, they may also include a very different kind of configuration. In all cases, the rules for derivation are relatively simple. With the table of standard forms and sufficient skill in the manipulation of the rules, we can set about evaluating derivatives with the greatest confidence. This confidence should be acquired in the course of the present chapter.

7.2 The power function and its derivative.

In elementary algebra we deal with expressions of the power type a^n, where n is called the *index* (or exponent) of the given *base* a. The meaning of the power varies, however, with the nature of the number n. For example, $a^2 = a \times a$, $a^{\frac{1}{2}} = \sqrt{a}$, $a^{-2} = \frac{1}{a^2}$ and $a^{-\frac{2}{3}} = \frac{1}{\sqrt[3]{a^2}}$.

In general terms, the power a^n is to be interpreted as follows:

(1) If n is a positive integer, a^n denotes the result of multiplying a by itself n times.

(2) If n is a positive fractional value, a^n denotes the positive value of a certain root: $a^{\frac{r}{q}} = \sqrt[q]{a^r}$.

(3) If n is a negative integral or fractional value, a^n denotes the reciprocal of the corresponding positive power: $a^{-\frac{r}{q}} = \frac{1}{\sqrt[q]{a^r}}$.

(4) If n is zero, a^n stands for unity: $a^0 = 1$.

(5) If n is an irrational number, the power a^n is more complicated in meaning. We can write n, in this case, as the limit of a sequence of integral or fractional numbers n_1, n_2, n_3, ... The power a^n is then defined as the limit of a^{n_r} as $r \to \infty$.

In all cases, the power a^n obeys the familiar "index laws" developed in elementary algebra.

One of the simplest of all functions is the *power function* $y=x^n$, where the index n is a fixed number. The variable x takes values continuously from the whole set of numbers except that only positive values can be considered in certain cases (such as $y=\sqrt{x}$). A distinctive characteristic of the power function is that both the function and its inverse are of the same type. For example, the power function $y=x^2$ gives the inverse function $x=\sqrt{y}$, which is a power function of y.

The derivative of x^n, for any value of n, is now required as a standard form. A number of simple cases can be taken first:

$$\frac{d}{dx}(x)=\lim_{h\to 0}\frac{(x+h)-x}{h}=\lim_{h\to 0}\frac{h}{h}=1,$$

$$\frac{d}{dx}(x^2)=\lim_{h\to 0}\frac{(x+h)^2-x^2}{h}=\lim_{h\to 0}\frac{2hx+h^2}{h}=\lim_{h\to 0}(2x+h)=2x,$$

$$\frac{d}{dx}(x^3)=\lim_{h\to 0}\frac{(x+h)^3-x^3}{h}=\lim_{h\to 0}\frac{3hx^2+3h^2x+h^3}{h}$$

$$=\lim_{h\to 0}(3x^2+3hx+h^2)=3x^2,$$

$$\frac{d}{dx}\left(\frac{1}{x}\right)=\lim_{h\to 0}\frac{1}{h}\left(\frac{1}{x+h}-\frac{1}{x}\right)=\lim_{h\to 0}-\frac{1}{x(x+h)}=-\frac{1}{x^2},$$

i.e. $$\frac{d}{dx}(x^{-1})=-x^{-2}.$$

$$\frac{d}{dx}\left(\frac{1}{x^2}\right)=\lim_{h\to 0}\frac{1}{h}\left\{\frac{1}{(x+h)^2}-\frac{1}{x^2}\right\}=\lim_{h\to 0}\frac{x^2-(x+h)^2}{hx^2(x+h)^2}$$

$$=\lim_{h\to 0}\frac{-2x-h}{x^2(x+h)^2}=-\frac{2}{x^3},$$

i.e. $$\frac{d}{dx}(x^{-2})=-2x^{-3},$$

$$\frac{d}{dx}(\sqrt{x})=\lim_{h\to 0}\frac{\sqrt{x+h}-\sqrt{x}}{h}=\lim_{h\to 0}\frac{(\sqrt{x+h}-\sqrt{x})(\sqrt{x+h}+\sqrt{x})}{h(\sqrt{x+h}+\sqrt{x})}$$

$$=\lim_{h\to 0}\frac{(x+h)-x}{h(\sqrt{x+h}+\sqrt{x})}=\lim_{h\to 0}\frac{1}{\sqrt{x+h}+\sqrt{x}}=\frac{1}{2\sqrt{x}},$$

i.e. $$\frac{d}{dx}(x^{\frac{1}{2}})=\tfrac{1}{2}x^{-\frac{1}{2}}.$$

A uniform law is now apparent. In each of the above cases the derivative of x^n is found to be of the form nx^{n-1}. This is, in fact, true for any given value of n whatever and, to provide a general proof, we need a result which will be established at a later stage (see 17.3, Ex. 1, below). The result can be expressed quite simply :

$$(1+x)^n = 1 + nx + \lambda x^2,$$

where λ is some quantity involving x which is finite even when $x \to 0$.

Then $$\frac{d}{dx}(x^n) = \lim_{h\to 0} \frac{(x+h)^n - x^n}{h} = \lim_{h\to 0} \frac{x^n}{h}\left\{\left(1+\frac{h}{x}\right)^n - 1\right\}$$

$$= \lim_{h\to 0} \frac{x^n}{h}\left\{1 + n\frac{h}{x} + \lambda\frac{h^2}{x^2} - 1\right\}$$

$$= \lim_{h\to 0} x^n\left(\frac{n}{x} + \lambda\frac{h}{x^2}\right) = x^n\frac{n}{x} = nx^{n-1}.$$

Hence, we have the *standard form* :

$$\frac{d}{dx}(x^n) = nx^{n-1} \quad \text{for any fixed value of } n.$$

This result can now be quoted whenever required. For example, we are entitled to write at once, not only the particular derivatives evaluated above, but also many others such as

$$\frac{d}{dx}(x^5) = 5x^4; \quad \frac{d}{dx}(x^{12}) = 12x^{11};$$

$$\frac{d}{dx}\left(\frac{1}{x^3}\right) = \frac{d}{dx}(x^{-3}) = -3x^{-4} = -\frac{3}{x^4};$$

$$\frac{d}{dx}(\sqrt[3]{x^2}) = \frac{d}{dx}(x^{\frac{2}{3}}) = \frac{2}{3}x^{-\frac{1}{3}} = \frac{2}{3\sqrt[3]{x}};$$

$$\frac{d}{dx}\left(\frac{1}{\sqrt{x}}\right) = \frac{d}{dx}(x^{-\frac{1}{2}}) = -\frac{1}{2}x^{-\frac{3}{2}} = -\frac{1}{2\sqrt{x^3}}.$$

Many rather different kinds of derivatives are thus included in the one standard form. The method of treating fractional and negative indices is to be noticed.

7.3 Rules for the evaluation of derivatives.

We come now to the second stage in the technique of derivation, to the framing of a set of rules for the derivation of combinations of functions. The first three rules relate to combinations which are sums (or differences), products and quotients respectively. In each of these rules, as set out below, it is assumed that u and v are two given functions of x with known derivatives at the point concerned.

RULE I. *The derivative of a sum or difference.*

The derivative of the sum (or difference) of two functions is the sum (or difference) of the separate derivatives:

$$\frac{d}{dx}(u+v)=\frac{du}{dx}+\frac{dv}{dx} \quad \text{and} \quad \frac{d}{dx}(u-v)=\frac{du}{dx}-\frac{dv}{dx}.$$

RULE II. *The derivative of a product.*

The derivative of the product of two functions is equal to the first function times the derivative of the second *plus* the second function times the derivative of the first:

$$\frac{d}{dx}(uv)=u\frac{dv}{dx}+v\frac{du}{dx}.$$

RULE III. *The derivative of a quotient.*

The derivative of the quotient of two functions is equal to the denominator times the derivative of the numerator *minus* the numerator times the derivative of the denominator, all divided by the square of the denominator:

$$\frac{d}{dx}\left(\frac{u}{v}\right)=\frac{v\dfrac{du}{dx}-u\dfrac{dv}{dx}}{v^2}.$$

As a special case:

$$\frac{d}{dx}\left(\frac{1}{u}\right)=-\frac{1}{u^2}\frac{du}{dx}.$$

Formal proofs of the three rules proceed directly from the definition of the derivative. The proof of the first rule is:

If $u=\phi(x)$ and $v=\psi(x)$ are the two given functions, then

$$\frac{d}{dx}(u+v)=\operatorname*{Lim}_{h\to 0}\frac{\{\phi(x+h)+\psi(x+h)\}-\{\phi(x)+\psi(x)\}}{h}$$

$$=\operatorname*{Lim}_{h\to 0}\frac{\phi(x+h)-\phi(x)}{h}+\operatorname*{Lim}_{h\to 0}\frac{\psi(x+h)-\psi(x)}{h}$$

$$=\frac{du}{dx}+\frac{dv}{dx},$$

and similarly for the difference of u and v. We use here the fact (see 4.5 above) that the limit of a sum (or difference) is the sum (or difference) of the separate limits. The proofs of the other two rules, though more complicated, can be given in essentially the same way.* These rules can also be deduced from the first rule by a method to be described at a later stage (see 10.2 below).

* See, for example, Hardy, *Pure Mathematics*, (3rd Ed., 1921), pp. 203-4.

Important deductions from Rules I and II concern the behaviour of constants in the process of derivation. A constant can be regarded as the (limiting) case of a function of x which does not change in value as x varies. A constant must thus have a zero rate of change, i.e. *the derivative of a constant is zero.* Rules I and II now give

$$\frac{d}{dx}(u+a)=\frac{du}{dx} \quad \text{and} \quad \frac{d}{dx}(au)=a\frac{du}{dx},$$

where u is any function of x and a is any constant. Hence, an additive constant disappears when the derivative is taken and a multiplicative constant remains unaffected by derivation. These facts are of very great service, as the examples of the following section show.

The three rules are set out above for the case of a combination of two functions only. But, if more than two functions appear in a combination, the rules can be applied several times in succession to give the derivative. It can be noticed, however, that the sum and difference rule extends at once to give the derivative of the " algebraic " sum of a number of functions as the similar " algebraic " sum of the derivatives of the separate functions. For example,

$$\frac{d}{dx}(u+v-w)=\frac{du}{dx}+\frac{dv}{dx}-\frac{dw}{dx},$$

where u, v and w are three functions of x with known derivatives.

The derivative of a product or quotient involving more than two functions is less easily obtained, as is shown by the following example: Taking the product (uvw) as (uv) times w, we obtain

$$\frac{d}{dx}(uvw)=uv\frac{dw}{dx}+w\frac{d}{dx}(uv)=uv\frac{dw}{dx}+w\left(u\frac{dv}{dx}+v\frac{du}{dx}\right)$$

$$=vw\frac{du}{dx}+uw\frac{dv}{dx}+uv\frac{dw}{dx}.$$

In an exactly similar way, we can show that

$$\frac{d}{dx}\left(\frac{uv}{w}\right)=\frac{1}{w^2}\left(vw\frac{du}{dx}+uw\frac{dv}{dx}-uv\frac{dw}{dx}\right).$$

It is possible to put the product and quotient rules in a somewhat different form suitable for the extension to the case of more than two functions. The product rule for the derivative of the function $y=uv$ can be divided through by $y=uv$ and written

$$\frac{1}{y}\frac{dy}{dx}=\frac{1}{u}\frac{du}{dx}+\frac{1}{v}\frac{dv}{dx}.$$

In the same way, the quotient rule for $y=\frac{u}{v}$ can be arranged

$$\frac{1}{y}\frac{dy}{dx}=\frac{1}{u}\frac{du}{dx}-\frac{1}{v}\frac{dv}{dx}.$$

Further, the derivative of $y=uvw$ obtained above becomes

$$\frac{1}{y}\frac{dy}{dx}=\frac{1}{u}\frac{du}{dx}+\frac{1}{v}\frac{dv}{dx}+\frac{1}{w}\frac{dw}{dx},$$

and the derivative of $y=\frac{uv}{w}$ becomes

$$\frac{1}{y}\frac{dy}{dx}=\frac{1}{u}\frac{du}{dx}+\frac{1}{v}\frac{dv}{dx}-\frac{1}{w}\frac{dw}{dx}.$$

We now have a common form of the product and quotient rules which can be extended, in an obvious way, to the case where any number of functions are involved. The meaning of this alternative expression of the rules will be made more clear later (see 10.2 below).

7.4 Examples of the evaluation of derivatives.

The examples below illustrate the way in which the rules for derivation are used in practice. The derivative of a fairly complicated function is to be obtained by several applications either of the same rule or of different rules. The method of dealing with constants in the derivation process is particularly to be noticed. It is also clear that some derivatives can be obtained in two or more different ways. The result is always the same and all that matters is to get it in some way or other.

Ex. 1. $\frac{d}{dx}(5x)=5\frac{d}{dx}(x)=5.$

Ex. 2. $\frac{d}{dx}\left(\frac{3}{x}\right)=3\frac{d}{dx}\left(\frac{1}{x}\right)=-\frac{3}{x^2}.$

Ex. 3. $\frac{d}{dx}\left(2x+3+\frac{1}{x}\right)=2\frac{d}{dx}(x)+\frac{d}{dx}\left(\frac{1}{x}\right)=2-\frac{1}{x^2}.$

Ex. 4. $\frac{d}{dx}(1+x^2)=\frac{d}{dx}(x^2)=2x.$

Ex. 5. $\frac{d}{dx}(x^2+3x-2)=\frac{d}{dx}(x^2)+3\frac{d}{dx}(x)=2x+3.$

Ex. 6. $\frac{d}{dx}(x^3-7x^2+5x+3)=\frac{d}{dx}(x^3)-7\frac{d}{dx}(x^2)+5\frac{d}{dx}(x)=3x^2-14x+5.$

Ex. 7. For any fixed values of the constants a, b, c, ... , we have

$$\frac{d}{dx}(ax+b)=a\,;\quad \frac{d}{dx}(ax^2+bx+c)=2ax+b\,;$$

$$\frac{d}{dx}(ax^3+bx^2+cx+d)=3ax^2+2bx+c\,;\ \text{and so on.}$$

Ex. 8. $\frac{d}{dx}\{x(1+x^2)\}=x\frac{d}{dx}(1+x^2)+(1+x^2)\frac{d}{dx}(x)=x(2x)+(1+x^2)$

$$=1+3x^2,$$

or $\frac{d}{dx}\{x(1+x^2)\}=\frac{d}{dx}(x+x^3)=\frac{d}{dx}(x)+\frac{d}{dx}(x^3)=1+3x^2.$

Ex. 9. $\frac{d}{dx}(1+x)^2=\frac{d}{dx}(1+x)(1+x)$

$$=(1+x)\frac{d}{dx}(1+x)+(1+x)\frac{d}{dx}(1+x)=2(1+x),$$

or $\frac{d}{dx}(1+x)^2=\frac{d}{dx}(1+2x+x^2)=2\frac{d}{dx}(x)+\frac{d}{dx}(x^2)=2(1+x).$

Ex. 10. $\frac{d}{dx}\{x(1+x)(1+2x)\}=x(1+x)\frac{d}{dx}(1+2x)+(1+2x)\frac{d}{dx}\{x(1+x)\}$

$$=2x(1+x)+(1+2x)(1+2x)$$
$$=2x+2x^2+1+4x+4x^2=1+6x+6x^2,$$

or $\frac{d}{dx}\{x(1+x)(1+2x)\}=\frac{d}{dx}(x+3x^2+2x^3)=1+6x+6x^2.$

Ex. 11. $\frac{d}{dx}\left(\frac{1}{3x+2}\right)=\frac{-\frac{d}{dx}(3x+2)}{(3x+2)^2}=-\frac{3}{(3x+2)^2}.$

Ex. 12. $\frac{d}{dx}\left(\frac{2x^2+1}{x}\right)=\frac{x\frac{d}{dx}(2x^2+1)-(2x^2+1)\frac{d}{dx}(x)}{x^2}$

$$=\frac{x(4x)-(2x^2+1)}{x^2}=\frac{2x^2-1}{x^2}$$

or $\frac{d}{dx}\left(\frac{2x^2+1}{x}\right)=\frac{d}{dx}\left(2x+\frac{1}{x}\right)=2-\frac{1}{x^2}=\frac{2x^2-1}{x^2}.$

Ex. 13. $$\frac{d}{dx}\left(\frac{1+x}{1-2x}\right)=\frac{(1-2x)\frac{d}{dx}(1+x)-(1+x)\frac{d}{dx}(1-2x)}{(1-2x)^2}$$

$$=\frac{(1-2x)+2(1+x)}{(1-2x)^2}=\frac{3}{(1-2x)^2}.$$

Ex. 14. $$\frac{d}{dx}\left\{\frac{x}{(x+1)(x+2)}\right\}$$

$$=\frac{(x+1)(x+2)-x(x+1+x+2)}{(x+1)^2(x+2)^2}=-\frac{x^2-2}{(x+1)^2(x+2)^2}.$$

Ex. 15. $$\frac{d}{dx}\left(\frac{ax}{x+b}-cx\right)=a\frac{d}{dx}\left(\frac{x}{x+b}\right)-c\frac{d}{dx}(x)$$

$$=a\frac{(x+b)-x}{(x+b)^2}-c=\frac{ab}{(x+b)^2}-c.$$

Ex. 16. $$\frac{d}{dx}\{xf(x)\}=x\frac{d}{dx}f(x)+f(x)\frac{d}{dx}(x)=xf'(x)+f(x),$$

$$\frac{d}{dx}\left\{\frac{f(x)}{x}\right\}=\frac{x\frac{d}{dx}f(x)-f(x)\frac{d}{dx}(x)}{x^2}=\frac{xf'(x)-f(x)}{x^2}.$$

where $f(x)$ is any single-valued function of x.

7.5 The function of a function rule.

The three rules we have given are not in themselves sufficient for the derivation of all the functions met with in ordinary mathematical analysis. On the contrary, there are many relatively simple functions the derivatives of which cannot be found by means of the three rules. Consider, for example, the function

$$y=\sqrt{2x^2-3},$$

which is a " mixture " of the simple quadratic function $(2x^2-3)$ and the simple square root function. On attempting to evaluate the derivative, however, it is found that the function does not break up into sums, differences, products or quotients of " standard form " functions. The " mixture ", in fact, is not as before. We have a function which is fundamentally different in nature and we need some method of evaluating its derivative.

The problem is solved by introducing a new variable :

$$u=2x^2-3.$$

Here, u is a function of x and its derivative is

$$\frac{du}{dx}=\frac{d}{dx}(2x^2-3)=4x.$$

The original function now appears as $y=\sqrt{u}$, i.e. as a simple function of the variable u with a known derivative *with respect to u* :

$$\frac{dy}{du}=\frac{d}{du}(u^{\frac{1}{2}})=\tfrac{1}{2}u^{-\frac{1}{2}}=\frac{1}{2\sqrt{u}}.$$

The function $y=\sqrt{2x^2-3}$ has been arranged, therefore, in such a way that we can say that y is a function of u, where u is a function of x. Both these functions have known derivatives. Can we now deduce the derivative of *y with respect to x*? The answer is provided by the very simple rule given below.

The way in which we have arranged the function $y=\sqrt{2x^2-3}$ shows that it can be described as a " function of a function of x ". It is clearly only one example of a wide range of such functions to which the following rule for derivation applies :

RULE IV. *The derivative of a function of a function.*

If y is a function of u where u is a function of x, then the derivative of y with respect to x is the product of the derivative of y with respect to u and the derivative of u with respect to x :

$$\frac{dy}{dx}=\frac{dy}{du}\cdot\frac{du}{dx}.$$

The formal proof of the rule proceeds :

If $y=f(u)$, where $u=\phi(x)$, then $y=f\{\phi(x)\}$ and

$$\frac{dy}{dx}=\lim_{h\to 0}\frac{f\{\phi(x+h)\}-f\{\phi(x)\}}{h}.$$

Let $\quad k=\phi(x+h)-\phi(x)\to 0,\ \text{as } h\to 0.$

So
$$\frac{dy}{dx}=\lim_{h\to 0}\frac{f(u+k)-f(u)}{k}\cdot\frac{\phi(x+h)-\phi(x)}{h}$$

$$=\lim_{k\to 0}\frac{f(u+k)-f(u)}{k}\cdot\lim_{h\to 0}\frac{\phi(x+h)-\phi(x)}{h}=\frac{dy}{du}\cdot\frac{du}{dx}.$$

We use here the result (see 4.5 above) that the limit of a product is the product of the separate limits.

The rule clearly extends to the case where there are several " intermediate" functions. For example, if y is a function of u where u is a function of v which, in its turn, is a function of x, then

$$\frac{dy}{dx}=\frac{dy}{du}\cdot\frac{du}{dv}\cdot\frac{dv}{dx}.$$

This follows from two successive applications of the simple function of a function rule given above.

The success of the function of a function rule in practice depends largely on the introduction of an "intermediate" function u in such a way that the derivative is most conveniently obtained. The method of breaking up functions for the application of the rule is illustrated by the following examples.

Ex. 1. $y=\sqrt{2x^2-3}$.

Write $y=\sqrt{u}$ where $u=2x^2-3$,

and so $$\frac{d}{dx}\sqrt{2x^2-3}=\frac{d}{du}\sqrt{u}\cdot\frac{d}{dx}(2x^2-3)=\frac{1}{2\sqrt{u}}\cdot 4x=\frac{2x}{\sqrt{2x^2-3}}.$$

Since u is only introduced for convenience of working, the final step is to get rid of this "intermediate" variable in terms of x.

Ex. 2. $y=(1+x)^2$,

i.e. $y=u^2$ where $u=1+x$.

So $$\frac{d}{dx}(1+x)^2=\frac{d}{du}(u^2)\cdot\frac{d}{dx}(1+x)=2u\,.\,1=2(1+x).$$

This result has been obtained before (7.4, Ex. 9) by other methods.

Ex. 3. $y=\dfrac{1}{3x+2}$,

i.e. $y=\dfrac{1}{u}$ where $u=3x+2$.

So $$\frac{d}{dx}\left(\frac{1}{3x+2}\right)=\frac{d}{du}\left(\frac{1}{u}\right)\cdot\frac{d}{dx}(3x+2)=-\frac{1}{u^2}\cdot 3=-\frac{3}{(3x+2)^2},$$

which has also been obtained before.

Ex. 4. $y=(ax+b)^n$,

i.e. $y=u^n$ where $u=ax+b$.

So $$\frac{d}{dx}(ax+b)^n=\frac{d}{du}(u^n)\cdot\frac{d}{dx}(ax+b)=nu^{n-1}\,.\,a=na(ax+b)^{n-1}.$$

Ex. 5. $y=\sqrt{\dfrac{1+x}{1-2x}}$,

i.e. $y=\sqrt{u}$ where $u=\dfrac{1+x}{1-2x}$.

Now $$\frac{d}{du}\sqrt{u}=\frac{1}{2\sqrt{u}}=\frac{1}{2}\sqrt{\frac{1-2x}{1+x}},$$

and $$\frac{d}{dx}\left(\frac{1+x}{1-2x}\right)=\frac{3}{(1-2x)^2}\quad (7.4, \text{Ex. } 13).$$

So $$\frac{d}{dx}\sqrt{\frac{1+x}{1-2x}}=\frac{1}{2}\sqrt{\frac{1-2x}{1+x}}\,\frac{3}{(1-2x)^2}=\frac{3}{2}\frac{1}{\sqrt{(1+x)(1-2x)^3}}.$$

Each of these five examples is an instance of one general form which can be derived at once from the function of a function rule:

$$\frac{d}{dx}(u^n)=nu^{n-1}\frac{du}{dx},$$

where u is any given function of x. In the case $n=-1$, we have

$$\frac{d}{dx}\left(\frac{1}{u}\right)=-\frac{1}{u^2}\frac{du}{dx},$$

which is the special case of the quotient rule already given.

7.6 The inverse function rule.

From the function of a function rule we can derive the fifth, and last, of the rules for derivation. This rule provides the derivative of the function inverse to a given function and it applies only if both the inverse functions are single-valued.

RULE V. *The derivative of an inverse function.*

The derivative of an inverse function is the reciprocal of the derivative of the original function:

$$\frac{dx}{dy}=\frac{1}{\dfrac{dy}{dx}},$$

provided that both functions are single-valued.

The proof is as follows. Suppose that the single-valued function $y=\phi(x)$ gives the single-valued inverse function $x=\psi(y)$. Then $\psi\{\phi(x)\}$ must equal x for any value of x. So

$$\frac{d}{dx}\psi\{\phi(x)\}=\frac{d}{dx}(x)=1.$$

But $$\frac{d}{dx}\psi\{\phi(x)\}=\frac{d}{dy}\psi(y)\cdot\frac{d}{dx}\phi(x)\quad \text{(by Rule IV)}$$

$$=\frac{dx}{dy}\cdot\frac{dy}{dx}.$$

So, the product of $\frac{dx}{dy}$ and $\frac{dy}{dx}$ is unity and the rule is proved.

As an illustration of the inverse function rule, we can take the case of the derivative of $\sqrt{x}$. If $y=\sqrt{x}$, then $x=y^2$. The inverse function rule then gives

$$\frac{d}{dx}(\sqrt{x})=\frac{1}{\frac{d}{dy}(y^2)}=\frac{1}{2y}=\frac{1}{2\sqrt{x}}.$$

This result agrees, of course, with that obtained from the standard form giving the derivative of x^n for any value of n.

The process of evaluating derivatives can now be regarded as complete. The five rules, taken in conjunction with a suitable table of standard forms, are sufficient for the derivation of all single-valued functions, no matter how complicated is their analytical expression. Even when functions of entirely new types are introduced, it is only necessary to extend the table of standard forms by including the derivatives of the simplest instances of the new function types, as obtained directly from the definition of the derivative. Examples of this extension of the table of standard forms will be given later.

7.7 The evaluation of second and higher order derivatives.

The practical method of obtaining the second and higher order derivatives of a function introduces nothing that is new. Having obtained the first derivative by means of the rules set out above, the second derivative is obtained by a further application of the rules, this time applied to the first derivative considered as a function of x. The third, fourth and higher derivatives are then obtained in succession in the same way.

It is possible, however, to extend the sum and product rules so that they apply directly to second and higher order derivatives. If u and v are two single-valued functions of x, we have

$$\frac{d^2}{dx^2}(u+v)=\frac{d}{dx}\left\{\frac{d(u+v)}{dx}\right\}=\frac{d}{dx}\left(\frac{du}{dx}+\frac{dv}{dx}\right)$$

$$=\frac{d}{dx}\left(\frac{du}{dx}\right)+\frac{d}{dx}\left(\frac{dv}{dx}\right)=\frac{d^2u}{dx^2}+\frac{d^2v}{dx^2}.$$

In general, for any value of r which is a positive integer,

$$\frac{d^r}{dx^r}(u \pm v) = \frac{d^r u}{dx^r} \pm \frac{d^r v}{dx^r}.$$

The extension of the product rule is more difficult. From

$$\frac{d}{dx}(uv) = u\frac{dv}{dx} + v\frac{du}{dx},$$

we have

$$\frac{d^2}{dx^2}(uv) = \frac{d}{dx}\left(u\frac{dv}{dx}\right) + \frac{d}{dx}\left(v\frac{du}{dx}\right)$$

$$= \left(u\frac{d^2v}{dx^2} + \frac{du}{dx}\frac{dv}{dx}\right) + \left(\frac{du}{dx}\frac{dv}{dx} + v\frac{d^2u}{dx^2}\right)$$

$$= u\frac{d^2v}{dx^2} + 2\frac{du}{dx}\frac{dv}{dx} + v\frac{d^2u}{dx^2}.$$

The third derivative is then found by a further derivation process as

$$\frac{d^3}{dx^3}(uv) = u\frac{d^3v}{dx^3} + 3\frac{du}{dx}\frac{d^2v}{dx^2} + 3\frac{d^2u}{dx^2}\frac{dv}{dx} + v\frac{d^3u}{dx^3}.$$

A general result for the rth derivative of the product is obtained, in this way, by repeated derivation :

$$\frac{d^r}{dx^r}(uv) = u\frac{d^rv}{dx^r} + r\frac{du}{dx}\frac{d^{r-1}v}{dx^{r-1}} + \frac{r(r-1)}{2}\frac{d^2u}{dx^2}\frac{d^{r-2}v}{dx^{r-2}} + \dots$$

$$+ \frac{r(r-1)}{2}\frac{d^{r-2}u}{dx^{r-2}}\frac{d^2v}{dx^2} + r\frac{d^{r-1}u}{dx^{r-1}}\frac{dv}{dx} + v\frac{d^ru}{dx^r}.$$

This is known as Leibnitz's Theorem and the coefficients of the successive terms on the right-hand side follow the well-known " binomial " law. From the point of view of the present development, however, only the first and second derivatives are normally required and these can always be obtained by direct application of the simple rules given above. A number of examples are given for illustration :

Ex. 1. $\frac{d}{dx}(x^4) = 4x^3$; $\frac{d^2}{dx^2}(x^4) = 12x^2$; $\frac{d^3}{dx^3}(x^4) = 24x$;

$$\frac{d^4}{dx^4}(x^4) = 24; \quad \frac{d^5}{dx^5}(x^4) = \frac{d^6}{dx^6}(x^4) = \dots = 0.$$

Ex. 2. $$\frac{d}{dx}(\sqrt{x}) = \tfrac{1}{2}x^{-\frac{1}{2}} = \frac{1}{2\sqrt{x}},$$

$$\frac{d^2}{dx^2}(\sqrt{x}) = \frac{1}{2}\frac{d}{dx}(x^{-\frac{1}{2}}) = \tfrac{1}{2}(-\tfrac{1}{2}x^{-\frac{3}{2}}) = -\frac{1}{4\sqrt{x^3}},$$

$$\frac{d^3}{dx^3}(\sqrt{x}) = -\frac{1}{4}\frac{d}{dx}(x^{-\frac{3}{2}}) = -\tfrac{1}{4}(-\tfrac{3}{2}x^{-\frac{5}{2}}) = \frac{3}{8\sqrt{x^5}}; \text{ and so on.}$$

Ex. 3. $$\frac{d}{dx}(x^n) = nx^{n-1}; \quad \frac{d^2}{dx^2}(x^n) = n(n-1)x^{n-2};$$

$$\frac{d^3}{dx^3}(x^n) = n(n-1)(n-2)x^{n-3}; \text{ and so on.}$$

The general result for the rth derivative of this standard form can be written down by inspection :

$$\frac{d^r}{dx^r}(x^n) = n(n-1)(n-2)\ldots(n-r+1)x^{n-r}.$$

The previous two examples are particular cases of this standard result. If n is a positive integer, as in the first example, then the derivatives after the nth order are all zero. If n has any other value, the process of finding successive derivatives can be extended indefinitely without producing zero derivatives.

Ex. 4. $$\frac{d}{dx}\left(\frac{1}{3x+2}\right) = -\frac{3}{(3x+2)^2}; \quad \frac{d^2}{dx^2}\left(\frac{1}{3x+2}\right) = \frac{18}{(3x+2)^3};$$

$$\frac{d^3}{dx^3}\left(\frac{1}{3x+2}\right) = -\frac{162}{(3x+2)^4}; \text{ and so on.}$$

Ex. 5. $$\frac{d}{dx}\sqrt{2x^2-3} = \frac{2x}{\sqrt{2x^2-3}};$$

$$\frac{d^2}{dx^2}\sqrt{2x^2-3} = \frac{2\sqrt{2x^2-3} - 2x\frac{d}{dx}\sqrt{2x^2-3}}{2x^2-3}$$

$$= \frac{2}{2x^2-3}\left(\sqrt{2x^2-3} - \frac{2x^2}{\sqrt{2x^2-3}}\right) = -\frac{6}{(\sqrt{2x^2-3})^3},$$

and similarly for higher order derivatives.

Ex. 6. If $f(x)$ denotes any single-valued function, then

$$\frac{d}{dx}\{xf(x)\} = xf'(x) + f(x),$$

and $$\frac{d^2}{dx^2}\{xf(x)\} = \frac{d}{dx}\{xf'(x)\} + \frac{d}{dx}f(x) = xf''(x) + 2f'(x).$$

Again, $$\frac{d}{dx}\left\{\frac{f(x)}{x}\right\}=\frac{xf'(x)-f(x)}{x^2},$$

and $$\frac{d^2}{dx^2}\left\{\frac{f(x)}{x}\right\}=\frac{1}{x^4}\left[x^2\frac{d}{dx}\{xf'(x)-f(x)\}-\{xf'(x)-f(x)\}\frac{d}{dx}(x^2)\right]$$

$$=\frac{1}{x^4}[x^2\{xf''(x)+f'(x)-f'(x)\}-2x\{xf'(x)-f(x)\}]$$

$$=\frac{x^2f''(x)-2xf'(x)+2f(x)}{x^3}.$$

EXAMPLES VII

Practical derivation

1. Write down, from the standard form, the derivatives of $\sqrt{x^3}$, $\frac{1}{x^5}$ and $\frac{1}{\sqrt[3]{x^2}}$.

2. Obtain the derivatives of $1+2x^2-3x^4$; $(1-x)^3$; $(x-2)(2x+1)$; $x+\frac{1}{\sqrt{x}}$; $x^2-\frac{1}{x^2}$; $\frac{\sqrt{x}}{2x+1}$; $\frac{1}{1-x^2}$; $\frac{x+1}{(x+2)(x+3)}$; $\frac{1+2x}{1-x+2x^2}$.

3. If a, b and c are constants, show that

$$\frac{d}{dx}\left(\frac{1}{ax+b}\right)=-\frac{a}{(ax+b)^2}\quad\text{and}\quad\frac{d}{dx}\left(\frac{1}{ax^2+bx+c}\right)=-\frac{2ax+b}{(ax^2+bx+c)^2}.$$

4. Show that $$\frac{d}{dx}\left(\frac{a_1x+b_1}{a_2x+b_2}\right)=\frac{a_1b_2-a_2b_1}{(a_2x+b_2)^2}.$$

5. If $f(x)=ax+b$, find the derivatives of $xf(x)$ and $\frac{f(x)}{x}$.

6. Verify that $1+x+x^2+\ldots+x^n=\frac{1-x^{n+1}}{1-x}$. By means of derivatives, deduce that

$$1+2x+3x^2+\ldots+nx^{n-1}=\frac{1-x^n}{(1-x)^2}-\frac{nx^n}{1-x}.$$

7. It is given that $y=1+u^2$ where $u=1-x^2$. Find y as a function of x. Write down the derivatives of y with respect to u and of y and u with respect to x. Verify that $\frac{dy}{dx}=\frac{dy}{du}\cdot\frac{du}{dx}$.

8. Find the derivatives of $\frac{1}{\sqrt{x+1}}$ and $\frac{1}{\sqrt{x+1}-\sqrt{x-1}}$.

9. Use the function of a function rule to find the derivatives of

$$\sqrt{a^2-x^2};\quad\sqrt{a^2+x^2}\quad\text{and}\quad\sqrt{\frac{a^2-x^2}{a^2+x^2}}.$$

Deduce the last derivative from the other two by the quotient rule.

10. From the function of a function rule find the derivatives of

$$\left(x+\frac{1}{x}\right)^3 \quad \text{and} \quad \left(\frac{2x+1}{2x-1}\right)^2.$$

Multiply out these powers and deduce the derivatives by using the sum, product and quotient rules only.

11. Find the derivative of each of the general expressions :

$$\frac{1}{ax+b};\ \sqrt{ax+b};\ \frac{1}{\sqrt{ax+b}};\ \frac{1}{ax^2+bx+c};\ \sqrt{ax^2+bx+c} \text{ and } \frac{1}{\sqrt{ax^2+bx+c}}.$$

12. If $f(x)$ is a single-valued function of x, express the derivatives of $\sqrt{f(x)}$ and its reciprocal in terms of the derivative of $f(x)$.

13. If x and y satisfy the relation $xy=a$, show that

$$\frac{dy}{dx}=-\frac{y}{x} \quad \text{and} \quad \frac{dx}{dy}=-\frac{x}{y},$$

so verifying the inverse function rule in this case. Consider the relation $xy^n=a$ in the same way.

14. If y is a single-valued function of x, show that $\frac{d}{dx}(x^2+y^2)=2\left(x+y\frac{dy}{dx}\right)$. Deduce that $\frac{dy}{dx}=-\frac{x}{y}$, if x and y satisfy the relation $x^2+y^2=1$. Verify by finding the derivative of $y=\sqrt{1-x^2}$.

15. Evaluate the second derivatives of $\frac{1}{\sqrt{x}}$ and $\sqrt{1-x^2}$.

16. Find the first and second derivatives of $\frac{1}{ax+b}$ and of $\frac{\alpha x+\beta}{ax+b}$. Deduce that successive derivatives of these functions differ only by a constant factor.

17. If $xy=a$, show that $x\frac{d^2y}{dx^2}+2\frac{dy}{dx}=0$. Show that the same result holds for $xy=a+bx$.

18. If $y=x^2+\frac{1}{x^2}$, show that $x^2\frac{d^2y}{dx^2}+x\frac{dy}{dx}-4y=0$. Does the same result hold for the relation $y=ax^2+\frac{b}{x^2}$?

19. Show that all derivatives of ax^2+bx+c of higher order than the second are zero.

20. If $f(x)$ is a single-valued function of x, obtain the third and fourth derivatives of $xf(x)$ in terms of the derivatives of $f(x)$. In general, deduce that

$$\frac{d^r}{dx^r}\{xf(x)\}=x\frac{d^r}{dx^r}f(x)+r\frac{d^{r-1}}{dx^{r-1}}f(x),$$

and verify by Leibnitz's Theorem.

21. Draw graphs of $y=x^3$ and $y=\sqrt[3]{x}$ for a range of positive and negative values of x and verify that each function is monotonic increasing. How is one graph to be obtained from the other? Find the derivative of $\sqrt[3]{x}$ and check that the tangent gradient to the curve $y=\sqrt[3]{x}$, as estimated from the graph, equals this derivative. Show that the derivative becomes infinite at $x=0$ and so that the tangent to the curve is Oy at this point.

22. Write x^2, x^3 and x^4 as products and deduce their derivatives from the derivative of x by the product rule. From the derivatives of x^2 and x^3, use the quotient rule to obtain the derivatives of $\frac{1}{x^2}$ and $\frac{1}{x^3}$, and the inverse function rule to obtain the derivatives of $\sqrt{x}$ and $\sqrt[3]{x}$. Find the derivatives of $x^{\frac{3}{2}}$ and $x^{-\frac{3}{2}}$ by the function of a function rule from the derivative of $\sqrt{x}$.

23. Generalise the results of the previous example to show how the standard form $\frac{d}{dx}(x^n)=nx^{n-1}$ can be deduced, for any rational value of n, from the rules of derivation and from the fact that $\frac{d}{dx}(x)=1$.

24. Find the derivative of $(a-bx)x^2$. If a hemispherical bowl of radius 10 inches is filled to a depth of x inches with water, it is known that the volume of water is $\pi(10-x)x^2$ cubic inches. Find the rate of increase of volume as the depth increases and an approximate value of the volume increase when the depth increases from 5 to 5·1 inches.

25. If water is poured into the bowl of the previous example at the rate of one cubic inch per second, show that the depth increases by $\frac{1}{\pi x(20-3x)}$ inches per second when the depth is x inches.

26. Find the equation of the tangent to the curve $y=ax+b+\frac{c}{x}$ at the point with abscissa x_1. Where is the tangent parallel to Ox?

27. A projectile travels $t(1+t^2)$ feet in t seconds. Show that it moves with increasing velocity and acceleration. What is the velocity after 10 seconds? Find an approximate value of the distance travelled during a small time Δt from the tth second.

28. After t seconds a body is $x=at+\frac{1}{2}bt^2$ feet from its starting-point. What are the velocity and acceleration then? Show that the velocity is $\sqrt{a^2+2bx}$ feet per second when x feet have been covered.

29. If the demand law is $p=\frac{a}{x}-c$, show that total revenue decreases as output increases, marginal revenue being a negative constant.

30. In the case of the demand law $p=(a-bx)^2$, show that the average and marginal revenue curves are both parabolas, the former lying above the latter. Show that marginal revenue falls to a negative value and then rises to zero as output increases. Plot the total, average and marginal revenue curves in the case of the demand of Examples V, 10.

31. Show that the demand law $p=a-bx^2$ gives parabolic average and marginal revenue curves of similar shape, both falling continuously as output increases. Where is marginal revenue zero?

32. What is the marginal revenue function for the demand $p=\sqrt{a-bx}$? Under the conditions of Examples V, 8, how many tenements must be rented per week before marginal revenue falls to zero?

33. Show that marginal revenue can always be expressed as $p+x\frac{dp}{dx}$. Deduce that the gradient of the demand curve is numerically equal to $\frac{p}{x}$ at the output where marginal revenue is zero. Check this result in the cases of the particular demand laws of the two previous examples.

34. The total cost function is $\Pi = \sqrt{ax+b}+c$. Find an expression for marginal cost and show that it decreases as output increases. Plot the marginal cost curve for the electricity output of Examples V, 17.

35. If $\Pi = ax\dfrac{x+b}{x+c}+d$ is the total cost of an output x, show that the marginal cost of the output is $a\left\{1+\dfrac{c(b-c)}{(x+c)^2}\right\}$. If $b>c$, deduce that marginal cost falls continuously as output increases. Draw the marginal cost curve of the tobacco manufacturer of Examples V, 20.

36. If a firm can produce two chocolate "lines" according to the conditions of Examples V, 25, interpret the value of the derivative $\dfrac{dy}{dx}$. At what output is a small decrease in the production of one line accompanied by an equal increase in the other line?

37. The output of a coal-mine is given in terms of the number of men working per shift according to the relation of Examples V, 27. Draw a graph showing variation of output. Find expressions for average and marginal output when x men per shift are worked and draw the corresponding curves. Show that both curves are parabolas and that average output only equals marginal output when the former has its greatest value.

CHAPTER VIII

APPLICATIONS OF DERIVATIVES

8.1 The sign and magnitude of the derivative.

WE have seen that the value of a derivative can be interpreted in two ways. The derivative of a single-valued function $f(x)$ measures the *rate of change* of the function and the *tangent gradient* of the curve $y=f(x)$ at the point in question. In the present chapter we base certain very important applications of the derivative on these interpretations of its value. It will be noticed, however, that our results depend largely upon diagrammatic " intuition " and are not rigorously established. The less strict development is sufficient at this stage, but it can be added that it is possible to provide formal analytical proofs of our results in all cases.*

The value of the derivative of $f(x)$ at the point $x=a$ is $f'(a)$. The meaning of the *sign* of $f'(a)$ is evident. If $f'(a)$ is positive, then the rate of change of $f(x)$ is positive, i.e. $f(x)$ increases as x increases through the value $x=a$. The tangent gradient of the curve $y=f(x)$ is positive and the tangent and curve slope upwards from left to right at the point with abscissa a. The converse properties hold if $f'(a)$ is negative. Hence,

(1) $f'(a)>0$ implies that $f(x)$ increases as x increases and that the curve $y=f(x)$ rises from left to right at the point $x=a$.

(2) $f'(a)<0$ implies that $f(x)$ decreases as x increases and that the curve $y=f(x)$ falls from left to right at the point $x=a$.

The *numerical magnitude* of the derivative $f'(a)$ then measures how rapidly the function $f(x)$ increases or decreases, and how steeply the curve $y=f(x)$ rises or falls, at the point $x=a$.

The results can be extended to indicate the nature of the function or curve over a whole range of values of the variable x. In order to

* Some of the analytical proofs are given at a later stage, when we consider the important result known as Taylor's series.

see when the function increases or decreases, or where the curve rises or falls, it is only necessary to examine the sign of the derivative of the function. For all ranges of values of x in which $f'(x)$ is positive, we know that $f(x)$ increases continuously as x increases and that the curve $y=f(x)$ rises continuously from left to right, and conversely for ranges in which $f'(x)$ is negative.

One case of special interest now presents itself, the case where the derivative is zero at a given point. If $f'(a)=0$, then $f(x)$ is neither increasing nor decreasing, and the curve $y=f(x)$ is neither rising nor falling, at the point $x=a$. The value of the function is momentarily stationary and the curve has a tangent parallel to Ox. The value of the function at such a point is called a *stationary value* and much of the following development is concerned with these values. Before proceeding, it is instructive to examine some particular functions and to obtain a general idea of the properties holding at their stationary values.

Ex. 1. $y=4x-x^2$.

Here $\frac{dy}{dx}=4-2x=-2(x-2)$.

The derivative is positive when $x<2$, zero when $x=2$ and negative when $x>2$. The value of the function increases at first, becomes stationary at $x=2$ and then decreases as x increases beyond this value. The corresponding curve is a parabola rising to a peak at the point where $x=2$, as shown in Fig. 47. In this case, we say that the function has a "maximum" value $y=4$ at the point where $x=2$.

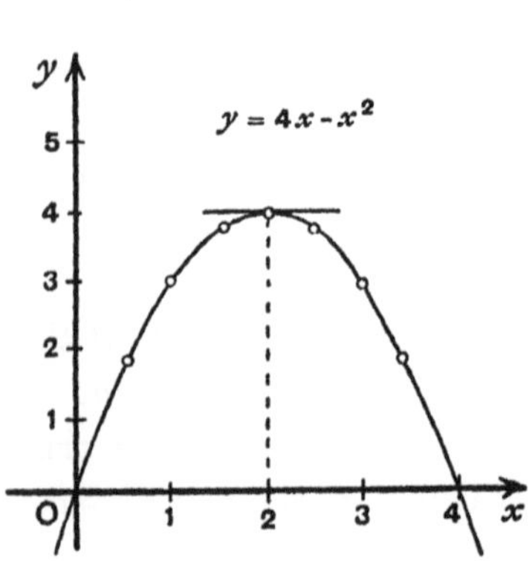

FIG. 47.

Ex. 2. $y=x^2-4x+8$.

Here $\frac{dy}{dx}=2x-4=2(x-2)$.

The sign of the derivative is exactly opposite to that found in the previous case. The value of the function is again stationary at $x=2$, but the function decreases as x increases up to the value 2 and increases as x increases beyond 2. The curve, graphed in Fig. 48, is a parabola with a lowest point at $x=2$. In this case, we say that the function has a "minimum" value $y=4$ at the point where $x=2$.

Ex. 3. $y = 6x - 3x^2 + \frac{1}{2}x^3$.

Here $\frac{dy}{dx} = 6 - 6x + \frac{3}{2}x^2 = \frac{3}{2}(x-2)^2$.

The derivative is positive at all points except that it becomes zero at the single point $x = 2$. The function is monotonic increasing but has a stationary value $y = 4$ at $x = 2$. The graph of the curve is shown in Fig. 49,

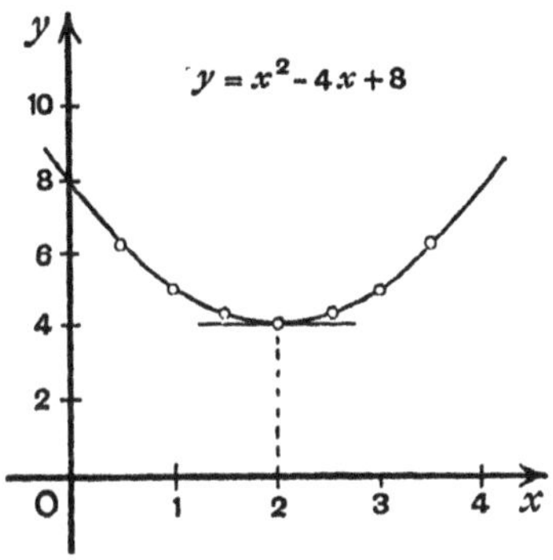

FIG. 48.

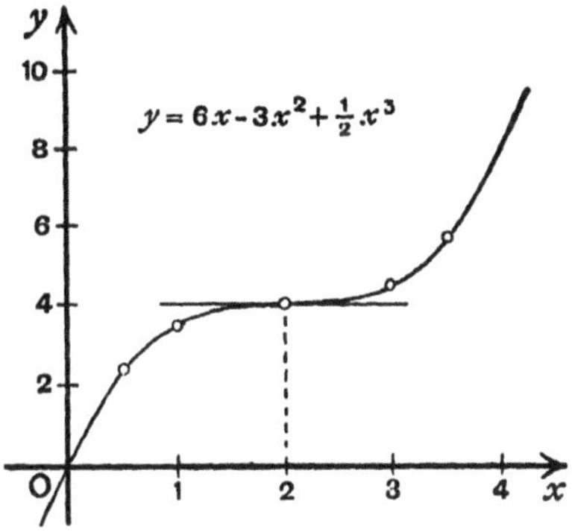

FIG. 49.

from which it is seen that the curve always rises except at the single point where the tangent is parallel to the axis Ox. The stationary value, in this case, is neither a " maximum " nor a " minimum ". The point concerned is an example of what is called a " point of inflexion ", a name derived from the fact that the curve crosses over the (horizontal) tangent as we pass through the point.

8.2 Maximum and minimum values.

It is now necessary to make more precise the meaning of the terms maximum and minimum values, which, we have seen, are connected with the stationary values of a function.

DEFINITION : The function $f(x)$ has a maximum (minimum) value at a point where the value of $f(x)$ is greater (less) than all values in the immediate neighbourhood of the point.

The maximum and minimum values together can be termed the *extreme values* of the function.

It is assumed, in the following development, that the function and its derivative are finite and continuous at all points. The corresponding curve is then smooth, being free of discontinuities and

" sharp points ". Fig. 50 shows a hypothetical, and rather artificial, curve of this nature, drawn for the purpose of indicating the various possibilities that can arise.

It is clear, in the first place, that a maximum or minimum value of $f(x)$ can occur only at a stationary point where the curve $y=f(x)$ has a horizontal tangent. If the tangent slopes upwards at any point, there are larger values of $f(x)$ immediately to the right of the point. If the tangent slopes downwards, there are larger values of $f(x)$ immediately to the left of the point. Neither of these cases is

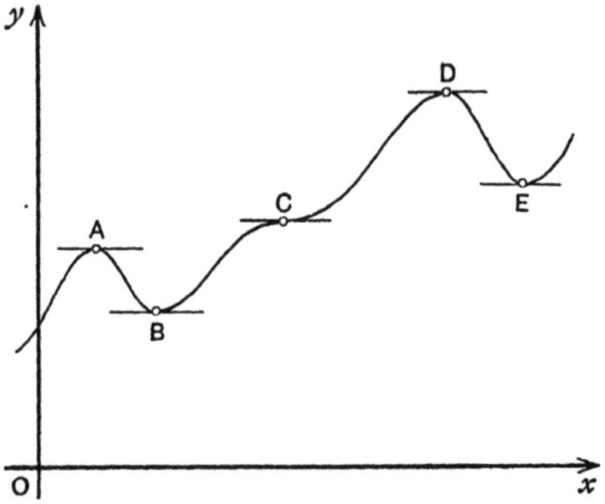

FIG. 50.

possible, by definition, at a point where $f(x)$ is a maximum. The tangent can, therefore, only be horizontal at a maximum point and (similarly) at a minimum point. Thus all the maximum and minimum values of a function are included amongst the stationary values.

In the second place, of all the points where the tangent to the curve $y=f(x)$ is horizontal, there are some points (such as A and D) giving maximum values and other points (such as B and E) giving minimum values of $f(x)$. But there remains the possibility of a third kind of point, such as the " inflexional " point C, where the function has neither a maximum nor a minimum value. Stationary values, therefore, include cases other than maximum and minimum values.

A method for distinguishing different kinds of stationary values is suggested by Fig. 50. The tangent gradient of the curve $y=f(x)$ is positive (the tangent being upward sloping) immediately to the left and negative (the tangent being downward sloping) immediately to the right of a point where $f(x)$ has a maximum value. Hence, the derivative $f'(x)$ changes from positive, through zero, to negative values as x increases through a value giving a maximum of $f(x)$. At a point where the function has a minimum value, the tangent gradient and derivative change sign in the opposite sense. Finally, at a point where there is a stationary value which is neither maximum nor minimum, the zero value of the tangent gradient and derivative does not mark a change in their sign at all; they have the same sign on each side of the point.

These results, indicated by inspection of the curve of Fig. 50, can be set out in precise analytical form :

CRITERION FOR MAXIMUM AND MINIMUM VALUES

(1) All maximum and minimum values of the single-valued function $f(x)$ are stationary values and occur where $f'(x)$ is zero.

(2) If $f'(x)$ changes in sign from positive to negative as x increases through a value a where $f'(a)=0$, then $f(a)$ is a maximum value of the function $f(x)$.

If $f'(x)$ changes sign from negative to positive under the same conditions, then $f(a)$ is a minimum value of the function.

If $f'(x)$ does not change sign under the same conditions, then $f(a)$ is neither a maximum nor a minimum value of the function.

There are thus two conditions to consider. The first is a "necessary" condition for an extreme value, while the second adds the "sufficient" condition which enables us to distinguish maximum, minimum and other stationary values. The conditions together make up "necessary and sufficient" conditions (see 8.9 below). We can notice an alternative, but less exact, way of expressing the necessary condition. If $y=f(x)$ has a stationary value (e.g. a maximum or minimum value) at a point, then y is momentarily steady in value and $\Delta y=0$ approximately for any small change in x from the point. Thus when treating the function at a maximum or minimum position, we can regard it as a constant for small variations about the position. In any case, the value of the function always changes very slowly from such a position.*

It must be emphasised that a maximum value of a function is not necessarily the "greatest" value of the function. A maximum value is greater than all *neighbouring* values but there may be still greater values of the function at more remote points. Similarly, a minimum value need not be the "smallest" value. There is only one greatest and only one least value of the function but there may be several alternating maximum and minimum values. This is clear from Fig. 50. Further, it is important to realise that *every* stationary value is not an extreme value and we should always be on the look-out for "inflexional" stationary values. Finally, our criterion may break down and fail to show a maximum or minimum value if the function or its derivative is not continuous. Fig. 51 illustrates the fact that a maximum value, occurring at a "sharp point" of the curve, may not be indicated by the criterion we have given.

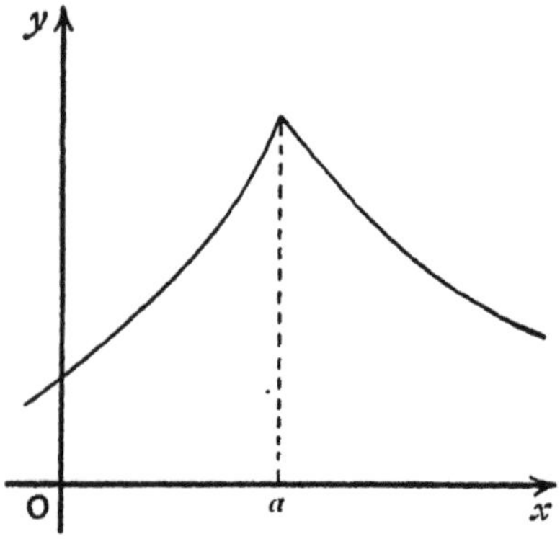

FIG. 51.

8.3 Applications of the second derivative.

The second derivative $f''(x)$ of a function, being the derivative of the first derivative $f'(x)$, measures the rate of increase or decrease of $f'(x)$, i.e. the rate of increase or decrease of the tangent gradient to the curve $y=f(x)$ as we pass through the point concerned. The sign of $f''(x)$ at any point $x=a$ provides some useful information. If $f''(a)$ is positive, then $f(x)$ is changing at an increasing rate as x increases through a and the tangent gradient to the curve $y=f(x)$ increases as we pass through the point with abscissa a. The tangent to the curve turns in the anticlockwise direction and the curve is convex when viewed from below at this point. Conversely, if $f''(a)$

* The device of keeping a maximised or minimised variable constant is often adopted by Wicksell. See, for example, *Lectures on Political Economy* (Ed. Robbins, 1934), Vol. I, p. 181.

is negative, then $f(x)$ changes at a decreasing rate as x increases, the tangent to the curve turns in the clockwise direction and the curve is concave from below at the point where $x=a$. These results, which are fully illustrated in Fig. 52, are independent of the value of the derivative $f'(a)$, of whether the tangent to the curve slopes upwards, downwards or is horizontal at the point where $x=a$.* Hence:

(1) $f''(a)>0$ implies that the function $f(x)$ changes at an increasing rate as x increases through the value a and that the curve $y=f(x)$ is convex from below at the point $x=a$.

(2) $f''(a)<0$ implies that the function $f(x)$ changes at a decreasing rate as x increases through the value a and that the curve $y=f(x)$ is concave from below at the point $x=a$.

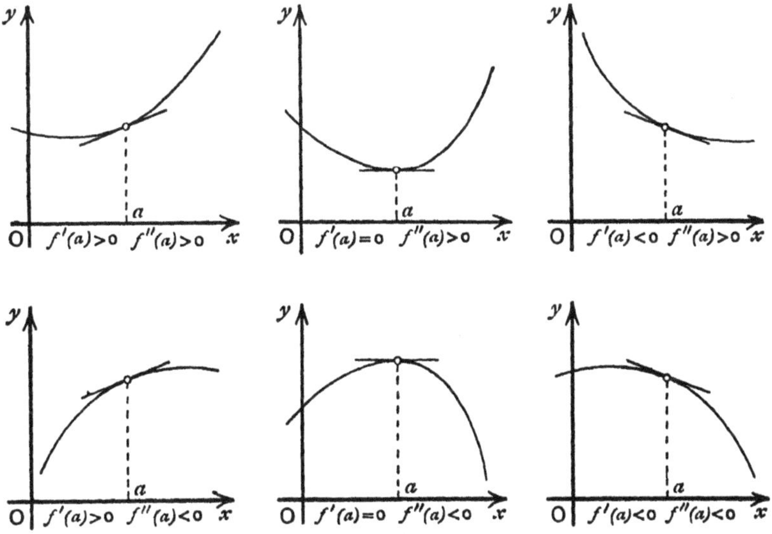

FIG. 52.

The numerical magnitude of $f''(a)$ then indicates how rapidly the value of $f(x)$ is "accelerating" and how great is the curvature of the curve $y=f(x)$ at the point $x=a$. All questions relating to the nature and extent of the curvature of a curve are answered by an

* But the results, which are obvious enough when $f(x)$ increases and the tangent slopes upwards, should be interpreted with care when $f(x)$ decreases and the tangent slopes downwards. When we say that $f'(a)<0$ and $f''(a)>0$ implies that $f(x)$ is decreasing and changing at an increasing rate at $x=a$, we mean that the *numerical* rate of decrease of $f(x)$ is getting less (and not greater) as x increases. A negative quantity, which is getting less in numerical magnitude, is increasing.

examination of the value of the second derivative of the function concerned.

An alternative criterion for extreme values of a function can be given in terms of the second derivative. It is assumed that the function $f(x)$ is finite, continuous and possesses continuous first and second derivatives. If $f'(a)$ is zero and $f''(a)$ negative, then $f'(x)$ must decrease through zero as x increases through a, i.e. the change in sign of $f'(x)$ is from positive to negative and $x=a$ must give a maximum value of the function. In the same way, we see that a minimum value of the function is obtained at $x=a$ when $f'(a)$ is zero and $f''(a)$ positive. Hence:

ALTERNATIVE CRITERION FOR MAXIMUM AND MINIMUM VALUES

(1) All maximum and minimum values of $f(x)$ occur where $f'(x)=0$.

(2) If $f'(a)=0$ and $f''(a)<0$, then $f(a)$ is a maximum value of the function. If $f'(a)=0$ and $f''(a)>0$, then $f(a)$ is a minimum value of the function.

The first condition is, as before, a "necessary" condition. The second is a "sufficient" condition indicating situations in which only a maximum (or only a minimum) value of $f(x)$ can occur. But it is not complete, i.e. both "necessary and sufficient" (see 8.9 below), since the case where $f''(a)$ is zero is not covered. This case includes all inflexional stationary values and may include other maximum or minimum values as well. The alternative criterion, though often more useful, is somewhat more limited in its scope than the first criterion.

8.4 Practical methods of finding maximum and minimum values.

In practical problems, we require the maximum or minimum values attained by one variable (y) as the value of another variable (x) is changed. If the variables are related by a single-valued function $y=f(x)$, we proceed:

(1) The derivative $f'(x)$ is obtained.

(2) The equation $f'(x)=0$ is solved to give a number of solutions $x=a$, $x=b$, $x=c$, ...

(3) Each solution (e.g. $x=a$) is taken in turn and examined.

Either: the change in sign of $f'(a+h)$ as h changes from negative to positive is determined. If the change is from positive to negative, $f(a)$ is a maximum value of $f(x)$. If the change is from negative to positive, $f(a)$ is a minimum value of $f(x)$. If there is no change of sign, $f(a)$ is an inflexional value of $f(x)$.

Or: $f''(a)$ is obtained and its sign determined. If $f''(a)$ is negative, $f(a)$ is a maximum value of $f(x)$. If $f''(a)$ is positive, $f(a)$ is a minimum value of $f(x)$.

If this process is carried out in the cases of the three simple functions quoted in 8.1 above, it is found that one stationary value $y=4$ occurs at $x=2$ in each case. This is a maximum value, a minimum value and an inflexional value in the three cases respectively. The following examples provide further illustrations.

Ex. 1. $\quad y=x^3-3x^2+5.$

Here $\quad \dfrac{dy}{dx}=3x^2-6x=3x(x-2); \quad \dfrac{d^2y}{dx^2}=6x-6=6(x-1).$

The stationary values of y occur where $3x(x-2)=0$, i.e. at $x=0$ and at $x=2$. The corresponding values of y are $y=5$ and $y=1$. To distinguish between maximum and minimum values,

$$\text{at } x=0, \quad \frac{d^2y}{dx^2}=-6<0;$$

$$\text{at } x=2, \quad \frac{d^2y}{dx^2}=6>0.$$

The function has a *maximum value* 5 at $x=0$ and a *minimum value* 1 at $x=2$. The graph of Fig. 56 below illustrates.

Ex. 2. $\quad y=x^4-4x^3+16x.$

Here $\quad \dfrac{dy}{dx}=4x^3-12x^2+16 \qquad \dfrac{d^2y}{dx^2}=12x^2-24x$

$\qquad =4(x-2)^2(x+1). \qquad\qquad =12x(x-2).$

There are stationary values $y=-11$ at $x=-1$ and $y=16$ at $x=2$. At $x=-1$, the value of the second derivative is 36. The function thus has a *minimum value* of -11 at this point. At $x=2$, the second derivative is zero and the nature of the stationary value here remains undecided. To settle this point, the first derivative at $x=2+h$ has value $4h^2(h+3)$. This is positive for all small (positive or negative) values of h. The function has an inflexional stationary value at $x=2$. The graph of Fig. 57 below illustrates.

Ex. 3. $$y=\frac{2x+1}{x-1}.$$

Here $$\frac{dy}{dx}=-\frac{3}{(x-1)^2},$$

which is negative for all values of x. The function has no stationary, and so no maximum or minimum, values. The curve representing this function is a rectangular hyperbola with asymptotes parallel to the axes, a curve which clearly has no tangent parallel to Ox.

Ex. 4. $$y=\frac{2-x}{x^2+x-2}.$$

Here $$\frac{dy}{dx}=\frac{x(x-4)}{(x^2+x-2)^2}.$$

There are stationary values $y=-1$ at $x=0$ and $y=-\frac{1}{9}$ at $x=4$. We now examine the sign of the derivative near these points:

$$\frac{dy}{dx}=\frac{h(h-4)}{(h^2+h-2)^2} \text{ at } x=h; \quad \frac{dy}{dx}=\frac{h(h+4)}{(h^2+9h+18)^2} \text{ at } x=4+h.$$

The first expression changes from positive to negative as h is given small values changing from negative to positive. The second expression changes in the opposite sense as h is varied similarly. The function thus has a *maximum value* -1 at $x=0$ and a *minimum value* $-\frac{1}{9}$ at $x=4$.

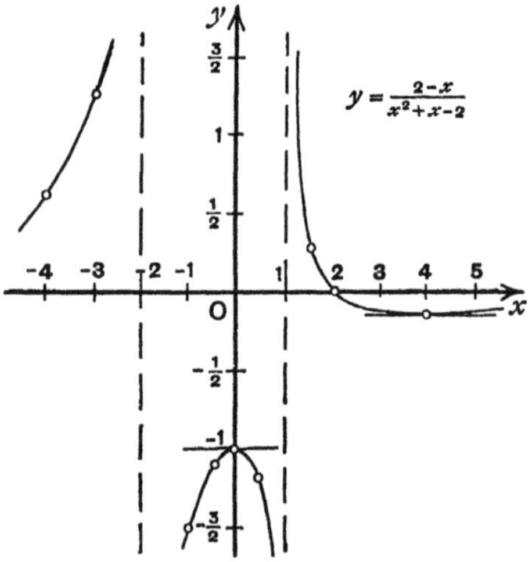

FIG. 53.

The curious feature of this case is that the maximum value of the function is smaller than the minimum value. This apparently paradoxical result is due to the fact that the function has infinite values, at $x=1$ and at $x=-2$. (At each of these values the denominator of y is zero.) The graph of the function, shown in Fig. 53, illustrates how the presence of infinities influences the maximum and minimum values.

Ex. 5. $$y=ax^2+bx+c.$$

Here $$\frac{dy}{dx}=2ax+b; \quad \frac{d^2y}{dx^2}=2a.$$

There is one stationary value $y = -\frac{b^2 - 4ac}{4a}$ at $x = -\frac{b}{2a}$. This value is a maximum if a is negative and a minimum if a is positive. The curve corresponding to this function is a parabola with axis vertical. The stationary value occurs at the vertex of the curve and is a maximum or minimum according to the direction of the axis as determined by the sign of a (see 3.4 above).

Ex. 6. $y = ax + b + \frac{c}{x}$, where a, b and c are positive constants and x is assumed to take only positive values.

Here $$\frac{dy}{dx} = a - \frac{c}{x^2}; \quad \frac{d^2y}{dx^2} = \frac{2c}{x^3}.$$

Since x is positive, there is only one stationary value and this occurs at the point $x = \sqrt{\frac{c}{a}}$. The second derivative is positive. So, the function has a *minimum value* $y = 2\sqrt{ac} + b$ at this point.

Ex. 7. $y = ax^2 - bx + c + \frac{d}{x}$, where a, b, c and d are positive constants and x is restricted to positive values.

Here $$\frac{dy}{dx} = 2ax - b - \frac{d}{x^2}; \quad \frac{d^2y}{dx^2} = 2a + \frac{2d}{x^3}.$$

The stationary values occur at points where the derivative is zero, i.e. where x satisfies the relation :

$$2ax^3 - bx^2 - d = 0.$$

It can be shown that this cubic equation has only one positive root, and that the value of this root is greater than $x = \frac{b}{3a}$. The second derivative is seen to be positive. The function has thus a single *minimum value* for the range of positive values of x.

Ex. 8. An open box consists of a square base with vertical sides and has a volume of 4 cubic feet. What are the dimensions of the box for a minimum surface area ?

Let x feet be the side of the square base and h feet the depth of the box. Then $hx^2 = 4$ since the volume is given as 4 cubic feet. From this we obtain h in terms of x. The surface area is

$$y = x^2 + 4hx = x^2 + \frac{16}{x} \quad \text{square feet.}$$

So, y is written as a function of x only and its minimum value is required for variation in x (which alone governs the dimensions of the box). The values of x for stationary values of y are given by

$$\frac{dy}{dx}=2x-\frac{16}{x^2}=0, \quad \text{i.e. } x^3=8.$$

Hence $x=2$ gives the only stationary value of y. Since

$$\frac{d^2y}{dx^2}=2+\frac{32}{x^3}=6>0, \quad \text{when } x=2,$$

the stationary value is a minimum. The minimum surface area is thus 12 square feet and the dimensions of the box are then: side of base $=2$ feet and depth of box $=1$ foot. The depth of the box is half the side of the square base.

8.5 A general problem of average and marginal values.

From a given single-valued function $f(x)$ at a point x, we derive the pair of values:

$$\text{Average value of } f(x)=\frac{f(x)}{x}; \quad \text{Marginal value of } f(x)=f'(x).$$

The average value is taken over the whole range from zero to the given value x; the marginal value refers to the " margin " at the given value x. In diagrammatic terms, if P is the point with abscissa x on the curve $y=f(x)$, then the average value of $f(x)$ is represented by the gradient of the " radius vector " OP and the marginal value by the gradient of the tangent at P. The application of these general concepts in the economic problems of demand and cost have been considered at an earlier stage (6.8 above).

It is now assumed, for convenience, that both x and $f(x)$ take only positive values. It is required to find the values of x which correspond to maximum or minimum values of the average $\frac{f(x)}{x}$. So

$$\frac{d}{dx}\left\{\frac{f(x)}{x}\right\}=\frac{xf'(x)-f(x)}{x^2},$$

$$\frac{d^2}{dx^2}\left\{\frac{f(x)}{x}\right\}=\frac{x^2f''(x)-2xf'(x)+2f(x)}{x^3}.$$

Stationary values of $\frac{f(x)}{x}$ occur where $xf'(x)-f(x)=0$, i.e. where

$$\frac{f(x)}{x}=f'(x).$$

At such a point, the value of the second derivative reduces to

$$\frac{d^2}{dx^2}\left\{\frac{f(x)}{x}\right\}=\frac{f''(x)}{x}$$

and, since x is taken as positive, this has the sign of $f''(x)$. The average value of $f(x)$ is thus stationary at any point where the average and marginal values of $f(x)$ are equal. The stationary value is a maximum if $f''(x)$ is negative at the point; it is a minimum if $f''(x)$ is positive at the point.

In diagrammatic terms, a stationary value of the average value of $f(x)$ is shown by a point P on the curve $y=f(x)$ where the tangent coincides with the radius vector OP, i.e. at a point where the tangent

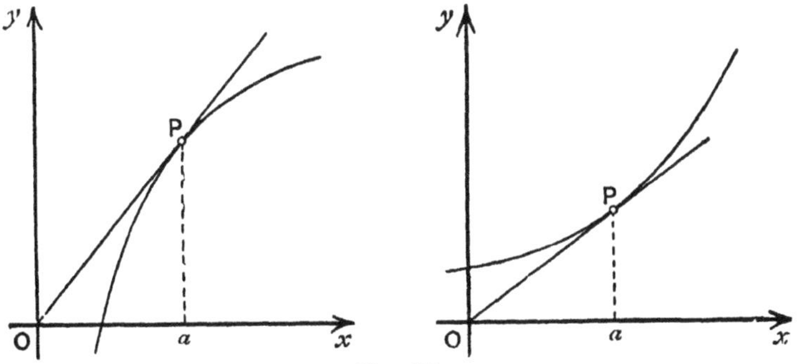

FIG. 54.

passes through the origin O. The average value is a maximum if the curve is concave from below at the point P; it is a minimum if the curve is convex from below at the point. These facts are clear from the two cases illustrated in Fig. 54.

8.6 Points of inflexion.

A single-valued function $y=f(x)$ is defined to have an inflexional value at a point where the corresponding curve crosses from one side of its tangent to the other. The point itself is described as a point of inflexion. The most important property of a point of inflexion is that it marks a change in curvature, the curve changing from convex to concave from below as we pass from left to right through the point, or conversely.* This property is clear from the inflexional

* If the function is single-valued and the curve smooth, then *all* changes of curvature occur at points of inflexion. For multi-valued functions, on the other hand, a change in curvature may occur where the curve "turns back on itself" and the tangent is vertical. Such a point is not a point of inflexion.

cases shown in Fig. 55. There are two classes of points of inflexion. A point of one class (as illustrated in the first two cases of Fig. 55) marks a change of curvature from convex to concave from below as we move from left to right along the curve. A point of the other class marks a change of curvature in the opposite sense (as shown in the second two cases of Fig. 55). The actual tangent at the point of inflexion is not restricted in any way ; it can slope upwards or downwards with any numerical gradient whatever. Upward and downward sloping tangents are shown in Fig. 55 for each of the two classes of inflexional points. Further, as limiting cases, the tangent can be parallel to Ox with zero gradient or parallel to Oy with infinite gradient.

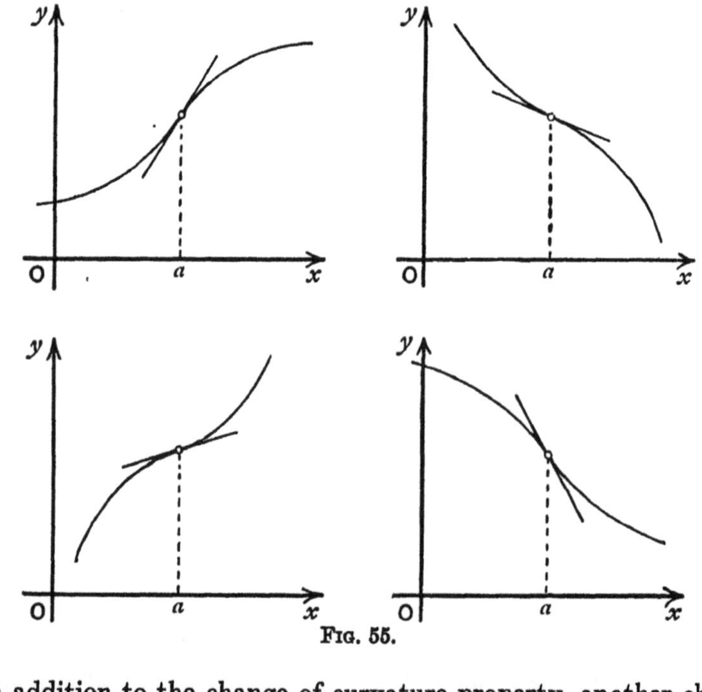

Fig. 55.

In addition to the change of curvature property, another characteristic of points of inflexion is evident from Fig. 55. A point of inflexion always corresponds to an extreme value of the tangent gradient of the curve. At a point of inflexion of the first class, the tangent gradient is a maximum, the gradient increasing as we move from the left towards the point and decreasing as we move to the right away from the point. At a point of inflexion of the second class, it is seen, in the same way, that the tangent gradient is a minimum.

Assuming that the single-valued function $f(x)$ is finite and continuous with continuous first and second derivatives, it is a simple matter to express the properties of an inflexional value of $f(x)$ in analytical terms. The derivative $f'(x)$, as the measure of the tangent gradient, must have an extreme value at any point of inflexion. It is necessary, therefore, that the second derivative $f''(x)$, being the derivative of $f'(x)$, has zero value at the point. Further, the value of $f''(x)$ must change in sign as x increases through the point concerned, the direction of the change determining the class to which the point of inflexion belongs. If $f''(x)$ changes from positive to negative, the derivative $f'(x)$ has a maximum value and the curvature changes from convex to concave from below, i.e. the point of inflexion is of the first class. The converse change of sign of $f''(x)$ indicates a point of inflexion of the other class. So

CRITERION FOR POINTS OF INFLEXION

(1) An inflexional value of the function $f(x)$ can only occur at a point where $f''(x)=0$.

(2) If $f''(a)=0$, and if $f''(x)$ changes in sign as x increases through the value a, then $f(a)$ is an inflexional value of the function $f(x)$. The direction of the change of sign of $f''(x)$ indicates the class of the point of inflexion.

The first condition is " necessary " for points of inflexion. The second adds the " sufficient " condition and, altogether, we have a criterion which is complete, i.e. " necessary and sufficient " (see 8.9 below).

If the function is assumed to have a continuous third derivative, an alternative form of the criterion can be given in which this derivative is used. If $f'''(a)$ is negative at a point where $f''(a)$ is zero, then $f'(x)$ is a maximum at the point $x=a$, i.e. we have a point of inflexion of the first class. Similarly, if $f'''(a)$ is positive at the point where $f''(a)$ is zero, we have a point of inflexion of the second class. Hence, if $f''(a)=0$ and $f'''(a)\neq 0$, there is a point of inflexion at $x=a$ and the sign of the non-zero third derivative indicates to which class the inflexional point belongs. This alternative form of the criterion is not complete (not " necessary and sufficient ") since the case where the third derivative is zero is not considered.

A point of inflexion, as we have remarked, is in no way dependent on the value assumed by the first derivative of the function at the point. It may happen, however, that the first derivative is zero at the point of inflexion, the value of the function being stationary as well as inflexional. The second, and incomplete, criterion for stationary values (8.3 above) can now be extended slightly by the addition of the condition :

If $f'(a)=f''(a)=0$ and if $f'''(a)\neq 0$, then $f(a)$ is a stationary and inflexional value of the function $f(x)$.

The criterion is still incomplete since it takes no account of cases where the third derivative is zero.*

As long as the third derivative of the function $f(x)$ is not zero at the point $x=a$, the following scheme indicates all the possible cases of stationary and inflexional values of $f(x)$:

	Stationary value		Inflexional value	
/		\ /		\
Extreme value		Stationary and inflexional value		Inflexional and non-stationary value
$f'(a)=0$ $f''(a)\neq 0$		$f'(a)=f''(a)=0$		$f'(a)\neq 0$ $f''(a)=0$

Two examples will illustrate the method of locating inflexional values in the cases of particular functions :

Ex. 1. $y=x^3-3x^2+5.$

Here $\dfrac{dy}{dx}=3x(x-2)$; $\dfrac{d^2y}{dx^2}=6(x-1)$; $\dfrac{d^3y}{dx^3}=6.$

There is thus only one inflexional value of the function, i.e. $y=3$ at $x=1$. The third derivative is positive and the second derivative changes sign from negative to positive as we pass from left to right through the point of inflexion. The point is thus an inflexion of the second class, the curvature of the curve changing from concave to convex and the tangent gradient (-3) being a minimum. Fig. 56 indicates the point of inflexion P and also the maximum and minimum points A and B on the curve representing this function.

Ex. 2. $y=x^4-4x^3+16x.$

Here $\dfrac{dy}{dx}=4(x-2)^2(x+1)$; $\dfrac{d^2y}{dx^2}=12x(x-2)$; $\dfrac{d^3y}{dx^3}=24(x-1).$

* The criterion is fully completed in 17.5 below, where extreme and inflexional values are determined by the first non-zero derivative of $f(x)$ of whatever order it may be.

There are two points of inflexion. At $x=0$ there is the inflexional value $y=0$; at $x=2$ there is the inflexional value $y=16$. At the first of these points the third derivative is negative. The point is an inflexion of the first class, the curvature changing from convex to concave and the tangent

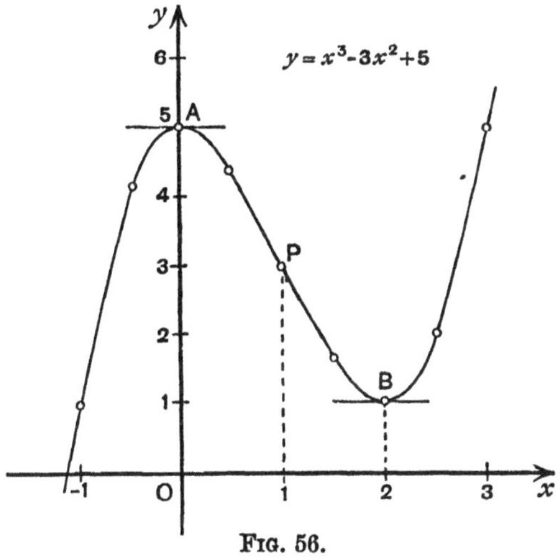

FIG. 56.

gradient (16) being a maximum. At the second of the points the third derivative is positive. The point is an inflexion of the second class, the curvature changing from concave to convex. Further, the first derivative is also zero at the point and we have here a stationary as well as an

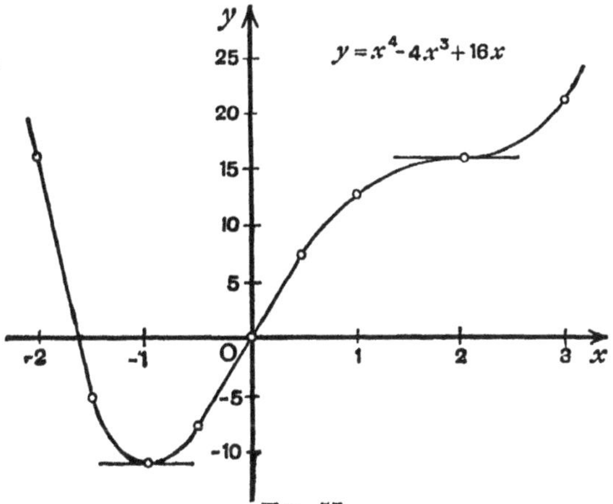

FIG. 57.

inflexional value of the function. The zero tangent gradient at the point is a minimum value of the tangent gradient. Fig. 57 indicates the graph of the curve representing the function and shows the two inflexional points in addition to the single minimum point.

8.7 Monopoly problems in economic theory.

Many of the problems of physics and other mathematical sciences can be reduced to the determination of positions of maximum or minimum. It is sufficient to instance the importance of maximum or minimum potential, energy, action and entropy. An exactly similar situation is found in certain branches of economic theory. In static problems, we find it convenient to assume that the individual consumer seeks the highest position on his "preference scale" consistent with market conditions, that the individual firm fixes output or price to produce the largest net revenue and organises its factors of production to give the largest output at a given cost or the smallest cost for a given output. Similar assumed principles, in more complicated forms, are to be found in dynamic problems. If functions of a single variable suffice to interpret the phenomena concerned, then the methods of the present chapter apply at once in the solution of our problems. Some simple examples, intended to illustrate the method, are given in the following paragraphs.*

As a first problem, suppose that a firm produces a good X under known cost conditions represented by the total cost function $\Pi = F(x)$. The demand of the firm's market for the good X is assumed to be known and represented by the demand function $x = \phi(p)$ or $p = \psi(x)$, where x is the demand at price p. Within the limits set by this demand relation, the firm is assumed to act as a monopolist with the object of maximising net revenue. Two alternative points of view can be taken. *Either*: the firm fixes its output and leaves the price to be determined by the demand conditions. *Or*: the firm fixes its price and the demand conditions determine the appropriate output. The analysis of the problem is different in the two cases but, as we shall see, the results obtained are effectively identical.

Assuming that the firm fixes output, then the price to clear any output x must be $p = \psi(x)$ as given by the demand conditions. The gross revenue from output x is $R = x\psi(x)$, the total cost is $\Pi = F(x)$ and the net revenue is $(R - \Pi)$ given as a function of x. The output

* For an account of the importance of maximum or minimum positions, particularly in relation to "loose indefinite relations" between economic variables, see Edgeworth, *Mathematical Psychics* (1881, reprinted 1932), pp. 1-15 and pp. 83-93.

x fixed by the firm for maximum net revenue must satisfy the two conditions :

$$\frac{d}{dx}(R-\Pi)=0 \quad \text{and} \quad \frac{d^2}{dx^2}(R-\Pi)<0.$$

The first condition is $\frac{dR}{dx}-\frac{d\Pi}{dx}=0$, i.e. $\frac{dR}{dx}=\frac{d\Pi}{dx}$. At the output for equilibrium, therefore, marginal revenue must equal marginal cost. The second condition, for a maximum as opposed to a minimum, is

$$\frac{d}{dx}\left\{\frac{d}{dx}(R-\Pi)\right\}=\frac{d}{dx}\left(\frac{dR}{dx}-\frac{d\Pi}{dx}\right)=\frac{d^2R}{dx^2}-\frac{d^2\Pi}{dx^2}<0,$$

i.e.

$$\frac{d^2R}{dx^2}<\frac{d^2\Pi}{dx^2}.$$

Hence, at the equilibrium output, marginal revenue must be increasing less rapidly than marginal cost. This second condition is

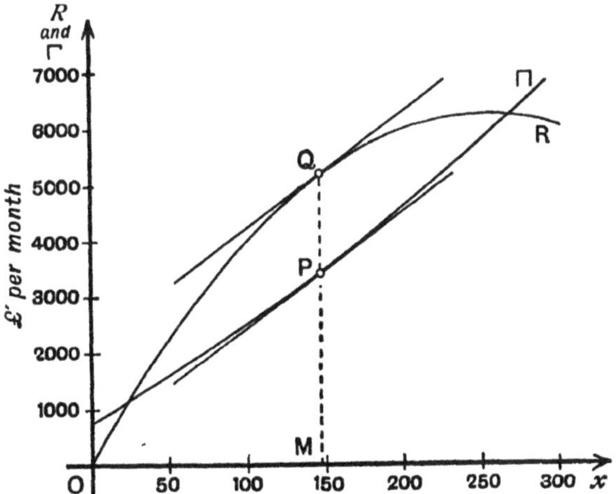

FIG. 58.

automatically satisfied if, for example, marginal revenue is decreasing while marginal cost is increasing as output increases from the value at which these marginal concepts are equal.

The position can be represented on a diagram in two different ways. The total revenue and cost curves can be drawn on one diagram, taking output x along the horizontal axis and revenue or cost along the vertical axis. Suppose that, on the two curves, points in the same vertical line can be found such that the tangents to the respective curves are parallel $\left(\frac{dR}{dx}=\frac{d\Pi}{dx}\right)$. The output common to such points is a monopoly equilibrium output provided only that the total revenue curve is less convex (or more concave) from below than the total cost curve $\left(\frac{d^2R}{dx^2}<\frac{d^2\Pi}{dx^2}\right)$.. When the total revenue and cost curves are of "normal" form (see 10.7-8 below), the position is illustrated by the curves of Fig. 58. The net revenue obtainable from any output is shown by the vertical distance of the total revenue curve above the total cost curve. This is clearly a maximum in the position PQ, the tangent to the total cost curve at P being parallel to the tangent to the total revenue curve at Q. The monopoly output OM is thus uniquely determined in this case and the monopoly price is read off as the gradient of OQ.

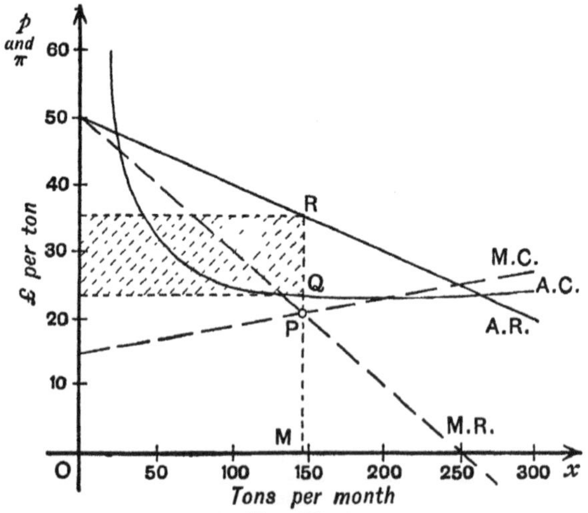

FIG. 59.

The monopoly situation can also be represented on a diagram showing the average and marginal revenue and cost curves referred to the same axes and scales. Fig. 59 exhibits these four curves as obtained from the total curves of Fig. 58. Three of the curves happen to be straight lines, but this is not an essential feature of the "normal" case here taken. The unique monopoly output is now given as the abscissa OM of the point P where the marginal curves intersect. The monopoly price is MR, the average cost of the monopoly output is MQ and the maximised net revenue is QR times OM, i.e. the rectangular area shaded. The condition for a maximum, as opposed to a minimum, is automatically satisfied here since the marginal revenue curve falls while the marginal cost curve rises.*

The analytical solution of the monopoly problem can be illustrated by assuming particular cost and demand functions of simple type appropriate to "normal" conditions. If the cost function is quadratic in form $\Pi = ax^2 + bx + c$ and if the demand function is the linear form $p = \beta - \alpha x$ (all the constants specified being positive), then the first condition of equality of marginal cost and revenue is

$$2ax + b = \beta - 2\alpha x,$$

i.e.

$$x = \frac{\beta - b}{2(a + \alpha)}.$$

There is a unique equilibrium output, provided that $\beta > b$. This last condition is simply that marginal revenue is greater than marginal cost at zero, or very small, output. If it is not satisfied, the firm never makes a positive profit and its losses are least when it produces nothing. The second condition for equilibrium $\left(\frac{d^2R}{dx^2} < \frac{d^2\Pi}{dx^2}\right)$ is always satisfied. Figs. 58 and 59 are drawn for cost and demand functions of these types. A sugar refinery produces an output of x tons per month at a total cost of £$(\frac{1}{50}x^2 + 15x + 800)$ and monopolises the sale of sugar on a market with a demand law $p = 50 - \frac{1}{10}x$, where £p is the price (per ton) of sugar. The monopoly equilibrium output is just under 150 tons per month and the monopoly price is approximately £35 per ton or rather under 4d. per lb.

Taking the second view of the monopoly problem, the firm fixes the price, and its output to meet the demand must be $x = \phi(p)$. Here

$$R = xp = p\phi(p) \quad \text{and} \quad \Pi = F(x) = F\{\phi(p)\}.$$

The net revenue at the price p is $(R - \Pi)$ and this is a maximum if

$$\frac{d}{dp}(R - \Pi) = 0 \quad \text{and} \quad \frac{d^2}{dp^2}(R - \Pi) < 0.$$

* It is to be noticed that, as far as our formal solution is concerned, there is no reason to suppose that net revenue is positive even when it is maximised. The total cost curve may lie completely above the total revenue curve. In this case, the firm either goes out of business or stays in and cuts its losses. Maximum net revenue is then minimum net loss.

The first condition is

$$\frac{d}{dp}\{p\phi(p)\}-\frac{d\Pi}{dp}=0,$$

i.e. $$\phi(p)+p\phi'(p)-\frac{d\Pi}{dx}\phi'(p)=0 \quad \left(\text{since } \frac{d\Pi}{dp}=\frac{d\Pi}{dx}\frac{dx}{dp}\right)$$

The equation to be solved for the equilibrium price is thus

$$\phi(p)+\left(p-\frac{d\Pi}{dx}\right)\phi'(p)=0.$$

The limiting condition for maximum net revenue is

$$\frac{d}{dp}\left\{\frac{d}{dp}(R-\Pi)\right\}<0,$$

i.e. $$\frac{d}{dp}\left\{\phi(p)+\left(p-\frac{d\Pi}{dx}\right)\phi'(p)\right\}<0,$$

i.e. $$\left\{2-\phi'(p)\frac{d^2\Pi}{dx^2}\right\}\phi'(p)+\left(p-\frac{d\Pi}{dx}\right)\phi''(p)<0.$$

Any value of p satisfying the above equation and inequality is a possible monopoly price and the corresponding output is $x=\phi(p)$.

The second analysis of the monopoly problem, which is due to Cournot, can be shown to lead to the same equilibrium price and output as is obtained in the first analysis. This fact can easily be checked in particular cases, e.g. with a quadratic cost function and a linear demand function. Or, in general terms, the above equation for the monopoly price gives

$$\frac{d\Pi}{dx}=\frac{\phi(p)+p\phi'(p)}{\phi'(p)}=\frac{\dfrac{dR}{dp}}{\dfrac{dx}{dp}}=\frac{\dfrac{dR}{dx}\dfrac{dx}{dp}}{\dfrac{dx}{dp}}=\frac{dR}{dx},$$

i.e. the condition of the first analysis that marginal cost and marginal revenue are equal is satisfied also in the condition of the second analysis.

8.8 Problems of duopoly.

The demand of a market for a good X is represented, as before, by the demand relation $p=\psi(x)$ connecting the price p and the demand x. The production of the good is shared between two duopolist firms selling at the same price p. The first duopolist produces an output x_1 at a total cost of $\Pi_1=F_1(x_1)$ and the second duopolist produces an output x_2 at a total cost of $\Pi_2 = F_2(x_2)$. The solution of the problem of the distribution of the market between the two duopolists depends entirely upon what is assumed about the reaction of one duopolist to any action on the part of the other.

It is assumed, in the simplest duopoly problem, that each duopolist expects the other to make no change in current output no matter what changes he makes in his own output. Subject to this expectation, each duopolist then aims at fixing his output for maximum net revenue. If outputs x_1 and x_2 are fixed by the two duopolists, the price of the good is determined by $p = \psi(x)$, where $x = x_1 + x_2$ is the total output. The net revenue of the first duopolist is $(x_1 p - \Pi_1)$ and, for this to be a maximum, x_1 must be chosen so that

$$\frac{d}{dx_1}(x_1 p - \Pi_1) = 0, \quad \text{i.e.} \quad \frac{d}{dx_1}(x_1 p) = \frac{d\Pi_1}{dx_1}.$$

This is the familiar equality between marginal revenue and marginal cost. The difficulty here is to express marginal revenue in a suitable form. Since $x_1 p = x_1 \psi(x)$, where $x = x_1 + x_2$, we have

$$\frac{d}{dx_1}(x_1 p) = \psi(x) + x_1 \frac{d}{dx}\psi(x) \frac{d}{dx_1}(x_1 + x_2) = \psi(x) + x_1 \psi'(x),$$

making use of the assumption that the first duopolist considers x_2 as fixed. Hence, for any given output x_2 of the second duopolist, the equation which determines the first duopolist's output x_1 is

$$\psi(x) + x_1 \psi'(x) = \frac{d\Pi_1}{dx_1}.$$

In the same way, given any output x_1 of the first duopolist, the second duopolist fixes his output x_2 so that

$$\psi(x) + x_2 \psi'(x) = \frac{d\Pi_2}{dx_2}.$$

These two equations are together sufficient, in general, to determine the outputs of the two duopolists. The total output and the price at which it is sold then follow at once.

The first equation gives the output of the first duopolist in terms of whatever output the second duopolist is producing, i.e. it gives x_1 as a function of x_2. It can be taken, under " normal " conditions, that an increase in x_2 results in a decrease in x_1 of smaller amount. The dependence of x_1 on x_2 can be represented by a " reaction curve " C_1 in the plane Ox_1x_2. The " normal " form of C_1 is shown in Fig. 60. The curve must be considered in relation to the axis Ox_2 and its gradient to this axis is negative but numerically less than unity. Similarly, the second equation gives x_2 as a function of x_1 and a second reaction curve C_2 is obtained. This curve is related to the axis Ox_1 and, in the " normal " case, its gradient to this axis is negative and numerically less than unity. The two curves C_1 and C_2 thus intersect in a single point P and the co-ordinates of P give the equilibrium outputs x_1 and x_2 of the two duopolists.

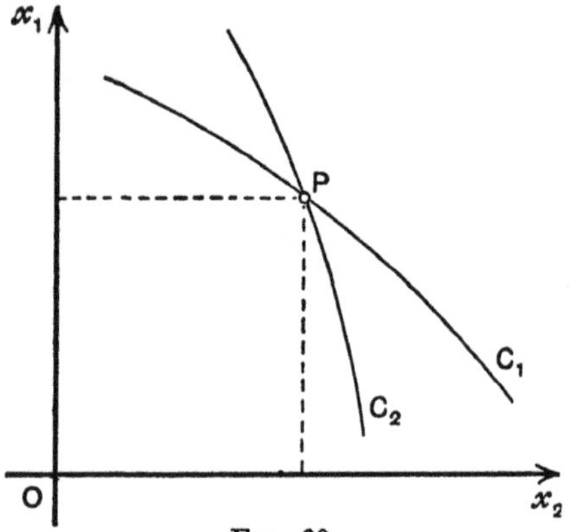

FIG. 60.

In the particular case where the duopolists have the same total cost function $\Pi = F(x)$, the equations giving their outputs are

$$\psi(x) + x_1\psi'(x) = F'(x_1) \quad \text{and} \quad \psi(x) + x_2\psi'(x) = F'(x_2).$$

The reaction curve C_1 when viewed from the axis Ox_2 is now of exactly the same form as the reaction curve C_2 viewed from the axis Ox_1. It follows that x_1 and x_2 must be equal at the point of intersection. Thus, as we expect, the total output is shared equally between the two duopolists: $x_1 = x_2 = \frac{1}{2}x$. The value of x is given by

$$\psi(x) + \tfrac{1}{2}x\psi'(x) = F'(\tfrac{1}{2}x).$$

Further, if each duopolist produces at constant total cost, the total output shared equally between them is given by

$$\psi(x) + \tfrac{1}{2}x\psi'(x) = 0,$$

i.e.
$$\psi(x) + \{\psi(x) + x\psi'(x)\} = 0,$$

i.e.
$$\psi(x) + \frac{d}{dx}\{x\psi(x)\} = 0.$$

The total output is such that the sum of the average and marginal revenue from the total demand $p = \psi(x)$ is equal to zero. If the average and marginal revenue curves are as shown in Fig. 61, the total output under duopoly is given by ON. This output can be compared easily with that under pure monopoly. The monopoly output of a single monopolist with constant total cost is such that marginal revenue is equal to marginal cost, i.e. is equal to zero. In Fig. 61, the monopoly output is OM, and this is less than the duopoly output ON.

In conclusion, we can indicate the way in which the general duopoly problem can be analysed. It is assumed that the first duopolist, when he varies his own output x_1, expects the second duopolist to react and vary his output x_2 according to some definite law

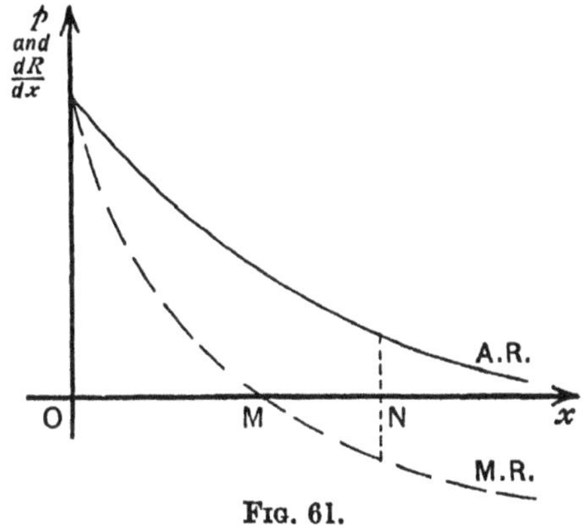

Fig. 61.

$x_2 = f(x_1)$. Thus, if he changes his output from a level x_1, he expects his rival's output to expand or contract at a rate indicated by the derivative $\frac{dx_2}{dx_1} = f'(x_1)$. Following Professor Frisch,* this derivative can be termed a " conjectural variation " and it may be positive or negative in value according to circumstances. For a maximum net revenue, the first duopolist's marginal cost must equal

$$\frac{d}{dx_1}(x_1 p) = \psi(x) + x_1 \frac{d}{dx}\psi(x)\frac{d}{dx_1}(x_1 + x_2) = \psi(x) + x_1\psi'(x)\left(1 + \frac{dx_2}{dx_1}\right).$$

The equation, which gives x_1 as a function of x_2 and determines the reaction curve C_1 of the first duopolist, is now

$$\psi(x) + x_1\psi'(x)\left(1 + \frac{dx_2}{dx_1}\right) = \frac{d\Pi_1}{dx_1}.$$

In the same way, it is assumed that the second duopolist expects the output of the first duopolist to vary according to a definite law $x_1 = g(x_2)$. The derivative $\frac{dx_1}{dx_2} = g'(x_2)$ is again the conjectural varia-

* Frisch, *Monopole-Polypole*, Nationaløkonomisk Tidsskrift, 1933.

tion. The equation giving x_2 as a function of x_1 and defining the reaction curve C_2 of the second duopolist is

$$\psi(x)+x_2\psi'(x)\left(1+\frac{dx_1}{dx_2}\right)=\frac{d\Pi_2}{dx_2}.$$

The two reaction curves, by their point or points of intersection, determine the duopoly distribution of output between the firms. Their forms depend on the nature of the conjectural variations of the two duopolists and, by making various assumptions about these conjectural variations, a whole series of duopoly problems can be defined and analysed. The problem will be elaborated at a later stage (13.9 below).*

8.9 A note on necessary and sufficient conditions.

In the above analysis we have had occasion to draw distinctions between " necessary " conditions, " sufficient " conditions and conditions which are " necessary and sufficient ". The following observations serve to explain more fully the nature of these distinctions and to illustrate their importance.

To start with a simple example, we can examine conditions under which a four-sided figure is a rectangle. First, if the figure is a rectangle, then one of its angles must be a right angle. This is a *necessary* condition. While all rectangles have a right angle for one angle, there are also other figures with the same property. If, however, one angle of the figure is a right angle and all sides are of the same length, then the figure must be a rectangle. This is a *sufficient* condition. All figures with the property stated are rectangles, but there are some rectangles (i.e. those not squares) which do not display the property. The condition is thus not complete, not necessary *and* sufficient. Finally, if the figure is a rectangle, then one angle is a right angle and opposite sides are of equal lengths. Conversely, if a figure has one angle a right angle and opposite sides of equal lengths, then the figure is a rectangle. We have here a *necessary and sufficient* condition; it is complete, including all rectangles and no figures other than rectangles.

In general, a necessary condition for a certain property is such

* The analysis given above, in the case where the conjectural variations are zero, is based on the work of Cournot. For the general problem, see Hicks, *The Theory of Monopoly*, Econometrica, 1935.

that, if the property holds, then the condition is satisfied. The condition is satisfied by all things with the property but may be satisfied also by things without the property. A sufficient condition is such that, if the condition is satisfied, then the property holds. The condition is satisfied by no thing without the property but may not be satisfied by some things with the property. A necessary and sufficient condition is one which holds in both the ways described—if the property holds, then the condition is satisfied; if the condition is satisfied, then the property holds. The condition is complete and includes all things with the property and no others.

An important instance of the distinctions here drawn is provided by the conditions for maximum and minimum values of a function $f(x)$ which has a finite and continuous derivative. A necessary condition for a maximum value at $x=a$ is that $f'(a)=0$. This condition is satisfied at all maximum positions but also at other positions (e.g. minimum or some inflexional positions). A sufficient condition for a maximum value is that $f'(a)=0$ and $f''(a)<0$. We have a maximum value whenever this condition is satisfied but some maximum values can occur even when the condition is not satisfied. A necessary and sufficient condition for a maximum value is that $f'(a)=0$ and that $f'(x)$ changes sign from positive to negative as x increases through the value a. A maximum value must satisfy this condition and we have a maximum value whenever the condition is satisfied.

EXAMPLES VIII

General applications of derivatives

1. Write down the derivative of $3x^3+3x^2+x-1$ and show that this function is monotonic increasing. If $y=ax^3+bx^2+cx+d$, express $\frac{dy}{dx}$ as a square plus a constant term. Deduce that the function is monotonic if $b^2<3ac$ and that it then increases or decreases according to the sign of a.

2. Show, by means of derivatives, that $y=\frac{1}{2x+1}$ and $y=\frac{1-x}{1+x}$ are both monotonic functions. Generalise by showing that $y=\frac{a_1x+b_1}{a_2x+b_2}$ is always a monotonic function. When does it increase and when decrease? Illustrate these results by considering the shape of the rectangular hyperbolas which represent the function.

3. Show that $y=x+\frac{1}{x}$ has one maximum and one minimum value and that the latter is larger than the former. Draw a graph to illustrate.

4. Show that $y=2x-1+\frac{1}{x}$ has a single minimum value and is positive for all positive values of x.

5. Find the maximum and minimum values of $y=x^3-3x-1$ and of $y=3x^4-10x^3+6x^2+5$. Illustrate graphically.

6. If $y=x^4-4x^3+6x^2-4x-3$, show that $\frac{dy}{dx}=4(x-1)^3$. Deduce, from the *first* criterion for maximum and minimum values, that y has a minimum value at $x=1$. Why does the *second* criterion fail to give the result in this case?

7. Show that each of the functions $y=\frac{1-x^2}{1+x^2}$ and $y=\sqrt{3-x^2}$ has a single maximum value. Draw rough graphs of the functions to illustrate.

8. Show that $y=x\sqrt{1+x}$ has a minimum and $y=-x\sqrt{1+x}$ a maximum value. Draw the graphs of the functions for $x>-1$ and deduce that the double-valued function $y^2=x^2(1+x)$ is continuous with a maximum and a minimum at the same value of x. Then consider $y=+\sqrt{x^2(1+x)}$ and $y=-\sqrt{x^2(1+x)}$ as the two single-valued branches of this function, showing that each branch is continuous but without a derivative at $x=0$. Illustrate the difficulty of defining derivatives for multi-valued functions.

9. Show that $y^2=x(x^2-1)$ can be divided into two single-valued branches, one with a maximum and the other with a minimum value. Deduce that the curve representing the double-valued function is continuous, defined only for certain ranges of x and shows a vertical tangent at three points.

10. Find the derivative of $y=\sqrt[3]{x^2}$ and show that it is infinite at $x=0$. Draw a graph of the function and indicate its behaviour in the neighbourhood of the origin. Deduce that y has a minimum value at the origin which is not a stationary value. Contrast this function and its graph with $y=\sqrt[3]{x}$ (see Examples VII, 21).

11. Show that the perimeter of a rectangle of area 16 square inches is least when the rectangle is a square of side 4 inches.

12. A rectangular area is to be marked off as a chicken run with one side along an existing wall. The other sides are marked by wire netting of which a given length is available. Show that the area of the run is a maximum if one side is made twice the other.

13. A cricket field consists of a rectangle with a semicircular area at each end. The perimeter is to be used as a quarter-mile running track. Find the dimensions of the field so that the area of the rectangular portion is the largest possible. (Take $\pi=3\cdot14159$.)

14. A tinned soup manufacturer uses tins which are circular cylinders closed top and bottom. Find the most economical dimensions of the tin (i.e. minimum surface area) when the volume is given. If the top and bottom of the tin are cut from square sheets and the surplus wasted, find the new dimensions for greatest economy.

15. Express the distance of O from a point on the line $2x+y=5$ as a function of the x-co-ordinate of the point. Find the point on the line nearest O and deduce that the shortest distance from O to the line is perpendicular to the line and of length $\sqrt{5}$. Generalise to show that the shortest distance from (x_1, y_1) to the line $ax+by+c=0$ is $\frac{ax_1+by_1+c}{\sqrt{a^2+b^2}}$.

16. Find, on the part of the rectangular hyperbola $xy=4$ in the positive quadrant, the point which is nearest to O and show that the shortest distance is perpendicular to the tangent at this point. What is the shortest distance of the point (0, 2) from the parabola $y=x^2$?

17. Show that the curve $y=2x-3+\frac{1}{x}$ is convex from below for all positive values of x. Is the same true of the curve $y=ax+b+\frac{c}{x}$?

18. Show that the curve $y=ax^3+bx^2+cx$ can have only one point of inflexion. If a is positive, show that the curvature changes from concave to convex from below as we pass through the inflexional point from left to right. Deduce that the point of inflexion is also a stationary point if $b^2=3ac$.

19. Show that the curve $y=\frac{2x}{x^2+1}$ has three points of inflexion separated by a maximum point and a minimum point. Verify these facts by drawing a graph of the curve.

20. Prove that the curve $y=x^3$ has a single stationary point which is a point of inflexion. Are there any other points of inflexion?

21. Show that the curve $y=\sqrt[3]{x}$ is convex from below for negative values of x and concave from below for positive values of x. Deduce that the origin is a point of inflexion. Why is this point not given by the criterion that the second derivative is zero? Check the result by considering the function as the inverse of $y=x^3$.

22. From the second derivative, verify that the rectangular hyperbola $xy=1$ is concave from below for negative values of x and convex from below for positive values of x. In what sense is $x=0$ a point of inflexion? Contrast this case with that of the previous example.

23. If $f(x)$ is a single-valued function of x, find where $\{xf(x)\}$ attains maximum and minimum values and interpret in terms of the curve $y=f(x)$. If x and $f(x)$ are both positive, show that $\{xf(x)\}$ can only be a maximum at a point where the curve $y=f(x)$ is downward sloping with a curvature less than a certain amount.

Economic applications of derivatives

24. Show that the demand curves $p=\frac{a}{x+b}-c$ and $p=(a-bx)^2$ are each downward sloping and convex from below. Do the same properties hold of the marginal revenue curves? Show further that, for each of the demand laws $p=\sqrt{a-bx}$ and $p=a-bx^2$, the demand and marginal revenue curves are downward sloping and concave from below.

25. Show that the demand curve $x=4a^3-3ap^2+p^3$, where a is a positive constant, and p is less than $2a$, is downward sloping with a point of inflexion. How does the curvature change?

26. With the aid of derivatives, check the positions of maximum total revenue obtained in the cases of the demand laws of Examples V, 2, 10 and 11.

27. It is given that a demand curve is convex from below $\left(\frac{d^2p}{dx^2}>0\right)$ at all points. Show that the marginal revenue curve is also convex from below

either if $\frac{d^3p}{dx^3}$ is positive *or* if $\frac{d^3p}{dx^3}$ is negative and numerically less than $\frac{3}{x}\frac{d^2p}{dx^2}$.
If the demand curve is always concave from below, does a similar property hold of the marginal revenue curve?

28. By examining the signs of certain derivatives, show that each of the total cost functions

$$\Pi = \sqrt{ax+b}+c \quad \text{and} \quad \Pi = ax\frac{x+b}{x+c}+d \quad (b>c)$$

gives average and marginal cost curves which fall continuously with increasing output.

29. From the sign of the second derivative, show that the transformation curve of Examples V, 25 is concave from below at all points.

30. If the supply of a good is related to its price by the law $x = a\sqrt{p-b}+c$, where a, b and c are positive constants, show that the supply curve is upward sloping and concave to the axis Op at all points. Illustrate with the case of Examples V, 23.

31. An indifference map is defined by the relation $(x+h)\sqrt{y+k}=a$, where h and k are fixed positive numbers and a is a positive parameter. By expressing y as a function of x and by finding derivatives, show that each indifference curve is downward sloping and convex from below.

32. If $\Pi = ax^2+bx+c$ is the cost function of a monopolist and if $p=\beta-\alpha x$ is the demand law, find the monopoly price and output when the monopolist is assumed to fix the price. Verify that this is the same result as when the monopolist fixes the output.

33. A radio manufacturer produces x sets per week at a total cost of £$(\frac{1}{25}x^2+3x+100)$. He is a monopolist and the demand of his market is $x=75-3p$, when the price is £p per set. Show that the maximum net revenue is obtained when about 30 sets are produced per week. What is the monopoly price? Illustrate by drawing an accurate graph.

34. If the manufacturer of the previous example, with the same costs, produces for a demand of $x=100-20\sqrt{p}$ sets per week, show that he should produce only 25 sets per week for maximum monopoly revenue. What is the monopoly price now?

35. In the case of Example 33, a tax of £k per set is imposed by the goverment. The manufacturer adds the tax to his cost and determines the monopoly output and price under the new conditions. Show that the price increases by rather less than half the tax. Find the decrease in output and monopoly revenue in terms of k. Express the receipts from the tax in terms of k and determine the tax for maximum return. Show that the monopoly price increases by about 33 per cent. when this particular tax is imposed.

36. Generalise the taxation problem of the previous example by finding the effects of a tax of k per unit of output when a monopolist's total cost is $\Pi=ax^2+bx+c$ and the demand law is $p=\beta-\alpha x$. Show that the tax brings in the maximum return when $k=\frac{1}{2}(\beta-b)$ and that the increase in monopoly price is always less than the tax.

37. If a monopolist has a total cost of $\Pi=ax^2+bx+c$ and if the demand law is $p=\beta-\alpha x^2$, show that the output for maximum revenue is

$$x=\frac{\sqrt{a^2+3\alpha(\beta-b)}-a}{3\alpha}.$$

How many sets per week should the radio manufacturer of Example 33 produce when the demand is $x = 10\sqrt{25 - p}$ sets per week?

38. The demand of a monopolist's market is $p = \beta - \alpha x$ and he produces an output of x units at a total cost of $\Pi = ax^3 - bx^2 + cx + d$. Show that the output for maximum net revenue is the positive root of

$$3ax^2 - 2(b - \alpha)x - (\beta - c) = 0.$$

Taking the cost conditions of the firm of Examples V, 18, and the demand law $p = 50 - \frac{5}{2}x$, show that the firm must produce $3\frac{1}{3}$ tons of its product per month for maximum monopoly revenue. What is the monopoly price per ton?

39. A firm with a total cost function $\Pi = F(x)$ sells on a perfectly competitive market, the market price being fixed at p. Show that the output of the firm for maximum net revenue is such that marginal cost equals p, provided that total costs are covered. Deduce that there is a supply relation for the firm, giving the output as a function of the market price p.

40. If $F(x) = ax^2 + bx + c$ in the previous example, show that the supply relation is linear. Show that p must exceed $b + 2\sqrt{ac}$ if total costs are to be covered but that, if only variable costs are to be covered, p need only exceed b. Illustrate the determination of the supply curve by drawing a diagram showing the average and marginal cost curves.

41. A sugar refinery has total cost equal to £$(\frac{1}{10}x^2 + 5x + 200)$ when x tons of sugar are produced per week. The fixed market price is £p per ton. What is the supply curve of the firm? What is the lowest price to cover total costs? At what price will 150 tons be produced?

42. A plant produces x tons of steel per week at a total cost of

$$£(\tfrac{1}{10}x^3 - 3x^2 + 50x + 300).$$

If the market price is fixed at £33 6s. 8d. per ton, show that the plant produces $16\frac{2}{3}$ tons per week.

43. In the problem of the previous example, show that the plant's output at the fixed market price of £p per ton is the root of

$$3x^2 - 60x + 10(50 - p) = 0,$$

which is greater than 10 tons per week. What is the smallest price for total costs to be covered? Show that the supply curve is

$$x = 10 + \tfrac{1}{3}\sqrt{30p - 600}$$

for values of p greater than this minimum amount. Connect the supply curve with the marginal cost curve and illustrate graphically.

44. The market demand for a good is given by $p = \beta - \alpha x$. The market is supplied by two duopolists with cost functions $\Pi_1 = a_1x_1^2 + b_1x_1 + c_1$ and $\Pi_2 = a_2x_2^2 + b_2x_2 + c_2$. Assuming that the " conjectural variations " are zero, show that the reaction curves are straight lines. Deduce the equilibrium output of each duopolist.

45. The duopolists of the previous example are radio manufacturers producing identical sets. The total cost of an output of x sets per week is £$(\frac{1}{25}x^2 + 3x + 100)$ in each case. When the price is £p per set, the market demand is $x = 75 - 3p$ sets per week. Show that the total equilibrium output is approximately 41 sets per week. Compare with the monopoly output of Example 33.

46. If the market demand of the previous example is $x = 10\sqrt{25 - p}$ sets per week, show that the reaction curve of the first duopolist is

$$x_1 = \tfrac{1}{3}(\sqrt{x_2^2 + 16x_2 + 6616} - 2x_2 - 4)$$

and similarly for the second duopolist. Draw a graph showing the two reaction curves and deduce that the total output is now approximately 32 sets per week. Compare with the monopoly output of Example 37.

47. A firm, with fixed plant, supplies of raw materials, etc., produces two goods X and Y in amounts related by the transformation function $y = f(x)$. The market prices of the two goods are fixed at p_x and p_y. Show that the outputs for maximum total revenue are such that $-f'(x)$ equals the ratio of p_x to p_y.

48. If the firm of the previous example monopolises the sale of both X and Y, the demand curves for X and Y being given and independent, show that the outputs for maximum total revenue are such that $-f'(x)$ equals the ratio of the marginal revenues from the X and Y demands.

49. A steel plant is capable of producing x tons per day of a low grade steel and y tons per day of a high grade steel, where $y = \dfrac{40 - 5x}{10 - x}$. If the fixed market price of low grade steel is half that of high grade steel, show that about $5\frac{1}{2}$ tons of low grade steel are produced per day for maximum total revenue.

50. The steel producer of the previous example monopolises the sale of both quality steels. If the prices of low and high grade steel are £p_x and £p_y per ton, the demands are $p_x = 20 - x$ and $p_y = 25 - 2y$. Find an equation giving the output x of low grade steel for maximum total revenue. Show, by a graphical method, that just under 6 tons of this steel are produced per day.

CHAPTER IX

EXPONENTIAL AND LOGARITHMIC FUNCTIONS

9.1 Exponential functions.

WE have been concerned, so far, with power functions and with the variety of functions that can be derived, in one way or another, from power functions. Such functions are of great practical importance and they serve to describe, accurately or approximately, many of the ways in which one variable depends on another. It is now convenient to extend the range of our function concept by the definition of an entirely new function type.

The power function is represented by the general form $y=x^n$, where n is any given number. A new function can be defined by the simple process of taking the base of the power as a fixed number and the index as variable. So, instead of writing a fixed power of a variable number, we write a variable power of a fixed number. The new function so obtained is called an *exponential function* and we write

$$y=a^x,$$

where a is the fixed base of the function. Since a power can be defined for all values of the index, the exponential function is a function of a continuous variable x.

It is found convenient, in general, to limit the value that can be allotted to the base a of an exponential function. If a is negative, then many values of a^x (e.g. when x is $\frac{1}{2}$ or $-\frac{1}{2}$) are not defined in terms of the real number system. To avoid this difficulty, we always take a as a *positive* number. If a is a positive fraction (between 0 and 1) and if we write $b=\frac{1}{a}>1$, then

$$a^x=\left(\frac{1}{b}\right)^x=\frac{1}{b^x}=b^{-x},$$

i.e. any power of a number less than unity can be reduced to a power of a number greater than unity, the sign of the index being reversed.

There can be no objection, therefore, if we take the base a as a number *greater than unity*. This is assumed throughout the following development.*

The graph of the exponential function can be plotted once the base is given a definite value. For the particular function $y=2^x$, a table of values of x and y can be obtained :

x	...	-2	$-\frac{3}{2}$	-1	$-\frac{1}{2}$	0	$\frac{1}{2}$	1	$\frac{3}{2}$	2	...
y	...	0·25	0·35	0·50	0·71	1	1·41	2	2·83	4	...

the value of y being taken, where necessary, to two decimal places. It is seen, from Fig. 62, that the points plotted from this table can

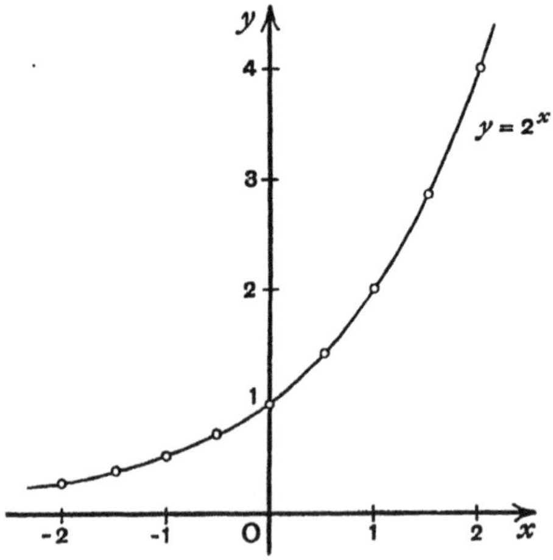

FIG. 62.

be joined by a smooth line which is the curve representing the function $y=2^x$. The graph shown is constructed for selected rational values of x only. But, for an irrational value of x, the value of 2^x is defined as the limit of a set of rational powers and the corresponding point must fit into the graph in a continuous way. The function $y=2^x$ is thus continuous.

The graphs of other examples of the exponential function type $y=a^x$ are of exactly similar shape. It will be seen later that the graph of one exponential function (e.g. $y=10^x$) can be obtained from

* Notice that the case $a=1$ is trivial. The exponential function is then the constant $y=1$.

that of any other (e.g. $y=2^x$) by a process of "stretching" or "contracting" in the direction of Ox. The general shape of the curve $y=a^x$ is always the same; its steepness or gradient varies with the value allotted to a and that is all.

We conclude that $y=a^x$ is a single-valued and continuous function. It is also seen that it is monotonic increasing, y increasing over the whole range of positive values as x increases from infinitely large negative to infinitely large positive values. It follows that, given any positive number p, we can find a unique value q so that $p=a^q$. An approximate value of q, for given values of a and p, can be read off the graph of the function $y=a^x$ as the abscissa corresponding to the ordinate p. We have, therefore, the important result that *any positive number can be expressed as a power of a given number greater than unity.*

It can be shown that, in general, irrational values of q correspond to rational values of p, and conversely. To illustrate the nature of the correspondence between p and q, we can take the useful base 10 and write, for example,

$$34=10^{1\cdot5315} \quad\text{and}\quad 7240=10^{3\cdot8597},$$

the indices being irrational numbers written correct to four decimal places. Again

$$10^{\frac{1}{3}}=2\cdot1528 \quad\text{and}\quad 10^{\frac{3}{2}}=31\cdot6228,$$

these numbers being irrational and given to four decimal places. The result we have just given leads us at once to the consideration of what are called "logarithms".

9.2 Logarithms and their properties.

The *logarithm* of a positive number p to the *base* a (>1) is defined as the index of that power of a which equals p. In symbols:

DEFINITION: If $p=a^q$, then $q=\log_a p$.

Since a logarithm is simply the index of the power to which the base must be raised to obtain the given number, it follows, from what has been said above, that any positive number has a unique logarithm to a definite base greater than unity. On the other hand, a negative number has no logarithm to any such base.

The logarithms of practical work have the convenient number 10 as their base. Since, to take two examples already quoted,

$$34=10^{1\cdot 5315} \quad \text{and} \quad 7240=10^{3\cdot 8597},$$

it follows that

$$\log_{10}34=1\cdot 5315 \quad \text{and} \quad \log_{10}7240=3\cdot 8597.$$

These logarithms are correct to four decimal places. The two notations, one in power form and the other in logarithmic form, are simply two ways of saying exactly the same thing.

The logarithms to the base 10 are called *common logarithms* and their values have been calculated and set out in tables of logarithms to four, five or more decimal places. Hence, to find the common logarithm of a given number, we have only to look up the number in the tables. The numbers 1·5315 and 3·8597 given above were, in fact, obtained in this way. No attempt is made here to give an account of the way in which common logarithms are used in practice to facilitate numerical work. There are many practical devices to remember and it is assumed that the technique is familiar or can be obtained by reference to a text-book on algebra. The general laws of logarithms, given below, provide the basis of the practical work.

Returning to logarithms to any base a greater than unity, two particular cases are derived from $a^0=1$ and $a^1=a$:

$$\log_a 1=0 \quad \text{and} \quad \log_a a=1.$$

Three general laws are obeyed by all logarithms :

LAW I. *The logarithm of a product.*

The logarithm of a product of two numbers is the sum of the separate logarithms :

$$\log_a (p_1 \cdot p_2)=\log_a p_1+\log_a p_2.$$

LAW II. *The logarithm of a quotient.*

The logarithm of a quotient of two numbers is the difference of the separate logarithms :

$$\log_a \left(\frac{p_1}{p_2}\right)=\log_a p_1-\log_a p_2.$$

LAW III. *The logarithm of a power.*

The logarithm of a power is the index times the logarithm of the base of the power :

$$\log_a (p^n)=n \log_a p.$$

The laws can be extended to the case of products or quotients of more than two numbers and they can be combined to give the logarithms of complicated products, quotients or powers. Since the logarithm of 1 is zero, we obtain a particular, and useful, case of Law II:

$$\log_a \left(\frac{1}{p}\right) = -\log_a p.$$

Notice, also, that the powers of p to which Law III applies include fractional powers. For example, we can write

$$\log_a \sqrt{p^3} = \log_a p^{\frac{3}{2}} = \tfrac{3}{2} \log_a p.$$

Since logarithms are indices, the laws of logarithms are simply translations of the index laws. For example, since indices are added when powers are multiplied, we expect that the logarithm of a product is the sum of the separate logarithms. Formal proofs of the three logarithm laws, based on the index laws, are as follows:

In Laws I and II, let

$$\log_a p_1 = q_1 \quad \text{and} \quad \log_a p_2 = q_2$$

so that $$p_1 = a^{q_1} \quad \text{and} \quad p_2 = a^{q_2}.$$

Then $$p_1 \cdot p_2 = a^{q_1} . a^{q_2} = a^{q_1+q_2},$$

and $$\frac{p_1}{p_2} = \frac{a^{q_1}}{a^{q_2}} = a^{q_1} . a^{-q_2} = a^{q_1-q_2}.$$

So $$\log_a (p_1 \cdot p_2) = q_1 + q_2 = \log_a p_1 + \log_a p_2,$$

and $$\log_a \left(\frac{p_1}{p_2}\right) = q_1 - q_2 = \log_a p_1 - \log_a p_2.$$

In Law III, let $\log_a p = q$ so that $p = a^q$.

Then $$p^n = (a^q)^n = a^{nq},$$

i.e. $$\log_a (p^n) = nq = n \log_a p.$$

The use of logarithms, both in theory and practice, is evident from the three laws. Expressions involving sums or differences are not easily treated by the use of logarithms. The logarithm of a sum or difference is not reducible and it must be stressed, in particular, that $\log_a (p_1 + p_2)$ is *not* equal to the sum of $\log_a p_1$ and $\log_a p_2$. On the other hand, by taking logarithms, an expression involving products and quotients is much simplified. The use of logarithms here replaces multiplication and division by the simpler processes of addition and subtraction. Further, logarithms are equally useful in dealing with powers and provide, for example, the only simple means of solving an equation in which the variable appears as an index. As a very simple instance, consider the equation :

$$3\,2^x - 4 = 0 \quad \text{or} \quad 2^x = \tfrac{4}{3}.$$

Here $\log(2^x) = \log(\tfrac{4}{3})$,

i.e. $x \log 2 = \log 4 - \log 3$,

i.e. $$x = \frac{\log 4 - \log 3}{\log 2}.$$

The logarithms can be to any base. Taking the base as 10 and looking up the logarithms concerned (to four decimal places) in tables, then

$$x = \frac{0{\cdot}6021 - 0{\cdot}4771}{0{\cdot}3010} = \frac{125}{301} = 0{\cdot}415,$$

correct to three decimal places. Other instances of equations most readily solved with the aid of logarithms arise in problems of compound interest (see 9.6 below).

Since the base of a logarithm can be chosen arbitrarily and for convenience, it is useful to have the following law which connects logarithms to one base with those to another base :

LAW IV. *The change of base in a logarithm.*

The logarithm of a number to the base a is the logarithm of the number to the base b times the logarithm of b to the base a :

$$\log_a p = \log_a b \log_b p.$$

As a particular case, put $p = a$. Then, since $\log_a a = 1$, we have

$$\log_a b \log_b a = 1, \quad \text{i.e. } \log_a b = \frac{1}{\log_b a}.$$

The change of base law can thus be written in two alternative ways :

$$\log_a p = \log_a b \log_b p = \frac{\log_b p}{\log_b a}.$$

The formal proof of the law proceeds :

Let $\log_b p = q$ so that $p = b^q$.

Then $\log_a p = \log_a (b^q) = q \log_a b = \log_a b \log_b p$.

The change of base law shows that the logarithms of a set of numbers to one base are simply constant multiples of the logarithms of the same set of numbers to another base. To change the base of a whole set of logarithms is only a matter of multiplying each logarithm by a constant. For example, logarithms to the base 2 can be obtained from common logarithms :

$$\log_2 p = (\log_2 10) \log_{10} p = \left(\frac{1}{\log_{10} 2}\right) \log_{10} p.$$

From tables of common logarithms, therefore, it is possible to construct tables giving logarithms to any base other than 10.

9.3 Logarithmic functions.

If the variable y is the value of the logarithm of the variable x to a given base a, we obtain the *logarithmic function*

$$y = \log_a x.$$

The relation between x and y defined by the logarithmic function is not, however, a new one. The logarithmic function, in fact, is simply the inverse of the exponential function. This follows from the definition of a logarithm ; if $y = \log_a x$, then $x = a^y$. Since the exponential function is single-valued, continuous and monotonic increasing ($a > 1$), the logarithmic function, as its inverse, possesses exactly similar properties.

For any definite value of a (>1), the graph of $y = \log_a x$ is the graph of the exponential function $y = a^x$ with the axes Ox and Oy transposed. Alternatively, the graph can be obtained directly from tables of logarithms. The graph of

$$y = \log_2 x$$

is shown in Fig. 63. It can be derived from the graph of $y = 2^x$, shown in Fig. 62, by transposing the axes.

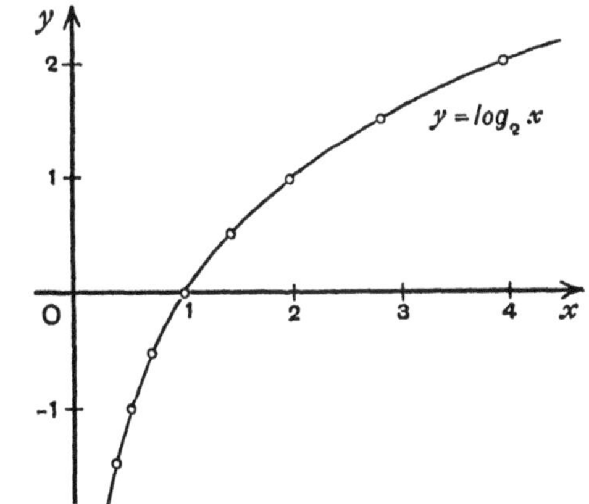

FIG. 63.

Logarithms to one base are constant multiples of logarithms to another base. It follows that the graph of $y = \log_b x$ can be derived from that of $y = \log_a x$ by multiplying all ordinates by a constant amount, i.e. by " stretching " or " contracting " the graph in the direction of Oy. All logarithmic functions have the same general shape when represented graphically ; the steepness of the graph varies with the value given to the base but the general shape of the graph remains. Further, since the graph of an exponential function is the graph of the corresponding logarithmic function with axes transposed, one exponential graph is obtainable from another exponential graph by a process of "stretching" or "contracting" in the direction of Ox. This fact was stated above; it has now been justified.

The graph of a logarithmic function shows the following properties. There is no logarithm of a negative number; the logarithm of a positive number less than unity is negative; the logarithm of the number 1 is zero; the logarithm of a number greater than unity is positive, increases and tends to infinity as the number increases. These properties are obtainable, of course, directly from the definition of a logarithm but they are particularly clear from the graph.

It is interesting to compare the three function types we have now considered. In their simplest forms, we can write

$$y=\log_\alpha x; \quad y=x^n; \quad y=a^x,$$

where α and a are greater than unity and where n is taken as positive. As x tends to infinity, so does the value of each function. But it can be shown that $\log_\alpha x$ tends to infinity more slowly than x^n and the latter, in its turn, tends to infinity more slowly than a^x. So, for a large value of x, the functions ascend in order of magnitude:

$$\log_\alpha x < x^n < a^x.$$

The proof of this statement can be given as follows.

If h is positive and if k is a positive integer, the Binomial Theorem (proved in text-books on algebra) gives

$$(1+h)^k=1+kh+\frac{k(k-1)}{2}h^2+\ldots+kh^{k-1}+h^k>1+kh.$$

Let $\sqrt{a}=1+h$ where h is positive and suppose that the number x lies between the integers k and $(k+1)$, i.e. $k \leqslant x < k+1$. Then

$$\sqrt{\frac{a^x}{x}}=\frac{(1+h)^x}{\sqrt{x}}>\frac{(1+h)^k}{\sqrt{k+1}}>\frac{1+kh}{\sqrt{k+1}}>\frac{kh}{\sqrt{2k}}.$$

Hence, $\frac{a^x}{x}$ is greater than $\frac{1}{2}h^2k$ and the latter tends to infinity as k, and so x, tends to infinity. So

$$\frac{a^x}{x}\to\infty \quad \text{as} \quad x\to\infty.$$

Now $$\frac{a^x}{x^n}=\left(\frac{a^{\frac{x}{n}}}{x}\right)^n=\left(\frac{1}{n}\frac{a^y}{y}\right)^n \text{ where } y=\frac{x}{n},$$

and this expression tends to infinity as y, and so as x, tends to infinity.

Finally $$\frac{x^n}{\log_a x}=n\frac{\alpha^z}{z} \text{ where } z=n\log_a x,$$

and this tends to infinity as z, and so as x, tends to infinity.

So $$\frac{a^x}{x^n}\to\infty \quad\text{and}\quad \frac{x^n}{\log_a x}\to\infty \quad\text{as}\quad x\to\infty,$$

and $\log_a x$, x^n and a^x must be in ascending order of magnitude if x is given any large value.

The result can be illustrated by comparing the three functions

$y=\log_2 x$; $y=x^2$; $y=2^x$

graphically. Fig. 64 shows the graphs of the three functions plotted on the same scales. It is clear that the graphs rise to the right at different rates and that

$$\log_2 x<x^2<2^x$$

for any large value of x.

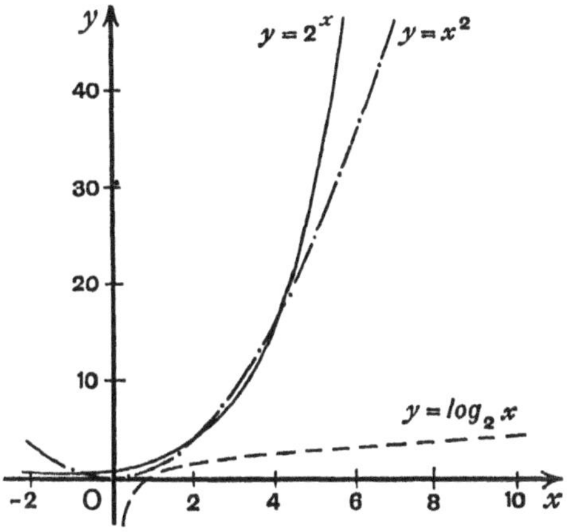

FIG. 64.

9.4 Logarithmic scales and graphs.

The introduction of logarithms enables us to extend the process of representing a function graphically by measuring the variables, not on the familiar " natural " or numerical scales, but on what are called " logarithmic " scales. The advantages and applications of this radical change in the graphical method are described in the present and following sections.

A variable x is measured according to some definite scale and a series of its values is represented by points on an axis Ox. The usual method is to take distances along Ox from a base point O as equal, or proportional, to the values of x plotted, a method which gives the *natural* scale for x. Now suppose that the points are plotted at distances along Ox which are equal, or proportional, to the logarithms of the values of x concerned, the point marked x on the axis being a distance $\log x$ from the base point. It is convenient, but not essential, to take logarithms to the base 10 from this purpose. A *logarithmic* scale for x is thus obtained.

The important characteristics of the logarithmic scale as compared with the natural scale are best introduced by considering some numerical examples. Of the two sequences of numbers

$$100, \quad 150, \quad 200, \quad 250, \quad 300, \quad \ldots,$$
$$100, \quad 150, \quad 225, \quad 337 \cdot 5, \quad 506 \cdot 25, \quad \ldots,$$

the first shows a regular increase of 50 units and the second a regular increase of 50 per cent. from one number to the next. On a natural scale, the points representing the first sequence appear at *equal* distances from each other and those representing the second sequence at *increasing* distances along the axis. The logarithms are

$$2, \quad 2 \cdot 176, \quad 2 \cdot 301, \quad 2 \cdot 398, \quad 2 \cdot 477, \quad \ldots,$$
$$2, \quad 2 \cdot 176, \quad 2 \cdot 352, \quad 2 \cdot 528, \quad 2 \cdot 704, \quad \ldots.$$

Hence, on a logarithmic scale, it is the second sequence that gives points at *equal* distances from each other and the first sequence shows points at *decreasing* distances along the axis. In the same way, the two decreasing sequences of numbers

$$100, \quad 80, \quad 60, \quad 40, \quad 20, \quad \ldots,$$
$$100, \quad 80, \quad 64, \quad 51 \cdot 2, \quad 40 \cdot 96, \quad \ldots,$$

show decreases of 20 units and of 20 per cent. respectively. The first sequence is represented by points at equal distances on a natural scale and at increasing distances (to the left) on a logarithmic scale. The second sequence corresponds to points at equal distances on a logarithmic scale and at decreasing distances (to the left) on a natural scale.

It appears, therefore, that equal distances between points on a natural scale indicate equal *absolute* changes in the variable, and equal distances between points on a logarithmic scale indicate equal *proportional* changes in the variable. This property is easily verified in general. If x_1, x_2, x_3 are values shown by points at equal distances on a natural scale, then

$$x_3 - x_2 = x_2 - x_1$$

and the variable increases by equal absolute amounts. The same property on a logarithmic scale implies that

$$\log x_3 - \log x_2 = \log x_2 - \log x_1,$$

i.e.
$$\log \frac{x_3}{x_2} = \log \frac{x_2}{x_1},$$

i.e.
$$\frac{x_3}{x_2} = \frac{x_2}{x_1},$$

and we have equal proportional changes in the variable. The obverse of this property is that equal proportional increases are shown on a natural scale by points at distances from the base point which increase more and more rapidly, while equal absolute increases are shown on a logarithmic scale by points at distances along the axis which increase more and more slowly.

If points are plotted from a function $y = f(x)$ by taking x along the horizontal axis Ox on a natural scale and by taking y along the vertical axis Oy on a logarithmic scale, we obtain the *semi-logarithmic graph* of the function. The graph shows a curve drawn through the plotted points and the varying height of the curve shows, not the variation of y with x, but the variation of $\log y$ with x. The curve is obtainable, of course, as the natural graph of $\log y$ as a function of x. If the curve is seen to rise through equal heights over certain ranges of values of x, then the value of y is subject, not to equal absolute increases, but to equal proportional or percentage increases. The use of semi-logarithmic graphs is thus clear. If we wish to compare percentage changes in the value of one variable as the other variable increases steadily, we plot the relation between the variables on a semi-logarithmic graph. If it is absolute changes that are important, we plot the relation on a natural graph.

Many statistical time-series are appropriately plotted as semi-logarithmic graphs. Time-intervals are taken, on a natural scale, on the horizontal axis. The variable of the time-series, to be plotted on the vertical axis, may be such that proportional changes in its value are important. Trade, consumption or employment figures for an increasing population are cases in point. In such cases, a logarithmic scale is used for the vertical axis and a semi-logarithmic graph is drawn. An example is given in the following section. Another application can be quoted in which the logarithmic scale is used for the independent variable measured along the horizontal axis. Suppose y represents the number of families with incomes over £x. Then the relation of y to x (the "cumulative" frequency distribution of incomes) is best plotted with y on a natural and x on a logarithmic scale. The characteristic that is most important is the variation of numbers for proportional changes in income.*

The logarithmic graphing of a function can be carried one stage further by taking both variables on logarithmic scales, in which case a *logarithmic graph* is obtained. The curve representing the function $y=f(x)$ now shows the variation of $\log y$ as $\log x$ changes and correlates proportional changes in y with proportional changes in x. So, if a curve on a logarithmic graph shows equal rises for equal moves to the right, then the result of increasing x by equal percentage amounts is to increase y by equal percentage amounts. The logarithmic graph is thus useful when proportional changes in both variables are important. For example, variables which are related for very small and, at the same time, for very large values are not easily shown on natural scales. Taking logarithmic scales, however, reduces the large variations to reasonable proportions while magnifying the small variations.†

Logarithmic graphs are extremely useful, though perhaps not well understood, in the statistico-economic field. Pareto's Income Law, for example, asserts that a logarithmic graph shows a straight line relation between x and y, where y is the number of persons with incomes over £x.‡ As income rises proportionally, the number of persons with that or higher income falls off in proportion. The main use of logarithmic graphs, however, is in the statistical correlation of economic variables by means of scatter diagrams (see 2.2 above). When we correlate, for example, the populations and unemployment rates in urban and rural areas of varying size, the variations are great and percentage changes are more significant than absolute changes. It is thus appropriate to plot the scatter diagram by logarithmic graphing. An example of a correlation scatter diagram

* See Gibrat, *Les Inégalités Économiques* (1931) and Allen and Bowley, *Family Expenditure* (1935), pp. 63-5 and Diagram XIII.

† This follows since $\log x$ increases more rapidly than x at first and then more slowly than x (see the graph of the logarithmic function).

‡ See Pareto, *Cours d'économie politique* (1897), Book 3, Chapter 1 and Bowley, *Elements of Statistics* (4th Ed. 1920), pp. 346-8.

plotted on logarithmic scales is given in the following section. Another important example of the use of a logarithmic scatter diagram arises in the statistical determination of demand or supply relations from the correlation of market data of prices and quantities demanded or supplied. The important thing here is the proportional change in price or quantity from one year to another, or in one year as compared with the " norm ".

9.5 Examples of logarithmic plotting.

The methods of logarithmic graphing, described above in general terms, are most clearly appreciated by giving actual examples of their use. The two cases considered in detail below illustrate very different applications of the methods.

Ex. 1. *Value of total imports, U.K.*, 1820-1897.

Centre year	Imports (£Mn.) average for 3 years	Centre year	Imports (£Mn.) average for 3 years
1821	31	1860	202
24	39	63	250
27	43	66	280
30	47	69	298
33	47	72	352
36	54	75	373
39	63	78	375
42	65	81	407
45	77	84	396
48	91	87	367
51	107	90	428
54	140	93	410
57	172	96	437

The variation in the value of U.K. imports, as given in the above table, is represented graphically in Fig. 65 by two different methods. The first graph takes a natural scale for the measurement of import values and the second graph a logarithmic scale. In each case, the years are measured along the horizontal axis on a natural scale and each import value is plotted at the centre of the averaged period.

It is seen that changes in import values are almost imperceptible, on the natural scale, in the early years and very marked in the later years. But this hides a very important fact. The *relative* changes from one year to another are roughly of the same importance in the early years as they are in the later. The logarithmic scale brings out this fact, the variations

IMPORTS, U.K., 1820-97.

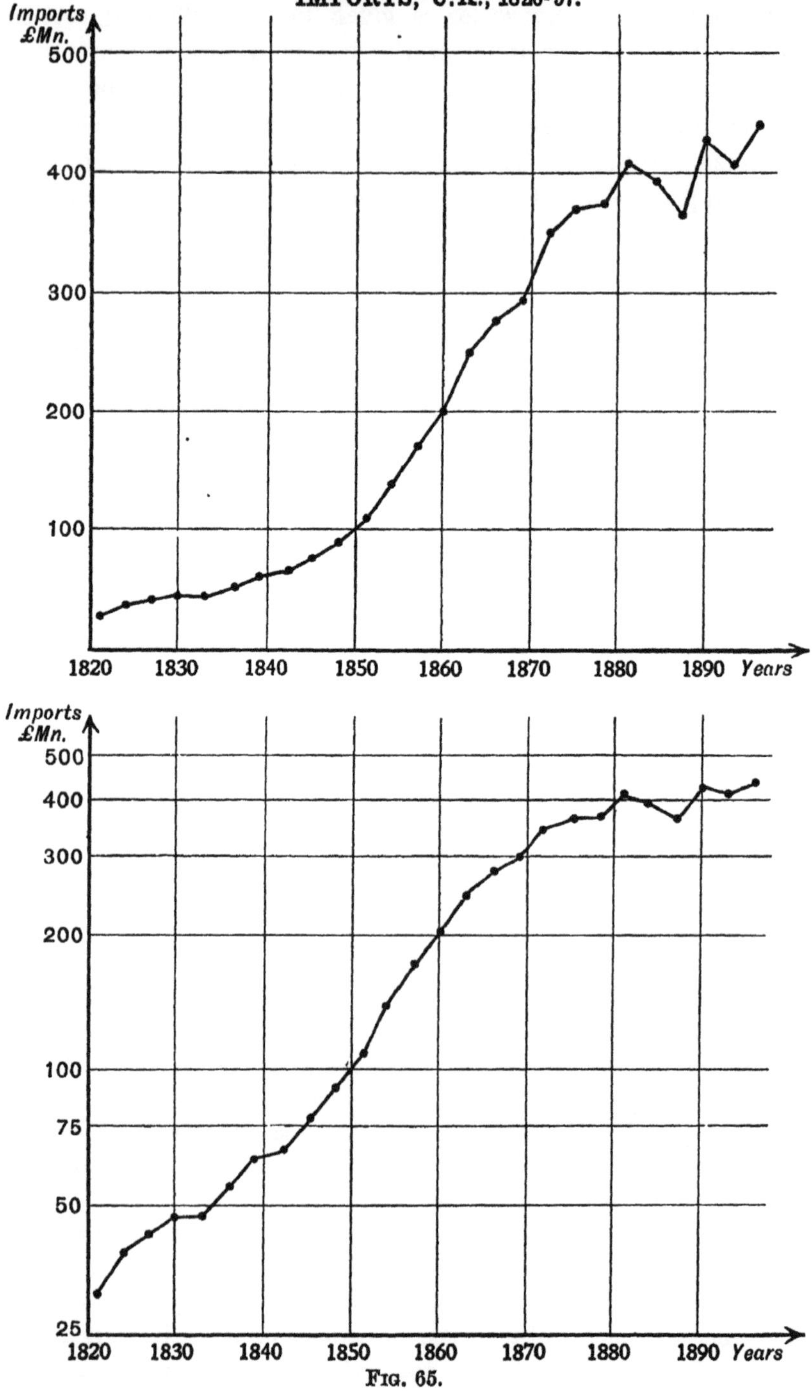

FIG. 65.

in the height of the second graph being of comparable magnitude at the beginning and end of the period. These variations represent relative changes in import values. For example, imports were valued at £65Mn. per year in the period round 1842 and £77Mn. in the period round 1845. The increase was £12Mn. or 18·5 per cent. of the earlier figure. In the two periods centred at 1887 and 1890, the increase was £61Mn. or 16·6 per cent. of the 1887 figure. The natural graph shows the second increase as roughly 5 times the first. The semi-logarithmic graph shows the second percentage increase as less than the first. For most purposes, the percentage increases are the more interesting and the semi-logarithmic graph is to be preferred.

Ex. 2. *Number of insured workers in employment, U.K.*, 1929-1935.

Industry	Employment per cent. increase (+) or decrease (−)	
	July 1929 to July 1932	July 1932 to July 1935
1. Coal-mining	− 29·6	+ 5·1
2. Bricks and tiles	− 8·0	+ 31·5
3. Pottery	− 23·1	+ 20·3
4. Iron and steel	− 40·1	+ 46·0
5. Shipbuilding	− 58·3	+ 37·3
6. Engineering, general	− 27·9	+ 19·9
7. ,, electrical	− 2·8	+ 11·2
8. ,, marine	− 57·4	+ 48·0
9. Electrical apparatus	+ 16·6	+ 31·7
10. Motors	− 13·8	+ 32·1
11. Cotton	− 26·8	− 1·3
12. Wool	− 15·5	+ 12·1
13. Silk	− 16·3	+ 24·5
14. Linen	− 34·9	+ 20·6
15. Leather	− 9·1	+ 14·1
16. Clothing	− 5·4	+ 3·8
17. Boots and shoes	− 8·2	+ 6·2
18. Food, etc.	− 1·8	+ 6·0
19. Building	− 17·7	+ 35·0
20. Public works	+ 31·6	− 13·2
21. Road transport	+ 6·6	+ 7·7
22. Distribution	+ 8·6	+ 4·7
23. Entertainment	+ 17·9	+ 25·8
24. Hotels	+ 4·4	+ 15·0
25. Local government	+ 10·9	+ 6·5
All insured industries	− 9·4	+ 11·9

The above table provides a picture of the decline in employment between 1929 and 1932 in various important industries and of the corresponding recovery between 1932 and 1935.* A graphical representation can be given by taking per cent. changes in the first period along the horizontal axis and per cent. changes in the second period along the vertical axis. One point is then plotted for each of the 25 industries, the height of the point indicating the per cent. change 1932-5 and the horizontal distance the per cent. change 1929-32. For convenience, the changes are referred to 100 as base in each case. Thus the decline in the coal-mining employment in the first period is represented by 70·4 (29·6 per cent. decrease) and the recovery in the second period by 105·1 (5·1 per cent. increase). The point representing the industry has co-ordinates (70·4, 105·1). It remains to determine whether natural or logarithmic scales are the more appropriate. It is clearly convenient to represent an industry which has recovered completely (1935 employment = 1929 employment) by a point with distance to the left of the 100 mark equal to height above the 100 mark. This is achieved by taking logarithmic scales. Suppose, for example, that employment falls to four-fifths its 1929 value in 1932 and then recovers completely by 1935. This implies a 20 per cent. decline and a 25 per cent. recovery in the two periods, i.e. the plotted point is (80,125). Now

$$\tfrac{80}{100} \cdot \tfrac{125}{100} = 1, \quad \text{i.e. } \log \tfrac{80}{100} + \log \tfrac{125}{100} = 0,$$

i.e.

$$\log 80 - \log 100 = -(\log 125 - \log 100).$$

So, on logarithmic scales, the point (80,125) is as much to the left of the 100 mark as it is above the 100 mark, as required.

A logarithmic scatter diagram of the data is shown in Fig. 66, the points representing the decline and recovery of the various industries. Most industries are shown by points in the N.W. quadrant, a decline followed by a recovery. The broken line, sloping downwards at 45°, corresponds to complete recovery, so that industries shown by points below the line have failed to recover completely and industries shown by points above the line have more than recovered their employment position. Six of the industries have increased employment in both periods, as shown by the points in the N.E. quadrant. The cotton industry, shown by a point in the S.W. quadrant, has a progressive decline in employment. Finally, Public Works employment has changed, as we should expect, in the opposite sense to the majority of industries and is shown by a point in the S.E. quadrant. Since the point is above the broken line, the later decline was not sufficient to reduce employment to the 1929 level.

* No correction has been made for the increase in the total insured population over the period 1929-35. The picture thus shows the recovery in a somewhat too favourable light.

It is to be noticed that we have marked our logarithmic scales according to equal intervals of the variable. In the vertical scale of Fig. 65, for example, the markings are at 10, 20, 30, ... (£Mn. of imports). The characteristic appearance of the logarithmic scale is then that the markings get closer and closer as we proceed up the scale.* Another point of importance is that a logarithmic scale avoids any difficulty about choice of zero mark or of units, a difficulty which is evident in using natural scales. In fact, a fixed distance on the logarithmic scale represents a given per cent. change no matter where we are on the scale and we have never to refer distances

NUMBER OF INSURED WORKERS IN EMPLOYMENT, U.K., 1929-35.

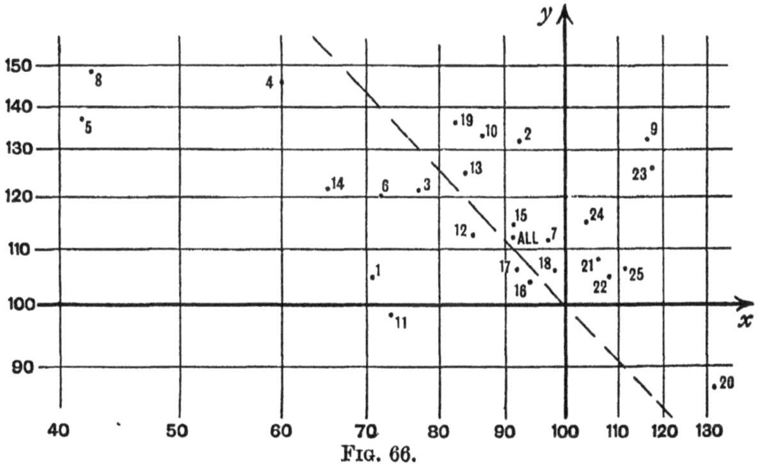

FIG. 66.

(as in the case of the natural scale) to the fixed origin or zero mark. The logarithmic scale, in this respect, has a clear advantage over the natural scale. A minor difficulty of the logarithmic scale is that only positive values of the variable can be plotted (since negative numbers have no logarithms). An adjustment can always be made, however, to avoid negative values, as is seen in Ex. 2 above, where negative percentages are converted to positive numbers below 100.

A functional relation between x and y can be represented, therefore, in three different ways on a diagram, as a natural, a semi-logarithmic or a logarithmic graph. The differences between the

* Special graph paper, marked in this way either for semi-logarithmic or for logarithmic graphing, is provided commerically. But ordinary graph paper can be used and adapted for the purpose by plotting distances directly from the logarithms.

three methods are well brought out by considering what relation is represented by a straight line in each case. On natural scales,

$$y = ax + b$$

is the relation represented by a straight line graph. If y is plotted on a logarithmic scale against x on a natural scale, the relation

$$y = ab^x$$

is represented by a straight line graph. For, we can write the relation as $\log y = \log a + x \log b$, i.e. $\log y$ is related linearly to x. Finally, on a logarithmic graph, the relation

$$y = ax^b$$

appears as a straight line. For, the relation can be written as $\log y = \log a + b \log x$ and $\log y$ is related linearly to $\log x$. So, an exponential function appears as a straight line on a semi-logarithmic graph and the line's gradient is the logarithm of the base of the exponential. A power function is shown as a straight line on a logarithmic graph and the gradient is the constant index of power. If y is the number of incomes over £x, for example, Pareto's Law is $y = \frac{a}{x^m}$, where a and m are constants. Plotting on a logarithmic graph, we obtain a downward sloping straight line with a numerical gradient equal to the constant m.

9.6 Compound interest.

A sum of £100 is invested and accumulates at compound interest at the rate of 4 per cent. per year. If interest is added yearly, then

$$£100 + £4 = £104 = £100(1{\cdot}04)$$

is the amount at the end of the first year. Again,

$$£104 + £\tfrac{4}{100}\,104 = £104(1{\cdot}04) = £100(1{\cdot}04)^2$$

is the amount at the end of two years. Similarly, £$100(1{\cdot}04)^3$ is the amount after three years, and so on. In general, if £y is the amount of the investment after x years, then

$$y = 100(1{\cdot}04)^x.$$

Suppose now that interest is added twice a year. Then, with a rate of 4 per cent. per year, 2 per cent. is added in each first half-year and another 2 per cent. in each second half-year. It follows, as before, that £100 amounts at the end of successive *half-years* to sums given by the sequence

$$£100(1{\cdot}02), \quad £100(1{\cdot}02)^2, \quad £100(1{\cdot}02)^3, \ldots .$$

Hence, at the end of x years, the amount is £y where

$$y = 100(1{\cdot}02)^{2x}.$$

These results can be generalised at once. If £a is invested at compound interest at $100r$ per cent. per year compounded yearly, then the amount after x years is £y where

$$y = a(1+r)^x.$$

If the interest is added n times a year, then

$$y = a\left(1+\frac{r}{n}\right)^{nx}.$$

In this result, which includes the previous one as a particular case, it is to be understood that x is a *discontinuous* variable, taking values which are multiples of $\frac{1}{n}$. Discontinuity is an essential feature of this compound interest problem.

The dependence of y on the parameters indicating the interest rate and the frequency of compounding interest is to be noticed. The amount y, after any period, is clearly larger the higher is the interest rate. Further, the amount is larger when interest is compounded twice a year than it is on yearly compounding. For,

$$(1+\tfrac{1}{2}r)^2 = 1 + r + \tfrac{1}{4}r^2 > 1 + r,$$

i.e.
$$a(1+\tfrac{1}{2}r)^{2x} > a(1+r)^x.$$

In general, the more frequently is interest added, the larger is the amount of a given sum at the end of any period.

The amount of an investment increases over time in what is called a " geometric progression ", each amount being a fixed multiple of the previous year's amount. Analytically, we can express this growth at an ever increasing rate by the exponential function $y = ab^{nx}$, where $b = 1 + \frac{r}{n}$ is a constant greater than unity. The growth is shown by the heights of successive points on a certain exponential graph, the points being spread out at equal distances along the horizontal time-axis. For example, the heights of points on the graph of $y = 2^x$ (shown in Fig. 62) at abscissa $x = 1, 2, 3, \ldots$

represent the amounts of £1 after successive years when interest is added yearly at 100 per cent. It is clear that semi-logarithmic graphing is appropriate to this problem. The compound interest growth curve is then a straight line with gradient $\log b = \log\left(1+\frac{r}{n}\right)$. The percentage rate of growth is a constant fixed by r and n.

A simple problem illustrates the way in which logarithms are to be used in dealing with compound interest growth:

A National Savings Certificate (1935 issue) costs 15s. and realises 20s. at the end of 10 years. Assuming that interest is added four times a year, it is required to find the rate of interest represented by this growth. If the interest rate is $100r$ per cent., then

$$y = 15(1+\tfrac{1}{4}r)^{4x}$$

is the amount (in shillings) after x years. But $y = 20$ when $x = 10$.

So, $$20 = 15(1+\tfrac{1}{4}r)^{40},$$

i.e. $$40\log(1+\tfrac{1}{4}r) = \log 20 - \log 15 = 0{\cdot}12494,$$

i.e. $$\log(1+\tfrac{1}{4}r) = 0{\cdot}0031235,$$

i.e. $$1+\tfrac{1}{4}r = 1{\cdot}00722.$$

So $100r = 2{\cdot}89$ approximately and the rate of interest per year is 2·9 per cent. correct to one decimal place.

The growth of an investment when interest is added at definite intervals is a function of a discontinuous variable. It remains to consider what meaning can be attached to a notion of growth at continuous compound interest. Our problem now is to examine the result of letting n, the number indicating the frequency of compounding interest, take larger and larger integral values.

To start with a simple case, £$\left(1+\frac{1}{n}\right)^n$ is the amount of £1 at the end of a year when interest is compounded at 100 per cent. per year n times in the year. The values of this expression for certain values of n are

n	1	10	100	1000	10,000
$\left(1+\frac{1}{n}\right)^n$	2	2·594	2·704	2·717	2·718

the values being given correct to three decimal places. It is clear that $\left(1+\frac{1}{n}\right)^n$ tends to a definite limit, in the neighbourhood of 2·718, as n tends to infinity. It can be shown, by rigid methods, that this is correct * and the limiting value is denoted by the letter e. So,

DEFINITION : $$e = \lim_{n\to\infty} \left(1+\frac{1}{n}\right)^n.$$

It is possible to show that e, which must be a pure number, is very similar to the familiar number π and cannot be expressed in fractional form or as a terminating decimal. The value of e can be found, however, correct to any given number of decimal places by giving n a sufficiently large value in $\left(1+\frac{1}{n}\right)^n$. So, to five decimal places,

$$e = 2{\cdot}71828.$$

Our definition of e is such that the amount of £1 at the end of one year, when the interest at 100 per cent. is added more and more frequently, approaches the value £e. This is only one of the many uses of the number e which is of very great importance in mathematical analysis.

Returning to the general case of compound interest growth,

$$a\left(1+\frac{r}{n}\right)^{nx} = a\left\{\left(1+\frac{r}{n}\right)^{\frac{n}{r}}\right\}^{rx} = a\left\{\left(1+\frac{1}{m}\right)^{m}\right\}^{rx} \to ae^{rx} \quad \text{as } n\to\infty,$$

since $\left(1+\frac{1}{m}\right)^m \to e$ as m, and so as $n=rm$, tends to infinity. Also, since the discontinuous variable x is a multiple of $\frac{1}{n}$, it tends to become less and less discontinuous as n increases. Hence, as n tends to infinity and interest is added more and more frequently, the compound interest formula tends to assume the form $y=ae^{rx}$ and the variable x tends to become continuous. We have now derived a concept of continuous compound interest as the result of a limiting process in which interest is compounded more and more often. Our result is:

The amount of £a after x years when interest is compounded continuously at the nominal rate of $100r$ per cent. per year is given by

$$y = ae^{rx}.$$

* See Hardy, *Pure Mathematics* (3rd Ed., 1921), p. 137.

This compound interest formula, dependent on a continuous variable x, is an abstract one. But, though interest can never be compounded continuously in actual practice, the formula can be taken as a convenient and approximate representation of the actual state of affairs when interest is compounded frequently.

9.7 Present values and capital values.

A sum of £a is due x years hence and the rate of interest and the frequency of its compounding are known. Then we can determine the sum to invest now so as to produce the given sum of £a at the end of x years. This sum is called the *present value*, or the *discounted value*, of £a available x years hence.

If interest is compounded once yearly at $100r$ per cent., then £y is the present value of £a available x years hence provided that

$$y(1+r)^x = a.$$

So

$$y = \frac{a}{(1+r)^x}.$$

Similarly, if interest is added n times a year at $100r$ per cent., we have

$$y = \frac{a}{\left(1+\frac{r}{n}\right)^{nx}}.$$

Finally, if interest is added continuously at $100r$ per cent., then

$$ye^{rx} = a,$$

i.e.

$$y = ae^{-rx}.$$

The present value £y of a given sum at a given date depends on the parameters r and n indicating the interest rate and the frequency of compounding. The present value is smaller, the higher is the interest rate and the more frequently is interest compounded.

The uses of a computation of present value are fairly obvious. If a dealer has a claim on £a in x years' time, he can sell his claim now to another person, not for the full £a, but for the present value of this sum calculated at the current rate of interest. Further, we can find the total present value of a claim on a whole series of sums due in successive years. The series

$$£a_0, \quad £a_1, \quad £a_2, \ \dots \ £a_m$$

may represent the values in the current and in m successive years of the crops obtainable from a piece of land, of the outputs of a given machine or plant or of the incomes due to a given individual. We have, in fact, a given "stream" of crop values, of output values or of incomes over time. The present value of the stream, if interest is added yearly at 100r per cent. per year, is given by the sum

$$a_0+\frac{a_1}{1+r}+\frac{a_2}{(1+r)^2}+\ldots+\frac{a_m}{(1+r)^m}.$$

This sum can be called the *capital value* of the land, machine or income stream in question. It represents the sum which must be invested now to produce incomes of £a_0, £a_1, £a_2, ... £a_m in successive years. It is to be noticed that the capital value of an output or income stream depends, not only on the items of the stream and on the number of years that the stream flows, but also on the interest rate that is taken. One and the same income stream has different capital values when different interest rates are current. Simple examples can be given to illustrate this fact.

Shares in a mining company are expected to produce dividends of £40, £32, £24, £16 and £8 in the present and in the four following years, and to be worth nothing thereafter. If interest is added once yearly at 5 per cent., the present or capital value of the holding is

$$£\left(40+\frac{32}{1{\cdot}05}+\frac{24}{1{\cdot}05^2}+\frac{16}{1{\cdot}05^3}+\frac{8}{1{\cdot}05^4}\right)=£112{\cdot}6,$$

the calculation being made with the aid of logarithms. A similar computation shows that the capital value is £116·2 if the interest rate is only 2½ per cent. These are the two sums which can be invested now to produce, at the respective interest rates, the given income stream over the five years.

Shares in a new trading company are expected to produce dividends of £1, £2, £4 and £6 in the present and in the three following years. In the fourth year, it is expected that the shares can be sold for £120. The capital value of this holding, taking interest as added once yearly at 5 per cent. per year, is given by

$$£\left(1+\frac{2}{1{\cdot}05}+\frac{4}{1{\cdot}05^2}+\frac{6}{1{\cdot}05^3}+\frac{120}{1{\cdot}05^4}\right)=£110{\cdot}4.$$

But, if the interest rate is 2½ per cent., the capital value is found to be £121·1. This holding is worth less than the previous one if the rate of interest is 5 per cent. but worth more if the rate of interest is 2½ per cent. The relative capital valuation of the two holdings is reversed by the change in the rate of interest.

In conclusion, we can consider a simplified economic problem in the field of capital and interest. Let £x be the income derived by an entrepreneur this year and £y his income next year. The relation between these two incomes is given by technical considerations as $y=f(x)$ (see 5.6 above). The entrepreneur wishes to arrange his resources to give incomes which correspond to the largest possible present value. If the rate of interest, compounded yearly, is fixed at $100r$ per cent., the present value of £x now and £y next year is

$$V=x+\frac{y}{1+r}=x+\frac{f(x)}{1+r}.$$

This is a maximum if $\frac{dV}{dx}=0$ and $\frac{d^2V}{dx^2}<0$, i.e. if

$$1+\frac{f'(x)}{1+r}=0 \quad \text{and} \quad \frac{f''(x)}{1+r}<0.$$

The first condition shows that r equals the expression $\{-f'(x)\}-1$ which, as the numerical gradient of the transformation curve $y=f(x)$ reduced by one, is called the marginal rate of return over cost.* At the incomes for maximum present value, therefore, this marginal rate must equal the fixed rate of interest. The second condition is that $f''(x)<0$, i.e. the transformation curve must be concave from below, a condition which is satisfied in the "normal" case.

9.8 Natural exponential and logarithmic functions.

The introduction of the number $e=\lim\limits_{n\to\infty}\left(1+\frac{1}{n}\right)^n=2{\cdot}71828\ldots$ provides us with the most important cases of the exponential and logarithmic functions. When the base of the function is taken as e, it is called a "natural" exponential or logarithmic function. The standard forms of the natural functions are $y=e^x$ and $y=\log_e x$ and the curves representing them are similar to those already indicated (in Figs. 62 and 63) for the base 2.

A more general form of the natural exponential function is

$$y=ae^{bx},$$

where a and b are constants. But this function can be derived quite easily from the standard form $y=e^x$. The method of derivation is

* See Fisher, *The Theory of Interest* (1930), pp. 159 *et seq.* and pp. 514-5. See also, Examples VI, 31.

best described in a particular case and in diagrammatic terms. The curve $y=e^{\frac{1}{2}x}$ is obtained by stretching the curve $y=e^x$ in the horizontal (Ox) direction so that each point on the former is twice as far from Oy as the point with the same height on the latter. This is illustrated in Fig. 67. The curve $y=2e^{\frac{1}{2}x}$ is then obtained from the curve $y=e^{\frac{1}{2}x}$ by stretching in the vertical (Oy) direction so that all ordinates are doubled, as shown in Fig. 67. Two stretching processes are thus needed to transform the standard curve $y=e^x$ into the curve $y=2e^{\frac{1}{2}x}$. In general, if a and b are positive, the curve $y=ae^{bx}$ can be derived from the curve $y=e^x$ in two stages. The standard exponential curve is first reduced in the Ox direction in the ratio $b:1$, i.e. contracted if $b>1$ and stretched if $b<1$. The curve so obtained is then expanded in the Oy direction in the ratio $a:1$, i.e. stretched if $a>1$ and contracted if $a<1$. As a result, the ordinate of the curve $y=ae^{bx}$ at abscissa $\frac{x}{b}$ is a times the ordinate of the curve $y=e^x$ at abscissa x.* Exactly similar remarks apply to the relation between the natural logarithmic function $y=a\log_e bx$, where a and b are constants, and the standard form $y=\log_e x$.

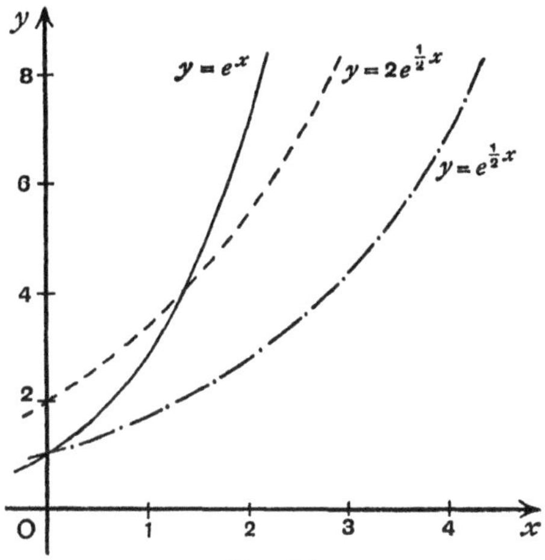

FIG. 67.

All problems of growth at compound interest added continuously are described by means of a natural exponential function. An investment of £a at $100r$ per cent. continuous compound interest increases over time according to the law $y=ae^{rx}$. The particular

* If the constants a and b are allowed to take negative values, a further modification of the standard exponential curve is needed. The curves $y=ae^{-bx}$ and $y=ae^{bx}$ differ only in that the abscissa $(-x)$ gives the same ordinate of one curve as the abscissa x of the other. One curve is the reflection of the other in Oy. One curve falls and the other rises from left to right. Similarly, the curve $y=-ae^{bx}$ is the reflection of the curve $y=ae^{bx}$ in Ox. Negative values of a and b, therefore, require the reflection of the standard exponential curve in one or other of the axes.

shape of the curve representing this growth depends, as we have seen, on the values of the constants a and r. The larger the interest rate, the steeper becomes the curve and the faster becomes the growth of the investment. The larger the original sum invested, the higher becomes the curve and the larger the amount of the investment at any time.

A somewhat different use of the curve $y=ae^{rx}$ can be noticed. Any point (x, y) in the positive quadrant of the plane represents a situation in which £y is available x years hence. If this point lies on the curve $y=ae^{rx}$, then the present or discounted value of this sum is £a if the interest is reckoned continuously at $100r$ per cent. The curve thus connects all points representing situations with the same present or discounted value at a given rate of interest and, for this reason, it can be described as a *discount curve*.

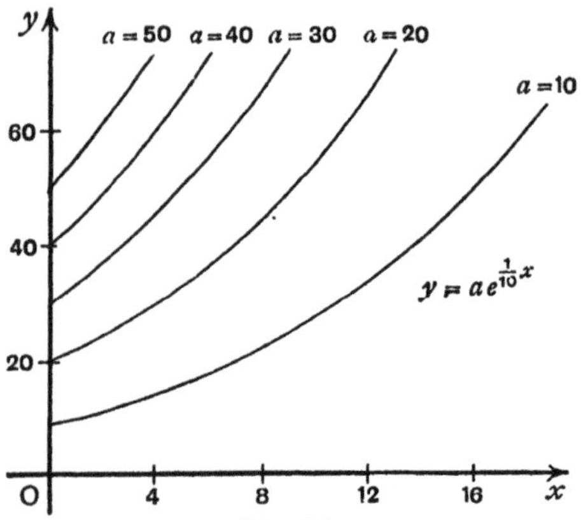

Fig. 68.

A system of discount curves $y=ae^{rx}$ is derived by fixing the value of r and taking a as a parameter. Certain curves of the system are shown in Fig. 68, in the case where the interest rate is 10 per cent. ($r=0{\cdot}1$). One curve of the system is obtained from any other by stretching or contracting in the vertical (Oy) direction and one curve passes through any given point in the positive quadrant of the plane. The discount curve system provides a simple means of comparing the present values of different sums available at different future dates. Suppose that two situations are given (certain sums available at certain dates) and represented by points in the plane Oxy. If the points are on the same discount curve, then the present values are equal and given by the parameter of the discount curve. If one point is on a higher discount curve than the other, then the present values are different and the first situation corresponds to the greater present value. It is necessary, of course, that the rate of interest should be known and fixed; the discount curve system is given only for one rate of discounting. The whole system changes when the rate is changed, the curves becoming steeper for larger rates of interest.* A discount curve $y=ae^{rx}$ becomes a straight line $\log y=\log a+rx$ when it is plotted on a semi-logarithmic graph in which the vertical scale for y is based on natural logarithms. The line slopes upwards with a gradient equal to the interest rate r. The discount curve system of Fig. 68 then reduces to a set of parallel straight lines.

In conclusion, it is easily shown that an exponential or logarithmic function to a base other than e can be expressed as a natural exponential or logarithmic function. If b is a positive number, the definition of a logarithm enables us to write

$$b=e^{\log_e b}.$$

The general exponential function $y=ab^{cx}$ then appears

$$y=ae^{(c\log_e b)x},$$

i.e. as a natural exponential function. Further, since

$$\log_b x=\frac{\log_e x}{\log_e b},$$

the general logarithmic function $y=a\log_b cx$ can be written

$$y=\frac{a}{\log_e b}\log_e cx,$$

i.e. as a natural logarithmic function.

There is, therefore, no need to consider exponential or logarithmic functions other than those to the natural base e. In practical work, it is convenient to take common logarithms and powers of the base 10. But, in theoretical work, it is always found preferable to use the natural base e. In this case, for example, the derivatives of the exponential and logarithmic functions appear in their simplest form (see 10.1 below). Natural exponential and logarithmic functions are thus assumed in all our theoretical work and, if we write a logarithm without specification of the base, it is to be understood that the base is e. The step from the practical logarithms to the base 10 to theoretical logarithms to the base e is, however, a very simple one :

$$\log_e x=\log_e 10\,.\,\log_{10}x=\frac{\log_{10}x}{\log_{10}e}=(0{\cdot}4343)\log_{10}x,$$

the numerical multiplier being given correct to four decimal places.

* Another system of discount curves is obtained when a is fixed and r is taken as a parameter. This sytem enables us to determine at what interest rate £y available in x years has a given present value of £a.

EXAMPLES IX

Exponential and logarithmic functions

1. Draw a graph of $y=10^x$. Read off the values of x for $y=0{\cdot}5$, $1{\cdot}2$ and $8{\cdot}9$ and check from tables of logarithms.

2. Use logarithmic tables to construct a graph of $y=\log_{10}x$. Add the graph of $y=x-2$ and so solve $\log_{10}x+2=x$ approximately.

3. Find, with the aid of logarithmic tables, approximate solutions of the equation $2^x \, . \, 3^{-x}=10$ and of the equation $2^{x^2}=7$.

4. Illustrate the fact that the logarithm of a non-prime number can be reduced to logarithms of primes by showing that

$$\log_{10}1176=3\log_{10}2+\log_{10}3+2\log_{10}7.$$

Check the result by looking up the logarithms in tables.

5. By expressing the left-hand side as a single logarithm, show that

$$\tfrac{1}{2}\log\tfrac{25}{9}-\log\tfrac{15}{4}+\tfrac{2}{3}\log\tfrac{27}{8}=0.$$

Express y explicitly in terms of x if it is given that

$$2\log y+\log(x-1)-\log(x+1)=0.$$

6. Express the logarithm of $\dfrac{x}{1+x}\sqrt{\dfrac{1+2x}{1+3x}}$ as sums and differences of logarithms of simpler expressions. Hence, find the value of this expression when $x=2{\cdot}4$.

7. Show that $(x+\sqrt{x^2-1})(x-\sqrt{x^2-1})=1$ and deduce that

$$\log(x+\sqrt{x^2-1})=-\log(x-\sqrt{x^2-1}).$$

8. Indicate why $\dfrac{1}{x}\log_2 x$ and $x^2\,3^{-x}$ both tend to zero as x tends to infinity.

Deduce that $\dfrac{x\log_2 x}{3^x}\to 0$ as $x\to\infty$.

9. *Yearly production of bricks, U.K.*, 1816-1849.

Year	Bricks Mn.	Year	Bricks Mn.	Year	Bricks Mn.	Year	Bricks Mn.
1816	673·0	1825	1948·8	1834	1152·4	1842	1271·9
17	701·7	26	1350·2	35	1349·3	43	1158·9
18	952·1	27	1103·3	36	1606·1	44	1420·7
19	1101·6	28	1078·8	37	1478·2	45	1820·7
20	949·2	29	1109·6	38	1427·0	46	2039·7
21	899·2	30	1091·3	39	1568·7	47	2193·8
22	1019·5	31	1125·4	40	1677·8	48	1461·0
23	1244·7	32	971·9	41	1423·8	49	1462·7
24	1463·2	33	1011·3				

Data from Shannon, *Bricks—A Trade Index*, 1785-1849, Economica, 1934.

Represent the above time-series on two graphs, one showing the number of bricks on a natural scale and the other on a logarithmic scale. Which graphical representation is to be preferred?

10. If y is the number of persons with incomes over £x, the following table is obtained for super-tax payers, U.K., 1911-2 :

x(£000)	5	10	15	20	25	35	45
y	11,554	4143	2114	1327	889	507	321

Plot the data on a logarithmic graph and show that the plotted points lie close to a downward sloping line. Deduce that Pareto's Income Law is approximately satisfied ($y = a/x^m$) and estimate the values of the constants a and m. (See Bowley, *Elements of Statistics*, 4th Ed. 1920, p. 347.)

11. Plot the price and consumption data of Examples I, 30 as a scatter diagram on logarithmic scales for both variables.

12. *Number of Insured Workers in Staffordshire, at July*, 1935, *and percentage Unemployed, at May*, 1936.

Area	Insured (000's) July 1935	Percentage unemployed May 1936	Area	Insured (000's) July 1935	Percentage unemployed May 1936
Audley	1·7	28·3	Rugeley	3·2	6·9
Biddulph	3·3	25·8	Smethwick	36·8	4·9
Bilston	16·8	11·2	Stafford	13·1	3·0
Brierley Hill	10·2	10·7	Stoke	120·7	16·7
Burton	18·4	8·8	Tamworth	8·3	6·1
Cannock	15·8	7·3	Tipton	15·7	7·8
Cheadle	5·1	27·3	Uttoxeter	2·7	3·1
Cradley Heath	16·1	9·8	Walsall	41·9	10·0
Darlaston	10·0	6·9	Wednesbury	11·6	12·0
Kidsgrove	3·3	50·5	W. Bromwich	22·7	7·6
Leek	11·1	14·3	Willenhall	11·8	8·4
Lichfield	2·5	6·5	Wolverhampton	55·3	11·1
Newcastle	16·3	20·3			

Plot a scatter diagram on logarithmic scales. Is there any evidence of a correlation between unemployment and the size of the area?

13. Given $\log_e 10 = 2{\cdot}3026$, use tables of common logarithms to plot a graph of $y = \log_e x$. Put $y = 1$ and read off the value of e.

14. Plot graphs of $y = e^{-x}$, $y = \frac{3}{2}e^{2x}$ and $y = 2e^{-\frac{1}{2}x}$. Show how these curves can be obtained from the curve $y = e^x$.

15. If $y = \frac{1}{2}(e^x - e^{-x})$, show that $e^{2x} - 2ye^x - 1 = 0$. Solve this quadratic equation in e^x to show that $x = \log(y + \sqrt{y^2+1})$. Why must the other root of the quadratic be neglected? In the same way, show that $y = \frac{1}{2}(e^x + e^{-x})$ and $x = \log(y + \sqrt{y^2 - 1})$ are inverse functions.

16. Show that $y = \dfrac{e^x - e^{-x}}{e^x + e^{-x}}$ and $x = \frac{1}{2}\log\dfrac{1+y}{1-y}$ are inverse functions.

17. Given $\log_{10}e = 0{\cdot}4343$, use tables of common logarithms to evaluate $y = xe^{-x}$ approximately for $x = 0, \frac{1}{2}, 1, \frac{3}{2}, 2, 3, 4$ and 5. Hence plot a graph of the curve $y = xe^{-x}$ for positive values of x. (This is one of Pearson's system of curves for fitting to statistical frequency distributions. See Elderton, *Frequency curves and correlation.*)

18. Using the method of the previous example, plot a graph of the curve $y = e^{-\frac{1}{2}x^2}$ for positive and negative values of x. (This is the *normal curve of error* of statistical theory.)

19. Indicate why $x^n e^{-x}$ and $x^n e^{-\frac{1}{2}x^2}$ must tend to zero as x tends to infinity for any finite value of n. From the graph of $y = xe^{-x}$, illustrate that $xe^{-x} \to 0$ as $x \to \infty$.

Compound interest problems

20. What is the amount of £100 after 5 years and what amount must be invested to realise £100 after 10 years when interest at the rate of $3\frac{1}{2}$ per cent. per year is added (*a*) yearly, (*b*) twice yearly, (*c*) continuously?

21. In how many years will an investment double itself when interest at $2\frac{1}{2}$ per cent. per year is added (*a*) yearly, (*b*) twice yearly, (*c*) continuously? What are the periods when the interest rate is 5 per cent.?

22. A National Savings Certificate costs 15s. and realises 20s. after 10 years. Find the rate of interest involved when it is added (*a*) yearly, (*b*) twice a year, (*c*) eight times a year, (*d*) continuously. Show that the nominal rate is smaller, the more frequently is interest added. Take $\log_{10}e = 0{\cdot}4342945$ and use Chambers' seven-figure logarithmic tables.

23. The certificate of the previous example produces 17s. 3d. after 5 years. What rate of interest, added yearly, does this represent? Is it less than the rate over the complete 10 years?

24. In a previous issue (1933), a certificate cost 16s. and realised 20s. after 8 years, 21s. 4d. after 10 years and 23s. after 12 years. Find the interest rate, added yearly, for each of these periods. Is it true, as claimed, that the present issue bears the same rate of interest as the 1933 issue over the 10 years period?

25. Interest at $100r$ per cent. compounded yearly is equivalent to interest at $100s$ per cent. compounded n times a year, a given sum producing the same amount after any period at the t ɔ rates. Show that $r = \left(1 + \frac{s}{n}\right)^n - 1$. Deduce that s is less than r by an amount equal to $\frac{3}{8}s^2$ approximately when s is small and $n = 4$.

26. If $100r$ per cent. compounded yearly and 100ρ per cent. compounded continuously are equivalent interest rates, show that $\rho = \log_e(1 + r)$. Plot a graph of this relation to show that ρ is always smaller than r but approximately equal to r when r is small.

27. A sinking fund is formed by investing £x at the end of each year for m years. Show that the final amount of the fund is

$$x + x(1+r) + x(1+r)^2 + \ldots + x(1+r)^{m-1}$$

when interest is added yearly at $100r$ per cent. per year. By writing the sum of this geometric progression (17.1 below), show that the fund will amount finally to £a if

$$x = \frac{ar}{(1+r)^m - 1}.$$

Show that approximately £79·5 must be set aside each year if the sinking fund (interest at 5 per cent.) is to replace a machine costing £1000 after 10 years. (See Fowler, *Depreciation of Capital*, 1934, p. 131.)

28. If the rate of interest (added yearly) will be $100r_1$ per cent., $100r_2$ per cent., ... $100r_m$ per cent. during the next m years, show that £a will amount to £$a(1+r_1)(1+r_2)\dots(1+r_m)$ at the end of the period. What is the present value of £b due in m years?

The interest rates in successive years from now will be 4 per cent., 3 per cent., $2\frac{1}{2}$ per cent., $2\frac{1}{2}$ per cent., $3\frac{1}{2}$ per cent., ... Find the amount of £100 after 5 years and the present value of £100 due in 4 years.

29. Find the present value of £100 due 10 years hence when interest at 2 per cent. per year is compounded (*a*) yearly, (*b*) continuously.

30. A mine-owner derives an income of £2000 this year and his income falls by £200 in each following year until no income results. Find the present value of the income stream when interest is added yearly at (*a*) 4 per cent. per year, (*b*) 5 per cent. per year.

31. Why can £$\frac{a}{r}$ be taken as the present value of an income stream of £a a year for ever? (Interest at $100r$ per cent. compounded yearly.)

32. A fir plantation brings its owner nothing this year or next year. In the two following years the incomes are £300 and £400 and the income is thereafter £500 a year for ever. What is the discounted value *at the beginning of the fifth year* of the constant income stream when interest is added yearly at (*a*) 4 per cent. per year, (*b*) 5 per cent. per year? Find the present value of the whole income stream now at the same interest rates. Compare these values with those of Example 30 above. (See Fisher, *The Theory of Interest*, 1930, pp. 133 *et seq.*)

33. Interest is added yearly at 3 per cent. per year. What is the present value of a perpetual income of £100 beginning two years from now? This income can be produced by investing £2000 in a business this year and £1200 next year. What is the present value of the investment? Is it a profitable investment?

34. Draw a graph of certain curves of the discount curve system when the rate of interest is fixed at 5 per cent. Use the graph to determine whether £220 due in 10 years has a larger or smaller present value than £150 due in 3 years, interest being added continuously at 5 per cent. per year.

35. If r is a parameter, what is the form of the curve system $y=100e^{rx}$ on a semi-logarithmic and on a natural graph? Draw certain curves of the system using a logarithmic scale for y. Hence estimate at what continuous rate of interest £200 due in 8 years has a present value of £100.

CHAPTER X

LOGARITHMIC DERIVATION

10.1 Derivatives of exponential and logarithmic functions.

THE introduction of the exponential and logarithmic functions makes it necessary to extend the list of standard derivative forms by the addition of the derivatives of the simplest of these functions, $y=e^x$ and $y=\log x$. The rules for derivation then apply, exactly as before, to give the derivatives of more complicated exponential and logarithmic functions and of combinations involving these functions.

Standard form derivatives are found from first principles, using the definition of the derivative and the properties of the function concerned. But the exponential and logarithmic functions are inverse to each other and we have only to find the derivative of one of them from first principles, the other derivative following from the inverse function rule. The derivative of the logarithmic function $y=\log x$ is found more easily from the definition :

$$\frac{\log(x+h)-\log x}{h}=\frac{1}{h}\log\left(\frac{x+h}{x}\right)=\frac{1}{x}\log\left(1+\frac{h}{x}\right)^{\frac{x}{h}}=\frac{1}{x}\log\left(1+\frac{1}{n}\right)^n,$$

where n denotes $\frac{x}{h}$ and tends to infinity as h tends to zero. But the expression $\left(1+\frac{1}{n}\right)^n$ tends to the limit e as n tends to infinity. So

$$\operatorname*{Lim}_{h\to 0}\frac{\log(x+h)-\log x}{h}=\frac{1}{x}\operatorname*{Lim}_{n\to\infty}\log\left(1+\frac{1}{n}\right)^n=\frac{1}{x}\log e=\frac{1}{x},$$

i.e. $$\frac{d}{dx}\log x=\frac{1}{x}.$$

The function inverse to $y=e^x$ is $x=\log y$ with derivative $\dfrac{dx}{dy}=\dfrac{1}{y}$.

Hence, the inverse function rule gives the derivative

$$\frac{d}{dx}e^x=\frac{1}{\frac{d}{dy}\log y}=y=e^x,$$

i.e.

$$\frac{d}{dx}e^x=e^x.$$

The derivative of the exponential or logarithmic function to a base a other than e can be deduced at once :

Since $$\log_a x=\log_a e\log_e x,$$

we have $$\frac{d}{dx}\log_a x=\log_a e\frac{d}{dx}\log_e x=\frac{1}{x}\log_a e.$$

Again $$a^x=(e^{\log_e a})^x=e^{x\log_e a},$$

and $$\frac{d}{dx}a^x=\frac{d}{du}e^u\frac{d}{dx}(x\log_e a)=e^u\log_e a,$$

where $u=x\log_e a$ and the function of a function rule is used. So

$$\frac{d}{dx}a^x=a^x\log_e a.$$

The derivative of the logarithm of any function of x or of any functional power of e can be deduced from the function of a function rule. If $u=f(x)$ is a single-valued function of x, then

$$\frac{d}{dx}\log u=\frac{d}{du}\log u\frac{du}{dx}=\frac{1}{u}\frac{du}{dx},$$

or $$\frac{d}{dx}\log f(x)=\frac{f'(x)}{f(x)}.$$

Similarly $$\frac{d}{dx}e^u=\frac{d}{du}e^u\frac{du}{dx}=e^u\frac{du}{dx},$$

or $$\frac{d}{dx}e^{f(x)}=f'(x)e^{f(x)}.$$

An important special case of the latter result is

$$\frac{d}{dx}e^{ax}=ae^{ax};$$

e.g. $$\frac{d}{dx}e^{-x}=-e^{-x}.$$

The list of *standard form derivatives*, as now constituted, is

$$\frac{d}{dx}x^n = nx^{n-1},$$

$$\frac{d}{dx}e^x = e^x, \qquad \frac{d}{dx}a^x = a^x \log_e a,$$

$$\frac{d}{dx}\log x = \frac{1}{x}, \qquad \frac{d}{dx}\log_a x = \frac{1}{x}\log_a e.$$

The following generalisations of the standard forms are also established :

$$\frac{d}{dx}u^n = nu^{n-1}\frac{du}{dx},$$

$$\frac{d}{dx}e^u = e^u\frac{du}{dx}, \qquad \frac{d}{dx}\log u = \frac{1}{u}\frac{du}{dx},$$

where u is any single-valued function of x.

The following examples illustrate the practical method of evaluating derivatives when the functions concerned involve exponential and logarithmic expressions :

Ex. 1. $\frac{d}{dx}e^{2x} = e^{2x}\frac{d}{dx}(2x) = 2e^{2x}.$

$$\frac{d}{dx}e^{x+1} = e^{x+1}\frac{d}{dx}(x+1) = e^{x+1}.$$

$$\frac{d}{dx}e^{1-2x} = e^{1-2x}\frac{d}{dx}(1-2x) = -2e^{1-2x}.$$

In general,

$$\frac{d}{dx}e^{ax+b} = ae^{ax+b}.$$

Ex. 2. $\frac{d}{dx}e^{x^2} = e^{x^2}\frac{d}{dx}(x^2) = 2xe^{x^2}.$

$$\frac{d}{dx}e^{-x^2} = e^{-x^2}\frac{d}{dx}(-x^2) = -2xe^{-x^2}.$$

$$\frac{d}{dx}e^{x^2+3x-2} = e^{x^2+3x-2}\frac{d}{dx}(x^2+3x-2) = (2x+3)\,e^{x^2+3x-2}.$$

In general,

$$\frac{d}{dx}e^{ax^2+bx+c} = (2ax+b)\,e^{ax^2+bx+c}.$$

Ex. 3. $\frac{d}{dx}x^2e^x = 2xe^x + x^2e^x = x(x+2)e^x.$

$$\frac{d}{dx}(x^2+3x-2)e^x = (2x+3)e^x + (x^2+3x-2)e^x = (x^2+5x+1)e^x.$$

In general,

$$\frac{d}{dx}(ax^2+bx+c)e^x = \{ax^2+(2a+b)x+(b+c)\}e^x.$$

Ex. 4. $\frac{d}{dx}\log(x+1) = \frac{1}{x+1}\frac{d}{dx}(x+1) = \frac{1}{x+1}.$

$$\frac{d}{dx}\log(1-2x) = \frac{1}{1-2x}\frac{d}{dx}(1-2x) = -\frac{2}{1-2x}.$$

In general,

$$\frac{d}{dx}\log(ax+b) = \frac{a}{ax+b}.$$

Ex. 5. $\frac{d}{dx}\log x^2 = \frac{1}{x^2}\frac{d}{dx}x^2 = \frac{1}{x^2}2x = \frac{2}{x}.$

$$\frac{d}{dx}\log(x^2+3x-2) = \frac{1}{x^2+3x-2}\frac{d}{dx}(x^2+3x-2)$$

$$= \frac{2x+3}{x^2+3x-2}.$$

In general,

$$\frac{d}{dx}\log(ax^2+bx+c) = \frac{2ax+b}{ax^2+bx+c}.$$

Ex. 6. $\frac{d}{dx}\log\left(\frac{1-2x}{1+x}\right) = \frac{d}{dx}\{\log(1-2x)-\log(1+x)\} = -\frac{2}{1-2x}-\frac{1}{1+x}$

$$= -\frac{3}{(1+x)(1-2x)}.$$

Ex. 7. $\frac{d}{dx}x^2\log x = 2x\log x + x^2\frac{1}{x} = x(1+2\log x).$

$$\frac{d}{dx}x^n\log x = x^{n-1}(1+n\log x).$$

Ex. 8. $\frac{d}{dx}(\log x)^2 = 2\log x\frac{d}{dx}\log x = \frac{2\log x}{x}.$

$$\frac{d}{dx}(\log x)^n = \frac{n(\log x)^{n-1}}{x}.$$

Ex. 9. Derivatives of the second and higher orders are to be obtained by successive applications of the derivation process. Thus

$$\frac{d}{dx}\log x=\frac{1}{x};\quad \frac{d^2}{dx^2}\log x=-\frac{1}{x^2};\quad \frac{d^3}{dx^3}\log x=\frac{2}{x^3};\text{ and so on.}$$

In general, the following results can be established:

$$\frac{d^r}{dx^r}x^n=n(n-1)(n-2)\dots(n-r+1)x^{n-r}.$$

$$\frac{d^r}{dx^r}e^x=e^x.$$

$$\frac{d^r}{dx^r}\log x=(-1)^{r-1}\frac{2.3.4\dots(r-1)}{x^r}.$$

10.2 Logarithmic derivation.

If $y=f(x)$ is any single-valued function of x, then

$$\frac{d(\log y)}{dx}=\frac{1}{y}\frac{dy}{dx},$$

an important result with many practical uses. The proportional change in the value of the function as x increases from x to $(x+h)$ is

$$\frac{f(x+h)-f(x)}{f(x)}.$$

The rate of proportional change in the value of the function at the point x is thus

$$\lim_{h\to 0}\frac{f(x+h)-f(x)}{hf(x)}=\frac{1}{f(x)}\lim_{h\to 0}\frac{f(x+h)-f(x)}{h}=\frac{1}{y}\frac{dy}{dx}=\frac{d(\log y)}{dx}.$$

The derivative $\frac{d(\log y)}{dx}=\frac{1}{y}\frac{dy}{dx}$ thus serves to measure the rate of proportional change of the function. In diagrammatic terms, just as $\frac{dy}{dx}$ measures the tangent gradient of the curve $y=f(x)$ plotted on natural scales, so $\frac{d(\log y)}{dx}$ measures the tangent gradient of the curve when plotted as a semi-logarithmic graph. In fact, whenever we consider proportional changes in the function $y=f(x)$ as x changes, we use a semi-logarithmic graph and the appropriate derivative $\frac{d(\log y)}{dx}$.

The logarithmic derivative provides the simplest method of

finding the derivative of a product or quotient. In general terms, if $y=uv$, where u and v are two single-valued functions of x, then

$$\log y=\log(uv)=\log u+\log v,$$

and $$\frac{d}{dx}\log y=\frac{d}{dx}\log u+\frac{d}{dx}\log v,$$

i.e. $$\frac{1}{y}\frac{dy}{dx}=\frac{1}{u}\frac{du}{dx}+\frac{1}{v}\frac{dv}{dx}.$$

Again, if $y=\frac{u}{v}$, we have

$$\log y=\log\left(\frac{u}{v}\right)=\log u-\log v,$$

and $$\frac{d}{dx}\log y=\frac{d}{dx}\log u-\frac{d}{dx}\log v,$$

i.e. $$\frac{1}{y}\frac{dy}{dx}=\frac{1}{u}\frac{du}{dx}-\frac{1}{v}\frac{dv}{dx}.$$

These two results are equivalent to the product and quotient rules for derivation given above and we have now established the rules with the use only of the sum or difference rule and the notion of logarithmic derivation.

A completely general form of the product and quotient rules can be given by adopting the same method of logarithmic derivation. If $y=\frac{u_1u_2u_3\ldots}{v_1v_2v_3\ldots}$, where u_1, u_2, u_3, ... and v_1, v_2, v_3, ... are any given single-valued functions of x, then

$$\log y=\log u_1+\log u_2+\log u_3+\ldots-\log v_1-\log v_2-\log v_3-\ldots.$$

So $$\frac{1}{y}\frac{dy}{dx}=\frac{1}{u_1}\frac{du_1}{dx}+\frac{1}{u_2}\frac{du_2}{dx}+\frac{1}{u_3}\frac{du_3}{dx}+\ldots-\frac{1}{v_1}\frac{dv_1}{dx}-\frac{1}{v_2}\frac{dv_2}{dx}-\frac{1}{v_3}\frac{dv_3}{dx}-\ldots.$$

Further, it is often more convenient, in practice, to evaluate a derivative by writing the logarithm of the function before proceeding to differentiate. In fact, we prefer to carry out the steps of the logarithmic derivation process rather than quote the rules for the derivatives of products or quotients. This remark holds particularly of the derivation of a function which involves complicated products, quotients or powers. The following examples illustrate :

Ex. 1. $y=\frac{x}{(x+1)(x+2)}$.

Here $$\log y = \log x - \log(x+1) - \log(x+2),$$

and $$\frac{1}{y}\frac{dy}{dx} = \frac{1}{x} - \frac{1}{x+1} - \frac{1}{x+2} = -\frac{x^2-2}{x(x+1)(x+2)},$$

i.e. $$\frac{dy}{dx} = \frac{-x}{(x+1)(x+2)}\,\frac{x^2-2}{x(x+1)(x+2)} = -\frac{x^2-2}{(x+1)^2(x+2)^2}.$$

Ex. 2. $y = x^2\sqrt{\dfrac{2x-1}{x+1}}$.

Here $$\log y = 2\log x + \tfrac{1}{2}\log(2x-1) - \tfrac{1}{2}\log(x+1),$$

and $$\frac{1}{y}\frac{dy}{dx} = \frac{2}{x} + \frac{1}{2}\frac{2}{2x-1} - \frac{1}{2}\frac{1}{x+1} = \frac{8x^2+7x-4}{2x(x+1)(2x-1)},$$

i.e. $$\frac{dy}{dx} = x^2\sqrt{\frac{2x-1}{x+1}}\,\frac{8x^2+7x-4}{2x(x+1)(2x-1)} = \frac{x(8x^2+7x-4)}{2(x+1)\sqrt{(x+1)(2x-1)}}.$$

Ex. 3. $y = x^n e^{-\frac{1}{2}(x-a)^2}$, where a and n are constants.

Here $$\log y = n\log x - \tfrac{1}{2}(x-a)^2,$$

and $$\frac{1}{y}\frac{dy}{dx} = \frac{n}{x} - (x-a) = \frac{1}{x}(n + ax - x^2).$$

i.e. $$\frac{dy}{dx} = (n + ax - x^2)x^{n-1}e^{-\frac{1}{2}(x-a)^2}.$$

10.3 A problem of capital and interest.

The exponential and logarithmic functions are important, as we have seen, in economic problems of capital and interest. A simple case can be considered here as an example of the type of analysis involved.* A good is produced at a given moment of time ($t=0$) at a fixed and known cost. The good is not sold at once but is stored for a variable length of time. The value of the good increases in a known way as time goes on, the selling price per unit being £$f(t)$ after t years. The function $f(t)$ is assumed to be known and to increase with t. As concrete, though necessarily highly simplified, examples of the kind of problem considered, we can quote the obvious cases of wine and timber. New wine is bought by a dealer and laid down for sale at a later time. Or a given piece of land is put under timber at a given time and the timber is to be cut for sale

* The development given here is based on the work of Wicksell, *Lectures on Political Economy*, Vol. I (Ed. Robbins, 1934), pp. 172-84, and of Fisher, *The Theory of Interest* (1930), pp. 161-5.

at a later time. The problem is to determine the optimum time for selling the wine or cutting the timber.

It is assumed that interest is added continuously at the fixed market rate of $100r$ per cent. over the periods concerned. Then, if £x denotes the present value per unit of the good sold after t years,

$$x = f(t)e^{-rt} \qquad \text{(see 9.7 above).}$$

It is assumed that the storage period of t years is chosen to maximise the net profit as discounted to the base time ($t=0$), i.e. to maximise the value of x since the cost of production is constant. The optimum storage period is given by

$$\frac{dx}{dt}=0 \quad \text{and} \quad \frac{d^2x}{dt^2}<0.$$

Now $$\log x = F(t) - rt \quad \text{where} \quad F(t) = \log f(t),$$

and $$\frac{1}{x}\frac{dx}{dt} = F'(t) - r,$$

$$\frac{d}{dt}\left(\frac{1}{x}\frac{dx}{dt}\right) = \frac{1}{x}\frac{d^2x}{dt^2} - \frac{1}{x^2}\left(\frac{dx}{dt}\right)^2 = F''(t).$$

The conditions for the optimum storage period are thus

$$F'(t) = r \quad \text{and} \quad F''(t) < 0.$$

The first condition is that $F'(t) = \frac{d}{dt}\log f(t) = \frac{f'(t)}{f(t)}$ is equal to the fixed rate of interest. At the optimum time of sale, therefore, the rate of proportional increase in the value of the good over time is equal to the market rate of interest. We find that it is the *marginal* variation of value that is the important factor. Further,

$$F''(t) = \frac{d^2}{dt^2}\log f(t) = \frac{d}{dt}\left\{\frac{f'(t)}{f(t)}\right\} < 0,$$

i.e. the rate of proportional increase of value must be decreasing at the optimum time of sale.

Fig. 69 shows the solution of the problem in diagrammatic terms. The curve represents the function $y=f(t)$ on a semi-logarithmic diagram, $\log y$ being taken against t. The zero point on the horizontal axis Ot is the time ($t=0$) when the initial cost of production is incurred and that on the vertical axis is taken, for convenience, as $y=a$, where £a is the fixed cost of production. The gradient of the tangent to the curve at the point P (at time t) is given by

$$\frac{d(\log y)}{dt}=\frac{f'(t)}{f(t)}.$$

Interest is reckoned continuously at the given rate of $100r$ per cent. per year and $y=xe^{rt}$ represents a system of discount curves, x being a parameter. Here, £x is the discounted value (at $t=0$) of £y available at time t. The system appears on the diagram as a set of parallel straight lines ($\log y = \log x+rt$) with gradient r and cutting the vertical axis at $y=x$. One line of the set passes through the point P and cuts the vertical at Q where

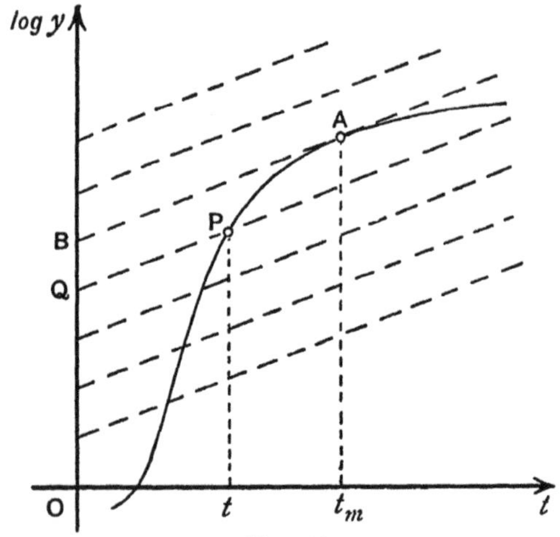

FIG. 69.

$$OQ=\log x-\log a=\log\frac{x}{a},$$

since the origin is at $y=a$. £x is now the discounted value of a unit of the good sold at time t and the optimum time of sale ($t=t_m$) is obtained when x or OQ is a maximum, i.e. it is given by the point A where a discount line touches the curve $y=f(t)$. At $t=t_m$, the tangent gradient of the curve equals the gradient of the discount line, i.e. $\frac{f'(t)}{f(t)}=r$. Further, for a genuine maximum, the curve must be concave from below, i.e. $\frac{d^2(\log y)}{dt^2}=\frac{d}{dt}\left\{\frac{f'(t)}{f(t)}\right\}<0$. This condition is satisfied in the "normal" case shown, the value of $f(t)$ increasing rapidly at first and then more slowly. The point B, where the tangent at A cuts the vertical axis, must lie above O if the discounted sale value of the good is to be greater than the cost incurred.

10.4 The elasticity of a function.

We have seen that $\frac{d(\log y)}{dx}$ measures the rate of proportional change of the function $y=f(x)$, the proportional change in y being related to the absolute change in x. Suppose, now, that we relate proportional changes in both variables. If x is changed from x to $(x+h)$, the proportional changes in x and y are $\frac{h}{x}$ and $\frac{f(x+h)-f(x)}{f(x)}$ respectively. The average proportional change in y per unit proportional change in x is

$$\frac{x}{f(x)}\frac{f(x+h)-f(x)}{h}.$$

If the derivative of the function exists, the rate of proportional change in y for proportional changes in x is

$$\operatorname*{Lim}_{h\to 0}\left\{\frac{x}{f(x)}\frac{f(x+h)-f(x)}{h}\right\}=\frac{xf'(x)}{f(x)}=\frac{x}{y}\frac{dy}{dx}.$$

Denoting the logarithms of x and y by u and v, we can write

$$v=\log y, \quad \text{where} \quad y=f(x), \quad \text{where} \quad x=e^u,$$

and
$$\frac{dv}{du}=\frac{dv}{dy}\frac{dy}{dx}\frac{dx}{du}=\frac{1}{y}\frac{dy}{dx}e^u=\frac{x}{y}\frac{dy}{dx}.$$

The rate of change of $y=f(x)$, when both changes are expressed in proportional terms, is thus measured by

$$\frac{d(\log y)}{d(\log x)}=\frac{x}{y}\frac{dy}{dx}.$$

This rate is termed the *elasticity* of the function at the point x and can be denoted by $\frac{Ey}{Ex}=\frac{E}{Ex}f(x)$.* Hence :

DEFINITION : The elasticity of the function $y=f(x)$ at the point x is the rate of proportional change in y per unit proportional change in x :

$$\frac{Ey}{Ex}=\frac{d(\log y)}{d(\log x)}=\frac{x}{y}\frac{dy}{dx}.$$

* The notation adopted here is that suggested by Champernowne, *A Mathematical Note on Substitution*, Economic Journal, 1935. An alternative notation is $E_x(y)=E_x\{f(x)\}$. No established notation for elasticities is in current use.

A logarithmic diagram is the appropriate representation of the function $y=f(x)$ when proportional changes in both x and y are considered. The gradient of the tangent to the curve shown on the diagram is then $\frac{d(\log y)}{d(\log x)}$. Hence, the elasticity of a function at various points is given by the tangent gradient of the corresponding curve when logarithmic scales are taken for both variables: the elasticity can be read off the logarithmic graph of the function just as the derivative is read off the natural graph.

The important property of the elasticity of a function is that it is a number which is independent of the units in which the variables are measured. This is clear since the elasticity is defined in terms of proportional changes which are necessarily independent of units. More formally, if units are changed, so that the new measures of x and y are $x'=\lambda x$ and $y'=\mu y$ respectively, then

$$\frac{x'}{y'}\frac{dy'}{dx'}=\frac{\lambda x}{\mu y}\frac{d(\mu y)}{d(\lambda x)}=\frac{\lambda x}{\mu y}\frac{\mu}{\lambda}\frac{dy}{dx}=\frac{x}{y}\frac{dy}{dx},$$

i.e. the elasticity of the function is unaltered.

10.5 The evaluation of elasticities.

Since the elasticity of a function is simply the derivative $\frac{dy}{dx}$ with a multiplicative factor $\frac{x}{y}$, which makes it independent of units, we can obtain elasticities from the corresponding derivatives. It is interesting, however, to translate the derivative rules into elasticity form and to examine the nature of the elasticities of simple functions. If u and v are single-valued functions of x, then

$$\frac{E(u\pm v)}{Ex}=\frac{u\frac{Eu}{Ex}\pm v\frac{Ev}{Ex}}{u\pm v},$$

$$\frac{E(uv)}{Ex}=\frac{Eu}{Ex}+\frac{Ev}{Ex},$$

$$\frac{E\left(\frac{u}{v}\right)}{Ex}=\frac{Eu}{Ex}-\frac{Ev}{Ex}.$$

Further, if y is a function of u, where u is a function of x, then

$$\frac{Ey}{Ex}=\frac{Ey}{Eu}\frac{Eu}{Ex}.$$

As an example of the method of proving these results, we have

$$\frac{E(uv)}{Ex}=\frac{x}{uv}\frac{d(uv)}{dx}=\frac{x}{uv}\left(v\frac{du}{dx}+u\frac{dv}{dx}\right)=\frac{x}{u}\frac{du}{dx}+\frac{x}{v}\frac{dv}{dx}=\frac{Eu}{Ex}+\frac{Ev}{Ex}.$$

The other results are established in a similar way.

It is to be noticed that the elasticity rules are simplest in the cases of product and quotient and more complicated for sum and difference. This completely reverses the position obtained in the case of the derivative rules. But it is no more than we expect since an elasticity is a derivative in which the variables are expressed as logarithms and since logarithms are designed to deal conveniently with products and quotients. Further, since the elasticity of a constant is zero,

$$\frac{E(u+a)}{Ex}=\frac{u}{u+a}\frac{Eu}{Ex} \quad \text{and} \quad \frac{E(au)}{Ex}=\frac{Eu}{Ex},$$

i.e. it is the multiplicative constant that disappears in an elasticity.

As examples of the elasticities of simple functions, we have :

$$\frac{E}{Ex}(ax+b)=\frac{ax}{ax+b}, \quad \frac{E}{Ex}(ax^{\alpha})=\alpha \quad \text{and} \quad \frac{E}{Ex}(ae^{\alpha x})=\alpha x.$$

Hence it is the power function, and not the linear function, which has a constant elasticity. If α is a positive constant, then the function with the same positive elasticity α at all points is $y=ax^{\alpha}$, and the function with the same negative elasticity $(-\alpha)$ at all points is $yx^{\alpha}=a$. In particular, the function $y=ax$ has elasticity 1 at all points and the function $xy=a$ has elasticity (-1) at all points. In diagrammatic terms, the function $y=ax^{\alpha}$ is represented by an upward sloping straight line with gradient α on logarithmic scales and the gradient measures the constant elasticity of the function. In the same way, the function $yx^{\alpha}=a$ is shown as a downward sloping line with numerical gradient α on logarithmic scales. The rectangular hyperbola $xy=a$ is a particular case, shown on a logarithmic diagram by a line sloping downward with gradient and elasticity numerically equal to unity.

The elasticity of a function varies, in general, with the value of x taken. Certain points may be found where the numerical value of the elasticity is unity and these points are of particular interest. If $\frac{E}{Ex}\{f(x)\}=1$ at a point x, then a proportional increase in x from this point gives rise to an equal proportional increase in $f(x)$. On the other hand, if $\frac{E}{Ex}\{f(x)\}>1$ at the point x, then a proportional increase in x results in a greater proportional increase in $f(x)$. Conversely, if $\frac{E}{Ex}\{f(x)\}<1$ at the point x, the proportional increase in $f(x)$ is less than that in x. Similar remarks apply when $\frac{E}{Ex}\{f(x)\}=-1$ at a point x, but here proportional increases in x correspond to proportional decreases in $f(x)$.

Further, at a stationary value of a function, the derivative, and so the elasticity, is zero. As particular cases of the product and quotient rules for elasticities, we have

$$\frac{E}{Ex}\{xf(x)\}=\frac{E}{Ex}\{f(x)\}+1\ ; \quad \frac{E}{Ex}\left\{\frac{f(x)}{x}\right\}=\frac{E}{Ex}\{f(x)\}-1.$$

It follows that a maximum or minimum value of the "total" expression $\{xf(x)\}$ can occur only at a point where the elasticity of $f(x)$ is -1 and that a maximum or minimum value of the "average" expression $\left\{\frac{f(x)}{x}\right\}$ can occur only where the elasticity of $f(x)$ is 1. Finally, it is clear, from the definition, that the elasticity of a function $f(x)$ is the ratio of the "marginal" value of $f(x)$ to the "average" value of $f(x)$ at the point in question, i.e. the ratio of $f'(x)$ to $\frac{f(x)}{x}$. Hence, at a point where the "average" value is a maximum or minimum and the elasticity is equal to 1, we have also that the "average" and "marginal" values of the function are equal (see 8.5 above). Points where the elasticity of a function is numerically equal to unity are thus of considerable importance. This is particularly evident, as we shall show, when we consider the application of the elasticity concept to the demand and cost problems of economic theory.

10.6 The elasticity of demand.

The market demand for a good can be represented, under certain conditions, by the monotonic decreasing function $x=\phi(p)$. The elasticity of this function defines the elasticity of demand at any price. Since its value is negative at all prices, it is found convenient to make it positive by the introduction of a negative sign. This step does not affect the elasticity concept in any material way. So :

$$\textit{Elasticity of demand } \eta = -\frac{p}{x}\frac{dx}{dp} = -\frac{d(\log x)}{d(\log p)}.$$

The value of η, which is independent of both price and quantity units, varies from point to point and always measures the rate of proportional decrease of demand for proportional increases in price from the price and demand in question.*

When we consider proportional changes in demand and price, it is a great advantage to plot the demand curve on a logarithmic diagram, taking both price and demand on a logarithmic scale. The tangent to the demand curve is then downward sloping with a numerical gradient equal to the elasticity of demand. For example, the demand law $x = ap^{-\alpha}$ is shown, on a logarithmic diagram, as a straight line with gradient $(-\alpha)$. The elasticity of demand, in this case, is constant and equal to α at all points.

If demand curves are drawn on natural scales, as is usual in economic works, it becomes more difficult to estimate and compare the elasticities at various points on the same or on different demand curves. It is tempting to estimate the elasticity from the gradient of the demand curve, to say that a demand curve steeply inclined to the price axis has a large elasticity. This is incorrect. Demand elasticity is the product of $\frac{p}{x}$ and $\left(-\frac{dx}{dp}\right)$ and its value cannot be judged solely from the second of these two factors. For example, the linear demand curve has a constant gradient but its elasticity is not constant. The elasticity, as is easily seen, decreases as the price decreases and the demand increases.

Two diagrammatic methods have been devised by Marshall for the purpose of estimating the elasticity of a demand curve drawn on

* The inverse demand function is $p = \psi(x)$ and the elasticity of price with respect to demand is $-\frac{x}{p}\frac{dp}{dx}$, the reciprocal of η. Some writers use the term *flexibility of price* to indicate the reciprocal concept (see Moore, *Synthetic Economics*, 1929, p. 38). Notice also that the elasticity of demand and flexibility of price are sometimes written without the negative sign (see Schultz, *Statistical Laws of Demand and Supply*, 1928).

natural scales.* The first method makes use of a simple geometrical property of the demand curve. With the notation of Fig. 70, tPT is the tangent at any point P on the demand curve and the elasticity of demand at P is

$$\eta=\frac{p}{x}\left(-\frac{dx}{dp}\right)=\frac{MP}{OM}\frac{MT}{MP}=\frac{MT}{OM}.$$

A well-known geometrical property of parallel lines gives $\frac{MT}{OM}=\frac{PT}{tP}$ (since MP is parallel to Op) and $\frac{PT}{tP}=\frac{ON}{Nt}$ (since NP is parallel to Ox). Hence:

$$\eta=\frac{MT}{OM}=\frac{ON}{Nt}=\frac{PT}{tP}.$$

We have thus three convenient length ratios as measures of the demand elasticity at any point on a demand curve.

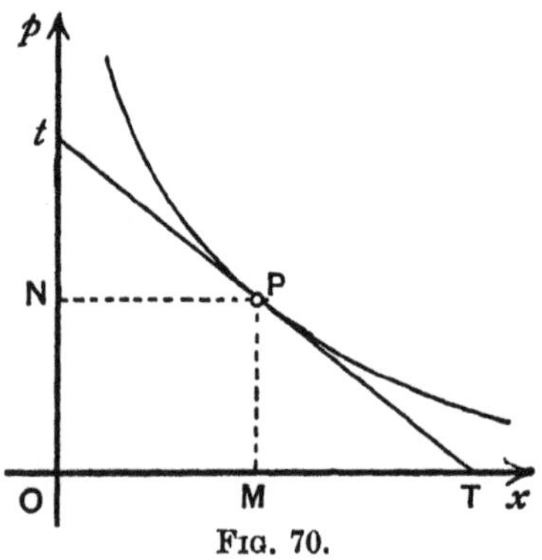

FIG. 70.

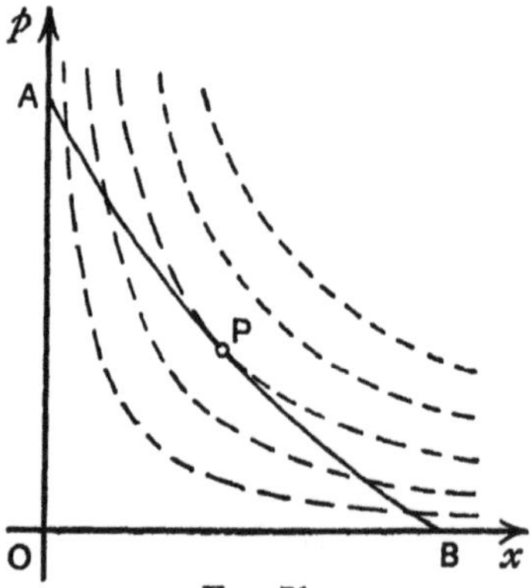

FIG. 71.

The second method depends on the fact that any curve with equation $xp=a$ is a demand curve with unit elasticity at all points. Such a curve can be called a *constant outlay curve*, since the outlay (xp) of consumers is constant at all prices. On logarithmic scales, the curves for various values of a form a set of parallel straight lines sloping downward with unit gradient. On natural scales, we have the system of rectangular hyperbolas shown in Fig. 71. Through each point P of a given demand curve AB, there passes one of the constant outlay curves. The elasticity of the demand curve AB at P can be compared with the unit elasticity of this constant outlay curve by comparison of the gradients of the two curves referred to

* Marshall, *Principles of Economics* (8th Ed. 1927), p. 839.

the price axis. This is possible since the other factor $\frac{p}{x}$ in the elasticity expression is the same for both curves at P. If the demand curve AB is steeper to the price axis than the constant outlay curve at P, then the demand elasticity is greater than unity at this point on AB. If AB is less steep than the constant outlay curve at P, then the demand elasticity of AB is less than unity. Finally, if AB touches the constant outlay curve at P (as shown in Fig. 71), then AB has unit elasticity of demand at P. Hence, the path of a given demand curve across the constant outlay curves gives us a good idea of the variation of demand elasticity along the demand curve.

10.7 Normal conditions of demand.

For any demand law $p=\psi(x)$, we have $R=xp=x\psi(x)$, and

$$\frac{dR}{dx}=\frac{d}{dx}(xp)=p+x\frac{dp}{dx}=p\left(1+\frac{x}{p}\frac{dp}{dx}\right).$$

But the elasticity of demand $\eta=-\frac{p}{x}\frac{dx}{dp}$ and so

$$\frac{dR}{dx}=p\left(1-\frac{1}{\eta}\right).$$

This is an important result. From it we deduce:

(1) If $\eta>1$ at a given price and demand, a small decrease in price results in a more than proportional increase in demand, marginal revenue $\frac{dR}{dx}$ is positive and total revenue increases as output (=demand) increases. This is the case of elastic demand.

(2) If $\eta=1$ at a given price and demand, the small decrease in price and increase in demand are proportionally equal, marginal revenue is zero and total revenue has a stationary (usually a maximum) value.

(3) If $\eta<1$ at a given price and demand, a small decrease in price is accompanied by a less than proportional increase in demand, marginal revenue is negative and total revenue decreases as output increases. This is the case of inelastic demand.

Our general idea of elastic and inelastic demands (see 5.4 above) is thus in agreement with the precise measure of demand elasticity.

We have assumed, in the normal case of demand, that price decreases are associated with demand increases, and conversely, i.e.

that the demand curve falls from left to right. A further property of demand is now assumed to hold in the normal case. It is taken that, as the demand increases, the elasticity of demand decreases continuously from values greater than unity at small demands to values less than unity at large demands. As demand increases it becomes continuously more inelastic. It follows that there is one definite demand $x=a$ where the demand elasticity is equal to unity. This normal property of the demand curve is illustrated by the curve *AB* of Fig. 71. The curve touches a constant outlay curve at one definite demand, is steeper to the price axis than the constant outlay curves for all smaller demands and less steep than the constant outlay curves for all larger demands.

In the normal case of demand, therefore, total revenue R increases with output at first ($\eta > 1$), reaches a definite maximum value at the

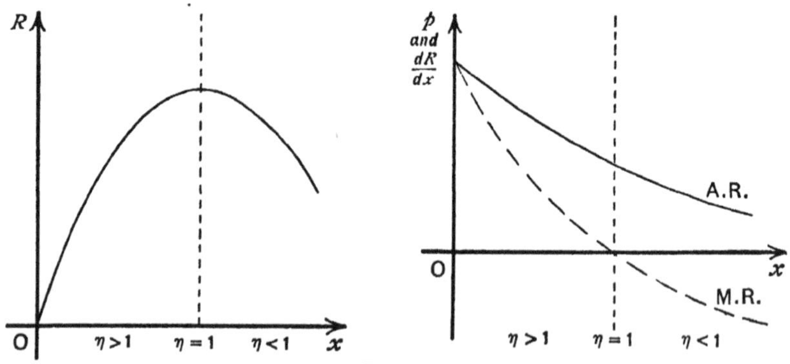

FIG. 72.

output $x=a$ ($\eta=1$) and then decreases as output increases further ($\eta<1$). Average revenue, of course, decreases continuously as output increases. Now, from the expression for $\frac{dR}{dx}$ above,

$$\text{marginal revenue} = \text{average revenue}\left(1-\frac{1}{\eta}\right).$$

Since η is positive, marginal revenue is less than average revenue at all outputs. Further, as output increases, η decreases and becomes unity at the output $x=a$. Hence, for outputs less than a, $\left(1-\frac{1}{\eta}\right)$ is positive and decreasing. So marginal revenue must decrease as output increases until it becomes zero at the output $x=a$ where $\eta=1$.

For further increases in output, marginal revenue is negative but need not decrease continuously. The normal forms of the total, average and marginal revenue curves are indicated in the two diagrams of Fig. 72. The diagrams must be consistent in the sense that the ordinates of the average and marginal revenue curves are respectively equal to the gradient of the vector from O and the gradient of the tangent at the corresponding point on the total revenue curve.

The linear demand law, $p=a-bx$, and the hyperbolic demand law, $p=\frac{a}{x+b}-c$, are both of the normal form here described. In the first case, $\eta=b\left(\frac{a}{x}-b\right)$ and, in the second case,

$$\eta=\frac{1}{a}\left(1+\frac{b}{x}\right)(1-bc-cx).$$

Each of these expressions decreases continuously as x increases. The total, average and marginal revenue curves in particular cases of linear and hyperbolic demands are graphed in Figs. 44 and 45 above. It is clear, from these graphs, that the demand curves are of normal form.

In conclusion, an interesting connection between the average and marginal revenue curves can be established.* If the tangent at any point P on the average revenue curve cuts the price axis in A and if A is joined to the point Q on the marginal revenue curve at the same output as P, then the gradient of AP referred to the price axis is twice that of AQ. In Fig. 73, PM and QN are perpendicular to the price axis. Then (10.6 above):

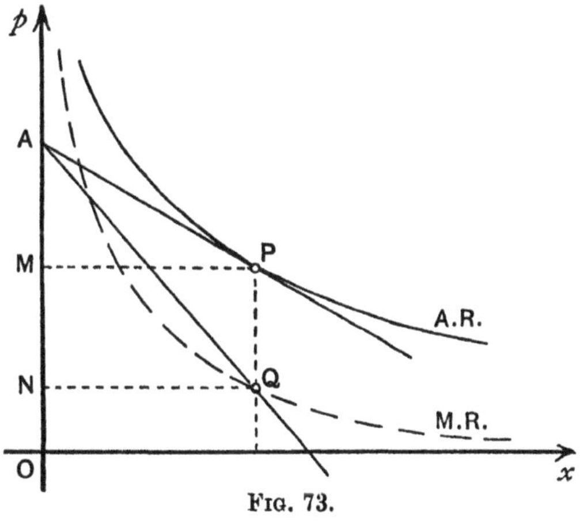

FIG. 73.

$$\eta=\frac{OM}{MA}=\frac{p}{MA}, \quad \text{i.e. } MA=\frac{p}{\eta},$$

* See Robinson, *The Economics of Imperfect Competition* (1933), pp. 29 *et seq.*

where η is the demand elasticity at P. But marginal revenue equals average revenue times $\left(1-\frac{1}{\eta}\right)$, i.e.

$$ON = OM\left(1-\frac{1}{\eta}\right),$$

and $$NM = OM - ON = \frac{OM}{\eta} = \frac{p}{\eta}.$$

So $$MA = NM = \tfrac{1}{2}NA.$$

The gradient of AP to the price axis is $\frac{MP}{MA} = \frac{NQ}{\frac{1}{2}NA} = 2\frac{NQ}{NA}$, i.e. twice the gradient of AQ to the price axis.

This result provides a method of tracing the marginal revenue curve from a given demand (or average revenue) curve. Select any point P on the demand curve, draw the tangent to cut the price axis in A, draw the line AQ with gradient to the price axis half that of AP and find the point Q on AQ with the same output as at P. Then Q is a point on the marginal revenue curve and the whole marginal revenue curve is traced by repeating the process for different points on the given demand curve.

10.8 Cost elasticity and normal cost conditions.

A less generally recognised, but extremely useful, application of the elasticity concept is found in the analysis of the cost problem (5.5 above). This application provides a striking contrast to the use of demand elasticity. The latter refers to the average quantity (average revenue) and enables us to determine, by relating its value to unity, whether the corresponding total quantity (total revenue) increases or decreases. The position is completely different in the cost problem. Here the total quantity (total cost) increases for all outputs and its elasticity is defined and used to deduce properties of the average quantity (average cost).

If a firm produces an output x at a total cost $\Pi = F(x)$, then

$$\textit{Elasticity of total cost } \kappa = \frac{x}{\Pi}\frac{d\Pi}{dx} = \frac{d(\log \Pi)}{d(\log x)}.$$

The value of κ is independent of both cost and output units and measures the proportional rate of increase of total cost for proportional increases in output from the level in question.* From the definition κ is equal to the ratio of marginal to average cost. Further, the elasticity of average cost $\pi=\frac{\Pi}{x}$ can be expressed:

$$\frac{x}{\pi}\frac{d\pi}{dx}=\frac{x^2}{\Pi}\frac{d}{dx}\left(\frac{\Pi}{x}\right)=\frac{x^2}{\Pi}\frac{1}{x^2}\left(x\frac{d\Pi}{dx}-\Pi\right)=\frac{x}{\Pi}\frac{d\Pi}{dx}-1=\kappa-1.$$

From these results, it follows that:

(1) If $\kappa<1$ at a given output, we have the case of increasing returns in which a small increase in output is obtained at a less than proportional increase in cost, average cost is greater than marginal cost and average cost decreases as output increases.

(2) If $\kappa=1$ at a given output, the case is one of constant returns in which the small increase in output is proportionally equal to the increase in cost, average cost equals marginal cost and average cost has a stationary (usually a minimum) value.

(3) If $\kappa>1$ at a given output, we have decreasing returns and the situation is exactly the reverse of that when $\kappa<1$.

It has been assumed, in the normal case of cost conditions, that total cost increases continuously from a positive value (representing overhead costs) as output increases from zero. A further property of the normal case is now added, the property that cost elasticity increases continuously from values less than unity at small outputs to values greater than unity at large outputs. Returns become increasingly unfavourable as output increases. There is one definite output $x=a$, where cost elasticity equals unity and returns cease to be increasing and become decreasing.†

It follows, in the normal case, that average cost falls with increasing output at first, reaches a minimum value at the output $x=a$ and then increases as output increases further. Marginal cost is less than average cost for outputs less than a and greater than average

* The elasticity of total cost is described by Professor Moore (*Synthetic Economics*, p. 77) as the "coefficient of relative cost of production" and its reciprocal as the "coefficient of relative efficiency of organisation". The reciprocal of κ has also been used by W. E. Johnson (*The Pure Theory of Utility Curves*, Economic Journal, 1913, p. 508) and by Professor Bowley (*The Mathematical Groundwork of Economics*, 1924, p. 32).

† As a slightly less severe normal assumption it can be taken simply that $\kappa=1$ at some definite output $x=a$, $\kappa<1$ for outputs $x<a$ and $\kappa>1$ for outputs $x>a$.

cost for outputs greater than a. At the output $x=a$, marginal cost is rising and equal to the (minimum) average cost. The normal forms of the total, average and marginal cost curves are shown in the two diagrams of Fig. 74. The forms of the average and marginal

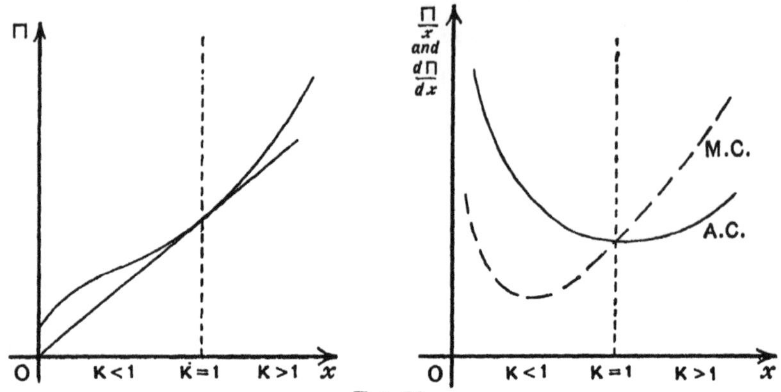

Fig. 74.

cost curves can be checked from the form of the total cost curve, and conversely. If P is the point on the total cost curve at a given output, then average and marginal cost are to be read off as the gradients of OP and of the tangent at P respectively.

The simplest cost function satisfying the normal cost conditions is the quadratic form

$$\Pi = ax^2 + bx + c$$

where a, b and c are positive constants. Here

$$\kappa = \frac{x(2ax+b)}{ax^2+bx+c}.$$

The derivative of κ can be shown to be positive, i.e. κ increases as x increases. The total, average and marginal cost curves in the case of a particular quadratic cost function are graphed in Fig. 46 and they are seen to be of normal form.

It may be convenient to take the total cost curve with a point of inflexion and the marginal cost curve with a minimum point at some output less than that at which average cost is a minimum. This is shown in Fig. 74. The quadratic cost function does not exhibit this additional property and the simplest normal cost function to do so is the cubic form

$$\Pi = ax^3 - bx^2 + cx + d$$

where a, b, c and d are positive constants ($b^2 < 3ac$). Here

$$\text{Average cost} = ax^2 - bx + c + \frac{d}{x},$$

$$\text{Marginal cost } \frac{d\Pi}{dx} = 3ax^2 - 2bx + c$$

and
$$\frac{d^2\Pi}{dx^2} = 6ax - 2b.$$

Since $\frac{d\Pi}{dx}$ is positive for all values of x (since $b^2 < 3ac$) and since $\frac{d^2\Pi}{dx^2} = 0$ at the single output $x = \frac{b}{3a}$, the total cost curve rises but has a point of inflexion at the output $x = \frac{b}{3a}$. The average cost curve has a single minimum value at an output greater than $x = \frac{b}{3a}$ (see 8.4, Ex. 7). The marginal cost curve is a parabola with a positive minimum value $\frac{3ac - b^2}{3a}$ at the output $x = \frac{b}{3a}$. The curves are thus of the normal form shown in Fig. 74.

As a final illustration of the elasticity concept, we can consider a problem very similar, in some respects, to the cost problem just discussed. The output of a firm, under given technical conditions, is determined by the amounts of the various factors of production employed. It is now assumed that the factors are always employed in the same fixed proportions. The output x of the good produced is then uniquely dependent upon the proportion λ by which the factors are increased ($\lambda > 1$) or decreased ($\lambda < 1$) from a given basic position. We can, therefore, take x as a function of λ and define

$$\textit{Elasticity of productivity } \epsilon = \frac{\lambda}{x}\frac{dx}{d\lambda} = \frac{d(\log x)}{d(\log \lambda)}.$$

If $\epsilon > 1$, we have the case of increasing returns in which a small proportionate increase in all factors employed results in a more than proportional increase in output. In the same way, $\epsilon = 1$ and $\epsilon < 1$ correspond to cases of constant and decreasing returns. In the normal case, ϵ decreases continuously as λ, and the amount of the factors employed, increases. It is to be noticed that the reciprocal of ϵ is analogous to κ in that it measures the elasticity of cost with respect to output, cost being expressed in terms of the amount of the factors used instead of in money terms. But ϵ is limited, as κ is not, to the case of factors used in fixed proportions.*

EXAMPLES X

Exponential and logarithmic derivatives

1. Obtain the derivatives of the functions: 2^x, e^{1-x}, $\frac{1}{\sqrt{e^x}}$, $e^{\frac{1}{x}}$, e^{1+x}, x^2e^{-x}, xe^{-x^2}, $\log(1-x)$, $\log(x+1)(x+2)$, $\log\sqrt{1-x^2}$, $\log(1+x^3)$, $\log\frac{x}{1-x}$, $\log\frac{1+x^2}{1-x^2}$ and $\log\frac{1}{\sqrt{x}}$.

2. By writing $e^{ax+b}=e^b(e^x)^a$, verify that $\frac{d}{dx}e^{ax+b}=ae^{ax+b}$.

3. Find the derivatives of $(ax+b)e^x$ and $(ax+b)e^{-x}$.

4. Show that $\frac{d}{dx}\{\frac{1}{2}(e^x+e^{-x})^2\}=e^{2x}-e^{-2x}$.

5. Show that $\frac{d}{dx}(x\log x)=1+\log x$, $\frac{d}{dx}\left(\frac{\log x}{x}\right)=\frac{1}{x^2}(1-\log x)$ and

$$\frac{d}{dx}\log(x+\sqrt{x^2-a^2})=\frac{1}{\sqrt{x^2-a^2}}.$$

6. Establish the general result

$$\frac{d}{dx}\log\frac{a_1x+b_1}{a_2x+b_2}=\frac{a_1b_2-a_2b_1}{(a_1x+b_1)(a_2x+b_2)}.$$

7. Find the second derivatives of e^{-x}, e^{ax+b}, $\log(ax+b)$, $x\log x$ and $x^2\log x$.

8. Show that $\frac{d^r}{dx^r}(xe^x)=(x+r)e^x$ and $\frac{d^r}{dx^r}(xe^{-x})=(-1)^r(x-r)e^{-x}$.

9. By logarithmic derivation, evaluate the derivatives of

$$x^2(x-1)^3, \quad \frac{(x+1)(x+2)}{(x-1)(x-2)} \quad \text{and} \quad x\sqrt{\frac{x+1}{x-1}}.$$

10. Find the derivatives of e^{ax^2+bx+c} and $(ax^2+bx+c)e^x$ by taking logarithms before derivation.

11. By logarithmic derivation, show that

$$\frac{d}{dx}(x^x)=x^x(1+\log x) \quad \text{and} \quad \frac{d}{dx}(x^ne^{ax+b})=(ax+n)x^{n-1}e^{ax+b}.$$

12. Find the maximum and minimum values of x^2e^x. Show that $(\log x-x)$ has only one maximum and $(x\log x)$ only one minimum value.

13. Show that the curve $y=xe^{-x}$ has one maximum point and one point of inflexion. Verify that y and its derivatives tend to vanish as $x\to\infty$. Illustrate from the graph of the curve (Examples IX, 17).

* The concept of elasticity of productivity (*Ergiebigkeitsgrad*) is used by Prof. Schneider (*Theorie der Produktion*, 1934). It describes one aspect of the "production function" which will be discussed at a later stage.

14. Show that, on the curve $y=e^{-\frac{1}{2}x^2}$, there are two points of inflexion separated by a maximum point. Hence, indicate the shape of the curve and describe its behaviour as $x \to \pm\infty$ (Examples IX, 18).

15. If $y=\frac{1}{2}(e^x+e^{-x})$, show that y and $\frac{d^2y}{dx^2}$ are always positive while $\frac{dy}{dx}$ changes sign at $x=0$. Deduce that the curve representing this function roughly resembles a parabola with a minimum at $x=0$. How can the curve be derived from the curves $y=e^x$ and $y=e^{-x}$?

16. If $y=\frac{1}{1+e^{-x}}$, show that $\frac{dy}{dx}=y(1-y)$ and $\frac{d^2y}{dx^2}=y(1-y)(1-2y)$. Deduce that y increases continuously between 0 and 1 as x increases. Show that the curve representing the function has a point of inflexion at $x=0$ and indicate its shape. (This is the *logistic curve*, see 16.2 (3), Ex. 3 below.)

17. The selling value of a good is £$f(t)$ after t years from the time when the fixed cost of production of £a was incurred. If £$f(t)$ represents $100r$ per cent. (reckoned continuously) on the outlay £a, find r as a function of t. It is assumed that the entrepreneur aims at maximising r. Show that the optimum storage period t and the (maximised) value of r are given by $f(t)=ae^{rt}$ and $f'(t)=rf(t)$. Illustrate diagrammatically and contrast this problem with that of 10.3.

18. The cost of planting a piece of land with timber is £272. The value of the timber after t years is £$100e^{\frac{1}{2}\sqrt{t}}$. Show that the present value of the timber is greatest (the rate of interest being 5 per cent. compounded continuously) if it is cut after 25 years. Show also that the maximum return (reckoned continuously) on outlay is $6\frac{1}{4}$ per cent. after 16 years. Hence illustrate the difference between the problem of 10.3 and that of the previous example. (Take $\log_e 2{\cdot}72=1$.)

19. It is assumed, at given prices of consumers' goods, that the utility of an individual consumer is measurable and dependent on his income £x. How is the marginal utility of income then defined? If it is known that marginal utility of income decreases towards zero as income increases, show that $u=\alpha\log\frac{x}{a}$ and $u=ax^{\alpha}$ are two possible forms for the utility function.

Elasticities and their applications

20. Show that the inverse function rule is of the same form for elasticities as for derivatives. Verify the rule by finding the elasticities of e^x and $\log x$ separately.

21. Evaluate the elasticities of xe^x, xe^{-x} and $x^a e^{-b(x+c)}$.

22. If ϵ is the elasticity of $f(x)$, then the elasticities of $xf(x)$ and $\frac{f(x)}{x}$ are $(\epsilon+1)$ and $(\epsilon-1)$ respectively. Check with $f(x)=ax^{\alpha}$.

23. Find the elasticity of demand η when the demand law is $x=\frac{20}{p+1}$ and $p=3$. Plot an accurate graph of the demand curve, draw the tangent at the point P where $p=3$ and locate the points T and t where it cuts the axes. Hence verify that $\eta=MT:OM=ON:Nt$, where PM and PN are perpendiculars to the axes.

24. If p cents per bushel is the price of corn and x bushels the yearly consumption per head of corn in the U.S.A., it is found that

p	50	55	60	65	70
x	30·4	28·4	26·7	25·3	24·0

By plotting on a logarithmic diagram, show that the elasticity of demand for corn can be taken as approximately constant for all prices.

25. The following table gives the average N.Y. price (p cents per lb.) and the yearly consumption (x th. short tons) of sugar, U.S.A., 1904-6:

Year	p	x
1904	4·8	3100
1905	5·25	2950
1906	4·5	3210

Plot on a logarithmic diagram and draw three parallel lines, one through each plotted point, so that the second line is equidistant from the other two. It is assumed that the elasticity of demand for sugar is constant and that the logarithmic demand line has shifted downwards by equal distances over the three years. Estimate the elasticity of demand for sugar. (See Pigou, *The Statistical Derivation of Demand Curves*, Economic Journal, 1930.)

26. If A.R. and M.R. denote the average and marginal revenue at any output, show that $\eta = \frac{\text{A.R.}}{\text{A.R.} - \text{M.R.}}$ at this output. Verify for the linear demand law $p = a - bx$.

27. Find the elasticity of demand in terms of x for each of the demand laws $p = \sqrt{a - bx}$, $p = (a - bx)^2$ and $p = a - bx^2$. Show that η decreases as x increases and find where η equals unity in each case.

28. If the demand law is $x = ae^{-bp}$, express demand elasticity and total, average and marginal revenue as functions of x. Show that the demand is of normal form. At what output is total revenue a maximum?

29. A monopolist radio manufacturer produces x sets per week at a total cost of £$(\frac{1}{25}x^2 + 3x + 100)$. The demand is $x = e^{-\frac{3p}{40}}$ sets per week when the price is £p per set. Plot the marginal cost and the marginal revenue curves on the same graph for outputs up to 50 sets per week and deduce that approximately 25 sets per week should be produced for maximum monopoly profit. (Take $\log_e 10 = 2{\cdot}3026$.)

30. If the demand law is $x = p^a e^{-b(p+c)}$, show that the demand increases as the price decreases, becoming large as the price approaches the value $\frac{a}{b}$. Find the elasticity of demand for any price greater than $\frac{a}{b}$. Is the demand of normal form?

31. If a firm produces an output x at a total cost of $\Pi = ax^2 + bx$, find an expression for κ, the elasticity of total cost. Show that κ is always greater than unity and increases as x increases.

32. Show that, for the total cost function $\Pi=\sqrt{ax+b}$, the elasticity of total cost increases but remains less than unity as x increases.

33. The total cost function of a firm is $\Pi=ax\dfrac{x+b}{x+c}+d$. Show that the average and marginal cost curves are of normal form if a, b, c and d are positive and if $d<a(c-b)$, the former having a minimum point and the latter rising continuously.

34. If $\Pi=ax^2\dfrac{x+b}{x+c}+d$ is the total cost function, where a, b, c and d are positive ($b<c$), express average and marginal cost as functions of x. Show that $\dfrac{d}{dx}\left(\dfrac{\Pi}{x}\right)$ increases with x, being negative for small x and positive for large x. Show, also, that $\dfrac{d\Pi}{dx}$ increases from zero to large values as x increases. Deduce that the cost curves are of normal form (see Examples V, 22).

35. The production (y) of a good Y depends on the amount (x) of a single variable factor of production X according to the law $y=ax^2(b-x)$, where a and b are positive constants. Express average and marginal product as functions of x and indicate that the forms of the corresponding curves can be taken as typical of the normal case of production (see Examples VII, 37).

36. The supply of a good is given by the law $x=a\sqrt{p-b}$, where p (which is greater than b) is the price and a and b are positive constants. Find an expression for e, the elasticity of supply as a function of price. Show that e decreases as price and supply increase and becomes unity at the price $2b$. What property of the supply curve holds at the point where $e=1$?

CHAPTER XI

FUNCTIONS OF TWO OR MORE VARIABLES

11.1 Functions of two variables.

OUR study of relations between variable numbers has been limited, so far, to the case of two variables, one of which is taken as a function of the other. This restriction can now be relaxed by defining relations involving many variables and by applying such relations in the interpretation of scientific phenomena. It is evident that all sciences are concerned with large numbers of inter-related variable quantities and that only by a process of severe simplification can functional relations between two variables be applied at all. For example, the volume of a gas depends on the pressure, on the temperature and on other factors. In economics, when an individual considers his purchases on a market, the demand for any good depends, not only on the price of the good, but also on his money income and on the prices of related goods. Again, to quote an everyday example, the sum of money extracted by a London taxi-driver from his passenger depends on the distance travelled, on the time taken for the journey and on the proportion expected by way of a tip.

The formal extension of the function concept presents no difficulty. Suppose, first, that we have three variables, x, y and z, each with its appropriate range of variation. The variables are related by an *implicit function* if the values taken by them are not arbitrary but connected in some definite way. In general, we write

$$f(x, y, z)=0.$$

Further, the variable z is an *explicit function* of the variables x and y if the values of z depend in a definite way on those allotted arbitrarily to x and y. In this case, we write

$$z=f(x, y).$$

A given implicit function between three variables gives rise to three distinct explicit functions; each variable can be taken as a function of the other two. For, when values are allotted arbitrarily to two of the variables, the functional relation determines the values of the third variable. When one particular explicit function is considered, say z as a function of x and y, the variable z is called the *dependent* and the variables x and y the *independent* variables.

The terms already used (2.4 above) to distinguish different kinds of functions extend at once. It is important to distinguish, in particular, between *single-valued* functions, to which the notation $z=f(x, y)$ particularly applies, and *multi-valued* functions. Of single-valued functions, we can note the class of monotonic functions where the dependent variable increases (or decreases) steadily as the independent variables increase. Again, the function $z=f(x, y)$ is said to be *symmetrical* in the variables x and y if an interchange of these variables leaves the function unaltered in form. Finally, the concepts of limits and continuity extend easily to the case of a single-valued function of two variables. If, for example, the function $z=f(x, y)$ approaches the limiting value λ as x approaches the value a at the same time that y approaches the value b, we write

$$f(x, y)\to\lambda \quad \text{as} \quad x\to a \text{ and } y\to b.$$

The function is continuous at the point $x=a$, $y=b$ if z takes the definite value $f(a, b)$ at the point and if z tends to the same value as x and y approach a and b respectively from either side. All ordinary functions are continuous except, perhaps, at a few isolated points where the variables assume infinite values.

The following examples show some common cases of functions of two variables and their grouping into function types:

Ex. 1. The implicit function $2x+3y-z-1=0$ is a particular case of the general linear functional relation

$$ax+by+cz+d=0.$$

Each of the explicit functions obtained here is single-valued. For example, from the particular function quoted, we have

$$z=2x+3y-1; \quad y=\tfrac{1}{3}(1-2x+z) \quad \text{and} \quad x=\tfrac{1}{2}(1-3y+z).$$

Ex. 2. The three explicit functions

$$z=3xy; \quad z=x^2+y^2 \quad \text{and} \quad x=x^2-2xy-y^2$$

are instances of the explicit and "homogeneous" quadratic type

$$z=ax^2+2hxy+by^2.$$

In the first of these instances, the other two explicit functions are also single-valued. In the other two instances, x and y are double-valued functions. For example, in the third case, we have

$$x = y \pm \sqrt{2y^2 + z} \quad \text{and} \quad y = -x \pm \sqrt{2x^2 - z}.$$

Ex. 3. The explicit function $z = \dfrac{x^2}{x - y + 1}$ appears in implicit form as

$$x^2 - xz + yz - z = 0,$$

and this is an instance of the implicit quadratic function type.

Here, y is also a single-valued function of x and z but x is a double-valued function of y and z. Another example of the same type is

$$x^2 + y^2 + z^2 = 16,$$

where each variable is a double-valued function of the other two. The general implicit quadratic function type can be written

$$ax^2 + by^2 + cz^2 + 2fyz + 2gxz + 2hxy + 2ux + 2vy + 2wz + d = 0,$$

where all the coefficients are constants.

Ex. 4. The explicit functions $z = e^{x^2 - y^2}$, $z = \log(x^2 + y^2)$ and $z = x^2 e^y$ illustrate the fact that exponential and logarithmic expressions can appear in functions of two variables. In the second case, for example, the expression $(x^2 + y^2)$ takes a definite value when values are allotted to x and y and the logarithm of this value then defines z.

11.2 Diagrammatic representation of functions of two variables.

If three axes $Oxyz$ are fixed to intersect at right angles at the origin O in space and if a definite scale of distance measurement is selected, then (from 1.9 above) a point P in space can be located by three co-ordinates (x, y, z). It will be assumed throughout the following development that Oz is drawn vertically upwards and that Ox and Oy are drawn, in a horizontal plane, in the W.E. and S.N. directions respectively. The co-ordinate z thus represents "heights" above or below the horizontal plane Oxy while the co-ordinates x and y represent distances E. and N. of the origin O in the plane Oxy. These conventions are adopted solely for ease in description; they are not essential to the argument.

From a given single-valued function $z = f(x, y)$, a table of values can be constructed by giving x and y arbitrarily selected values and by finding the corresponding values of z. To each pair of values of x and y there corresponds a point Q in the plane Oxy ($OM = x$ and

$ON = y$). A perpendicular QP is then erected of length equal to the value of z from the table (Fig. 75). The point P (x, y, z) is thus obtained in space with co-ordinates related by the given function. The whole table provides a finite " cluster " of such points in space and, if the variables x and y are continuous, there is no limit to the number of points that can be obtained. The infinite cluster of points makes up what is called a *surface* in space. The surface has the property that it is cut by lines parallel to Oz in no more than a single point the height of which above or below Oxy represents the value of the function $z = f(x, y)$ for the values of x and y concerned. An easily visualised example of such a surface is provided by an open hemispherical bowl resting with its rim in a horizontal plane.

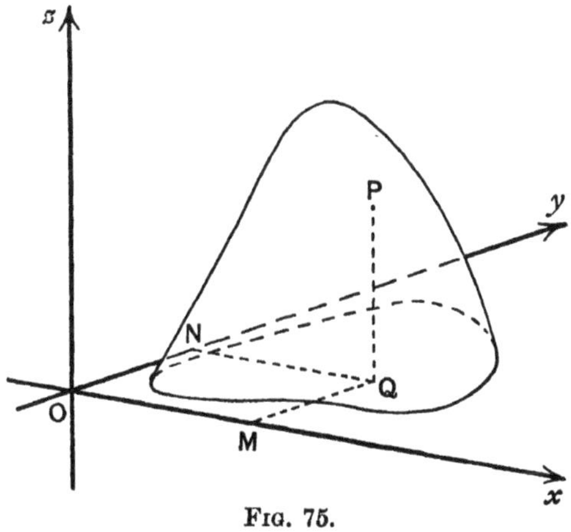

FIG. 75.

More generally, if $f(x, y, z) = 0$ is a given implicit function defining z as a multi-valued function of continuous variables x and y, we can still represent it by a surface drawn in space. The heights of various points on the surface above Oxy represent values of z obtained in the function from given values of x and y but, since the function is multi-valued, the surface can be cut in more than a single point by a parallel to Oz. A sphere can be taken as a simple example of the kind of surface obtained when z is a double-valued function of x and y.

To a functional relation between three variables x, y and z, therefore, there corresponds a surface referred to axes $Oxyz$ in space. Conversely, if axes and a scale of distance measurement are fixed, to any surface in space there corresponds a definite functional relation between three variables, the *equation* of the surface. The concept of a surface, it should be noticed, is not limited to the case of a surface without " jumps " or " gaps " and with an equation which is a continuous function of the variables.

We can construct an actual three-dimensional model of any surface of which the equation is given (e.g. the surface $z=x^2+y^2$). Axes Oxy are drawn in a plane board and a set of points is selected and marked to correspond to a set of values of x and y and to cover the plane fairly uniformly. At each point, a vertical pin is erected with height equal to the value of z given by the function. As more and more pins are erected, their heads are seen to describe a definite surface, the surface representing the function. Such a model is the three-dimensional equivalent of the graph of a function of one variable but, on account of the labour involved, its construction is rarely a practical proposition.

Again, if a function type is defined with the aid of certain parameters (e.g. the type $z=ax^2+2hxy+by^2$), then there corresponds a whole class of surfaces in space. Each surface of the class corresponds to the function obtained by giving definite values to the parameters. The systematic study of function types and surface classes is simply an extension of analytical geometry from two to three dimensions. For example, it is found that the class of *planes* in space is represented by the linear function type and that a class of surfaces known as *paraboloids* corresponds to the function type $z=ax^2+by^2$. It is not proposed, however, to elaborate this interesting study here.

11.3 Plane sections of a surface.

Without constructing an actual model of the surface in three dimensions, it is difficult to visualise the form of a surface corresponding to a given function. This difficulty can be overcome, to a large extent, by a method of great practical and theoretical service, the method of " plane sections ". The essentials of the method are contained in certain expedients familiar in everyday life. An architect, for example, seldom constructs a " working model " of a house he has designed; he draws ground and floor plans (horizontal sections) and cross-sections or elevations (vertical sections). Again, motorists and " hikers " do not carry relief models of Devonshire to guide them about that county; they refer to Ordnance maps which reduce the country to plane representations by means of a system of contours or horizontal sections. The methods underlying these and other practical devices are not difficult to generalise.

A given plane cuts a surface in points lying on a curve called a *plane section* of the surface. Of all the possible sections of the surface, the most convenient are the horizontal sections by planes perpendicular to Oz and the vertical sections by planes perpendicular to Oxy. Each of these sets of sections can be considered in turn.

A horizontal section of the surface $z=f(x, y)$ is a curve lying in the surface at a given and uniform height above the Oxy plane. Taking

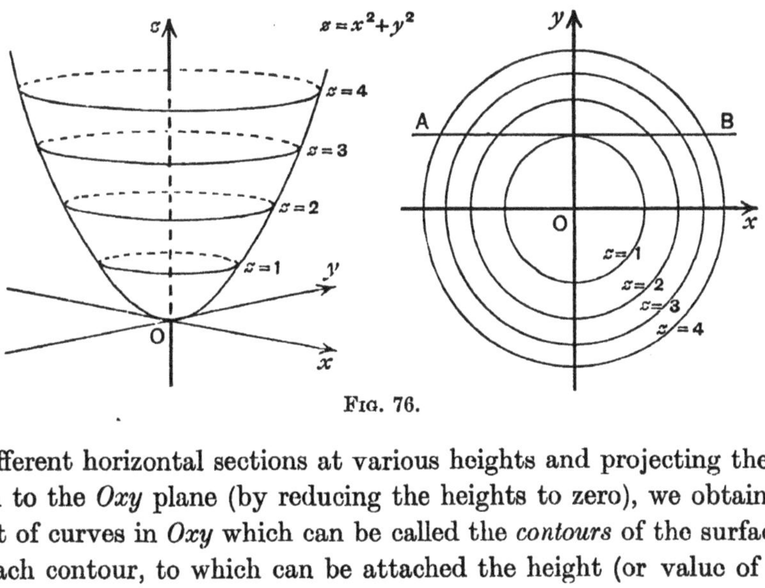

FIG. 76.

different horizontal sections at various heights and projecting them on to the Oxy plane (by reducing the heights to zero), we obtain a set of curves in Oxy which can be called the *contours* of the surface. Each contour, to which can be attached the height (or value of z) which defines it, shows the variation of x and y for the given value of z. If the variables are continuous, the set of contours consists of a limitless number of curves all included in the general equation

$$f(x, y) = \text{constant}$$

by allotting different values to the constant. In Fig. 76 are shown certain contours of the "paraboloid" surface with equation $z=x^2+y^2$, at heights $z=1$, 2, 3 and 4. The contours are here concentric circles with centre at the origin.

The shape of the surface, i.e. the variation of z with x and y, is shown in a very convenient way by the contour system. As the independent variables change, a point (x, y) moves in the plane Oxy and the variation of z (the height of the surface) is seen by tracing how the point moves in relation to the contours. The value of z increases, remains fixed or decreases according as the point crosses from lower to higher contours, remains on one contour or crosses from higher to lower contours. Further, the values of z attached to the contours crossed indicate the actual changes in z. If, as is usual, the contours are for values of z at constant intervals, then z changes more rapidly and the surface is steeper when the contours are closer together. A curve across the contour system (e.g. the line AB of Fig. 76) implies a definite change in x and y and the contours show how the corresponding value of z changes. The curve, in fact, is the projection on to Oxy of an actual path on the surface. The markings of roads, railways and other features of a country on a contoured Ordnance map are made on this principle.

Vertical sections of the surface can be defined in a similar way. Fixing a value of y, we have a section of the surface by a plane perpendicular to Oy showing the variation of z as x changes only. Such a section, which is conveniently projected on to the plane Oxz, is an "elevation" of the surface showing the up and down movement of a path on the surface in the W.E. direction. The section can be traced, at least roughly, from the contour map

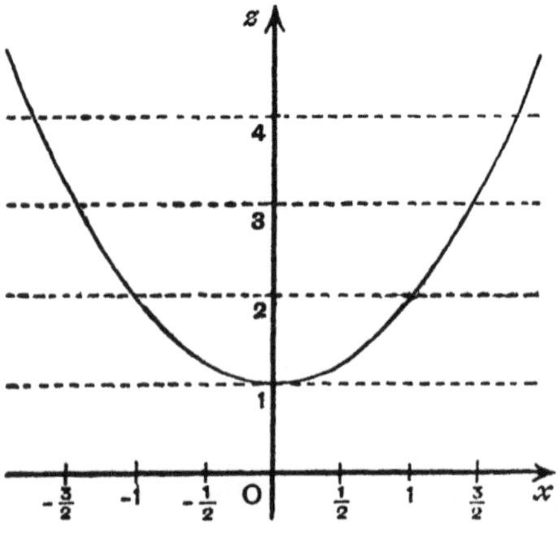

FIG. 77.

already obtained. The curve of Fig. 77 is the elevation of the surface $z = x^2 + y^2$ for $y = 1$, i.e. the parabola $z = 1 + x^2$. It can be traced by observing how the line AB $(y = 1)$ crosses the contours of Fig. 76. By taking different planes perpendicular to Oy, a whole set of vertical sections can be obtained in the plane Oxz, all sections being included in the general equation

$$z = f(x, b)$$

for different values of the parameter b. Other systems of vertical sections can be derived by taking a set of planes perpendicular to Ox (on each of which x has a constant value) or by taking a set of vertical planes through Oz and at various angles to Ox (on each of which x and y vary in proportion).

A system of sections by planes perpendicular to one or other of the axes shows, therefore, the related variation of two of the variables of the function for a given value of the third variable. Further, by arranging the sections of the system in order and relating them, we can even allow for the variation of the third variable. This idea of holding one variable constant while observing the variation of the other two is at least a first step in the examination of the complex variations of the complete function; it lies behind most of the analytical and diagrammatic expedients which are described below.

11.4 Functions of more than two variables.

There is little difficulty in completing the extension of the function concept and notation to the general case where any number of variables is taken. An implicit function relating four variables x, y, z and u is denoted by $f(x, y, z, u) = 0$ and implies that one variable is a (single-valued or multi-valued) function of the other three. Further, to denote u as an explicit function of three independent variables x, y and z, we write, in general, $u = f(x, y, z)$. This notation applies particularly, but not exclusively, to the case where the function is single-valued.

More generally, a functional relation between n variables

$$f(x_1, x_2, x_3, \dots x_n) = 0$$

gives one variable as a function of the other $(n-1)$ variables. Or, if a variable y is given as an explicit function of n independent variables $x_1, x_2, x_3, \dots x_n$, then we write

$$y = f(x_1, x_2, x_3, \dots x_n).$$

The different variables related, or the different independent variables, are here conveniently represented by a single letter x with different suffixes. It is easy to see, in this notation, the number of the variables concerned. The notation should not be confused, however, with the practice we have sometimes adopted of adding suffixes to indicate particular values of one variable.

Pure geometry as an abstract study is just as possible in four or more dimensions as in two or three dimensions, and the correspondence between the functions of analysis and the concepts of geometry persists no matter how many variables are related. But actual diagrammatic or graphical representations are no longer possible when more than three variables are involved. We should need to fix more than three axes intersecting at right angles and, unfortunately, the resources of concrete space are limited to three axes or dimensions. In the present development, therefore, we consider first the analysis of three-variable relations with concrete diagrammatic representations used as illustrations. We can then extend the results in a purely formal and analytical manner to the general case where many variables are related. Diagrammatic illustration finds no place in this extension, nor is it needed. Analysis, as we shall see, finds it little more difficult to deal with the general case than with the three-variable case.

It is sometimes convenient, however, to use certain geometrical terms even in the formal analysis of functions of many variables. The use of the terms is merely descriptive and made by analogy with the three-variable case. So, a set of values of n variables

$$(x_1, x_2, x_3, \dots x_n)$$

is called a " point " in n-dimensional space referred to n mutually perpendicular axes, and any relation between the variables is described as a " hyper-surface ", a " locus " of a " point " in n dimensions. In particular, the " hyper-surface " representing a linear relation between the variables is termed a " flat ", a concept which includes a straight line (two dimensions) and a plane (three dimensions) as particular and concrete cases.

11.5 Non-measurable variables.

It was seen (1.5 above) that many properties of scientific phenomena can be described by the measurable quantities with which our analysis has been concerned. There remain other properties capable of order without the additive character necessary for direct measurement and described by ordered but non-measurable magnitudes. We have yet to investigate whether, and under what conditions, mathematical analysis can be applied in such cases.

A clue to the solution of our problem is to be found in the use of a contour map of a function. As x and y vary continuously, a function $z=f(x, y)$ increases, remains constant or decreases according to the way in which the variable point (x, y) crosses the contour map of the function. There is no reference here to actual numerical changes in z; only the *order* of the values of z and of the contours is needed. It is possible to proceed to trace actual numerical changes in z by examining how close together the contours are as we cross them. But this second and quantitative use of the contour map is quite independent of the first use. It should be possible, therefore, to take a non-measurable magnitude as dependent on measurable variables and represent it by a contour map used only in the first of the ways described. This is the point we must now develop.

Suppose that the level assumed by a non-measurable magnitude depends uniquely on two measurable quantities x and y. As x and y vary, therefore, we can observe simply whether the magnitude increases, decreases or remains at the same level. A set of numbers z is now associated with the ordered set of levels of the magnitude so that z and the magnitude increase, decrease or remain constant together. Then z must be some function of x and y, $z=f(x, y)$, a function which is, however, by no means unique. The magnitude can be represented just as well by any other variable number or function which changes always in the same direction as $z=f(x, y)$ when x and y vary. For example, the variation of the magnitude might be indicated by $\{f(x, y)\}^2$, by $e^{f(x,y)}$ or by $\log f(x, y)$. In general, the dependence of the magnitude on the variables x and y can be expressed:

$$\zeta=F(z) \quad \text{where} \quad z=f(x, y),$$

i.e.

$$\zeta=F\{f(x, y)\}.$$

Here, $z=f(x, y)$ denotes *any one variable number* which indicates the variation of the magnitude and ζ then denotes *any other variable number*, or the whole class of possible variable numbers, indicating the magnitude's variation. $F(z)$ is any function restricted only by the fact that $F'(z)$ must be positive, so that ζ and z move always in the same direction.

In diagrammatic terms, the variable magnitude is shown by the height above *Oxy* of any one of a large number of surfaces. The points on the different surfaces for a given pair of values of x and

y are at all kinds of heights, but they move up and down together as x and y vary. In particular, if x and y vary so that the height of one surface is unaltered, then the same is true of all surfaces. The different surfaces have one feature in common; they all have identically the same contour map. The values of ζ attached to the contours of the map change according to the choice of the function F, but this affects in no way the form and order of the curves of the map. Since $\zeta = F(z)$ is constant if z is, the equation of the contour map can be written, in a form quite independent of F, as

$$f(x, y) = \text{constant}.$$

For example, the same non-measurable magnitude can be shown by

$$z = x^2 + y^2, \quad z = (x^2 + y^2)^2, \quad z = e^{x^2+y^2} \quad \text{or} \quad z = \log(x^2 + y^2)$$

and the contour map of each function is the same system of circles

$$x^2 + y^2 = \text{constant}$$

shown in Fig. 76. The value of z attached to the various circles changes from one function to another but that is all.

The dependence of a given non-measurable magnitude on two variable quantities x and y is described by the function $F\{f(x, y)\}$, where $f(x, y)$ is any one form of the dependence and where F is any monotonic increasing function. The dependence can be represented, without ambiguity, by a system of curves in the plane Oxy defined by the relation $f(x, y) = \text{constant}$. As x and y vary, the path of the point (x, y) across the curve system tells us whether the magnitude increases, decreases or remains constant, and the curve system must be used only in this non-quantitative sense. It follows that mathematical analysis can be applied, subject to the limitations indicated, in problems concerned with non-measurable magnitudes. The development given here can be extended, of course, to cases where the magnitude depends on more than two measurable quantities.

11.6 Systems of equations.

It has been shown (2.9 above) that two equations provide, in general, a determinate solution for two variables. This result can now be taken up and generalised. If one relation

$$f_1(x, y) = 0$$

is given between two variables x and y, there is an indefinitely large number of pairs of values of x and y possible and the solution of the equation for x and y is said to be "indeterminate". If two relations

$$f_1(x, y)=0 \quad \text{and} \quad f_2(x, y)=0$$

are given, there is, in general, only a finite number of possible pairs of values of x and y and the solution of the equations is then "determinate". The solution may be unique (as in the case of two linear equations) or it may be multiple (as in the case of a linear and a quadratic equation). If we are given three relations

$$f_1(x, y)=0, \quad f_2(x, y)=0 \quad \text{and} \quad f_3(x, y)=0$$

then the first two equations give a determinate set of solutions for x and y and these do not, in general, also satisfy the third equation. There is, in fact, no solution consistent with the three equations. The same result holds, *a fortiori*, if more than three relations between the variables are given. In diagrammatic terms, one equation is shown by a curve in the plane Oxy and the co-ordinates of any one of the infinite number of points on the curve satisfy the equation. Two equations are represented by a pair of curves which intersect, in general, in only a finite and determinate number of points. The co-ordinates of any point of intersection provide a solution of the equations. Three equations are represented by three curves and, since there is in general no single point lying on all curves, the equations have no solution.

Very similar results hold of a system of equations in three variables. If the number of equations is less than three, an infinite number of sets of values of the variables is possible and the solution of the equations is indeterminate. If there are three equations, the number of sets of values of the variables is finite (unique or multiple) and the solution of the equations is determinate. If there are more than three equations, then the solutions of any three of them do not, in general, satisfy the others and the equations have no consistent solution. These solutions are supported by considering the three dimensional surfaces which represent the equations.

A general result is now evident. A system of equations is given between n variables. Then:

(1) If there are less than n equations, the system is indeterminate in the sense that an infinite number of sets of values of the variables is possible.

(2) If there are exactly n equations, the system is determinate in the sense that only a finite number of sets of values of the variables is possible, the solutions of the equations.

(3) If there are more than n equations, there is no one set of values of the variables consistent with all equations.

One restriction on the generality of these results must be noticed. It is essential that the equations of the system should be *mutually consistent* with each other and *independent* of each other. In fact, it is not permissible to include an equation in the system *either* if it cannot hold if one of the others does *or* if it automatically holds if the others do. For example, the equation $2x+2y-4z=5$ cannot be included in the same system as $x+y-2z=3$; they are not consistent. Nor can the equation $2x+2y-4z=6$ be included with $x+y-2z=3$; they are not independent equations and the one tells us no more about the variables than the other. Hence, before the solution of a system of equations can be examined, we must see whether the system is mutually consistent and independent. It often happens, for example, that the number of equations in a given system is one more than the number of variables, so that it would appear that no solution exists. On examination, however, it is found that one of the equations is derivable from (i.e. not independent of) the others. When this equation is discarded from the system, the number of equations equals the number of variables and the system is determinate.*

The solution of a system of equations can also be approached from a different angle. One equation between n variables determines one variable as a function (not necessarily single-valued) of the other $(n-1)$; two equations determine two variables as functions of the other $(n-2)$; and so on. In general, if there are m equations in n variables $(m<n)$, any m of the variables can be determined as functions of the other $(n-m)$. Then, if further equations between the variables are found, the m selected variables can be eliminated by the substitution of the functions already obtained, and the new forms of the equations involve only $(n-m)$ variables. Thus each

* This case often arises in the analysis of economic equilibrium. See, for example, Pareto, *Manuel d'économie politique* (2nd Ed. 1927), pp. 591-3 and pp. 610-5; Bowley, *The Mathematical Groundwork of Economics* (1924), pp. 20-2 and p. 51.

equation of a system can be used to determine and eliminate one of the variables in terms of the others. As long as the number of equations is less than the number of variables, there remain some variables which cannot be determined and eliminated and to which values can be allotted at will. The equations are indeterminate. But when the number of equations and variables is equal, we can eliminate the variables one by one until a single equation in one variable is left to complete a determinate solution. For example,

$$x+y+z+u-1=0 \quad \text{and} \quad 2x-y-2z+u+1=0$$

are two equations from which we find

$$z=\tfrac{1}{3}(2+x-2y) \quad \text{and} \quad u=\tfrac{1}{3}(1-4x-y).$$

If, now, a third equation is given

$$x^2+y^2+z^2+u^2=35,$$

we have on substitution an equation in the variables x and y only:

$$13x^2+2xy+7y^2-2x-5y-15=0.$$

From the system of three equations, two have been used to eliminate two variables and the other gives an equation in the remaining variables.

11.7 Functions of several variables in economic theory.

It is clear that functions of many variables must play an important part in any precise and general interpretation of economic phenomena. This fact has been implicitly recognised already when we found that functions of one variable could only be applied when the problems were simplified and abstraction made of the inter-relations of the many variable quantities involved. We are now in a position to reconsider the problems in a more general light.

To generalise the problem of demand, we suppose that n consumers' goods $X_1, X_2, X_3, \ldots X_n$ are sold at uniform prices $p_1, p_2, p_3, \ldots p_n$ on a competitive market consisting of a fixed number of consumers with given tastes and incomes. Then the amount x_r of any one good X_r demanded by the market is uniquely dependent on the prices of all the goods on the market. We can thus write

$$x_r=\phi_r(p_1, p_2, p_3, \ldots p_n)$$

as the demand function for the good X_r, a function which, for convenience, can be assumed continuous in all the variables.

The number of the variables overcrowds our picture of market demand. It is possible, however, to select a few of the prices according to the particular aspect of the problem considered and to assume that all the other prices are fixed. In particular, we can study the inter-relations of the demands for two goods X_1 and X_2 by assuming that the prices of all other goods are fixed. Then

$$x_1 = \phi_1(p_1, p_2) \quad \text{and} \quad x_2 = \phi_2(p_1, p_2)$$

are the demand functions, each dependent on the two variable prices. Each function can be shown as a demand surface with heights above the horizontal plane Op_1p_2 representing the varying demand for the good. The vertical sections of such a surface are particularly interesting. The section of the surface $x_1 = \phi_1(p_1, p_2)$ by any plane perpendicular to Op_2 (on which p_2 has a fixed value) is an ordinary demand curve showing the variation of x_1 as p_1 varies. There is one such demand curve for each fixed price p_2 of the other good and the whole system of demand curves shows the way in which demand shifts as the price p_2 is changed. All these demand curves are downward sloping in the normal case. The section of the surface by a plane perpendicular to Op_1 is a curve showing the variation of x_1 as p_2 varies for a given value of p_1. At any point where this section is upward sloping, a rise in the price of X_2 results in a rise in the demand for X_1 and the goods can be called " competitive ", at least in a rough sense. If the section is downward sloping, the converse holds and the goods can be called " complementary " at the prices concerned.

It is often convenient to assume, as an approximation for certain ranges of the prices, that the demand functions are of definite types. The following are examples of demand functions of relatively simple type in the case of two related goods :

$$(1)\quad x_1 = a_1 - a_{11} p_1 + a_{12} p_2, \qquad x_2 = a_2 + a_{21} p_1 - a_{22} p_2.$$

$$(2)\quad x_1 = \frac{a_1}{p_1 + a_{11}} + a_{12} p_2, \qquad x_2 = \frac{a_2}{p_2 + a_{22}} + a_{21} p_1.$$

$$(3)\quad x_1 = a_1 p_1^{-a_{11}} p_2^{a_{12}}, \qquad x_2 = a_2 p_1^{a_{21}} p_2^{-a_{22}}.$$

$$(4)\quad x_1 = p_1^{-a_{11}} e^{a_{12} p_2 + a_1}, \qquad x_2 = p_2^{-a_{22}} e^{a_{21} p_1 + a_2}.$$

In the normal case of demand, the constants a_{11} and a_{22}, as well as a_1 and a_2, can be taken as positive. But the constants a_{12} and a_{21} can be of either sign. The goods are " competitive " if they are both positive and " complementary " if they are both negative. It can be noticed that the demand functions (3) and (4) become linear if the demands and either both or one of the prices are taken on logarithmic scales.

A slightly different presentation can be given to show the demand of an individual consumer instead of that of the whole market. If a consumer has given tastes, his demand for a good X_r is a function of his income μ and of all the prices:

$$x_r = \phi_r(\mu, p_1, p_2, p_3, \dots p_n).$$

As a problem of particular interest, we can trace the effect on the demand for X_r of changes in income and in the price of X_r. If all other prices are taken as fixed, we have a demand surface

$$x_r = \phi_r(\mu, p_r),$$

showing the demands for X_r for various incomes and prices. The section of the surface by a plane perpendicular to $O\mu$ is an ordinary demand curve for a fixed income level. As the income level is changed, the section shifts in position. We can thus observe the shifting of demand consequent upon the variation of individual income. On the other hand, if the price of the good is fixed, the variation of demand as income increases is shown by the section of the surface by a plane perpendicular to the axis Op_r.

A generalised cost problem provides a second example of the use of functions of several variables. If a firm produces different goods or different qualities of the same kind of good, and if the technique of production and the supply conditions of the factors of production are given, then definite amounts of the various goods can be obtained at a determinate minimal total cost:

$$\Pi = F(x_1, x_2, x_3, \dots x_n),$$

where $x_1, x_2, x_3, \dots x_n$ are the amounts of the goods $X_1, X_2, X_3, \dots X_n$ produced. If only two goods X and Y are produced, the cost function $\Pi = F(x, y)$ can be represented by a cost surface with heights above the horizontal plane Oxy showing the costs of producing various combinations of outputs. The vertical sections of the surface represent the cost of producing a variable output of one good in conjunction with a fixed output of the other good. A particular form of the joint cost function, such as

$$\Pi = ax^2 + 2hxy + by^2 + 2gx + 2fy + c,$$

can be assumed for convenience in certain cases.

The transformation functions of 5.6 above can also be generalised in an obvious way. The technique of production and the productive resources of a given firm or country are fixed. Then, if $x_1, x_2, x_3, \ldots x_n$ are the amounts of various goods $X_1, X_2, X_3, \ldots X_n$, that can be produced in a given year, or the incomes obtainable in a series of years, there must be some relation between these variables:

$$F(x_1, x_2, x_3, \ldots x_n)=0.$$

For, given the outputs desired of all goods but one, or the incomes desired in all years but one, then technical considerations determine the amount of the remaining output or income. When there are three goods or incomes, we have a transformation surface

$$F(x, y, z)=0$$

referred to axes Ox, Oy and Oz along which amounts of the goods or incomes are measured. In the normal case, we can take the surface as downward sloping and convex to the origin at all points. A simple case of the normal type of transformation function is

$$ax^2+by^2+cz^2=d,$$

where a, b, c and d are positive constants.

11.8 The production function and constant product curves.

An important problem, to which only incidental reference has so far been made,* concerns the conditions under which factors of production are combined in the production of a given good by a firm or industry. With given technical conditions of production, the amount of a good X produced depends uniquely on the amounts of the variable productive factors, $A_1, A_2, A_3, \ldots A_n$ used. If x is the amount produced when amounts $a_1, a_2, a_3, \ldots a_n$ of the factors are employed, we can write the *production function*

$$x=f(a_1, a_2, a_3, \ldots a_n).$$

It will be assumed, throughout the present development, that the factors of production are continuously divisible and that the productive process is continuously variable. The production function is then a continuous function of continuous variables. It is to be noticed that our presentation of the problem does not exclude the

* See 10.8 above.

possibility of the variable factors being used with certain fixed factors of production. A definite production function can be assumed in all cases, no matter how many factors are fixed and how many variable. Only the appropriate form of the function depends on considerations relating to the nature, the variability or otherwise, of the factors.

If there are two variable factors A and B, the production function $x=f(a, b)$ can be represented by a *production surface* referred to axes of which Oa and Ob are taken horizontally and Ox vertically. The method of plane sections is now of great advantage. The contours of the production surface consist of a system of curves in the plane Oab which can be termed *constant product curves* and are defined by

$$f(a, b) = \text{constant}.$$

One curve of the system, corresponding to a given value x_1 of the constant, includes all points (a, b) representing amounts of the factors giving a definite product x_1. The curves form a continuous and non-intersecting system covering the positive quadrant of the plane Oab in such a way that one, and only one, curve of the system passes through each point. As the amount of the factors used are changed in any way, the corresponding point (a, b) moves in the plane Oab and its path across the constant product curves determines the resulting variation of the product obtainable.

The vertical section of the production surface by a plane perpendicular to Ob (on which the value of b is fixed) is a curve showing the variation of product as various amounts of the factor A are used with a given amount of the factor B. Thus

$$x=f(a, b_1)=\phi_1(a), \quad x=f(a, b_2)=\phi_2(a), \quad \ldots$$

are the equations of the vertical sections for fixed amounts $b_1, b_2 \ldots$ of the factor B. They are different "elevations" of the production surface. A similar set of sections or "elevations" is obtained by fixing the value of a and taking planes perpendicular to Oa. A third set of vertical sections, corresponding to "elevations" of the surface of a different kind, can also be used to advantage. The sections are obtained by taking planes through Ox and at various angles to Oa and Ob, and they show curves giving the variation of product as the factors are varied in proportion. Thus, if a fixed point on one of the sections corresponds to amounts a_1 and b_1 of the factors, then any other point on the section is given by $a = \lambda a_1$ and $b = \lambda b_1$, where λ is a variable proportion. The equation of the section is

$$x = f(\lambda a_1, \lambda b_1) = \phi(\lambda),$$

a function of the variable λ (see 10.8 above). Different sections of the complete set are obtained by allotting different values to the constants a_1 and b_1.

In the "normal" case of production, the production surface can be subjected to certain restrictions apart from that of continuity. It can be assumed, in the simplest "normal" case, that product can only be maintained, when less of the factor B is used, by increasing the use of the factor A. Further, as the substitution of the factor

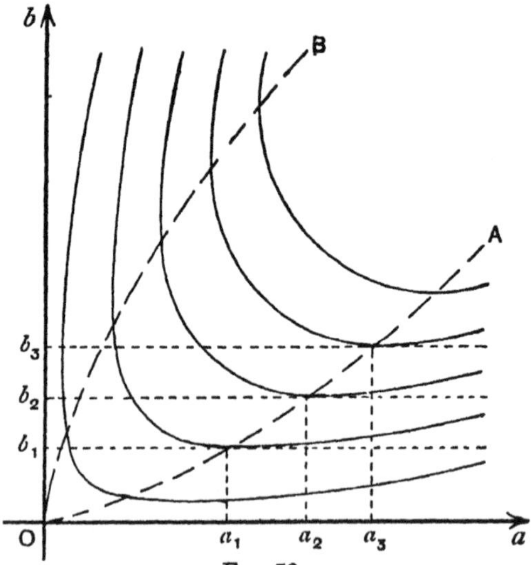

FIG. 78.

A for the factor B proceeds, increasingly larger additions of A are needed to compensate (i.e. to maintain product) for a given reduction of B.* The constant product curves are then downward sloping and convex to the origin at all points of the positive quadrant of Oab, i.e. they form a system of curves similar in form to that of the "normal" indifference map described in 5.7 above.

In a more general "normal" case, the properties assumed above apply only to certain ranges of employment of the factors while, outside these ranges, product is maintained only by increasing the use

* This is the principle of "increasing marginal rate of substitution" (see 13.7 below).

of both factors. The form of the constant product curve system is then that shown in Fig. 78. In the area of Oab between the curves OA and OB, the curves are downward sloping and convex to the origin; outside this area, the curves turn back and slope upwards. Here, when one factor is used in much greater proportion than the other, an increase in the factor's use requires more of the other factor even to maintain product.

The normal form of the vertical sections of the production surface can be traced from that of the constant product curves. If a line distant b_1 from the axis Oa is drawn in the plane Oab, the way in which it cuts the constant product curves shows the variation of product

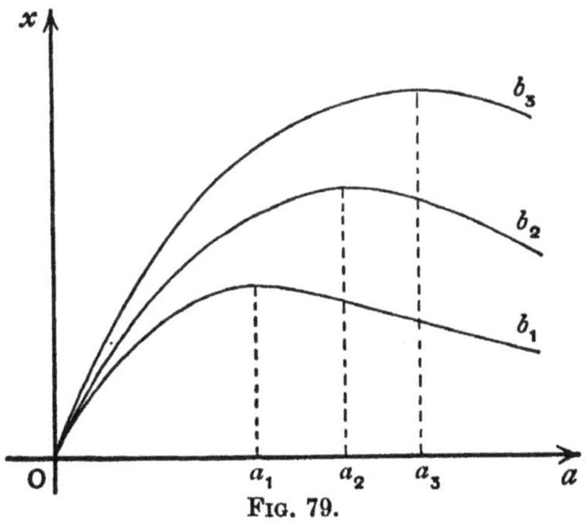

FIG. 79.

on the section perpendicular to Ob corresponding to the fixed amount b_1 of the factor B. In the simpler normal case, the line cuts higher and higher constant product curves from left to right, i.e. the vertical section rises as a increases. In the more general normal case of Fig. 78, the product increases as a increases at first, attains a maximum value when a equals a certain value a_1 (where the line cuts the curve OA and touches a constant product curve) and then decreases as a increases further. The vertical sections of the production surface, for fixed amounts b_1, b_2, b_3, ... of the factor B, are of the form shown in Fig. 79. Each section has a " peak " product corresponding to a combination of factors shown by a point on the curve OA of Fig. 78. The greater the fixed amount of B used, the more of A do we need before we obtain the peak product and the greater is the peak product. The curve OA of Fig. 78 is thus of great interest;

it corresponds to a " ridge line " of the production surface and indicates the peak product obtainable from any given use of the factor B. The vertical sections of the production surface for fixed use of the factor A are similar in form and lead to the definition of the curve OB of Fig. 78 as another " ridge line " of the surface.*

A particular production function of the simpler normal type is

$$x = Aa^{\alpha} b^{1-\alpha},$$

where α is a positive fraction.† The constant product curve, for the given product $x = x_1$, is given by

$$a^{\alpha} b^{1-\alpha} = \frac{x_1}{A},$$

i.e.
$$b = \frac{A'}{a^m},$$

where $A' = \left(\frac{x_1}{A}\right)^{\frac{1}{1-\alpha}}$ and $m = \frac{\alpha}{1-\alpha}$ are positive constants. So

$$\frac{db}{da} = -\frac{mA'}{a^{m+1}} < 0 \quad \text{and} \quad \frac{d^2b}{da^2} = \frac{m(m+1)A'}{a^{m+2}} > 0,$$

and the constant product curves are downward sloping and convex to the origin at all points. In the particular case where $\alpha = \frac{1}{2}$ the curves are rectangular hyperbolas with Oa and Ob as asymptotes. Further, the vertical section for the fixed amount b_1 of the factor B is

$$x = A''a^{\alpha}, \qquad (A'' = Ab_1^{1-\alpha}).$$

Since α is a positive fraction, the section rises from left to right at a decreasing rate and never attains a " peak " value. Notice that

$$\log x = \log A + \alpha \log a + (1 - \alpha) \log b.$$

On taking logarithmic scales for all variables, the production surface becomes a plane and the constant product curves straight lines.

A production function of the more general normal type is

$$x = 2Hab - Aa^2 - Bb^2,$$

where A, B and H are positive constants such that $H^2 > AB$. It can be shown that the constant product curves are of the form of Fig. 78

* The normal form of the vertical sections is described by Knight, *Risk, Uncertainty and Profit* (1921), pp. 100-1.

† See, for example, Douglas, *The Theory of Wages* (1934).

with OA and OB as straight lines (see Examples XIII, 32). The vertical section for which $b = b_1$ has equation

$$x = -Aa^2 + 2Hb_1a - Bb_1^2,$$

i.e. $$\left(x - \frac{H^2 - AB}{A} b_1^2\right) = -A\left(a - \frac{H}{A} b_1\right)^2,$$

i.e. the section is a parabola with axis vertical. The peak product $\frac{H^2 - AB}{A} b_1^2$ is obtained when the amount $\frac{H}{A} b_1$ of the factor A is used with the fixed amount b_1 of the factor B. It is to be noticed that a zero (or negative) product is obtained if the amount of the factor A used is less than $\frac{b_1}{A}(H - \sqrt{H^2 - AB})$ or greater than $\frac{b_1}{A}(H + \sqrt{H^2 - AB})$. For intermediate amounts of the factor, the product rises from zero

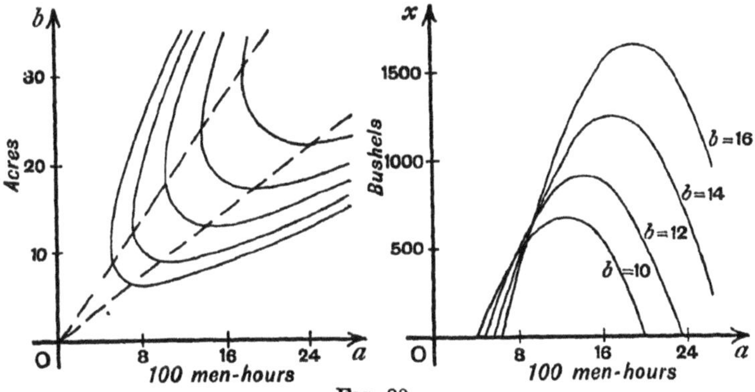

FIG. 80.

to a peak and then falls to zero again. Fig. 80 shows certain constant product curves and vertical sections when the product of wheat in bushels is

$$x = 2(12ab - 5a^2 - 4b^2),$$

when $100a$ men-hours of labour are employed on b acres of land. It appears, for example, that no wheat is obtainable from 10 acres when less than 400 men-hours are employed and that the wheat product rises to a maximum of 640 bushels when 1200 men-hours are employed.

11.9 The utility function and indifference curves.

An individual consumer has a scale of preferences for two consumers' goods X and Y represented, according to the conditions of 5.7 above, by the indifference map with equation

$$\phi(x, y) = \text{constant}.$$

Denote $$u = \phi(x, y),$$

a continuous function of two continuous variables. The original indifference map is then the contour map of the surface representing this function. The variable u takes a constant value on any one indifference curve and increases as we move from lower to higher indifference curves. Hence, as the purchases of the individual change, the value of u increases, remains constant or decreases according as the change leaves the individual better off, indifferent or worse off, i.e. the value of u indicates the level of preference or the utility of the purchases x and y to the individual consumer.

The concept of preference or utility, however, is an ordered but non-measurable one and the remarks of 11.5 above are of direct application. The function $u = \phi(x, y)$ is only one way of indicating the dependence of utility on the purchases. In general, if $F(u)$ is any monotonic increasing function of u, then

$$U = F(u) = F\{\phi(x, y)\}$$

serves as an indicator of utility and can be termed the general *index-function of utility.** It is quite immaterial which utility function, e.g. $u = \phi(x, y)$, $u = \log \phi(x, y)$, ... , is used to represent the dependence of utility, as a non-measurable magnitude, on the individual's purchases. The surfaces representing the various functions, referred to axes of which Ou is vertical, are different but rise and fall together and have the same contour map, the indifference map of the individual as originally defined.

The problem is generalised by assuming that the individual has a definite preference scale for the goods $X_1, X_2, X_3, \ldots X_n$ appearing in his budget, a scale described by the indifference map

$$\phi(x_1, x_2, x_3, \ldots x_n) = \text{constant}.$$

For a given value of the constant, this relation connects all purchases which are indifferent to the individual while, for increasing values of the constant, we move from one set of indifferent purchases to another set at a higher level of preference. If there are three goods

* This is a translation of the term " fonction-indice " introduced by Pareto, *Economie mathématique* (Encyclopédie des sciences mathématiques), p. 596.

X, Y and Z, the indifference map is represented by a system of surfaces in $Oxyz$ space which, in their normal form, fit into each other as a series of shells, all downward sloping and convex to the origin. For more than three goods, no diagrammatic representation is possible, but the indifference map can be described, by analogy, as a system of hyper-surfaces in many-dimensional space. The individual's utility is again a non-measurable concept dependent on the purchases according to the general index-function of utility

$$U = F(u) = F\{\phi(x_1, x_2, x_3, \dots x_n)\},$$

where $F(u)$ is any function such that $F'(u) > 0$. The particular utility function $u = \phi(x_1, x_2, x_3, \dots x_n)$ is only one of many possible forms.

A similar representation applies when we consider, not the purchases of the individual of different goods at one point of time, but the flow of his purchases or income over time. If $x_1, x_2, x_3, \dots x_n$ represent incomes in successive years according to the conditions of 5.8 above, we have a definite indifference map

$$\psi(x_1, x_2, x_3, \dots x_n) = \text{constant},$$

and we can write a corresponding index-function of utility

$$U = F(u) = F\{\psi(x_1, x_2, x_3, \dots x_n)\},$$

where $F(u)$ is any monotonic increasing function.

As an approximate representation of a normal indifference map for two goods or incomes, we can sometimes take the particular form

$$u = (x+a)^\alpha (y+b)^\beta,$$

where a, b, α and β are positive constants. Any monotonic increasing function of u, for example the function

$$u' = \log u = \alpha \log(x+a) + \beta \log(y+b),$$

serves equally well as an index of utility. The indifference map, common to all forms of this particular utility function, is shown in Fig. 25 in the case where $\alpha = \beta = 1$ and $a = 2$, $b = 1$. It is seen to be of normal form. A rather different utility function

$$u = ax + by + c\sqrt{xy}$$

also corresponds to a normal case of an indifference map for two goods or incomes. Any monotonic increasing function of u, such as

$$u' = e^u = e^{ax+by+c\sqrt{xy}},$$

serves equally well as an index of utility. The corresponding indifference map is shown in Fig. 26 in the case where $a = b = 1$ and $c = \sqrt{2}$.

EXAMPLES XI

Functions of two or more variables

1. Express the function $z = -\frac{xy}{x+y}$ in implicit form and show that each variable is a single-valued function of the other two of the same form in each case. Deduce that the sections of the surface representing the function by planes perpendicular to an axis are rectangular hyperbolas. In what sense is this a symmetrical function?

2. Illustrate graphically the form of the contours, and of the other sections, of the surface $z = \sqrt{xy}$.

3. Show that the sections of the surface $x + y + 2z = 3$ by planes perpendicular to an axis are parallel straight lines and deduce that the surface is a plane. Locate the points where the plane cuts the axes. Generalise to show that $ax + by + cz + d = 0$ is always the equation of a plane.

4. Examine the sections of the surface $x^2 + y^2 + z^2 = a^2$ and show that it is a sphere of radius a.

5. Show that $x^2 + y^2 - z^2 = 0$ is the equation of a circular cone with its axis along Oz.

6. Show that the contours of the surface $x^2 + y^2 + z^2 - 2xz - 2yz = 0$ form a system of circles and deduce that the surface is a cone.

7. By a geometrical construction, show that $\left(\frac{x_1 + x_2}{2}, \frac{y_1 + y_2}{2}, \frac{z_1 + z_2}{2}\right)$ is the mid-point between two given points (x_1, y_1, z_1) and (x_2, y_2, z_2) referred to rectangular axes in space. Show that the mid-point between any two points on $x + y + z + 1 = 0$ is also in the surface and deduce that this is the equation of a plane.

8. If (x_1, y_1, z_1) and (x_2, y_2, z_2) are two given points referred to rectangular axes in space, show that the distance between the points is

$$\sqrt{(x_1 - x_2)^2 + (y_1 - y_2)^2 + (z_1 - z_2)^2}.$$

Deduce that the equation of a sphere with centre (a, b, c) and radius r is

$$(x - a)^2 + (y - b)^2 + (z - c)^2 = r^2.$$

9. Show that any point on the surface $x^2 + y^2 + z^2 - 2z = 0$ is such that its distance from O is twice its distance from Oxy.

10. By examining the sections of the surface by various planes, show that $z = x^2 + y^2$ is a surface obtained by revolving the parabola $z = x^2$ about the axis Oz in the plane Oxz.

11. Show that $z = f(x^2 + y^2)$ is a surface obtained by revolving the curve $z = f(x^2)$ about the axis Oz in the plane Oxz. Illustrate by considering the surfaces $z = x^2 + y^2$, $z = e^{x^2+y^2}$ and $z = \log(x^2 + y^2)$.

12. If $z = \frac{x^2}{x - y}$, show that $z \to \pm\infty$ as $x \to a$ and $y \to a$, where a is any constant. Deduce that the surface represented by the equation has "infinities" at all points above the line $x - y = 0$ in the plane Oxy. Check from the vertical sections of the surface.

13. Show that each section of the surface $z=(x+y)e^{-(x+y)}$ by planes perpendicular to Oy has a single maximum point and that the sections are the same curve translated varying distances parallel to Ox.

14. Show that, for each of the functions

$$z=\sqrt[3]{x^2y},\ z=\sqrt{x^2-y^2}\ \text{and}\ z=\frac{x^2+y^2}{x+y},$$

the value of z changes from z to λz when the values of x and y are changed in proportion from x to λx and from y to λy. Deduce that the surface representing any of these functions has a straight line section by a vertical plane through O and a given point on the surface.

15. For any of the functions of the previous example, show that $z=x\phi(r)$, where $r=\dfrac{y}{x}$, and that $z=y\psi(s)$, where $s=\dfrac{x}{y}$.

16. If $\phi(x)$ and $\psi(y)$ are two functions of single variables, show that the sections of the surface $z=\phi(x)+\psi(y)$ by planes perpendicular to Ox or Oy are of the same shape but variable height, and that similar sections of the surface $z=\phi(x)\psi(y)$ consist of a curve "stretched" or "contracted" by a variable amount in the direction Oz.

17. In what sense can it be said that the function $u=x^2+y^2+z^2$ has contours consisting of a set of concentric spheres in $Oxyz$ space?

18. Show that $x+y+z=1$, $x+3y-z=3$ and $2x-y+z=1$ have a unique solution in x, y and z. Illustrate by drawing the planes which represent these equations.

19. Have the equations $x+2y-z=1$, $2x-y+z=3$ and $x+7y-4z=0$ any solution? Verify that one equation can be deduced from the other two, i.e. that the equations are not independent.

20. Eliminate z and u from the equations

$$2xy+z-u+1=0,\quad x^2+y^2-2z+u=2\quad\text{and}\quad 4x^2y^2+z^2-u^2+1=0$$

and obtain an equation in x and y only.

Economic functions and surfaces

21. If £z is the present value of £y available x years hence, interest being reckoned continuously at $100r$ per cent., show that $z=ye^{-rx}$. Show that the contours of the surface $z=ye^{-rx}$, for a given value of r, form a system of discount curves. What are the vertical sections?

22. The demand for tea is $x_1=40\dfrac{p_2}{p_1}$ and for coffee $x_2=10\dfrac{p_1}{p_2}$ thousand lbs. per week, where p_1 and p_2 are the respective prices of tea and coffee in pence per lb. At what relative prices of tea and coffee are the demands equal? Draw a graph to show the shifts of the demand curve for tea when the price of coffee increases from 2s. to 2s. 6d. and to 3s. per lb.

23. Show that $x_1=a_1\dfrac{p_2}{p_1}$ and $x_1=\dfrac{a_1}{p_1}e^{p_2}$ are two simple examples of a demand law for a good X_1 in competition with a good X_2 and that

$$x_1=\frac{a_1}{p_1p_2}\ \text{and}\ x_1=\frac{a_1}{p_1}e^{-p_2}$$

are corresponding laws when X_1 and X_2 are complementary.

24. The demands x_1 and x_2 of a consumer for two goods at market prices p_1 and p_2 are related by the condition $\frac{a_1+b_1x_1}{p_1}=\frac{a_2+b_2x_2}{p_2}$, where the a's and the b's are constants. If $\mu=x_1p_1+x_2p_2$ is the consumer's income, find the demands as functions of μ, p_1 and p_2. If the prices are fixed, show that each demand is a linear function of μ.

25. It is known that $x=Aa^{\alpha}b^{1-\alpha}$ is the product (in bushels) of wheat when $100a$ men-hours of labour are employed on b acres of land. Find the constants A and α, given that 1500 bushels can be obtained from 100 acres when 10,000 men-hours of labour are employed and that 2120 bushels can be obtained from the same area when 20,000 men-hours of labour are employed. Draw a graph of the variation of product as varying amounts of labour are applied to 100 acres. What is the product when 20,000 men-hours of labour are employed on 150 acres?

26. The production function is $x=Aa^{\alpha}b^{\beta}$, where A, α and β are constants. If the factors are increased in proportion, show that the product increases in greater or less proportion according as $(\alpha+\beta)$ is greater or less than unity. How is this property shown on a vertical section of the production surface through O and a given point on the surface? What is the special property of the case $\alpha=1-\beta$?

27. The following data are taken from Douglas, *The Theory of Wages*:

Manufacturing Industries, U.S.A., 1900-20.

x = Day's index of physical volume of manufactures (1899 = 100).
a = Average number of wage-earners in manufacture (1899 = 100).
b = Volume of fixed capital in manufacture (1899 = 100).

Year	x	a	b	Year	x	a	b	Year	x	a	b
1900	101	105	107	1907	151	138	176	1914	169	149	244
01	112	110	114	08	126	121	185	15	189	154	266
02	122	118	122	09	155	140	198	16	225	182	298
03	124	123	131	10	159	144	208	17	227	196	335
04	122	116	138	11	153	145	216	18	223	200	366
05	143	125	149	12	177	152	226	19	218	193	387
06	152	133	163	13	184	154	236	20	231	193	407

Find the value of x given each year by the production function $x=a^{\frac{3}{4}}b^{\frac{1}{4}}$ and express as a percentage of the actual value of x. Show that the deviation never exceeds 10 per cent. and find the average percentage deviation (neglecting signs). Plot a graph on a logarithmic scale of the variation of the actual and estimated values of x over the whole period.

28. It is given that $x=40\,\frac{12ab-5a^2-4b^2}{a+b}$ bushels of wheat are obtained when $100a$ men-hours of labour are employed on b acres of land. If 10 acres are cultivated, draw a graph to show the variation of product as a varies. Show that the greatest product is obtained from this area when approximately 1050 men-hours of labour are employed.

29. Consider vertical sections of the production surface

$$x=\frac{2Hab-Aa^2-Bb^2}{Ca+Db}$$

and show that a proportional increase in the factors results in an equal proportional increase in product.

30. Show that the production surfaces

$$x=2Hab-Aa^2-Bb^2 \quad \text{and} \quad x=\sqrt{2Hab-Aa^2-Bb^2}$$

have constant product curves of identical form but that the variation of product from one curve to another is different. How does product increase in the two cases as the factors are varied in proportion from any combination?

31. If a and b are positive constants, show that the utility function

$$u=(x-a)^2+(y-b)^2$$

gives an indifference map of normal form for purchases x and y in the ranges $0 \leqslant x \leqslant a$, $0 \leqslant y \leqslant b$, the curves being arcs of concentric circles.

32. Examine the utility function $u=\dfrac{x+a}{c-\sqrt{y+b}}$, where a, b and c are positive constants, and show that the indifference map is a set of parabolic arcs and of normal form for certain ranges of values of the purchases x and y.

CHAPTER XII

PARTIAL DERIVATIVES AND THEIR APPLICATIONS

12.1 Partial derivatives of functions of two variables.

IN a function of two variables $z=f(x, y)$, the two variables x and y can be varied in any way quite independently of each other. In particular, one of the variables can be allotted a fixed value and the other allowed to vary. The function, in such a case, reduces to a function of a single variable only. Two functions can be obtained in this way, z as a function of x only (y fixed) and z as a function of y only (x fixed). The derivative of each of these functions can be defined at any point and evaluated according to the familiar technique. The derivatives so obtained are called the "partial derivatives" of the function $z=f(x, y)$, the term "partial" implying that they are defined only for very special variations of the independent variables. One partial derivative follows when x is varied and y kept constant, the other when y is varied and x kept constant.

If the function $z=f(x, y)$ is single-valued, its partial derivative with respect to x at the point (x, y) is the limiting value of the ratio $\dfrac{f(x+h, y)-f(x, y)}{h}$ as the arbitrary increment h tends to zero and the partial derivative is written *either* as $\dfrac{\partial z}{\partial x}=\dfrac{\partial}{\partial x}f(x, y)$, *or* as $z_x'=f_x'(x, y)$. The notations are similar to those for ordinary derivatives. But, to indicate that we have the *partial* derivative with respect to x (i.e. that the other variable y is regarded as fixed), we use the symbol "∂" instead of the previous "d" in the first notation and add a suffix x in the second notation. An exactly similar definition and notation can be given for the other partial derivative. Hence:

DEFINITION: The partial derivative of $z=f(x, y)$ with respect to x at the point (x, y) is

$$\frac{\partial z}{\partial x}=\frac{\partial}{\partial x}f(x, y)=z_x'=f_x'(x, y)=\operatorname*{Lim}_{h\to 0}\frac{f(x+h, y)-f(x, y)}{h}$$

and the partial derivative with respect to y at the same point is

$$\frac{\partial z}{\partial y}=\frac{\partial}{\partial y}f(x,\,y)=z_y'=f_y'(x,\,y)=\operatorname*{Lim}_{k\to 0}\frac{f(x,\,y+k)-f(x,\,y)}{k}.$$

It is sometimes convenient to use slightly different notations. For example, $\frac{\partial f}{\partial x}$ and $\frac{\partial f}{\partial y}$ can be used instead of $\frac{\partial z}{\partial x}$ and $\frac{\partial z}{\partial y}$. Further, when no ambiguity is likely to arise, it is possible to use f_x and f_y as simple notations for the two partial derivatives.

The meaning of partial derivatives is most clearly seen when they are interpreted in diagrammatic terms. In Fig. 81, P is the point on

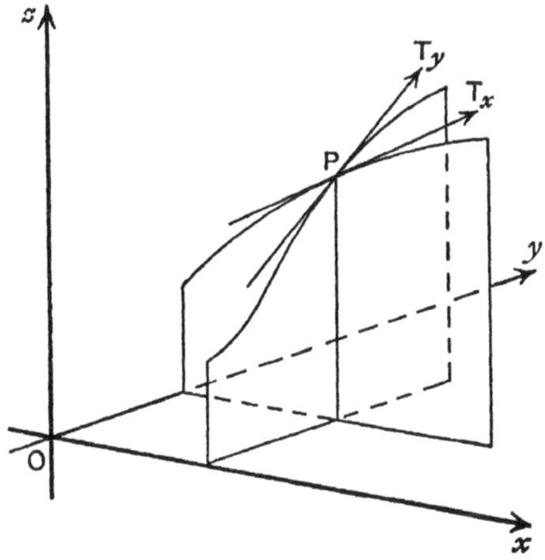

FIG. 81.

the surface $z=f(x,\,y)$ defined by the values x and y of the independent variables. Two vertical sections of the surface can be drawn through P, one perpendicular to Oy and the other perpendicular to Ox. The former is the curve on the surface passing through P in the W.E. direction and showing the variation of z as x varies. On this section, the value of y retains throughout the constant value allotted initially at P. The tangent PT_x to the section at P has gradient measured by the derivative of z as a function of x (y constant), i.e. by the partial derivative evaluated at $(x,\,y)$. Hence, $\frac{\partial z}{\partial x}$ is to be interpreted as the gradient of the W.E. section of the surface at P or, more shortly, as the W.E. gradient of the surface at P. In the same way, the value of $\frac{\partial z}{\partial y}$ at (x, y) measures the gradient of PT_y, the tangent at P to the vertical section of the surface perpendicular to Ox, i.e. the S.N. gradient of the surface at P. Hence:

The partial derivatives $\frac{\partial z}{\partial x}$ and $\frac{\partial z}{\partial y}$ measure the gradients of the surface $z = f(x, y)$ in two perpendicular directions at the point P defined by (x, y), the directions in the planes perpendicular to Oy and Ox respectively.

It is to be noticed that we have not yet obtained a measure of the gradient of the surface in any other direction.

The remarks (6.2 above) on the meaning of an ordinary derivative apply, with obvious modifications, to partial derivatives. One additional point, however, must be stressed. The value of $\frac{\partial z}{\partial x}$ at a point (x, y) depends, not only on the value of x from which we start in the limiting variation defining the partial derivative, but also on the value of y which is kept constant during the limiting process. The partial derivative, as the W.E. gradient of the surface $z = f(x, y)$ at the point P, varies in value when the point P is moved in any way, i.e. when either or both of the variables x and y are varied. Similar remarks apply to the partial derivative $\frac{\partial z}{\partial y}$. Hence, the value of each partial derivative at a point (x, y) is a function of both x and y. To stress this point, we can refer to the *partial derived functions* and use the particularly appropriate notations $f_x'(x, y)$ and $f_y'(x, y)$.

No essentially new idea is involved in the definition of the partial derivatives and they can be evaluated exactly as ordinary derivatives. The variable other than the one directly concerned must, of course, be treated as a constant in the actual process of derivation. In particular, the rules for finding the partial derivatives of composite functions are similar to those for derivatives. The modified form of the function of a function rule, however, merits separate notice. It can be expressed:

If z is a single-valued function of u where u is a single-valued function of x and y, then z is a function of x and y with the partial derivatives

$$\frac{\partial z}{\partial x}=\frac{dz}{du}\frac{\partial u}{\partial x} \quad \text{and} \quad \frac{\partial z}{\partial y}=\frac{dz}{du}\frac{\partial u}{\partial y}.$$

In particular, if u is any single-valued function of x and y, we have

$$\frac{\partial}{\partial x}u^n=nu^{n-1}\frac{\partial u}{\partial x};\quad \frac{\partial}{dx}e^u=e^u\frac{\partial u}{\partial x} \quad \text{and} \quad \frac{\partial}{\partial x}\log u=\frac{1}{u}\frac{\partial u}{\partial x}$$

and similar forms for the partial derivatives with respect to y.

The practical technique of partial derivation is fully illustrated by the following examples :

Ex. 1. $\frac{\partial}{\partial x}(2x-3y+1)=2$; $\frac{\partial}{\partial y}(2x-3y+1)=-3$.

Ex. 2. $\frac{\partial}{\partial x}(x^2+2xy-y^2)=2(x+y)$; $\frac{\partial}{\partial y}(x^2+2xy-y^2)=2(x-y)$.

Ex. 3. For any fixed values of the coefficients $a, b, c, \ldots$,

$$\frac{\partial}{\partial x}(ax+by+c)=a ; \quad \frac{\partial}{\partial y}(ax+by+c)=b ;$$

$$\frac{\partial}{\partial x}(ax^2+2hxy+by^2+2gx+2fy+c)=2(ax+hy+g) ;$$

$$\frac{\partial}{\partial y}(ax^2+2hxy+by^2+2gx+2fy+c)=2(hx+by+f).$$

Ex. 4.

$$\frac{\partial}{\partial x}\left(\frac{x^2}{x-y+1}\right)=\frac{1}{(x-y+1)^2}\left\{(x-y+1)\frac{\partial}{\partial x}(x^2)-x^2\frac{\partial}{\partial x}(x-y+1)\right\}$$

$$=\frac{2x(x-y+1)-x^2}{(x-y+1)^2}=\frac{x(x-2y+2)}{(x-y+1)^2};$$

$$\frac{\partial}{\partial y}\left(\frac{x^2}{x-y+1}\right)=x^2\frac{\partial}{\partial y}\left(\frac{1}{x-y+1}\right)=-\frac{x^2}{(x-y+1)^2}\frac{\partial}{\partial y}(x-y+1)$$

$$=\frac{x^2}{(x-y+1)^2}.$$

Ex. 5. $\frac{\partial}{\partial x}(x^2+y^2)^2=4x(x^2+y^2)$; $\frac{\partial}{\partial y}(x^2+y^2)^2=4y(x^2+y^2)$.

$$\frac{\partial}{\partial x}e^{x^2+y^2}=2xe^{x^2+y^2} ; \quad \frac{\partial}{\partial y}e^{x^2+y^2}=2ye^{x^2+y^2},$$

$$\frac{\partial}{\partial x}\log(x^2+y^2)=\frac{2x}{x^2+y^2} ; \quad \frac{\partial}{\partial y}\log(x^2+y^2)=\frac{2y}{x^2+y^2}.$$

Ex. 6. $\frac{\partial}{\partial x}(x^2e^y)=2xe^y$; $\frac{\partial}{\partial y}(x^2e^y)=x^2e^y$.

Ex. 7. If $z = x^2\sqrt{\dfrac{x+y}{x-y}}$, then

$$\log z = 2\log x + \tfrac{1}{2}\log(x+y) - \tfrac{1}{2}\log(x-y).$$

So
$$\frac{1}{z}\frac{\partial z}{\partial x} = \frac{2}{x} + \frac{1}{2}\frac{1}{x+y} - \frac{1}{2}\frac{1}{x-y} = \frac{2x^2 - xy - 2y^2}{x(x^2-y^2)},$$

i.e.
$$\frac{\partial z}{\partial x} = \frac{x(2x^2 - xy - 2y^2)}{\sqrt{(x+y)(x-y)^3}}.$$

Similarly
$$\frac{\partial z}{\partial y} = \frac{x^3}{\sqrt{(x+y)(x-y)^3}}.$$

Ex. 8. The definition of the *partial elasticities* of a function of two variables $z = f(x, y)$ is similar to that of the elasticity of a function of one variable:

$$\frac{Ez}{Ex} = \frac{x}{z}\frac{\partial z}{\partial x} = \frac{\partial(\log z)}{\partial(\log x)} \quad \text{and} \quad \frac{Ez}{Ey} = \frac{y}{z}\frac{\partial z}{\partial y} = \frac{\partial(\log z)}{\partial(\log y)}.$$

So, if $z = x^{\alpha}y^{\beta}$, where α and β are constants, then

$$\log z = \alpha \log x + \beta \log y$$

and
$$\frac{Ez}{Ex} = \alpha \quad \text{and} \quad \frac{Ez}{Ey} = \beta.$$

12.2 Partial derivatives of the second and higher orders.

The two partial derivatives of a function $z = f(x, y)$ are themselves functions of x and y. The partial derivation process can thus be repeated and we obtain two partial derivatives from each of the partial derivatives of z, i.e. four *second-order partial derivatives* of the original function z. By an extension of the notation, we can write the new second-order partial derivatives as

$$\frac{\partial^2 z}{\partial x^2} = \frac{\partial}{\partial x}\left(\frac{\partial z}{\partial x}\right); \quad \frac{\partial^2 z}{\partial y\,\partial x} = \frac{\partial}{\partial y}\left(\frac{\partial z}{\partial x}\right); \quad \frac{\partial^2 z}{\partial x\,\partial y} = \frac{\partial}{\partial x}\left(\frac{\partial z}{\partial y}\right) \quad \text{and} \quad \frac{\partial^2 z}{\partial y^2} = \frac{\partial}{\partial y}\left(\frac{\partial z}{\partial y}\right).$$

Alternatively, if $f_x'(x, y)$ and $f_y'(x, y)$ denote the (first-order) partial derivatives, the second-order partial derivatives appear as

$$f''_{xx}(x, y); \quad f''_{yx}(x, y); \quad f''_{xy}(x, y) \quad \text{and} \quad f''_{yy}(x, y)$$

or, more shortly, as

$$f_{xx}; \quad f_{yx}; \quad f_{xy} \quad \text{and} \quad f_{yy}.$$

The order of the suffixes indicates the order in which the partial derivations are carried out. Thus, $\dfrac{\partial^2 z}{\partial y\,\partial x} = f''_{yx}(x, y) = f_{yx}$ is the second-order partial derivative obtained by partial derivation first with respect to x and then with respect to y.

It is to be noticed that the two "cross" partial derivatives $\frac{\partial^2 z}{\partial y\,\partial x}$ and $\frac{\partial^2 z}{\partial x\,\partial y}$ are quite distinct in meaning and there is no *a priori* reason to assume that they have the same value at any point. A result can be established, however, stating that the two "cross" partial derivatives are identical in value, provided that certain conditions relating to the continuity of the function are satisfied.* This result, known as Young's Theorem, will not be proved here but it is verified below in the cases of two particular functions. It is, in fact, safe to assume that, for any ordinary continuous function, the order of partial derivation is immaterial :

$$\frac{\partial^2 z}{\partial y\,\partial x}=\frac{\partial^2 z}{\partial x\,\partial y}.$$

The following examples illustrate :

Ex. 1. From $\frac{\partial}{\partial x}\left(\frac{x^2}{x-y+1}\right)=\frac{x(x-2y+2)}{(x-y+1)^2}$, we derive

$$\frac{\partial^2}{\partial x^2}\left(\frac{x^2}{x-y+1}\right)=\frac{1}{(x-y+1)^4}$$

$$\times\{(2x-2y+2)(x-y+1)^2-x(x-2y+2)2(x-y+1)\}=\frac{2(y-1)^2}{(x-y+1)^3},$$

$$\frac{\partial^2}{\partial y\,\partial x}\left(\frac{x^2}{x-y+1}\right)=\frac{1}{(x-y+1)^4}$$

$$\times\{(-2x)(x-y+1)^2+x(x-2y+2)2(x-y+1)\}=-\frac{2x(y-1)}{(x-y+1)^3}.$$

From $\frac{\partial}{\partial y}\left(\frac{x^2}{x-y+1}\right)=\frac{x^2}{(x-y+1)^2}$, we derive

$$\frac{\partial^2}{\partial x\,\partial y}\left(\frac{x^2}{x-y+1}\right)=\frac{1}{(x-y+1)^4}\{2x(x-y+1)^2-x^2 2(x-y+1)\}$$

$$=-\frac{2x(y-1)}{(x-y+1)^3},$$

$$\frac{\partial^2}{\partial y^2}\left(\frac{x^2}{x-y+1}\right)=x^2\frac{\partial}{\partial y}\left\{\frac{1}{(x-y+1)^2}\right\}=\frac{2x^2}{(x-y+1)^3}.$$

* See Courant, *Differential and Integral Calculus*, Vol. II (1936), pp. 55-7.

Ex. 2. Since $\frac{\partial}{\partial x}\log(x^2+y^2)=\frac{2x}{x^2+y^2}$ and $\frac{\partial}{\partial y}\log(x^2+y^2)=\frac{2y}{x^2+y^2}$,

we have
$$\frac{\partial^2}{\partial x^2}\log(x^2+y^2)=\frac{2(x^2+y^2)-4x^2}{(x^2+y^2)^2}=-\frac{2(x^2-y^2)}{(x^2+y^2)^2},$$
$$\frac{\partial^2}{\partial y\,\partial x}\log(x^2+y^2)=2x\left\{-\frac{2y}{(x^2+y^2)^2}\right\}=-\frac{4xy}{(x^2+y^2)^2},$$
$$\frac{\partial^2}{\partial x\,\partial y}\log(x^2+y^2)=2y\left\{-\frac{2x}{(x^2+y^2)^2}\right\}=-\frac{4xy}{(x^2+y^2)^2},$$
$$\frac{\partial^2}{\partial y^2}\log(x^2+y^2)=\frac{2(x^2+y^2)-4y^2}{(x^2-y^2)^2}=\frac{2(x^2-y^2)}{(x^2+y^2)^2}.$$

Ex. 3. The partial derivatives of a function $\phi(x, y)$ are ϕ_x and ϕ_y. It is given that the partial derivatives of the ratio of ϕ_x to ϕ_y, considered as a function of x and y, are respectively negative and positive:
$$\frac{\partial}{\partial x}\left(\frac{\phi_x}{\phi_y}\right)<0 \quad\text{and}\quad \frac{\partial}{\partial y}\left(\frac{\phi_x}{\phi_y}\right)>0.$$

It is required to express these inequalities in terms of the first and second-order partial derivatives of $\phi(x, y)$.* We have
$$\frac{\partial}{\partial x}\left(\frac{\phi_x}{\phi_y}\right)=\frac{1}{\phi_y^2}\left(\phi_y\frac{\partial}{\partial x}\phi_x-\phi_x\frac{\partial}{\partial x}\phi_y\right)=\frac{\phi_y\phi_{xx}-\phi_x\phi_{xy}}{\phi_y^2},$$
$$\frac{\partial}{\partial y}\left(\frac{\phi_x}{\phi_y}\right)=\frac{1}{\phi_y^2}\left(\phi_y\frac{\partial}{\partial y}\phi_x-\phi_x\frac{\partial}{\partial y}\phi_y\right)=-\frac{\phi_x\phi_{yy}-\phi_y\phi_{xy}}{\phi_y^2}.$$
The inequalities are, therefore, equivalent to the forms
$$\phi_y\phi_{xx}-\phi_x\phi_{xy}<0 \quad\text{and}\quad \phi_x\phi_{yy}-\phi_y\phi_{xy}<0.$$

Each of the four second-order partial derivatives is a function of x and y and, by a further partial derivation process, we obtain eight partial derivatives of the third order. We can then proceed to partial derivatives of the fourth and higher orders. The notations we have given clearly extend to partial derivatives of order higher than the second. Young's Theorem can also be extended to show that the order of partial derivation is immaterial in a partial derivative of any order whatever. Partial derivatives of the third and higher orders are, however, of little practical use and we shall seldom have need of them in the present development.

* The problem is taken from Pareto, *Manuel d'économie politique* (2nd Ed. 1927), p. 575, where the mathematical steps given here are omitted. The function $\phi(x, y)$ is a utility function of an individual for two goods X and Y.

12.3 The signs of partial derivatives.

The definition of partial derivatives shows that $\frac{\partial z}{\partial x}$ and $\frac{\partial z}{\partial y}$ must measure the rates of change of the function $z=f(x, y)$ as x varies (y constant) and as y varies (x constant) from the point (x, y). Further, as we have seen, these partial derivatives measure the gradients of the surface $z=f(x, y)$ in two perpendicular directions. The meaning of the signs assumed by the partial derivatives is now clear. We have, for the partial derivative with respect to x:

(1) If $\frac{\partial z}{\partial x}>0$ at the point (a, b), then the function $z=f(x, y)$ increases as x increases from the value a (y remaining equal to the value b) and the surface $z=f(x, y)$ rises from W. to E. at the point where $x=a$ and $y=b$.

(2) If $\frac{\partial z}{\partial x}<0$ at the point (a, b), then the function decreases and the surface falls in the direction named above.

The sign of $\frac{\partial z}{\partial y}$ at the point (a, b) is to be interpreted in a similar way and refers to the variation of y from the value b (x remaining equal to the value a) and to the path on the surface through the point where $x=a$ and $y=b$ in the S.N. direction. The signs of the partial derivatives thus determine, at various points, the increasing or decreasing nature of the function, the rising or falling nature of the surface, in two particular and important directions. There remain, for later detailed consideration (14.1 below), the cases where one or both partial derivatives assume zero values.

The values of the two "direct" second-order partial derivatives of $z=f(x, y)$ have also quite simple interpretations. The partial derivative $\frac{\partial^2 z}{\partial x^2}$ measures the rate of change of $\frac{\partial z}{\partial x}$ as x increases from the point in question (y remaining constant), i.e. it measures the "acceleration" of the function $z=f(x, y)$ as x increases and y remains constant. In diagrammatic terms, $\frac{\partial^2 z}{\partial x^2}$ measures the rate of change of the W.E. gradient of the surface $z=f(x, y)$ as we move from the point concerned in the same W.E. direction, i.e. it determines the curvature of the surface in the W.E. direction. So:

(1) If $\frac{\partial^2 z}{\partial x^2} > 0$ at the point (a, b), then the function $z = f(x, y)$ changes at an increasing rate as x increases from the value a (y remaining equal to b) and the surface $z = f(x, y)$ is convex from below in the W.E. direction at the point (a, b).

(2) If $\frac{\partial^2 z}{\partial x^2} < 0$ at the point (a, b), then the function changes at a decreasing rate and the surface is concave from below in the direction named above.

The interpretation of the value and sign of $\frac{\partial^2 z}{dy^2}$ is exactly similar.

It is more difficult to interpret the value of the "cross" second-order partial derivative. The value of $\frac{\partial^2 z}{\partial x \, \partial y}$ at any point measures both the rate of change of $\frac{\partial z}{\partial x}$ as y increases (x remaining constant) and the rate of change of $\frac{\partial z}{\partial y}$ as x increases (y remaining constant). In diagrammatic terms, the value indicates the way in which the W.E. gradient of the surface $z = f(x, y)$ changes as we move in the S.N. direction, and the way in which the S.N. gradient changes as we move in the W.E. direction. So, if $\frac{\partial^2 z}{\partial x \, \partial y}$ is positive at a point P on the surface, the W.E. gradient increases as we move N. from P and the S.N. gradient of the surface increases as we move E. from P. It is difficult to visualise this state of affairs. For example, if the surface rises to the right and falls to the left as we look N. from P, then the positive sign of $\frac{\partial^2 z}{\partial x \, \partial y}$ implies that the surface tends to rise *more* rapidly to the right and to fall *more* steeply to the left as we actually move N. from P. The S.N. path on the surface through P moves on an increasingly precipitous incline to right and left. The opposite result holds if $\frac{\partial^2 z}{\partial x \, \partial y}$ is negative at P. These properties of the surface are quite different from, and independent of, the separate curvatures of the surface in the W.E. and S.N. directions.

One important point must be stressed at this stage. The various partial derivatives of a function serve to indicate the rising and falling nature and the curvature of the surface representing the function—but only in two particular directions, the " fundamental " directions in planes perpendicular to Ox and Oy. Nothing is said about the other directions on the surface, directions in which both x and y vary. For example, the fact that both $\frac{\partial^2 z}{\partial x^2}$ and $\frac{\partial^2 z}{\partial y^2}$ are positive at a point P on the surface $z=f(x, y)$ implies that the surface is convex from below in the W.E. direction and in the S.N. direction at P. It does *not* imply that the surface is also convex from below in any other direction (say the S.W. to N.E. direction) at P. In fact, it is quite possible that there is a " dip " in the surface between the two fundamental directions, so that it appears concave from below in the S.W. to N.E. direction.

The results obtained above clearly extend to allow for the cases where any or all of the variables are measured on logarithmic, instead of on natural, scales. For example,

$$\frac{\partial (\log z)}{\partial x}=\frac{1}{z}\frac{\partial z}{\partial x}$$

measures the rate of proportional change of $z=f(x, y)$ for actual changes in x (y constant) and is shown by the W.E. gradient of the surface obtained when z is taken on a logarithmic and x and y on natural scales. Again, the partial elasticity

$$\frac{Ez}{Ex}=\frac{\partial (\log z)}{\partial (\log x)}=\frac{x}{z}\frac{\partial z}{\partial x}$$

measures the proportional change in z for proportional changes in x (y constant) and is shown by the gradient in the W.E. direction of the surface obtained when all variables are plotted on logarithmic scales.

12.4 The tangent plane to a surface.

The simplest two-dimensional curve is a straight line, represented by a linear equation in two variables. The simplest surface in three dimensions is the plane and we can show that it is represented, as we expect, by a linear equation in three variables. Geometrical considerations (similar to those of 3.1 above) establish the simple result that, if two points in space have co-ordinates (x_1, y_1, z_1) and (x_2, y_2, z_2) referred to rectangular axes $Oxyz$, then the mid-point between them has co-ordinates $\left(\frac{x_1+x_2}{2}, \frac{y_1+y_2}{2}, \frac{z_1+z_2}{2}\right)$. Suppose, now, that (x_1, y_1, z_1) and (x_2, y_2, z_2) are any two points on the surface (whatever its form may be) corresponding to the linear equation

$$ax+by+cz=d$$

where a, b, c and d are constants. Then

$$ax_1+by_1+cz_1=d \quad \text{and} \quad ax_2+by_2+cz_2=d.$$

Adding, and dividing through by 2, we find

$$a\left(\frac{x_1+x_2}{2}\right)+b\left(\frac{y_1+y_2}{2}\right)+c\left(\frac{z_1+z_2}{2}\right)=d,$$

i.e. the mid-point of the two selected points also lies on the surface. This is true of any selected pair of points on the surface and it follows

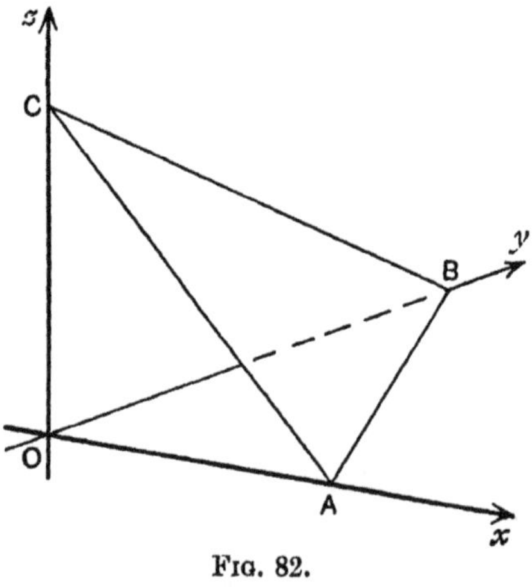

FIG. 82.

that the latter must be a plane. The general linear equation in three variables thus represents a plane. The actual location in space of a plane whose equation is given is easily determined by finding the points A, B and C where the plane cuts the axes (see Fig. 82). On the axis Ox, we have $y=z=0$. Substituting in the equation of the plane, $ax=d$. So

$$OA=\frac{d}{a}.$$

Similarly $$OB=\frac{d}{b} \quad \text{and} \quad OC=\frac{d}{c}.$$

The cases where one (or more) of the constants a, b and c has a zero value clearly correspond to planes parallel to one of the axes.

A plane is given as passing through the fixed point (x_1, y_1, z_1). If the equation of the plane is

$$ax+by+cz=d,$$

then $$ax_1+by_1+cz_1=d.$$

It follows that d can be eliminated so that the equation of the plane appears

$$a(x-x_1)+b(y-y_1)+c(z-z_1)=0,$$

or $$z-z_1=-\frac{a}{c}(x-x_1)-\frac{b}{c}(y-y_1) \quad (c\neq 0).$$

Hence $$\frac{\partial z}{\partial x}=-\frac{a}{c} \quad \text{and} \quad \frac{\partial z}{\partial y}=-\frac{b}{c}$$

are the gradients of the plane (referred to the horizontal plane Oxy) in the two fundamental directions perpendicular to Oy and to Ox respectively. The following result is thus obtained :

The equation of the plane passing through the point (x_1, y_1, z_1) with gradients α and β in the fundamental directions is

$$z-z_1=\alpha(x-x_1)+\beta(y-y_1).$$

In general; a plane through a given point P on the surface $z=f(x, y)$ cuts the surface in a curved section on which the point P lies. If it is possible to vary the position of the plane so that the part of the section including P encloses a smaller and smaller area and finally tends to close down on P itself, then the limiting position of the plane can be described as the *tangent plane* to the surface at P.* Suppose that the tangent plane exists at P and that a section of the surface and its tangent plane is taken by a plane through P perpendicular

* The possibilities are rather more complicated than this brief statement indicates. It may happen, of course, that no such limiting plane exists at all and the surface may have no tangent plane at a point where there is a "sharp point" or an "edge" of the surface. Further, it may happen that the section tends, not to a point, but to a straight line through P. The limiting plane is then a tangent plane touching the surface at all points on the straight line through P. In any case, it should be noticed that the section of the surface by the tangent plane can consist of a curve as well as the isolated point P. Only one part of the section need close down on the point P.

to Oy. We obtain, from the surface, the curved path in the W.E. direction and, from the tangent plane, the tangent PT_x to the path at P (see Fig. 81). It follows that the tangent plane, containing the line PT_x, has gradient in the direction of the plane perpendicular to Oy given by the value of $\frac{\partial z}{\partial x}$ at P. In the same way, the gradient of the tangent plane in the other fundamental direction is given by the value of $\frac{\partial z}{\partial y}$ at P. If the point P has co-ordinates (x_1, y_1, z_1), the fundamental gradients of the tangent plane at P can be denoted:

$$\left(\frac{\partial z}{\partial x}\right)_1 = f_x'(x_1, y_1) \quad \text{and} \quad \left(\frac{\partial z}{\partial y}\right)_1 = f_y'(x_1, y_1).$$

So:

The equation of the tangent plane at the point (x_1, y_1, z_1) on the surface $z = f(x, y)$ is

$$z - z_1 = \left(\frac{\partial z}{\partial x}\right)_1 (x - x_1) + \left(\frac{\partial z}{\partial y}\right)_1 (y - y_1).$$

This is an obvious extension of the form given for the tangent line to a two-dimensional curve (6.5 above).

As an example, the paraboloid surface $z = x^2 + y^2$ gives

$$\frac{\partial z}{\partial x} = 2x \quad \text{and} \quad \frac{\partial z}{\partial y} = 2y.$$

The tangent plane at (x_1, y_1, z_1) has equation

$$z - z_1 = 2x_1 (x - x_1) + 2y_1 (y - y_1),$$

i.e.
$$z = 2xx_1 + 2yy_1 + z_1 - 2(x_1^2 + y_1^2).$$

Since the point must lie on the surface, we have $z_1 = x_1^2 + y_1^2$ and

$$z + z_1 = 2xx_1 + 2yy_1$$

is the equation of the tangent plane.

An interesting result can now be deduced. If $\frac{\partial z}{\partial x}$ and $\frac{\partial z}{\partial y}$ are both positive at a point P on the surface $z = f(x, y)$, then the tangent plane at P is positively inclined to the plane Oxy in the two fundamental directions, and so in any direction in which x and y increase together. Hence, in the particular case in which $\frac{\partial z}{\partial x}$ and $\frac{\partial z}{\partial y}$ are both positive, the function $z=f(x, y)$ increases whenever x and y both increase. The converse result holds in the particular case when both partial derivatives are negative.

12.5 Partial derivatives of functions of more than two variables.

The definition of partial derivatives extends readily to cases of functions of more than two variables. If $u=f(x, y, z)$ is a single-valued function of three variables, then a function of one variable is obtained whenever two of the variables are given fixed values. This can be done in three different ways and we have u as a function of x only (y and z fixed), of y only (z and x fixed) and of z only (x and y fixed). Hence, we can define three partial derivatives, $\frac{\partial u}{\partial x}, \frac{\partial u}{\partial y}$ and $\frac{\partial u}{\partial z}$, of the function at any point (x, y, z), e.g.

$$\frac{\partial u}{\partial x}=\operatorname*{Lim}_{h\to 0}\frac{f(x+h, y, z)-f(x, y, z)}{h}.$$

Then three second-order partial derivatives are obtained from each of the first-order partial derivatives, making a group of nine in all. For all ordinary continuous functions, however, Young's Theorem asserts that the order of partial derivation is immaterial and only three of the six "cross" second-order partial derivatives are distinct. So we have only three "direct" partial derivatives $\left(\frac{\partial^2 u}{\partial x^2}, \frac{\partial^2 u}{\partial y^2} \text{ and } \frac{\partial^2 u}{\partial z^2}\right)$ and three "cross" partial derivatives $\left(\frac{\partial^2 u}{\partial x\,\partial y}, \frac{\partial^2 u}{\partial y\,\partial z} \text{ and } \frac{\partial^2 u}{\partial z\,\partial x}\right)$. The alternative notations for partial derivatives also extend, in an obvious way, to this three-variable case. Finally, if necessary, we can obtain partial derivatives of higher order than the second by further derivation processes.

In the general case of a single-valued function of n variables, $y=f(x_1, x_2, \ldots x_n)$, there are n first-order partial derivatives

$$\frac{\partial y}{\partial x_1}, \frac{\partial y}{\partial x_2}, \ldots \frac{\partial y}{\partial x_n}.$$

Each corresponds to the variation of y as one of the independent variables changes, the other $(n-1)$ variables remaining constant. There are then n^2 second-order partial derivatives but this number is reduced, in the case of ordinary continuous functions, since Young's

Theorem shows that the order of partial derivation is immaterial. There are, in fact, n " direct " partial derivatives

$$\frac{\partial^2 y}{\partial x_1^2}, \frac{\partial^2 y}{\partial x_2^2}, \ldots \frac{\partial^2 y}{\partial x_n^2}$$

and $\frac{1}{2}n(n-1)$ " cross " partial derivatives

$$\frac{\partial^2 y}{\partial x_1 \partial x_2}, \frac{\partial^2 y}{\partial x_1 \partial x_3}, \ldots \frac{\partial^2 y}{\partial x_1 \partial x_n}, \frac{\partial^2 y}{\partial x_2 \partial x_3}, \ldots \frac{\partial^2 y}{\partial x_2 \partial x_n}, \ldots .$$

The analytical interpretation of the partial derivatives extends at once. The first-order partial derivatives, evaluated at a given point, measure the rates of change of the function as one of the variables increases from the given value, the other variables remaining constant. Thus, if $\frac{\partial u}{\partial x}$ is positive at the point (a, b, c), then the function $u = f(x, y, z)$ increases as x increases from a, y and z having the fixed values b and c respectively. Further, the " direct " second-order partial derivatives measure the various " accelerations " of the function, one of the variables increasing from the given value and the others remaining constant. It is not possible, of course, to give any concrete diagrammatic interpretation of partial derivatives when there are more than two independent variables.

12.6 Economic applications of partial derivatives.

Under the conditions of 11.7 above, the market demand for any good X_r is a function of all market prices:

$$x_r = \phi_r(p_1, p_2, \ldots p_n) \qquad (r = 1, 2, \ldots n).$$

The partial derivatives of this function indicate the variations of demand as one of the prices varies, other prices remaining constant. The partial derivative $\frac{\partial x_r}{\partial p_r}$, which must be negative in the normal case, shows the rate at which the demand for X_r decreases as its price increases. It is usually convenient to put this rate into elasticity form—the partial elasticity of demand for X_r with respect to its price:

$$\eta_{rr} = -\frac{\partial(\log x_r)}{\partial(\log p_r)} = -\frac{p_r}{x_r}\frac{\partial x_r}{\partial p_r}.$$

This expression, independent of demand or price units, shows the rate of proportional decrease of demand for proportional increases in price. It is an extension of the ordinary demand elasticity obtained when demand is regarded as dependent only on the price of the good concerned. The ordinary demand elasticity is defined only when the prices of other goods are fixed and known; if these prices change, the whole demand law shifts and the elasticity must be evaluated anew. But our generalised demand law and the demand elasticity η_{rr} obtained from it automatically allow for these shifts in demand. In fact, η_{rr} is a function, not only of the price p_r, but also of all other prices; its value is thus seen to change when *any* of the prices change.

The partial derivative $\frac{\partial x_r}{\partial p_s}$ measures the rate of change in the demand for X_r when the price of another good X_s is increased, and it is best considered in relation to the corresponding partial derivative $\frac{\partial x_s}{\partial p_r}$, measuring the rate of change in the demand for X_s for increases in the price of X_r. If both partial derivatives are positive, then the demand for X_r increases with the price of X_s and the demand for X_s increases with the price of X_r. The goods X_r and X_s are then *competitive*, using this term in a broad sense. If both partial derivatives are negative, the demand for one good changes in the sense opposite to that of the price of the other good and the goods can be called, in the same broad way, *complementary* goods.* It is often convenient, again, to use the elasticity forms

$$\eta_{rs} = -\frac{\partial(\log x_r)}{\partial(\log p_s)} = -\frac{p_s}{x_r}\frac{\partial x_r}{\partial p_s}; \quad \eta_{sr} = -\frac{\partial(\log x_s)}{\partial(\log p_r)} = -\frac{p_r}{x_s}\frac{\partial x_s}{\partial p_r},$$

i.e. the partial elasticities of demand for one good with respect to the price of another good.

If the demand laws for two goods X_1 and X_2 are linear:

$$x_1 = a_1 - a_{11}p_1 + a_{12}p_2; \quad x_2 = a_2 + a_{21}p_1 - a_{22}p_2,$$

then the price elasticities of demand are

$$\eta_{11} = a_{11}\frac{p_1}{x_1}; \quad \eta_{12} = -a_{12}\frac{p_2}{x_1}; \quad \eta_{21} = -a_{21}\frac{p_1}{x_2} \quad \text{and} \quad \eta_{22} = a_{22}\frac{p_2}{x_2}$$

and all depend on the prices assumed for both goods.

* There remains the possibility that the partial derivatives are of opposite signs. In this case, the goods, on the present definition, are neither competitive nor complementary. See Schultz, *Interrelations of Demand*, Journal of Political Economy, 1933, and 19.7 below.

On the other hand, if the demand laws are of the form

$$x_1 = a_1 p_1^{-a_{11}} p_2^{a_{12}}; \quad x_2 = a_2 p_1^{a_{21}} p_2^{-a_{22}},$$

then we have linear relations when all variables are taken on logarithmic scales and the price elasticities of demand are constants:

$$\eta_{11} = a_{11}; \quad \eta_{12} = -a_{12}; \quad \eta_{21} = -a_{21} \quad \text{and} \quad \eta_{22} = a_{22}.$$

It is to be noticed that the two elasticities of demand for one good with respect to the price of the other are of the same sign, in each of the above cases, only if a_{12} and a_{21} are of the same sign.

The production function $x = f(a_1, a_2, a_3, \ldots a_n)$ shows the dependence of the output of a good X on the amounts of the variable factors $A_1, A_2, A_3, \ldots A_n$ used in the production. Suppose that a

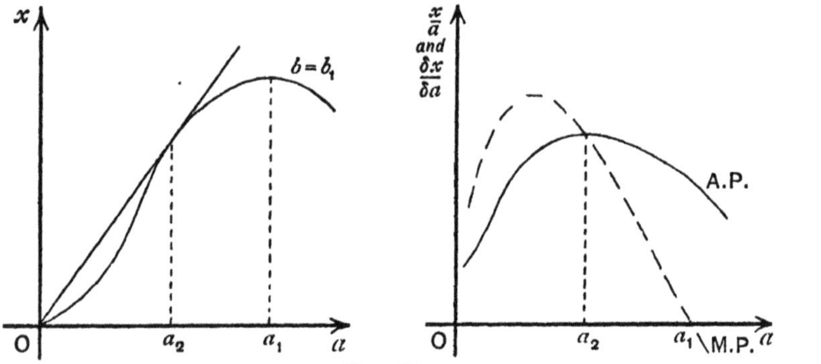

FIG. 83.

particular combination $(a_1, a_2, a_3, \ldots a_n)$ of the factors is used and that all factors except A_1 are regarded as fixed in amount. The ratio $\frac{x}{a_1}$ (the amount of product per unit of the factor A_1) can be called the *average product* and the partial derivative $\frac{\partial x}{\partial a_1}$ the *marginal product* of the factor A_1 at the combination $(a_1, a_2, a_3, \ldots a_n)$. The latter measures the rate of increase of output as the factor A_1 is increased, the amounts of other factors remaining unchanged. Both average and marginal products are functions of $a_1, a_2, a_3, \ldots a_n$, i.e. depend on the grouping of the factors considered. Average and marginal products of the other factors can be defined in similar ways.

With two variable factors A and B, average and marginal products can be shown diagrammatically on the vertical sections of the production surface $x = f(a, b)$. The section of the surface by the plane

$b=b_1$ shows the variation of product for a variable use of the factor A with a fixed amount b_1 of the factor B. If P (a, b_1) is a point on this section, then the average product of A at (a, b_1) is shown by the gradient (referred to Oa) of the line OP and the marginal product of A by the gradient (to Oa) of the tangent PT. In the normal case of production (see 11.8 above), the section, which can be called the product curve for the factor A, takes the form shown in the first diagram of Fig. 83. A second diagram can now be constructed showing the variation of average and marginal products (see Fig. 83). Average product increases to a maximum at a definite point $a=a_2$ where the tangent to the product curve passes through the origin. Average product equals marginal product at this position. Further, at some larger amount $a=a_1$, product is a maximum and marginal product zero. All three curves depend on the fixed amount b_1 of the factor B used and change in position and form when this amount b_1 is changed.

In the particular case where the production function is

$$x=2Hab-Aa^2-Bb^2 \qquad (H^2>AB),$$

then $$\frac{x}{a}=2Hb-Aa-B\frac{b^2}{a} \quad \text{and} \quad \frac{\partial x}{\partial a}=2(Hb-Aa).$$

So $$\frac{\partial}{\partial a}\left(\frac{x}{a}\right)=B\frac{b^2}{a^2}-A \quad \text{and} \quad \frac{\partial^2}{\partial a^2}\left(\frac{x}{a}\right)=-2B\frac{b^2}{a^3}<0.$$

Hence, for the fixed amount b_1 of the factor B, average product of A is a maximum when $a=\sqrt{\frac{B}{A}}\,b_1$, and both average and marginal products are then equal to $2(H-\sqrt{AB})b_1$. In this case, the "optimum" use of the factor A and the maximised average product increase in proportion to the fixed amount of the factor B used. The marginal product curve for A, when $b=b_1$, is a downward sloping straight line cutting Oa where $a=\frac{H}{A}b_1$, product here being a maximum. The average and marginal product curves are of the normal form of Fig. 83.

If $u=\phi(x_1, x_2, x_3, \ldots x_n)$ is one form of an individual's utility function for consumers' goods $X_1, X_2, X_3, \ldots X_n$, it might appear that the partial derivatives $\frac{\partial u}{\partial x_1}, \frac{\partial u}{\partial x_2}, \frac{\partial u}{\partial x_3}, \ldots \frac{\partial u}{\partial x_n}$ represent the "marginal utilities" of the various goods to the individual at a set of purchases $(x_1, x_2, x_3, \dots x_n)$. For example, $\frac{\partial u}{\partial x_1}$ measures the rate of increase of utility when the purchase of X_1 is increased, the purchases of other goods remaining unchanged. But we have seen that utility is a non-measurable magnitude to be indicated in general by

$$U = F(u) = F\{\phi(x_1, x_2, x_3, \dots x_n)\}$$

where $F(u)$ is any function such that $F'(u) > 0$. Can any meaning be attached, therefore, to increments of utility and so to "marginal utility"? We have

$$\frac{\partial U}{\partial x_1} = F'(u)\frac{\partial u}{\partial x_1}; \quad \frac{\partial U}{\partial x_2} = F'(u)\frac{\partial u}{\partial x_2}; \quad \frac{\partial U}{\partial x_3} = F'(u)\frac{\partial u}{\partial x_3}; \dots \frac{\partial U}{\partial x_n} = F'(u)\frac{\partial u}{\partial x_n},$$

i.e. marginal utilities involve the arbitrary function F and lose their definiteness when utility is regarded as non-measurable. But

$$\frac{\partial U}{\partial x_1} : \frac{\partial U}{\partial x_2} : \frac{\partial U}{\partial x_3} : \dots : \frac{\partial U}{\partial x_n} = \frac{\partial u}{\partial x_1} : \frac{\partial u}{\partial x_2} : \frac{\partial u}{\partial x_3} : \dots : \frac{\partial u}{\partial x_n}$$

and the *ratios* of the marginal utilities are definite concepts quite independent of the non-measurability of utility. This fact will be developed in the following chapter (13.7 below).

Suppose, for example, that one form of the utility function is

$$u = (x+a)^\alpha (y+b)^\beta,$$

where only two goods X and Y are considered by the individual. Then

$$\frac{\partial u}{\partial x} = \alpha (x+a)^{\alpha-1}(y+b)^\beta \quad \text{and} \quad \frac{\partial u}{\partial y} = \beta (x+a)^\alpha (y+b)^{\beta-1},$$

i.e. $$\frac{\partial u}{\partial x} : \frac{\partial u}{\partial y} = \frac{\alpha}{x+a} : \frac{\beta}{y+b}.$$

Another form of the same utility function is

$$u' = \alpha \log(x+a) + \beta \log(y+b),$$

whence $$\frac{\partial u'}{\partial x} = \frac{\alpha}{x+a} \quad \text{and} \quad \frac{\partial u'}{\partial y} = \frac{\beta}{y+b}.$$

The ratio of marginal utilities, $\frac{\alpha}{x+a} : \frac{\beta}{y+b}$, is thus perfectly definite.

12.7 Homogeneous functions.

When the variables x and y are increased or decreased in a fixed proportion from given values, the corresponding increase or decrease in the function $z=f(x, y)$ may be in greater, in equal or in less proportion. In the very special case where $z=f(x, y)$ increases or decreases always in the *same* proportion as x and y, the function is said to be homogeneous of the first degree, or to be *linear and homogeneous*. Such a function has the property that its value is doubled whenever the values of the independent variables are simultaneously doubled from any position whatever. And a similar result holds for any ratio other than that of doubling. Hence :

Definition : $z=f(x, y)$ is a linear homogeneous function if $f(\lambda x, \lambda y)=\lambda f(x, y)$ for any point (x, y) and for any value of λ whatever.

To quote particular examples, each of the following function types is linear and homogeneous :

(1) $z=ax+by$, (2) $z=ax^{\alpha}y^{1-\alpha}$,

(3) $z=\sqrt{ax^2+2hxy+by^2}$, (4) $z=\dfrac{ax^2+2hxy+by^2}{cx+dy}$

where the coefficients, a, b, c, ... , and the index α are constants. The fact that z changes proportionally for proportional changes in x and y can be verified in each case.

If P is a given point on the surface representing a linear homogeneous function, then any point with co-ordinates proportional to those of P also lies on the surface. The points (x, y, z), $(\frac{1}{2}x, \frac{1}{2}y, \frac{1}{2}z)$ and $(2x, 2y, 2z)$ all lie on the surface if one of them does. But all points of this nature lie on the straight line OP joining the origin to the given point P. In fact, the line joining O to any point on a linear homogeneous surface lies entirely in the surface. The surface is completely described by lines passing through the origin and is a particular case of what is called a *ruled surface*. It also follows that the tangent plane at P touches the surface, not only at P, but also at all points on the line OP. A cone with vertex at the origin is an easily visualised example of the type of surface corresponding to a linear homogeneous function.

The sections of a linear homogeneous surface by planes perpendicular to an axis also reflect its special properties. If P is a point (x, y) on the contour by the plane $z=z_1$, then the point Q $(2x, 2y)$ must lie on the contour by the plane $z=2z_1$. But the points P and Q lie on a line through O so that $OQ=2OP$. Hence, any radius through O in the plane Oxy cuts the z_1 and $2z_1$ contours in points of which the second is twice as far from O as the first. The contour $2z_1$ is of exactly the same shape as the contour z_1 but radially double in size. A similar result holds for any pair of contours :

All contours of the linear homogeneous surface are similar in shape and vary in size radially according to the ratios of the fixed values of z defining the contours.

Any one contour is a radial " projection " of any other and the whole system of contours can be derived given any one contour. Further, the tangents to the various contours at points where they are cut by a given radius through O must all be parallel.

Exactly similar results hold of sections of the linear homogeneous surface by planes perpendicular to Ox or Oy. The various sections of such a system are radial projections of any one section of the system and vary in size according to the fixed values of x or y defining the sections. Also, from the ruled surface property, the vertical section of the linear homogeneous surface by a plane through O and a point P on the surface consists of a straight line.

The linear homogeneous function is the simplest case of a wider class of homogeneous functions. More generally, if

$$f(\lambda x, \lambda y)=\lambda^r f(x, y)$$

for any point (x, y) and for any value of λ, then the function $z=f(x, y)$ is homogeneous of the rth degree. For example, $z=\dfrac{a_1x+b_1y}{a_2x+b_2y}$ is homogeneous of zero degree, $z=ax^2+2hxy+by^2$ is homogeneous of the second degree and $z=ax^\alpha y^\beta$ is homogeneous of degree $(\alpha+\beta)$. The diagrammatic properties of a general homogeneous function can be illustrated in the quadratic case $(r=2)$. The vertical section of a quadratic homogeneous surface by a plane through O and a given point on the surface is always a parabola with vertex at O and axis along Oz. The contours of the surface are radial projections of each other and vary in size according to the *square roots* of the values of z

defining them. A similar property holds also of the vertical sections of the surface by planes perpendicular to Ox or Oy.

The definition of homogeneous functions extends without difficulty to the case of functions of more than two variables. The function $y=f(x_1, x_2, x_3, \ldots)$ is homogeneous of the rth degree if

$$f(\lambda x_1, \lambda x_2, \lambda x_3, \ldots)=\lambda^r f(x_1, x_2, x_3, \ldots)$$

for any point $(x_1, x_2, x_3, \ldots)$ and for any value of λ. The case $r=1$ gives the linear homogeneous function, and here the proportional change in the values of all the independent variables produces an equal proportional change in the value of y.

12.8 Euler's Theorem and other properties of homogeneous functions.

If $z=f(x, y)$ is a linear homogeneous function of two variables, then the following properties can be shown to hold at any point (x, y) whatever:

(1) The function can be written in either of the forms

$$z=x\phi\left(\frac{y}{x}\right)=y\psi\left(\frac{x}{y}\right),$$

where ϕ and ψ are some functions of a single variable.

(2) The partial derivatives $\frac{\partial z}{\partial x}$ and $\frac{\partial z}{\partial y}$ are functions of the ratio of x to y only.

(3) *Euler's Theorem*: $\quad x\frac{\partial z}{\partial x}+y\frac{\partial z}{\partial y}=z.$

(4) The direct second-order partial derivatives are expressed in terms of the cross second-order partial derivative

$$\frac{\partial^2 z}{\partial x^2}=-\frac{y}{x}\frac{\partial^2 z}{\partial x\,\partial y} \quad \text{and} \quad \frac{\partial^2 z}{\partial y^2}=-\frac{x}{y}\frac{\partial^2 z}{\partial x\,\partial y}.$$

The proofs of these results can be set out formally:

Since $\quad f(\lambda x, \lambda y)=\lambda f(x, y)\quad$ for any value of λ,

we have $\quad f\left(1, \frac{y}{x}\right)=\frac{1}{x}f(x, y)\quad$ for $\lambda=\frac{1}{x}$,

i.e. $\quad z=xf\left(1, \frac{y}{x}\right)=x\phi\left(\frac{y}{x}\right),$

since $f\left(1, \frac{y}{x}\right)$ is a function of $\frac{y}{x}$ only. Similarly, for $\lambda=\frac{1}{y}$,

$$z=yf\left(\frac{x}{y}, 1\right)=y\psi\left(\frac{x}{y}\right).$$

Finding the partial derivative of $z=x\phi\left(\frac{y}{x}\right)$ with respect to x:

$$\frac{\partial z}{\partial x}=\phi\left(\frac{y}{x}\right)+x\phi'\left(\frac{y}{x}\right)\frac{\partial}{\partial x}\left(\frac{y}{x}\right)=\phi\left(\frac{y}{x}\right)-\frac{y}{x}\phi'\left(\frac{y}{x}\right)$$

where $\phi'\left(\frac{y}{x}\right)$ denotes the derivative of the single variable function $\phi\left(\frac{y}{x}\right)$ with respect to $\frac{y}{x}$. Further,

$$\frac{\partial z}{\partial y}=x\phi'\left(\frac{y}{x}\right)\frac{\partial}{\partial y}\left(\frac{y}{x}\right)=\phi'\left(\frac{y}{x}\right).$$

Hence, both $\frac{\partial z}{\partial x}$ and $\frac{\partial z}{\partial y}$ appear as functions of the ratio $\frac{y}{x}$ only.

Now, $$x\frac{\partial z}{\partial x}+y\frac{\partial z}{\partial y}=x\phi\left(\frac{y}{x}\right)-y\phi'\left(\frac{y}{x}\right)+y\phi'\left(\frac{y}{x}\right)=x\phi\left(\frac{y}{x}\right)=z,$$

which is Euler's Theorem. Finally, to prove the last result, we notice that Euler's Theorem holds for any values of x and y whatever, i.e. $x\frac{\partial z}{\partial x}+y\frac{\partial z}{\partial y}$ is *identically* equal to z. This identity is maintained no matter how x and y are varied and it follows that the partial derivatives of one expression are equal to the corresponding partial derivatives of the other.* Hence:

$$\frac{\partial}{\partial x}\left(x\frac{\partial z}{\partial x}+y\frac{\partial z}{\partial y}\right)=\frac{\partial z}{\partial x},$$

i.e. $$\frac{\partial z}{\partial x}+x\frac{\partial^2 z}{\partial x^2}+y\frac{\partial^2 z}{\partial x\,\partial y}=\frac{\partial z}{\partial x},\qquad \text{i.e.}\quad \frac{\partial^2 z}{\partial x^2}=-\frac{y}{x}\frac{\partial^2 z}{\partial x\,\partial y}.$$

A similar result is obtained by partial derivation with respect to y.

* The equation of the derivatives of each side of an identity is a device of wide application (see, e.g., 14.8 below). Before applying the device, it is essential to establish that the relation used holds identically for all the variables with respect to which the derivation is carried out. A relation holding only for certain values of the variables cannot be treated in this way.

Two examples make this clear. For all values of x, we have

$$(1+x)^2=1+2x+x^2.$$

Hence $$\frac{d}{dx}(1+x)^2=\frac{d}{dx}(1+2x+x^2)=2(1+x).$$

But $$x^2+2=3x$$

holds only for the particular values $x=1$ and $x=2$. Hence

$$\frac{d}{dx}(x^2+2)\neq\frac{d}{dx}(3x),$$

as can be verified at $x=1$ or $x=2$.

The truth of Euler's Theorem is also clear in diagrammatic terms. The tangent plane (12.4 above) at the point (x_1, y_1, z_1) on the surface $z=f(x, y)$ has equation

$$z - z_1 = \left(\frac{\partial z}{\partial x}\right)_1 (x - x_1) + \left(\frac{\partial z}{\partial y}\right)_1 (y - y_1).$$

It passes through the origin if $x=y=z=0$ satisfies the equation,

i.e. if
$$z_1 = \left(\frac{\partial z}{\partial x}\right)_1 x_1 + \left(\frac{\partial z}{\partial y}\right)_1 y_1.$$

Euler's Theorem asserts that this condition is satisfied at all points on a linear homogeneous surface. The tangent plane thus passes through the origin at all points, a fact which agrees with the ruled surface property of linear homogeneous surfaces.

The four properties generalise to the case of a function $z=f(x, y)$ which is homogeneous of degree r:

(1) $z = x^r \phi\left(\frac{y}{x}\right) = y^r \psi\left(\frac{x}{y}\right)$.

(2) $\frac{\partial z}{\partial x}$ and $\frac{\partial z}{\partial y}$ are homogeneous of degree $(r-1)$.

(3) $x\frac{\partial z}{\partial x} + y\frac{\partial z}{\partial y} = rz$.

(4) $x^2\frac{\partial^2 z}{\partial x^2} + 2xy\frac{\partial^2 z}{\partial x\,\partial y} + y^2\frac{\partial^2 z}{\partial y^2} = r(r-1)z$.

The first of these extended results is established, exactly as before, by using the definition of a homogeneous function. The second and third results then follow by derivation. The fourth result is obtained from the third result, which is identically true for all values of x and y, by finding the separate partial derivatives:

$$\frac{\partial}{\partial x}\left(x\frac{\partial z}{\partial x} + y\frac{\partial z}{\partial y}\right) = r\frac{\partial z}{\partial x},$$

i.e.
$$x\frac{\partial^2 z}{\partial x^2} + y\frac{\partial^2 z}{\partial x\,\partial y} = (r-1)\frac{\partial z}{\partial x}.$$

Similarly
$$x\frac{\partial^2 z}{\partial x\,\partial y} + y\frac{\partial^2 z}{\partial y^2} = (r-1)\frac{\partial z}{\partial y}.$$

Multiplying by x and y respectively and adding,

$$x^2\frac{\partial^2 z}{\partial x^2} + 2xy\frac{\partial^2 z}{\partial x\,\partial y} + y^2\frac{\partial^2 z}{\partial y^2} = (r-1)\left(x\frac{\partial z}{\partial x} + y\frac{\partial z}{\partial y}\right) = r(r-1)z.$$

Finally, the properties hold, in modified and extended forms, for homogeneous functions of more than two variables. It is not necessary to set out the properties in the completely general case here; the nature of the extensions is sufficiently obvious.*

12.9 The linear homogeneous production function.

The case where the production function of a good X with respect to variable factors $A_1, A_2, A_3, \ldots A_n$ is linear and homogeneous is one of particular interest. From the definition and results (1) and (2) of 12.8 above, the case is characterised by the fact that a proportional increase in *all* factors leads to a proportional increase in product and leaves the average and marginal product of each factor unaltered. We have, in fact, the case of " constant returns to scale " where only the *relative* amounts of the factors used is important and not the actual scale of production. For example, if wheat is produced with land and labour under constant returns to scale, then the wheat product is (e.g.) doubled when twice the number of men are employed on twice the area of land. Further, the product of wheat per man or per acre depends only on the number of men employed per acre, as does the marginal product of wheat per man or acre.

In the case of two factors A and B and constant returns to scale, the production surface is subject to restrictions additional to the normal ones already indicated (11.8 above). The surface is ruled by straight lines through the origin and any section through Ox consists of a straight line. The constant product curves in the plane Oab are now radial projections of each other and vary in size according to the constant products which define them. In particular, any radius through O cuts the curves in points where the tangents are parallel. So, in the normal case of Fig. 78, the curves OA and OB, marking the points with tangents parallel to an axis, reduce to straight lines. The sections of the production surface by planes perpendicular to (e.g.) Ob are also radial projections of each other and, in the normal case of Fig. 79, the maximum points all lie on a straight line through O, i.e. as the amount used of the factor B increases, the maximum product obtainable by varying A increases in proportion. Further, there is one line through O which touches all sections, i.e. the

* See Examples XII, 24-6, and 18.5 below.

maximised value of the average product of A is the same no matter what fixed amount of B is used.

Euler's Theorem states that, for any combination of factors,

$$x = a\frac{\partial x}{\partial a} + b\frac{\partial x}{\partial b},$$

and this can be illustrated diagrammatically (Fig. 84). P is the point on the production surface for the combination (a, b) of factors and the tangent PQ to the section perpendicular to Ob is drawn. Then

$$\frac{\partial x}{\partial a} = \text{gradient of } PQ = \frac{RP}{QR} = \frac{RP}{OM}, \quad \text{i.e.} \quad a\frac{\partial x}{\partial a} = RP.$$

So $$b\frac{\partial x}{\partial b} = x - a\frac{\partial x}{\partial a} = MP - RP = MR$$

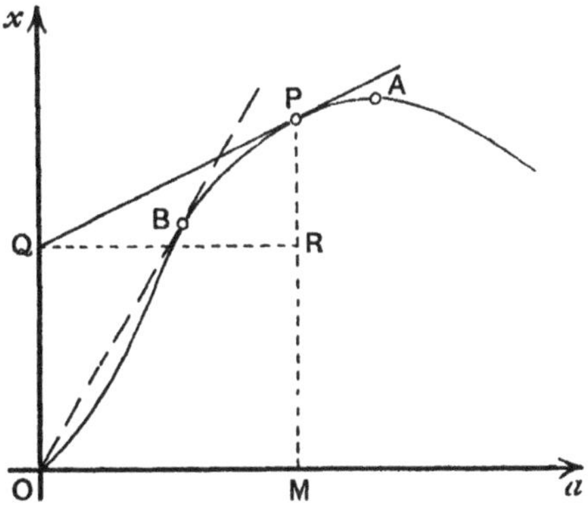

FIG. 84.

The total product MP is divided, $MR = b\frac{\partial x}{\partial b}$ and $RP = a\frac{\partial x}{\partial a}$. At the point A on the section where total product is a maximum, $\frac{\partial x}{\partial a} = 0$ and so $\frac{\partial x}{\partial b} = \frac{x}{b}$. At such a point, the average product of B is a maximum on the other section perpendicular to Oa. Similarly, at the point B on the section where the average product of A is a maximum, $\frac{\partial x}{\partial a} = \frac{x}{a}$ and so $\frac{\partial x}{\partial b} = 0$, i.e. the total product is a maximum on the other section perpendicular to Oa. There is a correspondence between the maximised total product for variation of one factor and the maximised average product for variation of the other factor.

The following are simple examples of linear homogeneous production functions corresponding to constant returns to scale :

$$(1)\ x=Aa^{\alpha}b^{1-\alpha}\,;\ (2)\ x=\frac{2Hab-Aa^2-Bb^2}{Ca+Db}\,;\ (3)\ x=\sqrt{2Hab-Aa^2-Bb^2}.$$

($H^2>AB$ in the two latter cases.) In each case, it is easily verified that the average products $\left(\frac{x}{a}\text{ and }\frac{x}{b}\right)$ and the marginal products $\left(\frac{\partial x}{\partial a}\text{ and }\frac{\partial x}{\partial b}\right)$ are functions of $\frac{b}{a}$ only, i.e. depend only on the ratio of the factors used. Case (1) corresponds to the simpler case of normal production and cases (2) and (3) to the full normal case illustrated by Figs. 78 and 79.

EXAMPLES XII

Partial derivatives

1. Find the first and second-order partial derivatives of each of the functions x^3+y^3-3xy; $\frac{1}{\sqrt{x^2+y^2}}$; e^{x-y} and $\log\frac{x}{x+y}$. Verify Young's Theorem that the order of partial derivation is immaterial in each case.

2. Evaluate the partial elasticities of x^2e^y and $(x+y)e^{x+y}$.

3. Obtain the partial derivatives of $\frac{a_1x+b_1y+c_1}{a_2x+b_2y+c_2}$.

4. By logarithmic derivation, evaluate the partial derivatives of

$$\frac{(x+y)(x+2y)}{(x-y)(x-2y)}$$

and show that $\frac{\partial}{\partial x}\{(x+y)^ne^{x+y}\}=\frac{\partial}{\partial y}\{(x+y)^ne^{x+y}\}=(x+y+n)(x+y)^{n-1}e^{x+y}$.

5. Show that
$$\frac{\partial^n}{\partial x^r\,\partial y^{n-r}}\{(x+y)e^{x+y}\}=(x+y+n)e^{x+y}$$
and that
$$\frac{\partial^n}{\partial x^r\,\partial y^{n-r}}\{(x-y)e^{x+y}\}=(x-y+2r-n)e^{x+y}.$$

6. Show that $z=\frac{x^2}{x-y}$ and its partial derivatives of all orders become infinite at any point where $x=y$.

7. If $z=f(u)$ where u is a function of x and y, show that

(*a*) $\frac{\partial z}{\partial x}=\frac{\partial z}{\partial y}$ if $u=x+y$; (*b*) $x\frac{\partial z}{\partial x}=y\frac{\partial z}{\partial y}$ if $u=xy$; (*c*) $x\frac{\partial z}{\partial x}+y\frac{\partial z}{\partial y}=0$ if $u=\frac{x}{y}$.

Verify in the case where $f(u)=\log u$.

8. If $z=\phi(u)+\psi(v)$, where $\phi(u)$ is a single-valued function of $u=x+ay$ and $\psi(v)$ is a single-valued function of $v=x-ay$, show that $\frac{\partial^2z}{\partial y^2}=a^2\frac{\partial^2z}{\partial x^2}$. Verify when $\phi(u)=u^2$ and $\psi(v)=e^v$.

9. Evaluate the first and second-order partial derivatives of

$$u=(x^2+2xy-y^2)e^z$$

and verify that the order of partial derivation is immaterial.

10. Show that the surface $z=\sqrt{xy}$ (x and y positive) rises and is concave from below in each of the fundamental directions.

11. If $\dfrac{\partial z}{\partial x}>0$ and $\dfrac{\partial z}{\partial y}<0$ at a point (x, y), show that z increases whenever x increases and y decreases from the point (x, y).

12. Find the equation of the plane passing through the points with co-ordinates (2, 1, 0), (−1, 0, 4), and (1, 2, 1) referred to rectangular axes. Where does the plane cut the axes?

13. If λ is a parameter, show that the equation $x+y+\lambda z=1$ represents a system of planes intersecting in a given line in the co-ordinate plane Oxy.

14. Show that the tangent plane at the point (x_1, y_1, z_1) on the surface $z=\sqrt{xy}$ has equation $xy_1+yx_1-2zz_1=0$.

15. Find the equation of the tangent plane at the point (2, −1, 2) on the surface $z=\dfrac{x}{x+y}$. Where does it cut the axes? Deduce that z decreases whenever x and y increase from the values $x=2$ and $y=-1$.

Homogeneous functions

16. Graph the contours of the surface $z=\sqrt{xy}$ given by $z=2$ and $z=4$ and show that the latter is radially double the size of the former.

17. Show that the contours of the surface $z=\dfrac{x^2+y^2}{x+y}$ form a system of circles which are radial projections of each other.

18. Show that the following are linear homogeneous functions :

$$z=\sqrt[3]{x^2y}\ ; \quad z=\sqrt{x^2+y^2}\ ; \quad z=\frac{x^3+y^3}{x^2+y^2}.$$

Express each in the form $z=x\phi\left(\dfrac{y}{x}\right)$. Obtain $\dfrac{\partial z}{\partial x}$ and $\dfrac{\partial z}{\partial y}$ and so verify Euler's Theorem in each case.

19. Verify property (4) of 12.8 in the case of $z=\sqrt{xy}$.

20. Show that $z=(a_1x+b_1y)^{\alpha}(a_2x+b_2y)^{1-\alpha}$ and $z=a_1x^{\alpha_1}y^{1-\alpha_1}+a_2x^{\alpha_2}y^{1-\alpha_2}$ are linear and homogeneous and verify Euler's Theorem in each case.

21. For each of the quadratic homogeneous functions $z=x^2-xy+2y^2$ and $z=\dfrac{x^3+y^3}{x-y}$, show that the partial derivatives are linear and homogeneous and verify that $x\dfrac{\partial z}{\partial x}+y\dfrac{\partial z}{\partial y}=2z$.

22. For the homogeneous function $z=ax^{\alpha}y^{\beta}$, show that

$$x\frac{\partial z}{\partial x}+y\frac{\partial z}{\partial y}=(\alpha+\beta)z \quad \text{and} \quad x^2\frac{\partial^2 z}{\partial x^2}+2xy\frac{\partial^2 z}{\partial x\,\partial y}+y^2\frac{\partial^2 z}{\partial y^2}=(\alpha+\beta)(\alpha+\beta-1)z.$$

23. If z is a function of x and y homogeneous of degree zero, show that z can be written as a function of $\frac{y}{x}$ only and that $\frac{\partial z}{\partial y} = -\frac{x}{y}\frac{\partial z}{\partial x}$. Verify by considering $z = \log\frac{x-y}{x+y}$ and $z = \frac{x^2 - xy + 2y^2}{2x^2 + xy + y^2}$.

24. If $y = f(x_1, x_2, x_3, \dots x_n)$ is homogeneous of degree r, show that

$$y = x_1^r \phi\left(\frac{x_2}{x_1}, \frac{x_3}{x_1}, \dots \frac{x_n}{x_1}\right),$$

where ϕ is some function of $(n-1)$ variables. Deduce that the partial derivatives are homogeneous of degree $(r-1)$ and that

$$x_1\frac{\partial y}{\partial x_1} + x_2\frac{\partial y}{\partial x_2} + x_3\frac{\partial y}{\partial x_3} + \dots + x_n\frac{\partial y}{\partial x_n} = ry.$$

25. If y is a linear homogeneous function of $x_1, x_2, \dots x_n$, show that

$$\frac{\partial^2 y}{\partial x_1^2} = -\frac{1}{x_1}\left(x_2\frac{\partial^2 y}{\partial x_1 \partial x_2} + x_3\frac{\partial^2 y}{\partial x_1 \partial x_3} + \dots + x_n\frac{\partial^2 y}{\partial x_1 \partial x_n}\right)$$

and similar expressions for the other direct partial derivatives.

26. Show that the relation between the second-order partial derivatives of a function of n variables, homogeneous of degree r, is

$$x_1^2\frac{\partial^2 y}{\partial x_1^2} + x_2^2\frac{\partial^2 y}{\partial x_2^2} + \dots + 2x_1x_2\frac{\partial^2 y}{\partial x_1 \partial x_2} + \dots = r(r-1)y.$$

Economic applications of partial derivatives and homogeneous functions

27. If the demand laws for two goods are given by (4) of 11.7, show that the "direct" price elasticities of demand are independent of the prices while the "cross" price elasticities are determined in sign by the constants a_{12} and a_{21} respectively.

28. The employment of $100a$ men-hours on b acres of land gives $x = 2(12ab - 5a^2 - 4b^2)$ bushels of wheat. Graph the average and marginal product curves for labour when 10 acres are cultivated. Compare with the product curve of Fig. 80.

29. Draw the average and marginal product curves for labour employed on (*a*) 10 acres and (*b*) 100 acres when $x = \frac{40}{a+b}(12ab - 5a^2 - 4b^2)$ is the wheat product for $100a$ men-hours on b acres. Verify that the maximum average product is the same in the two cases, ten times as much labour being required in the second case as compared with the first.

30. For the linear homogeneous production function $x = \frac{2Hab - Aa^2 - Bb^2}{Ca + Db}$, show that the average and marginal products of the factors depend only on the ratio of the factors and verify that the product is always a times the marginal product of A plus b times the marginal product of B.

31. The production function $x = \sqrt{2Hab - Aa^2 - Bb^2}$ is linear and homogeneous; show that the maximum value of the average product of A is a constant $\sqrt{\frac{H^2 - AB}{B}}$ independent of the fixed amount of B used.

32. The production function is $x = Aa^{\alpha}b^{\beta}$ where $\alpha + \beta < 1$. Show that there are decreasing returns to scale and deduce that the total product is greater than a times the marginal product of A plus b times the marginal product of B.

33. If a men are employed in planting b acres with timber, the amount of timber cut after t years is $x = f(a, b, t)$. What meanings can be attached to $\frac{\partial x}{\partial a}$, $\frac{\partial x}{\partial b}$ and $\frac{\partial x}{\partial t}$? If $x = Aa^{\alpha}b^{1-\alpha}t^{\beta}$, where α and β are fixed positive fractions, show that there are constant returns to land and labour after a fixed time t and that, for given employments of land and labour, the timber product increases, but in a decreasing proportion, as time goes on.

34. Find the ratio of the marginal utilities for two goods when the utility function is $u = ax + by + c\sqrt{xy}$. Verify that the same result is obtained when the utility function is written $u' = \log(ax + by + c\sqrt{xy})$.

35. If $U = F(u)$ is the index-function of utility where $u = axy$, show that (*a*) $\frac{\partial^2 U}{\partial x^2} = 0$, when $F(u) = u$; (*b*) $\frac{\partial^2 U}{\partial x^2} < 0$, when $F(u) = \sqrt{u}$; and (*c*) $\frac{\partial^2 U}{\partial x^2} > 0$, when $F(u) = e^u$.

36. If $u = \phi(x, y, z, \ldots)$ is one form of the utility function for goods $X, Y, Z, \ldots$ and if $U = F(u)$ is the index-function of utility, show that

$$\frac{\partial^2 U}{\partial x^2} = F'(u)\frac{\partial^2 u}{\partial x^2} + F''(u)\left(\frac{\partial u}{\partial x}\right)^2,$$

and similarly for the other second-order partial derivatives. Deduce that the sign of each of these derivatives varies, in general, with the choice of the arbitrary function F. Illustrate with the results of the previous example.

CHAPTER XIII

DIFFERENTIALS AND DIFFERENTIATION

13.1 The variation of a function of two variables.

THERE is only one way in which the value of a function of a single variable can change, i.e. by changing the value of the independent variable. The variation of the function is then adequately described by means of the derivative. The rate of change of $y=f(x)$ from any point x is measured by $f'(x)$ and $\Delta y=f'(x)\Delta x$ is the approximate increment in y for an arbitrary small increment Δx in x.

The expression of the variation of a function of two variables is much less simple. The value of $z=f(x, y)$ changes when x changes (y remaining constant) at a rate measured by $\frac{\partial z}{\partial x}$. Hence, if $\Delta_x z$ denotes the increment in z due to an arbitrary small increment Δx in the variable x from a point (x, y), then

$$\Delta_x z=\frac{\partial z}{\partial x}\Delta x \text{ approximately.}$$

In the same way, if $\Delta_y z$ denotes the increment in z due to an arbitrary small increment Δy in the variable y from the point (x, y), then

$$\Delta_y z=\frac{\partial z}{\partial y}\Delta y \text{ approximately.}$$

But these are only two special ways in which the value of the function can change and there remains the important problem of expressing the variation of the function when the independent variables vary together in any way whatever. A single partial derivative is not sufficient here and an addition to our mathematical equipment is needed.

It is assumed that the function $z=f(x, y)$ possesses continuous partial derivatives at the point (x, y). If h and k denote arbitrary increments in the variables x and y from the point (x, y), the corresponding increment in the value of the function is

$$\begin{aligned}\Delta z &= f(x+h, y+k) - f(x, y) \\ &= \frac{f(x+h, y+k) - f(x, y+k)}{h} h + \frac{f(x, y+k) - f(x, y)}{k} k.\end{aligned}$$

But, from the definition of a partial derivative,

$$\begin{aligned}\frac{f(x+h, y+k) - f(x, y+k)}{h} &\to \frac{\partial}{\partial x} f(x, y+k) \quad \text{as } h \to 0, \\ &\to \frac{\partial}{\partial x} f(x, y) \qquad \text{as } k \to 0,\end{aligned}$$

since the partial derivative is continuous. It follows that

$$\frac{f(x+h, y+k) - f(x, y+k)}{h} = \frac{\partial}{\partial x} f(x, y) + \epsilon, \text{ where } \epsilon \to 0 \text{ as } h \text{ and } k \to 0.$$

Again,

$$\frac{f(x, y+k) - f(x, y)}{k} \to \frac{\partial}{\partial y} f(x, y) \qquad \text{as } k \to 0,$$

i.e. $$\frac{f(x, y+k) - f(x, y)}{k} = \frac{\partial}{\partial y} f(x, y) + \eta, \text{ where } \eta \to 0 \text{ as } k \to 0.$$

Hence, writing the partial derivatives in an alternative notation,

$$\Delta z = \left(\frac{\partial z}{\partial x} + \epsilon\right) h + \left(\frac{\partial z}{\partial y} + \eta\right) k, \text{ where } \epsilon \text{ and } \eta \to 0 \text{ as } h \text{ and } k \to 0.$$

The expressions ϵ and η must be small when both h and k are small. So, changing the notation for the increments in x and y, we have:

The increment in the function $z = f(x, y)$ corresponding to arbitrary small increments Δx and Δy in x and y is approximately

$$\Delta z = \frac{\partial z}{\partial x} \Delta x + \frac{\partial z}{\partial y} \Delta y.$$

This is a result of the first importance. The increment in z when x varies alone is represented approximately by $\frac{\partial z}{\partial x} \Delta x$ and the increment when y varies alone approximately by $\frac{\partial z}{\partial y} \Delta y$. It is now established that the increment in z *when x and y vary together* is expressed approximately by the *sum* of the two increments in the simple directions. The two partial derivatives can be used, therefore, to describe the general variation of a function of two variables. But it is clearly convenient to have a precise symbolism for this general variation and a technique for operating upon the symbols. The concept of a " differential " and the process of " differentiation " are introduced to serve just this purpose.

Our fundamental result can be written in a variant form as

$$f(a+h,\ b+k)=f(a,\ b)+hf_x'(a,\ b)+kf_y'(a,\ b)$$

approximately when h and k are small. Hence if $x=a$ and $y=b$ are inserted in $f(x,\ y)$ instead of the correct values $x=a+h$ and $y=b+k$, then the error in the value of the function is approximately

$$hf_x'(a,\ b)+kf_y'(a,\ b).$$

Again, putting $x=a+h$, $y=b+k$, we have

$$f(x,\ y)=f(a,\ b)+(x-a)\,f_x'(a,\ b)+(y-b)\,f_y'(a,\ b)$$

approximately when $(x-a)$ and $(y-b)$ are small. This result enables us to replace $f(x,\ y)$ by an approximate linear expression in x and y for small ranges of values about the values $x=a$, $y=b$ (cf. 6.4 above).

13.2 The differential of a function of two variables.

The increment in the value of a function $z=f(x,\ y)$ for small increments Δx and Δy in the independent variables can be reduced to two parts, a precise expression, $\frac{\partial z}{\partial x}\Delta x+\frac{\partial z}{\partial y}\Delta y$, which is of the same order of smallness as the increments Δx and Δy themselves, and an expression, $\epsilon\Delta x+\eta\Delta y$, depending on values ϵ and η which are small when Δx amd Δy are small. The second expression is thus of the order of the squares of Δx and Δy. The precise and first-order part of Δz is termed the *differential* of the function $z=f(x,\ y)$:

Definition : The differential of a function $z=f(x,\ y)$ with continuous partial derivatives at a point $(x,\ y)$ is

$$dz=\frac{\partial z}{\partial x}\Delta x+\frac{\partial z}{\partial y}\Delta y,$$

where Δx and Δy denote arbitrary increments in the independent variables from the point $(x,\ y)$.

For convenience, we now denote the arbitrary increment Δx by

dx and call it the differential of the independent variable x. Similarly, Δy is written dy and called the differential of y. The differential of the dependent variable z is then defined in terms of the independent differentials by the formula

$$dz = \frac{\partial z}{\partial x} dx + \frac{\partial z}{\partial y} dy.$$

It must always be remembered, in using this formula, that dx and dy are no more than arbitrary increments in the independent variables. An alternative notation for the differential of $z = f(x, y)$ is

$$df(x, y) = f_x'(x, y) dx + f_y'(x, y) dy = f_x dx + f_y dy.$$

The interpretation of the differential of a function follows from the definition. The statement that $dz = \frac{\partial z}{\partial x} dx + \frac{\partial z}{\partial y} dy$ implies that the increment in the value of $z = f(x, y)$ is *approximately*

$$\Delta z = \frac{\partial z}{\partial x} \Delta x + \frac{\partial z}{\partial y} \Delta y$$

when *small* increments Δx and Δy are allotted to x and y. The differential dz replaces the increment Δz in much the same way that the derivative replaces the average rate of change. So $z = f(x, y)$ tends to increase (or to decrease) whenever x and y are varied by small amounts so that the corresponding differential dz is positive (or negative).

The expression $dz = \frac{\partial z}{\partial x} dx + \frac{\partial z}{\partial y} dy$ is often called the "complete" differential of the function $z = f(x, y)$ and from it can be derived two partial differentials. When x is varied by an amount dx while y is held constant, the differential becomes

$$dz = \frac{\partial z}{\partial x} dx$$

and the partial derivative $\frac{\partial z}{\partial x}$ is thus the ratio of the differential dz (y held constant) to the differential dx. Similarly, the partial derivative $\frac{\partial z}{\partial y}$ is the ratio of the differential dz (x held constant) to the differential dy. A partial derivative can be regarded as the ratio of two differentials.

It can be noticed that the definition of a differential applies automatically to the case of a function of one variable. If $y=f(x)$ has a continuous derivative $f'(x)$, then the differential of y is

$$dy=f'(x)\,dx$$

for an arbitrary increment dx in x.* This implies simply that

$$\Delta y=f'(x)\,\Delta x \quad \text{approximately}$$

when the arbitrary increment Δx is small. It follows that the derivative $f'(x)$ is the ratio of the differential dy to the differential dx. The notation $\frac{dy}{dx}$ is now very useful since it can be regarded either as a derivative or as a ratio of differentials. The differential notation, however, adds nothing to the derivative notation in the case of a function of one variable. It is only when at least two independent variables are involved that the differential notation becomes essential in the description of variation.

13.3 The technique of differentiation.

The process of obtaining the differential of a function is called *differentiation*. The most obvious method of differentiating $z=f(x, y)$ is to evaluate the partial derivatives and to substitute in the fundamental formula $dz=\frac{\partial z}{\partial x}dx+\frac{\partial z}{\partial y}dy$. But other methods are often more convenient. It can be shown quite easily that the rules obeyed by differentials are exactly similar to those obeyed by derivatives. If u and v are two functions of x and y, then

$$d(u \pm v)=du \pm dv\,; \quad d(uv)=u\,dv+v\,du\,; \quad d\left(\frac{u}{v}\right)=\frac{v\,du-u\,dv}{v^2}.$$

Further, if $z=f(u)$ where u is a function of x and y, then

$$dz=f'(u)\,du.$$

So $$d(u^n)=nu^{n-1}\,du\,; \quad d(e^u)=e^u\,du\,; \quad d(\log u)=\frac{du}{u}.$$

As examples of the method of proof, we have

$$\begin{aligned} d(uv) &= \frac{\partial}{\partial x}(uv)\,dx+\frac{\partial}{\partial y}(uv)\,dy \\ &= \left(u\frac{\partial v}{\partial x}+v\frac{\partial u}{\partial x}\right)dx+\left(u\frac{\partial v}{\partial y}+v\frac{\partial u}{\partial y}\right)dy \\ &= u\left(\frac{\partial v}{\partial x}dx+\frac{\partial v}{\partial y}dy\right)+v\left(\frac{\partial u}{\partial x}dx+\frac{\partial u}{\partial y}dy\right)=u\,dv+v\,du. \end{aligned}$$

* The derivative $f'(x)$ is sometimes called the "differential coefficient"; it is the coefficient of the differential dx in the differential of the function $y=f(x)$.

Again, if $z=f(u)$, then

$$\frac{\partial z}{\partial x}=f'(u)\frac{\partial u}{\partial x} \quad \text{and} \quad \frac{\partial z}{\partial y}=f'(u)\frac{\partial u}{\partial y}$$

and
$$dz=\frac{\partial z}{\partial x}dx+\frac{\partial z}{\partial y}dy=f'(u)\left(\frac{\partial u}{\partial x}dx+\frac{\partial u}{\partial y}dy\right)=f'(u)\,du.$$

A good practical method of differentiation is to split up the given function into groupings of functions $u, v, w, \ldots$, each of which involves only one of the variables x and y. The differential of the function is expressed in terms of the differentials of $u, v, w, \ldots$ by the above rules. The latter differentials are then given by ordinary single-variable derivatives, e.g. the standard forms

$$d(x^n)=nx^{n-1}\,dx; \quad d(e^x)=e^x\,dx; \quad d(\log x)=\frac{dx}{x},$$

and similar results in the variable y. Finally, it is to be noticed that the result $d(\log u)=\dfrac{du}{u}$ provides a method of logarithmic differentiation. For example, if u, v and w are functions of x and y and if $z=\dfrac{uv}{w}$, then on taking the logarithm

$$\log z=\log u+\log v-\log w$$

and
$$\frac{dz}{z}=\frac{du}{u}+\frac{dv}{v}-\frac{dw}{w}.$$

The practical technique of differentiation is illustrated by the following examples. Each result can be checked by using the partial derivatives already obtained (12.2 above).

Ex. 1. $z=x^2+2xy-y^2$.

Here
$$\begin{aligned} dz &= d(x^2)+2d(xy)-d(y^2) \\ &= 2x\,dx+2(x\,dy+y\,dx)-2y\,dy \\ &= 2(x+y)\,dx+2(x-y)\,dy. \end{aligned}$$

Ex. 2. $z=\dfrac{x^2}{x-y+1}$.

Here
$$\begin{aligned} dz &= \frac{(x-y+1)\,d(x^2)-x^2d(x-y+1)}{(x-y+1)^2} \\ &= \frac{(x-y+1)\,2x\,dx-x^2(dx-dy)}{(x-y+1)^2} \\ &= \frac{x(x-2y+2)\,dx+x^2\,dy}{(x-y+1)^2}. \end{aligned}$$

Ex. 3. $z = \log(x^2 + y^2)$.

So $z = \log u$ where $u = x^2 + y^2$,

and $$dz = \frac{du}{u} = \frac{d(x^2+y^2)}{x^2+y^2} = 2\frac{x\,dx + y\,dy}{x^2+y^2}.$$

Ex. 4. $z = x^2\sqrt{\dfrac{x+y}{x-y}}$.

So $\log z = 2\log x + \frac{1}{2}\log(x+y) - \frac{1}{2}\log(x-y)$,

and $$\frac{dz}{z} = 2\frac{dx}{x} + \frac{1}{2}\frac{d(x+y)}{x+y} - \frac{1}{2}\frac{d(x-y)}{x-y}$$

$$= \frac{4(x^2-y^2)\,dx + x(x-y)(dx+dy) - x(x+y)(dx-dy)}{2x(x^2-y^2)}$$

$$= \frac{(2x^2 - xy - 2y^2)\,dx + x^2\,dy}{x(x^2-y^2)},$$

i.e. $$dz = \frac{x}{\sqrt{(x+y)(x-y)^3}}\{(2x^2 - xy - 2y^2)\,dx + x^2\,dy\}.$$

13.4 Differentiation of functions of functions.

The differential $dz = \dfrac{\partial z}{\partial x}dx + \dfrac{\partial z}{\partial y}dy$ is defined for a function of two *independent* variables, dx and dy being then arbitrary increments. But exactly the same formula can be shown to give the differential of a function in which the variables x and y are not independent but *dependent on some other set of variables*. In this case, dx and dy are no longer arbitrary increments; they are proper differentials depending on the arbitrary increments in the actual independent variables defining x and y. The proof of this extension of the differential formula is too involved to be given here.* It can be assumed to hold at least for ordinary continuous functions. Hence,

The differential of the function $z = f(x, y)$ is given by

$$dz = \frac{\partial z}{\partial x}dx + \frac{\partial z}{\partial y}dy$$

whether the variables x and y are independent or not.

The extended formula is of importance in the evaluation of the differentials, and hence of the derivatives, of various types of functions of functions. The simplest case of a function of a function

* See, for example, Phillips, *A Course of Analysis* (1930), pp. 229-31.

has already been considered. If $z=f(u)$ where u is a function of x and y, then we have seen that

$$dz=f'(u)\,du$$

and $$\frac{\partial z}{\partial x}=f'(u)\frac{\partial u}{\partial x} \quad \text{and} \quad \frac{\partial z}{\partial y}=f'(u)\frac{\partial u}{\partial y}.$$

We can now pass to the two most useful cases of more complicated functions of functions.

In the first case, z is a given function of two variables x and y which are themselves given functions of a single independent variable t. Hence, z is a function of t and the derivative $\frac{dz}{dt}$ is required. Write $z=f(x, y)$ with partial derivatives $\frac{\partial z}{\partial x}$ and $\frac{\partial z}{\partial y}$. Further, write $x=\phi(t)$ with derivative $\frac{dx}{dt}=\phi'(t)$ and $y=\psi(t)$ with derivative $\frac{dy}{dt}=\psi'(t)$. Then

$$dz=\frac{\partial z}{\partial x}dx+\frac{\partial z}{\partial y}dy,$$

where $$dx=\phi'(t)dt \quad \text{and} \quad dy=\psi'(t)dt.$$

So $$dz=\left\{\frac{\partial z}{\partial x}\phi'(t)+\frac{\partial z}{\partial y}\psi'(t)\right\}dt,$$

i.e. $$\frac{dz}{dt}=\frac{\partial z}{\partial x}\phi'(t)+\frac{\partial z}{\partial y}\psi'(t).$$

The ratio of the differentials $dz:dt$ is to be interpreted at once as the derivative of z as a function of t. With a small notational change, the derivative can be written

$$\frac{dz}{dt}=\frac{\partial z}{\partial x}\frac{dx}{dt}+\frac{\partial z}{\partial y}\frac{dy}{dt}.$$

This result is a very simple one and, in practice, all we need do is to write the fundamental formula $dz=\frac{\partial z}{\partial x}dx+\frac{\partial z}{\partial y}dy$, divide through by the differential dt and interpret the ratios of differentials as derivatives according to the definition of the problem. As a particular case, if $z=f(x, y)$, where y is a given function of x, then

$$\frac{dz}{dx}=\frac{\partial z}{\partial x}+\frac{\partial z}{\partial y}\frac{dy}{dx}.$$

In the second case, z is a given function of two variables x and y which are themselves functions of two independent variables u and v. It is required to find the partial derivatives of z considered as a function of u and v. Write $z=f(x, y)$ where $x=\phi(u, v)$ and $y=\psi(u, v)$, all functions having known partial derivatives.

Then $$dz=\frac{\partial z}{\partial x}dx+\frac{\partial z}{\partial y}dy,$$

where $$dx=\frac{\partial x}{\partial u}du+\frac{\partial x}{\partial v}dv \quad \text{and} \quad dy=\frac{\partial y}{\partial u}du+\frac{\partial y}{\partial v}dv.$$

So $$dz=\left(\frac{\partial z}{\partial x}\frac{\partial x}{\partial u}+\frac{\partial z}{\partial y}\frac{\partial y}{\partial u}\right)du+\left(\frac{\partial z}{\partial x}\frac{\partial x}{\partial v}+\frac{\partial z}{\partial y}\frac{\partial y}{\partial v}\right)dv.$$

The partial derivative of z with respect to u is the ratio of the differentials dz and du when v is kept constant, i.e. it is the first bracket in the above expression for dz. The partial derivative of z with respect to v is obtained, in the same way, as the second bracket in the expression for dz. Hence,

$$\frac{\partial z}{\partial u}=\frac{\partial z}{\partial x}\frac{\partial x}{\partial u}+\frac{\partial z}{\partial y}\frac{\partial y}{\partial u} \quad \text{and} \quad \frac{\partial z}{\partial v}=\frac{\partial z}{\partial x}\frac{\partial x}{\partial v}+\frac{\partial z}{\partial y}\frac{\partial y}{\partial v}.$$

These results are again very simple. In practice, we divide the formula $dz=\frac{\partial z}{\partial x}dx+\frac{\partial z}{\partial y}dy$ through by either du or dv and interpret the resulting differential ratios as partial derivatives.

The basic form of the differential formula is thus capable of dealing with various types of functions of functions; it is merely a matter of making the interpretation appropriate to the case under consideration. This is a striking illustration of the flexibility of the differential notation. Our results are still limited, however, to the case of single-valued functions and it remains to extend the application of differentials in the construction of a theory of multi-valued and implicit functions.

13.5 Differentiation of implicit functions.

Two variables x and y are related by a given implicit relation $f(x, y)=0$. In general, we now have y as a multi-valued function of x and x as a multi-valued function of y. The simplest method of dealing with such a relation is to introduce a third variable z which takes its value from $f(x, y)$ for any values of x and y whatever. The single-valued function $z=f(x, y)$ has the given implicit function as its zero contour by $z=0$, i.e. the given implicit function can be studied by relating those values of x and y which make z equal to zero in the explicit function $z=f(x, y)$.

As x and y vary in any way, independently or not, the variation of $z=f(x, y)$ is given by the complete differential

$$dz=f_x\,dx+f_y\,dy$$

where f_x and f_y are the partial derivatives of $f(x, y)$. If (x, y) are values making $z=0$ and if dx and dy are variations from these values keeping $z=0$, then $dz=0$ and

$$f_x\,dx+f_y\,dy=0$$

is the relation between the differential increments satisfying the given implicit function $f(x, y)=0$. Hence,

$$\frac{dy}{dx}=-\frac{f_x}{f_y} \quad \text{and} \quad \frac{dx}{dy}=-\frac{f_y}{f_x}.$$

Starting from a given point (x, y) satisfying the relation $f(x, y)=0$ and remaining on one branch of y as a function of x, let Δy be the increment in y corresponding to a given increment Δx in x. Then, by the definition of differentials, the ratio $dy:dx$ written above is the approximate and limiting value of the ratio $\Delta y:\Delta x$ when Δx is small. Hence, the ratio $dy : dx$ can be interpreted as the derivative of y as a function of x given by $f(x, y)=0$, taking only one branch of this function through (x, y). In this way, the notion of a derivative is extended to apply to multi-valued, as well as to single-valued, functions. The ratio $dx : dy$ is interpreted, in the same way, as the derivative of x as a function of y given by the same implicit function. Hence,

The implicit function $f(x, y)=0$ gives y as a function of x with derivative $\frac{dy}{dx}=-\frac{f_x}{f_y}$ at the point (x, y), and it gives x as a function of y with derivative $\frac{dx}{dy}=-\frac{f_y}{f_x}$ at the point (x, y).

The two functions are inverse to each other and it is seen that one derivative is the reciprocal of the other.

One point must now be stressed most strongly. The derivatives $\frac{dy}{dx}$ and $\frac{dx}{dy}$, being expressed in terms of the partial derivatives of

$f(x, y)=0$, are themselves *functions of both x and y.* The derivatives differ, therefore, from those of single-valued functions and the reason for this is not difficult to see. If y is a single-valued function of x, then it is sufficient to fix a value of x, the corresponding value of y and of the derivative $\frac{dy}{dx}$ being uniquely determined. The derivative of a single-valued function of x depends only on x. But, if y is a multi-valued function of x, it is not sufficient to fix a value of x since this still leaves several equally possible values of y at choice. The branch of the function, and hence the derivative $\frac{dy}{dx}$, can only be determined when the starting-point is specified as regards both x and y. The derivative of a multi-valued function is essentially dependent on both x and y.

These remarks can be illustrated by the simple relation

$$x^2+y^2=16.$$

Here $\qquad f(x, y)=x^2+y^2-16 \quad$ and $\quad f_x=2x,\ f_y=2y.$

So $\qquad \frac{dy}{dx}=-\frac{x}{y}\quad$ at the point (x, y).

The values (x, y) must, of course, satisfy the given relation. In this case, y is a double-valued function of x and it is possible to separate the two branches :

$$y=\sqrt{16-x^2} \quad \text{and} \quad y=-\sqrt{16-x^2}.$$

On the positive branch, we have the single-valued function

$$y=+\sqrt{16-x^2},$$

which has derivative, at any value of x, given by

$$\frac{dy}{dx}=-\frac{x}{\sqrt{16-x^2}}=-\frac{x}{y}.$$

On the negative branch, $y=-\sqrt{16-x^2}$ with derivative

$$\frac{dy}{dx}=\frac{x}{\sqrt{16-x^2}}=-\frac{x}{y}.$$

For a given value of x, therefore, there are two different values of $\frac{dy}{dx}$, one for each branch of the function. But both values are included in the general result $\frac{dy}{dx}=-\frac{x}{y}$; the implicit form of the function and of the derivative expression has thus great advantages even when the branches of the function can be separated.

The derivative of an implicit function given by a relation $f(x, y)=0$ is evaluated at once from the partial derivatives of the expression $f(x, y)$. In practice, however, it is often better to differentiate $f(x, y)=0$ as it stands according to the method indicated in the following examples.

Ex. 1. The relation $x^2+y^2-2x+4y+1=0$ gives y as a double-valued function of x. From

$$f(x, y)=x^2+y^2-2x+4y+1,$$

we have $$f_x=2(x-1) \quad \text{and} \quad f_y=2(y+2).$$

So $$\frac{dy}{dx}=-\frac{x-1}{y+2}.$$

Alternatively, differentiating the relation as it stands,

$$\frac{d}{dx}(x^2)+\frac{d}{dx}(y^2)-2\frac{d}{dx}(x)+4\frac{d}{dx}(y)=0,$$

i.e. $$2x+2y\frac{dy}{dx}-2+4\frac{dy}{dx}=0,$$

i.e. $$\frac{dy}{dx}=-\frac{x-1}{y+2} \text{ as before.}$$

Ex. 2. The relation $x^3+y^3-3xy=0$ gives y as a triple-valued function of x. From $f(x, y)=x^3+y^3-3xy$, we have

$$f_x=3x^2-3y \quad \text{and} \quad f_y=3y^2-3x.$$

So $$\frac{dy}{dx}=-\frac{x^2-y}{y^2-x}.$$

Alternatively $$\frac{d}{dx}(x^3)+\frac{d}{dx}(y^3)-3\frac{d}{dx}(xy)=0,$$

i.e. $$3x^2+3y^2\frac{dy}{dx}-3\left(y+x\frac{dy}{dx}\right)=0,$$

which gives the same value of the derivative.

The second derivative of an implicit function is defined and obtained from the first derivative. If $f(x, y)=0$ gives y as a function of x, then from $\dfrac{dy}{dx}=-\dfrac{f_x}{f_y}$, we obtain

$$\frac{d^2y}{dx^2}=-\frac{d}{dx}\left(\frac{f_x}{f_y}\right)=-\left\{\frac{\partial}{\partial x}\left(\frac{f_x}{f_y}\right)+\frac{\partial}{\partial y}\left(\frac{f_x}{f_y}\right)\frac{dy}{dx}\right\}.$$

But $\quad \frac{\partial}{\partial x}\left(\frac{f_x}{f_y}\right)=\frac{f_{xx}f_y-f_{xy}f_x}{f_y^2}$ and $\frac{\partial}{\partial y}\left(\frac{f_x}{f_y}\right)=\frac{f_{xy}f_y-f_{yy}f_x}{f_y^2}.$

Substituting these expressions and the value of $\frac{dy}{dx}$, we have

$$\frac{d^2y}{dx^2}=-\left\{\frac{f_{xx}f_y-f_{xy}f_x}{f_y^2}-\frac{f_{xy}f_y-f_{yy}f_x}{f_y^2}\frac{f_x}{f_y}\right\}$$

$$=-\frac{1}{f_y^3}(f_{xx}f_y^2-2f_{xy}f_xf_y+f_{yy}f_x^2).$$

The second derivative thus involves the first and second-order partial derivatives of $f(x, y)$ and is a function of x and y. For example, the relation $x^2+y^2=16$ gives $\frac{dy}{dx}=-\frac{x}{y}$ and

$$\frac{d^2y}{dx^2}=-\frac{d}{dx}\left(\frac{x}{y}\right)=-\frac{1}{y^2}\left\{y\frac{d}{dx}(x)-x\frac{d}{dx}(y)\right\}$$

$$=-\frac{1}{y^2}\left(y-x\frac{dy}{dx}\right)=-\frac{1}{y^2}\left(y+\frac{x^2}{y}\right)=-\frac{x^2+y^2}{y^3}=-\frac{16}{y^3}.$$

In diagrammatic terms, the implicit function $f(x, y)=0$ is represented by a curve in the plane Oxy. In general, one branch of the curve passes through a point (x, y) satisfying the relation and there is a definite tangent to the branch at the point. The tangent gradient, referred to Ox, is then given by the value of the derivative

$$\frac{dy}{dx}=-\frac{f_x}{f_y} \text{ at } (x, y).$$

Further, the curvature of the branch at the point is indicated by the value of $\frac{d^2y}{dx^2}$ at (x, y). Our previous discussion of the applications of derivatives now extends at once to the case where we have a curve with an implicit and multi-valued equation. It must be remembered, however, that the results refer only to a single branch of the curve, the branch passing through a specified point (x, y).

The equation of the tangent at a point (x_1, y_1) on the curve $f(x, y)=0$ is readily obtainable. The tangent is the line passing through the point (x_1, y_1) with gradient to Ox given by the value of $\frac{dy}{dx}=-\frac{f_x}{f_y}$ at this point. The tangent equation is thus

$$y-y_1=\left(-\frac{f_{x_1}}{f_{y_1}}\right)(x-x_1),$$

i.e. $$f_{x_1}(x-x_1)+f_{y_1}(y-y_1)=0,$$

where f_{x_1} and f_{y_1} are the values of the partial derivatives of $f(x, y)$ at the point (x_1, y_1). This is a generalised form of the tangent equation previously obtained (6.5 above).

For example, in the case of the circle with equation

$$x^2+y^2-2x+4y+1=0,$$

we have $$f_{x_1}=2(x_1-1) \quad \text{and} \quad f_{y_1}=2(y_1+2).$$

The tangent at (x_1, y_1) on the circle has equation

$$(x_1-1)(x-x_1)+(y_1+2)(y-y_1)=0,$$

i.e. $$(x_1-1)x+(y_1+2)y=x_1^2+y_1^2-x_1+2y_1.$$

Since the point (x_1, y_1) lies on the circle, we have

$$x_1^2+y_1^2-2x_1+4y_1+1=0,$$

i.e. $$x_1^2+y_1^2-x_1+2y_1=x_1-2y_1-1.$$

The equation of the tangent is thus

$$(x_1-1)x+(y_1+2)y=x_1-2y_1-1.$$

13.6 The differential of a function of more than two variables.

The definition of a differential extends at once to the case of a function of more than two variables. The variation of $u=f(x, y, z)$ as the independent variables vary is given by the differential

$$du=\frac{\partial u}{\partial x}dx+\frac{\partial u}{\partial y}dy+\frac{\partial u}{\partial z}dz.$$

This implies that, if Δx, Δy and Δz are arbitrary small increments in the independent variables, then the increment in u is approximately

$$\Delta u=\frac{\partial u}{\partial x}\Delta x+\frac{\partial u}{\partial y}\Delta y+\frac{\partial u}{\partial z}\Delta z.$$

The complete differential is again the sum of the separate (approximate) variations due to changes in x only, in y only and in z only. The differential formula again extends to the case where x, y and z are no longer independent variables and it can be used to give the derivatives of functions of functions of various types.

The variation of variables x, y and z connected implicitly by some relation $f(x, y, z)=0$ must satisfy

$$f_x dx+f_y dy+f_z dz=0,$$

where f_x, f_y and f_z are the partial derivatives of $f(x, y, z)$. Regarding z as a multi-valued function of x and y defined by this relation, we obtain the partial derivative $\frac{\partial z}{\partial x}$ as the value of the ratio $dz : dx$ for y constant. Putting $dy=0$ in the above equation, we have

$$f_x dx + f_z dz = 0,$$

i.e. $$\frac{\partial z}{\partial x} = \left(\frac{dz}{dx}\right)_{y \text{ constant}} = -\frac{f_x}{f_z}.$$

Similarly $$\frac{\partial z}{\partial y} = \left(\frac{dz}{dy}\right)_{x \text{ constant}} = -\frac{f_y}{f_z}.$$

In diagrammatic terms, a three-dimensional surface with equation in implicit form $f(x, y, z)=0$ has gradients (referred to Oxy) in the two fundamental directions given by the above ratios of partial derivatives. The tangent plane at (x_1, y_1, z_1) thus has equation

$$z - z_1 = \left(-\frac{f_{x_1}}{f_{z_1}}\right)(x - x_1) + \left(-\frac{f_{y_1}}{f_{z_1}}\right)(y - y_1) \quad \text{(see 12.5)}$$

i.e. $$f_{x_1}(x - x_1) + f_{y_1}(y - y_1) + f_{z_1}(z - z_1) = 0.$$

More generally, for a function $y=f(x_1, x_2, \ldots x_n)$ of n independent variables, we define

$$dy = \frac{\partial y}{\partial x_1} dx_1 + \frac{\partial y}{\partial x_2} dx_2 + \ldots + \frac{\partial y}{\partial x_n} dx_n.$$

The complete differential contains as many additive terms as there are independent variables. The interpretation and application of this formula are exactly similar to those already discussed in the simpler cases above.

13.7 The substitution of factors in production.

The production function $x=f(a, b)$ for two variable factors of production gives a system of constant product curves

$$f(a, b) = \text{constant}.$$

In the normal case, one curve of the system passes through each point (a, b) of the positive part of the plane Oab and is downward sloping and convex to the origin (at least over a certain area of the plane). Differentiating and denoting marginal products by $f_a = \frac{\partial x}{\partial a}$ and $f_b = \frac{\partial x}{\partial b}$,

$$f_a da + f_b db = 0$$

is the (approximate) relation between increments da and db in the factors along the constant product curve through (a, b). This relation holds for any point (a, b) we care to select. Hence, the tangent gradient of the constant product curve through (a, b) is

$$\frac{db}{da} = -\frac{f_a}{f_b}$$

when referred to the axis Oa. In the normal case and over the relevant area of Oab, this gradient is negative and its numerical value is termed the *marginal rate of substitution* of the factor B for the factor A in the production of the good X :

$$r = -\frac{db}{da} = \frac{f_a}{f_b}.$$

The value of r depends on the combination of the factors considered, i.e. is a function of both a and b. It represents the additional amount of the factor B, from the given combination of factors, necessary to maintain product unchanged when a small unit reduction is made in the use of the factor A.

Since the constant product curves are convex to the origin in the normal case, the value of r must increase as b increases (and a decreases) along the constant product curve. The convexity condition is thus the expression of the principle of increasing marginal rate of substitution, of the assumption that it becomes increasingly more difficult to substitute B for A as the substitution proceeds. The interesting point now is to determine how fast r increases, i.e. to measure the "elasticity" of substitution. For any change from (a, b) along the constant product curve, $d\left(\frac{b}{a}\right)$ represents the increase (or decrease) in the use of B as compared with that of A and $dr = d\left(\frac{f_a}{f_b}\right)$ the corresponding increase (or decrease) in the marginal rate of substitution. The ratio of these differentials, expressed in proportional terms to make them independent of units of measurement, is defined as the *elasticity of substitution* between the factors at the combination of factors considered. Hence,

DEFINITION : The elasticity of substitution between A and B is

$$\sigma = \frac{\frac{a}{b} d\left(\frac{b}{a}\right)}{\frac{1}{r} dr},$$

where the differentials correspond to a variation along the constant product curve through (a, b).

In diagrammatic terms, σ appears as the ratio of the relative increase in the gradient of OP to the relative increase in the tangent gradient of the constant product curve at P as the point P moves along a constant product curve in the plane Oab.

The value of σ can be written in terms of the partial derivatives of r or of the production function itself. We have

$$d\left(\frac{b}{a}\right)=\frac{a\,db-b\,da}{a^2} \quad \text{and} \quad dr=\frac{\partial r}{\partial a}da+\frac{\partial r}{\partial b}db.$$

But $db=-\frac{f_a}{f_b}da=-r\,da$, and so

$$d\left(\frac{b}{a}\right)=-\frac{ar+b}{a^2}da \quad \text{and} \quad dr=-\left(r\frac{\partial r}{\partial b}-\frac{\partial r}{\partial a}\right)da.$$

Hence,
$$\sigma=\frac{r}{ab}\frac{ar+b}{r\frac{\partial r}{\partial b}-\frac{\partial r}{\partial a}}.$$

Evaluating $\frac{\partial r}{\partial a}=\frac{\partial}{\partial a}\left(\frac{f_a}{f_b}\right)$ and $\frac{\partial r}{\partial b}=\frac{\partial}{\partial b}\left(\frac{f_a}{f_b}\right)$ in terms of the first and second-order partial derivatives of $f(a, b)$, it is found that

$$\sigma=\frac{f_a f_b(af_a+bf_b)}{abT},$$

where
$$T=-(f_{aa}f_b^2-2f_{ab}f_af_b+f_{bb}f_a^2).$$

From this last result, it appears that, although σ has been defined for the substitution of B for A, the same value is obtained when the substitution is of A for B. The elasticity of substitution is symmetrical with respect to the two factors.

The curvature of the constant product curve at (a, b) is

$$\frac{d^2b}{da^2}=\frac{d}{da}\left(\frac{db}{da}\right)=-\frac{d}{da}(r)=-\left(\frac{\partial r}{\partial a}+\frac{\partial r}{\partial b}\frac{db}{da}\right)=r\frac{\partial r}{\partial b}-\frac{\partial r}{\partial a}.$$

Hence, σ is a positive multiple of the reciprocal of $\frac{d^2b}{da^2}$, i.e. it is positive and inversely proportional to the curvature of the constant product curve. The larger is the value of σ, the flatter is the constant product curve and the more slowly does the marginal rate of substitution increase as B is substituted for A. The magnitude of σ is thus an indication of the ease with which product can be maintained by substituting B for A. There are two limiting cases. If A and B are perfect substitutes, so that product is maintained by increasing

B in proportion as A is decreased, then the constant product curve is a straight line, $\frac{d^2b}{da^2}$ is zero and σ is infinite. If A and B are entirely incapable of substitution, being needed in a fixed proportion, then an increase in *one* of the factors from this proportion must leave product unchanged. The constant product curve has a right angle at the point concerned, $\frac{d^2b}{da^2}$ is infinite and σ is zero. As σ increases from zero to infinity, substitution between the factors becomes increasingly easier.

So, the elasticity of substitution σ varies from one combination of factors to another, is independent of the units in which factors and product are measured, is a symmetrical relation between the factors, is positive for all normal combinations of factors and varies from zero to infinity according to the ease with which one factor can be substituted for the other in production.

When the production function is linear and homogeneous (constant returns to scale), σ becomes of simpler form. Since

$$f_{aa} = -\frac{b}{a}f_{ab} \quad \text{and} \quad f_{bb} = -\frac{a}{b}f_{ab} \quad \text{(12.8 (4) above)}$$

we have $\quad T = \frac{f_{ab}}{ab}(a^2 f_a{}^2 + 2abf_a f_b + b^2 f_b{}^2) = \frac{f_{ab}}{ab}(af_a + bf_b)^2$

and $\quad \sigma = \frac{f_a f_b}{(af_a + bf_b)f_{ab}} = \frac{f_a f_b}{xf_{ab}}$ (by Euler's Theorem).

Using the alternative notation for partial derivatives,

$$\sigma = \frac{\frac{\partial x}{\partial a}\frac{\partial x}{\partial b}}{x\frac{\partial^2 x}{\partial a\,\partial b}}.$$

When there are constant returns to scale, σ is inversely proportional to the cross second-order partial derivative of the (linear and homogeneous) production function.* In the particular case where the production function is $x = Aa^{\alpha}b^{1-\alpha}$, it is easily seen that $r = \frac{\alpha}{1-\alpha}\frac{b}{a}$ and that $\sigma = 1$ at all points.

* It was in this form that σ was first defined by Hicks, *The Theory of Wages* (1932), pp. 117 and 245. See, also, Robinson, *The Economics of Imperfect Competition* (1933), pp. 256 and 330, Lerner and Kahn, *Notes on the Elasticity of Substitution*, Review of Economic Studies, 1933.

13.8 Substitution in other economic problems.

Exactly similar definitions can be given for the marginal rate and elasticity of substitution for an individual consumer's scale of preferences. A system of indifference curves for two goods X and Y is defined by the equation

$$\phi(x, y) = \text{constant}.$$

The tangent gradient to the curve through the point (x, y) is

$$\frac{dy}{dx} = -\frac{\phi_x}{\phi_y}.$$

In the normal case, the curves are downward sloping and convex to the origin. The marginal rate of substitution of the good Y for the good X in consumption, defined as

$$R = -\frac{dy}{dx} = \frac{\phi_x}{\phi_y},$$

is thus positive and increases as the substitution of Y for X proceeds. It is to be noticed that R is the ratio of the "marginal utilities" of X and Y and independent of the form we assume for the individual's utility function (see 12.8 above). But there is no need to consider "marginal utility" at all; the expression R is sufficient for the description of the indifference map. The elasticity of substitution between X and Y is then defined as in the production case, taking values between zero and infinity according to the ease with which X and Y can be substituted in consumption to maintain a given level of indifference.*

A similar analysis applies for a consumer's preference scale for incomes in two years. If x and y are the amounts of this and next year's incomes, we have an indifference curve system given by

$$\psi(x, y) = \text{constant}.$$

The marginal rate of time-preference is defined as the numerical gradient of the indifference curve through (x, y), reduced by unity †

$$R = \left(-\frac{dy}{dx}\right) - 1 = \frac{\psi_x}{\psi_y} - 1.$$

* See Hicks and Allen, *A Reconsideration of the Theory of Value*, Economica, 1934.

† The reduction of $\left(-\frac{dy}{dx}\right)$ by 1 is made since we wish to define a zero time-preference as the case where an equal addition to next year's income compensates for a given reduction of this year's income, i.e. where $\left(-\frac{dy}{dx}\right) = 1$. This is only a matter of the scale on which time-preference is measured.

The elasticity of substitution between the two incomes is then defined exactly as before.

The same concepts are also of use in other problems, e.g. in problems involving the notion of a transformation function (see 11.7 above). If two goods are produced with given resources so that

$$F(x, y)=0$$

is the relation between the outputs x and y, then the expression $\left(-\frac{dy}{dx}\right)=\frac{F_x}{F_y}$ gives the marginal rate of substitution of Y production for X production. Similarly, if x and y are this year's and next year's incomes obtainable from given resources, the expression written is the marginal rate of return over cost. The corresponding elasticities of substitution then follow as before except that a negative sign must be added, the transformation curves being concave, and not convex, to the origin in the plane Oxy.

13.9 Further consideration of duopoly problems.

In the duopoly problem with conjectural variations zero, as analysed in 8.8 above, the reaction curve C_1 of the first duopolist gives x_1 as a function of x_2 defined by

$$f(x_1, x_2)=\psi(x)+x_1\psi'(x)-\frac{d\Pi_1}{dx_1}=0. \quad (\text{where } x=x_1+x_2)$$

The equilibrium outputs are then determined by the intersection of C_1 and the similar curve C_2 for the second duopolist. For stable equilibrium, we require that C_1 be downward sloping with a gradient (to the axis Ox_2) less than unity and similarly for C_2. It is now possible to investigate the implications of these conditions.

Differentiating the above implicit relation,

$$\frac{dx_1}{dx_2}=-\frac{f_{x_2}}{f_{x_1}}=-\frac{\psi'(x)+x_1\psi''(x)}{2\psi'(x)+x_1\psi''(x)-\dfrac{d^2\Pi_1}{dx_1{}^2}}=-\frac{1}{1+\rho},$$

where
$$\rho=-\frac{\dfrac{d^2\Pi_1}{dx_1{}^2}-\psi'(x)}{\psi'(x)+x_1\psi''(x)}.$$

Hence, $\frac{dx_1}{dx_2}$ is negative and numerically less than unity provided that ρ is positive. It is assumed, in the normal case, that the demand function $p = \psi(x)$ is decreasing and that marginal cost $\frac{d\Pi_1}{dx_1}$ is increasing. It follows that the numerator of ρ is positive. The condition we require then reduces to

$$\psi'(x) + x_1\psi''(x) < 0,$$

which is satisfied *either* if $\psi''(x)$ is negative *or* if $\psi''(x)$, though positive, is less that $\frac{1}{x_1}\{-\psi'(x)\}$. These alternative conditions apply to the convexity of the demand curve and one or other is satisfied in many cases of demand laws, e.g. in the case of the linear demand law $p = \beta - \alpha x$.

A rather more complex duopoly problem is the following. The first duopolist is assumed to act as before, expecting his rival's output to be unaffected by his own actions. The reaction curve C_1

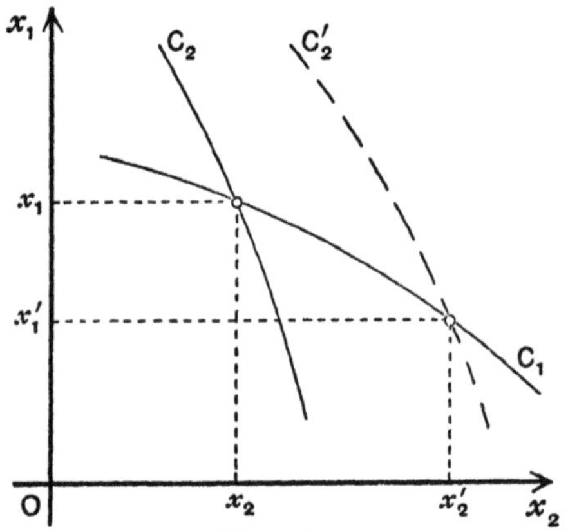

FIG. 85.

is then obtained with gradient $\frac{dx_1}{dx_2}$ as given above with value lying between -1 and 0 in the normal case. The second duopolist is assumed to forecast correctly the effect of his own actions on his rival's output, *i.e.* his conjectural variation is $\frac{dx_1}{dx_2}$ as obtained from C_1. The second reaction curve gives x_2 as a function of x_1 defined by

$$\psi(x) + x_2\psi'(x)\left(1 + \frac{dx_1}{dx_2}\right) = \frac{d\Pi_2}{dx_2}. \quad \text{(where } x = x_1 + x_2\text{)}$$

Since $\psi'(x)<0$ and $0<1+\frac{dx_1}{dx_2}<1$, the left-hand side of this equation is greater than the form $\{\psi(x)+x_2\psi'(x)\}$ in the case where his conjectural variation is zero. Hence, for any given x_1, $\frac{d\Pi_2}{dx_2}$ must be greater than before and, since this is assumed to be an increasing function of x_2, the value of x_2 must be greater than before. The new reaction curve C_2' is thus farther to the right than the old curve C_2, Ox_2 being horizontal as in Fig. 85. If x_1 and x_2 are the old equilibrium outputs, and x_1' and x_2' the new, it follows that $x_1'<x_1$ and $x_2'>x_2$. The second duopolist gains by being "alive" to his rival's reactions. Further, since C_1 has a numerical gradient (to Ox_2) less than unity, the sum of x_1' and x_2' is greater than the sum of x_1 and x_2, at least for a small shift in C_2. The joint output is greater, and the common price charged by the duopolists smaller, than in the previous case.

EXAMPLES XIII

Differentiation

1. If h and k are increments in x and y from the point (x, y), find the increment in $z=x^2+2xy-y^2$ in terms of x, y, h and k. What is the difference between this increment and dz? Show that the difference is of the second order in h and k.

2. Write down the equation of the tangent plane at (a, b, c) on the surface $z=f(x, y)$ and show that the increment in the height of the plane when increments dx and dy are given to x and y from (a, b, c) is dz. Hence illustrate the approximation of dz to the increment in z.

3. Evaluate the differentials of $z=x^3+y^3-3xy$; $z=\sqrt{x^2+y^2}$; $z=e^{x-y}$; $z=\log\frac{x}{x+y}$.

4. Obtain the differential of $z=(x+y)(x-y)$ by logarithmic differentiation and check by differentiating $z=x^2-y^2$ directly.

5. Take logarithms and differentiate $z=(x+y)\sqrt{x-y}$ and $z=(x+y)e^{x-y}$.

6. Find the differential of $u=(x^2+2xy-y^2)e^z$.

7. Show that $f(x, y)=f(0, 0)+xf_x'(0, 0)+yf_y'(0, 0)$ approximately when x and y are small. Deduce that, for small values of x and y,

$$\frac{1}{1+x-y}=1-x+y \quad \text{approximately.}$$

Find the difference between these expressions in terms of x and y.

8. Show that $\frac{x}{x+y}=\frac{1}{4}(x-y+2)$ approximately when x and y are each nearly equal to 1. What error is involved in the use of this approximate value when $x=1{\cdot}2$ and $y=0{\cdot}9$?

9. The volume of a right-circular cone of base radius x and height y is $\frac{1}{3}\pi x^2 y$. If x is estimated as a with an error of h and y as b with an error of k, show that the error in the estimate of the volume is $\frac{1}{3}\pi a(2bh+ak)$ approximately.

10. For the cone of the previous example, show that the proportional increase in volume is approximately $2\frac{\Delta x}{x}+\frac{\Delta y}{y}$ when x and y are increased by Δx and Δy. Deduce that a 1 per cent. increase in base radius increases volume by the same percentage amount as a 2 per cent. increase in height.

11. If $pv=at$ is the relation between the pressure (p), volume (v) and temperature (t) of a gas, a being a constant, show that the proportional increase in volume is approximately equal to the difference between the (small) proportional increases in temperature and pressure. Deduce that a simultaneous 1 per cent. increase in temperature and pressure leave the volume approximately unaltered.

12. From the expression for dz, find $\frac{dz}{dt}$ when $z=x^2+y^2$ where $x=1+t$ and $y=1-t$. Express z as a function of t and check your result by direct derivation.

13. Evaluate $\frac{dz}{dt}$ if $z=\frac{1}{x+y}$, where $x=e^t$ and $y=e^{-t}$.

14. Find $\frac{\partial z}{\partial u}$ and $\frac{\partial z}{\partial v}$ if $z=\log(x-y)$, where $x=\frac{u}{v}$ and $y=\frac{v}{u}$. Express z as a function of u and v and check by direct partial derivation.

15. If $z=f(x, y)$, where $x=a+\alpha t$ and $y=b+\beta t$, show that

$$\frac{dz}{dt}=\alpha\frac{\partial z}{\partial x}+\beta\frac{\partial z}{\partial y} \quad \text{and} \quad \frac{d^2z}{dt^2}=\alpha^2\frac{\partial^2 z}{\partial x^2}+2\alpha\beta\frac{\partial^2 z}{\partial x\,\partial y}+\beta^2\frac{\partial^2 z}{\partial y^2}.$$

16. If $z=f(x, y)$, where $x=au+\alpha v$ and $y=bu+\beta v$, show that

$$\frac{\partial z}{\partial u}=a\frac{\partial z}{\partial x}+b\frac{\partial z}{\partial y} \quad \text{and} \quad \frac{\partial z}{\partial v}=\alpha\frac{\partial z}{\partial x}+\beta\frac{\partial z}{\partial y}.$$

Generalise by finding the partial derivatives of $f(x, y, z, \ldots)$ as a function of u and v if $x=au+\alpha v$, $y=bu+\beta v$, $z=cu+\gamma v$, ...

17. Find $\frac{dz}{dt}$ when $z=f(x, y)$, where x and y are functions of t and show that

$$\frac{d^2z}{dt^2}=\frac{\partial^2 z}{\partial x^2}\left(\frac{dx}{dt}\right)^2+2\frac{\partial^2 z}{\partial x\,\partial y}\frac{dx}{dt}\frac{dy}{dt}+\frac{\partial^2 z}{\partial y^2}\left(\frac{dy}{dx}\right)^2+\frac{\partial z}{\partial x}\frac{d^2x}{dt^2}+\frac{\partial z}{\partial y}\frac{d^2y}{dt^2}.$$

18. A function $f(x, y)$ is homogeneous of the rth degree if $f(u, v)=\lambda^r f(x, y)$ where $u=\lambda x$ and $v=\lambda y$. Differentiate each side with respect to λ, put $\lambda=1$ and show that

$$x\frac{\partial f}{\partial x}+y\frac{\partial f}{\partial y}=rf(x, y) \qquad \text{(Euler's Theorem)}$$

and

$$x^2\frac{\partial^2 f}{\partial x^2}+2xy\frac{\partial^2 f}{\partial x\,\partial y}+y^2\frac{\partial^2 f}{\partial y^2}=r(r-1)f(x, y).$$

19. If $z=f(x, y)$, where $x=\lambda x_0$ and $y=\lambda y_0$ vary in a fixed proportion, consider z as a function of λ and show that $\frac{dz}{d\lambda}=\frac{1}{\lambda}\left(x\frac{\partial z}{\partial x}+y\frac{\partial z}{\partial y}\right)$. If $\epsilon=\frac{\lambda}{z}\frac{dz}{d\lambda}$ is the elasticity of $z=f(x, y)$ for proportional variation in x and y, show that $x\frac{\partial z}{\partial x}+y\frac{\partial z}{\partial y}=\epsilon z$. If the function is homogeneous of degree r, show that $\epsilon=r$ and so deduce Euler's Theorem.

20. Find $\frac{dy}{dx}$ and $\frac{d^2y}{dx^2}$ from the implicit function $(x-a)^2+(y-b)^2=c^2$.

21. If $ax^2+2hxy+by^2+2gx+2fy+c=0$, show that $\frac{dy}{dx}=-\frac{ax+hy+g}{hx+by+f}$ and evaluate the second derivative.

22. Find $\frac{dy}{dx}$ and $\frac{d^2y}{dx^2}$ when $x^3+y^3-3xy=0$. Show that y is a maximum when $x=\sqrt[3]{2}$. Interpret in terms of the graph of the function.

23. If $x^3+y^3+z^3-3xy=0$ defines z as a function of x and y, find $\frac{\partial z}{\partial x}$ and $\frac{\partial z}{\partial y}$. Deduce the second-order partial derivatives of z.

24. Given $f(x, y, z)=0$, find the first and second-order partial derivatives of z as a function of x and y.

25. If $x^2+y^2-2ax+1=0$ defines y as a function of x, a being a constant, show that $2xy\frac{dy}{dx}+x^2-y^2-1=0$.

26. Show that the tangent line at (x_1, y_1) on the curve $ax^2+by^2=c$ is $axx_1+byy_1=c$, and that the tangent plane at (x_1, y_1, z_1) on the surface $ax^2+by^2+cz^2=d$ is $axx_1+byy_1+czz_1=d$.

27. Explain the meaning of each of the six partial derivatives, of one variable in terms of another, obtainable from the implicit function $f(x, y, z)=0$. Express each in terms of the partial derivatives of f and deduce that

$$\frac{\partial z}{\partial y}\cdot\frac{\partial y}{\partial x}\cdot\frac{\partial x}{\partial z}=-1.$$

28. Two relations $\phi(x, y, z)=0$ and $\psi(x, y, z)=0$ define y and z as functions of x. Differentiate each relation and obtain expressions for $\frac{dy}{dx}$ and $\frac{dz}{dx}$ in terms of the partial derivatives of ϕ and ψ.

29. Differentiate the relations $xy+yz-z^2=0$ and $x^2+y-z=0$ to give $\frac{dy}{dx}$ and $\frac{dz}{dx}$. Solve the equations to obtain y and z explicitly in terms of x and obtain the derivatives directly.

Economic applications of differentials

30. If $x=f(a, b)$ is a linear homogeneous production function, show that the marginal rate of substitution between the factors depends only on the ratio $b:a$ of the factors used. Verify in the particular cases

$$x=\sqrt{2Hab-Aa^2-Bb^2} \quad \text{and} \quad x=\frac{2Hab-Aa^2-Bb^2}{Ca+Db}.$$

31. Show that the property of the previous example is also true of the quadratic homogeneous production function and illustrate with the function

$x=2Hab-Aa^2-Bb^2$. Is the property true for any homogeneous production function, e.g. for $x=Aa^{\alpha}b^{\beta}$?

32. If the production function is $x=2Hab-Aa^2-Bb^2$, show that at any point (a, b) on a constant product curve

$$\frac{db}{da}=-\frac{Hb-Aa}{Ha-Bb} \quad \text{and} \quad \frac{d^2b}{da^2}=\frac{(H^2-AB)x}{(Ha-Bb)^3}.$$

Deduce that the first derivative is negative and the second positive provided that $\frac{H}{B}a>b>\frac{A}{H}a$. Deduce that the constant product curves are of normal form in the area between the lines $Hb-Aa=0$ and $Ha-Bb=0$.

33. Show that, for the production function $x=\sqrt{2Hab-Aa^2-Bb^2}$,

$$\sigma=\frac{Hx^2}{(H^2-AB)ab}-1.$$

34. Show that the production function $x=Aa^{\alpha}b^{\beta}$ gives $\sigma=1$ for all combinations of factors.

35. Show that the elasticity of productivity ϵ (10.8 above) is the elasticity of the vertical section of the production surface through the axis Ox and the given point on the surface. Use the results of Example 19 above to show that, for any production function,

$$a\frac{\partial x}{\partial a}+b\frac{\partial x}{\partial b}=\epsilon x.$$

36. If $u=\log(x+a)^{\alpha}(y+b)^{\beta}$ is one form of a utility function, find the marginal rate of substitution between the goods X and Y and deduce that the elasticity of substitution is

$$\sigma=1+\frac{b\alpha x+a\beta y}{(\alpha+\beta)xy}.$$

37. If $ax^2+by^2=\text{constant}$ is the transformation function for two goods X and Y, show that the marginal rate of substitution of Y production for X production is $\frac{ax}{by}$ and that the elasticity of substitution is always unity.

38. Two goods X and Y are produced jointly with the aid of two factors A and B. Technical conditions define a relation $f(x, y, a, b)=0$, so that (e.g.) the production of Y is determined if the amounts of the factors used and the production of X are given. Show how a and b can be used as parameters to describe the shifting of a transformation curve between X and Y production. What do the partial derivatives $\frac{\partial y}{\partial x}$, $\frac{\partial y}{\partial a}$ and $\frac{\partial y}{\partial b}$ represent?

39. If, in the problem of the previous example, technical conditions give two production functions $x=\phi(a, b)$ and $y=\psi(a, b)$, show that

$$\left(\frac{da}{dx}\right)_{y \text{ constant}}=\frac{\psi_b}{\phi_a\psi_b-\phi_b\psi_a}$$

and interpret the meaning of this derivative.

40. Of the duopolists of Examples VIII, 45, the first acts (as before) on the assumption that the second does not change output while the second correctly estimates the first's output changes consequent upon his own changes. Show that the total equilibrium output is now approximately 44 sets per week. How is this output divided?

CHAPTER XIV

PROBLEMS OF MAXIMUM AND MINIMUM VALUES

14.1 Partial stationary values.

In our discussion of the partial derivatives of a function of two variables (12.4 above), we postponed consideration of the cases where one or both of the derivatives assume zero values. Suppose now that $\frac{\partial z}{\partial x}=0$ and $\frac{\partial z}{\partial y}\neq 0$ for a function $z=f(x, y)$ at a point (a, b). Then z has a stationary value for variation in x from the value a, y retaining the constant value b. The sign of $\frac{\partial^2 z}{\partial x^2}$ at (a, b) determines whether z is a maximum or a minimum.* The vertical section of the surface $z=f(x, y)$ which passes through P (where $x=a$ and $y=b$) in the W.E. direction has a maximum or minimum point at P.

In general, the condition $\frac{\partial z}{\partial x}=0$ is an implicit relation between x and y represented by some curve in the plane Oxy. Points on the surface $z=f(x, y)$ above this curve correspond to maximum or minimum values of z in the W.E. direction. Some of these points may correspond to maximum values $\left(\text{satisfying the condition } \frac{\partial^2 z}{\partial x^2}<0\right)$ and make up what can be called a *ridge line* of the surface in the W.E. direction. Others of the points may correspond to minimum values $\left(\frac{\partial^2 z}{\partial x^2}>0\right)$ and define a *trough line* of the surface in the W.E. direction. The curve in Oxy giving the ridge and trough lines of the surface is clearly located on the contour map of the

* Throughout the present chapter, the possibility that stationary values can include points of inflexion, though not overlooked, will not be taken into account.

surface. The tangent gradient of the contour through the point (x, y) is

$$\frac{dy}{dx} = -\frac{\frac{\partial z}{\partial x}}{\frac{\partial z}{\partial y}},$$

which is zero whenever $\frac{\partial z}{\partial x}=0$. The ridge and trough lines of the surface thus lie above the locus of points on the contour map where the tangents are parallel to the axis Ox. For example, in the case shown in Fig. 87 below, the line AB across the contour map corresponds to a ridge line of the surface in the W.E. direction.

A similar analysis holds if $\frac{\partial z}{\partial x} \neq 0$ and $\frac{\partial z}{\partial y}=0$ at any point. The function then has a partial maximum or minimum value $\left(\text{according to the sign of } \frac{\partial^2 z}{\partial y^2}\right)$ for variation of y alone. The ridge and trough lines of the corresponding surface are then obtained in the S.N. direction. They lie above the points on the contour map where the tangents to the contours are parallel to Oy. The line CD across the contour map of Fig. 87, for example, represents a ridge line of the surface in the S.N. direction.

The analysis extends without difficulty to functions of more than two variables. In general, if $y=f(x_1, x_2, \ldots x_n)$, there are n different kinds of partial stationary values, each corresponding to the vanishing of one of the n partial derivatives of the function. Further, though the analysis is expressed in terms of explicit functions, it applies equally well to functions written in implicit form.

14.2 Maximum and minimum values of a function of two or more variables.

If both the partial derivatives of the function $z=f(x, y)$ are zero at the point (a, b), then z has a stationary value for variation in x alone and for variation in y alone from (a, b). Further,

$$dz=\frac{\partial z}{\partial x}dx+\frac{\partial z}{\partial y}dy=0$$

for any values of dx and dy, i.e. z has a stationary value for any variation whatever from the point (a, b). Such a point is said to define a *stationary value* of the function. Ignoring the possibility of inflexional points, there are three cases to distinguish :

(1) The value of z is a maximum for any variation of x and y from (a, b), the surface $z=f(x, y)$ having a " peak " at the point where $x=a$ and $y=b$.

(2) The value of z is a minimum for any variation, the surface having a " hollow " with lowest point where $x=a$ and $y=b$.

(3) The value of z is a maximum for some variations and a minimum for other variations from (a, b), the surface having a " saddle point " where $x=a$ and $y=b$, i.e. a point where there is a " pass " over a " ridge " of the surface.

The three cases are illustrated in Fig. 86.

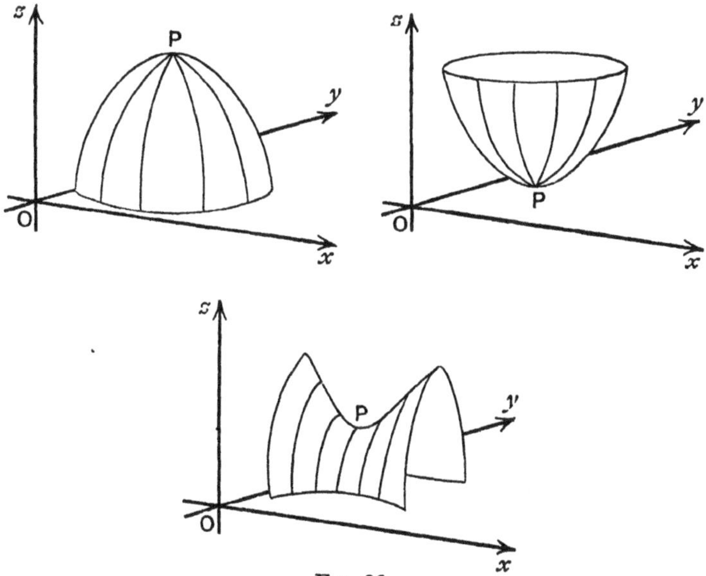

FIG. 86.

Stationary values, occurring where the tangent plane to the surface is horizontal (parallel to Oxy), are located on the contour map of the surface at points where the locus of points with tangents parallel to Ox $\left(\frac{\partial z}{\partial x}=0\right)$ cuts the locus of points with tangents parallel to Oy $\left(\frac{\partial z}{\partial y}=0\right)$. There is a stationary value wherever the ridge or trough line in the W.E. direction cuts the similar line in the S.N. direction. The lie of the contours around such a point determines the nature of the stationary value. If the contours tend to close down to a single point as their height increases or decreases, then the stationary value is a maximum value of type (1) or a minimum value of type (2). If a contour crosses itself at the point and there is no tendency to close down on the point, then the stationary value occurs at a saddle point of type (3). The cases are fully illustrated by the particular contour maps of Figs. 87 and 88.

In more rigid analytical terms, the function $z=f(x, y)$ has a *maximum value* at (a, b) if z is a maximum for any variation of x and y from the values a and b respectively. A *minimum value* of z is defined similarly. Maximum and minimum values together make up the *extreme values* of the function. It is assumed that the function and its partial derivatives are finite and continuous.* A maximum (minimum) value can occur only if the function is a maximum (minimum) in each of the two fundamental directions in which x and y vary alone, i.e. only if the two partial derivatives are separately zero and the function has a stationary value. To distinguish maximum from minimum values, or either from other stationary values, is a matter of some difficulty. If $\frac{\partial^2 z}{\partial x^2}$ and $\frac{\partial^2 z}{\partial y^2}$ are both negative at a stationary point, then z must be a maximum in each of the fundamental directions and it is *likely* that z is a maximum in all directions. Similarly, if $\frac{\partial^2 z}{\partial x^2}$ and $\frac{\partial^2 z}{\partial y^2}$ are both positive at such a point, then z is likely to have a minimum value for all directions of variation. But these results are by no means certain. We must exclude the possibility that z behaves differently in some directions as compared with the fundamental directions. It is shown below (14.7, Ex. 3 and 19.1) that

$$\frac{\partial^2 z}{\partial x^2}\frac{\partial^2 z}{\partial y^2} > \left(\frac{\partial^2 z}{\partial x\,\partial y}\right)^2$$

is a condition sufficient to exclude this possibility. Hence:

* As in the one-variable case, a maximum or minimum value of $z=f(x, y)$ can occur at a point where the partial derivatives are not continuous. Such a point is not shown by the criterion below.

CRITERION FOR MAXIMUM AND MINIMUM VALUES

(1) All maximum and minimum values of $z=f(x, y)$ occur where $dz=0$ for all variations of x and y, i.e. where

$$\frac{\partial z}{\partial x}=\frac{\partial z}{\partial y}=0.$$

(2) If $\frac{\partial^2 z}{\partial x^2}$ and $\frac{\partial^2 z}{\partial y^2}$ are negative and such that $\frac{\partial^2 z}{\partial x^2}\frac{\partial^2 z}{\partial y^2}>\left(\frac{\partial^2 z}{\partial x\, \partial y}\right)^2$ at a point $x=a$, $y=b$, then $f(a, b)$ is a maximum value of the function. If $\frac{\partial^2 z}{\partial x^2}$ and $\frac{\partial^2 z}{\partial y^2}$ are positive subject to the same inequality at a point $x=a$, $y=b$, then $f(a, b)$ is a minimum value of the function.

The first is a necessary condition. The second condition when added to the first is a sufficient condition and defines situations in which only a maximum (or only a minimum) value can occur. But the condition is not complete; it is not necessary and sufficient. It is possible to show that, whenever $\frac{\partial^2 z}{\partial x^2}\frac{\partial^2 z}{\partial y^2}<\left(\frac{\partial^2 z}{\partial x\, \partial y}\right)^2$, then the stationary point concerned is a saddle point. But the case where $\frac{\partial^2 z}{\partial x^2}\frac{\partial^2 z}{\partial y^2}=\left(\frac{\partial^2 z}{\partial x\, \partial y}\right)^2$ is still open and the stationary value concerned may be a maximum or minimum value or it may not.

The criterion holds, in exactly the same form, for functions given implicitly. If $f(x, y, z)=0$ defines z as a function of x and y, we have only to write

$$\frac{\partial z}{\partial x}=-\frac{f_x}{f_z} \quad \text{and} \quad \frac{\partial z}{\partial y}=-\frac{f_y}{f_z}$$

and the criterion applies. The necessary condition for a maximum or minimum value of z reduces to $f_x=f_y=0$.

The criterion of maximum and minimum values can be extended to functions of more than two variables, but only as regards the necessary condition at this stage. All the maximum and minimum values of $y=f(x_1, x_2, \dots x_n)$ occur at points where $dy=0$ for all variations in the independent variables, i.e. where

$$\frac{\partial y}{\partial x_1}=\frac{\partial y}{\partial x_2}=\dots=\frac{\partial y}{\partial x_n}=0.$$

Similarly, if $f(x_1, x_2, \dots x_n, y)=0$ defines y as a function of the other variables, then the maximum and minimum values of y occur where

$$f_{x_1}=f_{x_2}=\ldots=f_{x_n}=0.$$

Sufficient conditions are given later (19.1).

An important feature of the results here given is that, since the differential of a maximised or minimised variable is zero at the point concerned, we can treat such a variable *as a constant* in the process of differentiation. Suppose that, in the relation

$$f(x_1, x_2, \ldots x_n, y, z)=0,$$

the variable y is to be maximised for variation in $x_1, x_2, \ldots x_n$, the variable z remaining fixed. Differentiating and putting $dy=0$ (since y is maximised) and $dz=0$ (since z is fixed), we have

$$f_{x_1}dx_1+f_{x_2}dx_2+\ldots+f_{x_n}dx_n=0$$

for any variations in the variables $x_1, x_2, \ldots x_n$. So

$$f_{x_1}=f_{x_2}=\ldots=f_{x_n}=0.$$

Exactly the same conditions obtain if z is maximised while y is held constant. There is no distinction, in differentiating, between variables which are maximised (or minimised) and variables which are held constant.

14.3 Examples of maximum and minimum values.

Maximum and minimum values of a function $z=f(x, y)$ are located in practice as follows. The equations $\frac{\partial z}{\partial x}=\frac{\partial z}{\partial y}=0$ are solved for x and y and each solution examined in turn. In the majority of cases, the sufficient conditions (involving the second-order partial derivatives) enable us to distinguish maximum and minimum values. In some cases where the conditions fail, we can examine the contour map of the function in the neighbourhood of the point concerned. And, even if the sufficient conditions apply, the contour map provides a useful illustration of the situation. The following examples illustrate the method :

Ex. 1. $z=x^2+y^2$.

Here $\frac{\partial z}{\partial x}=2x$ and $\frac{\partial z}{\partial y}=2y$.

There is one stationary value ($z=0$) occurring where $x=y=0$ and this value must be a *minimum* since all other values of z are positive. The sufficient conditions are found to support this, since

$$\frac{\partial^2 z}{\partial x^2}=\frac{\partial^2 z}{\partial y^2}=2 \quad\text{and}\quad \frac{\partial^2 z}{\partial x\,\partial y}=0.$$

The contour map of the function has trough lines along Ox and Oy and is shown in Fig. 76 above.

Ex. 2. $z=2x+y-x^2+xy-y^2$.

Here $\frac{\partial z}{\partial x}=-2x+y+2$ and $\frac{\partial z}{\partial y}=x-2y+1$

and stationary values of z occur where

$$2x-y-2=0 \quad\text{and}\quad x-2y+1=0.$$

There is a single stationary value ($z=\frac{7}{3}$), where $x=\frac{5}{3}$ and $y=\frac{4}{3}$.

Since $\frac{\partial^2 z}{\partial x^2}=-2,\quad \frac{\partial^2 z}{\partial x\,\partial y}=1$ and $\frac{\partial^2 z}{\partial y^2}=-2,$

the sufficient conditions show that $z=\frac{7}{3}$ is a maximum value. The contour

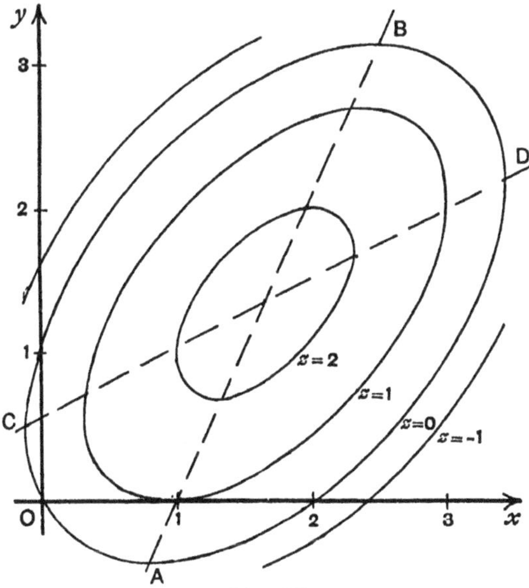

FIG. 87.

map of the function is shown in Fig. 87, rising contours closing down on the point where $x=\frac{5}{3}$, $y=\frac{4}{3}$. The line AB ($2x-y-2=0$) is a ridge line in the W.E. direction and the line CD ($x-2y+1=0$) is a ridge line in the S.N. direction.

Ex. 3. $z = 4x^2 - xy + y^2 - x^3.$

Here $\dfrac{\partial z}{\partial x} = 8x - y - 3x^2 ; \quad \dfrac{\partial z}{\partial y} = -x + 2y ;$

$$\frac{\partial^2 z}{\partial x^2} = 2(4 - 3x) ; \quad \frac{\partial^2 z}{\partial x \, \partial y} = -1 \quad \text{and} \quad \frac{\partial^2 z}{\partial y^2} = 2.$$

Stationary values of z are given by

$$3x^2 - 8x + y = 0 \quad \text{and} \quad x = 2y.$$

On solving (e.g. by eliminating y), we find two stationary values, $z = 0$ where $x = y = 0$, and $z = \frac{125}{16}$ where $x = \frac{5}{2}$, $y = \frac{5}{4}$. The first is a *minimum* value of z since, when $x = y = 0$, we have

$$\frac{\partial^2 z}{\partial x^2} = 8 > 0 ; \quad \frac{\partial^2 z}{\partial y^2} = 2 > 0 \quad \text{and} \quad \frac{\partial^2 z}{\partial x^2} \frac{\partial^2 z}{\partial y^2} = 16 > \left(\frac{\partial^2 z}{\partial x \, \partial y} \right)^2 = 1.$$

But when $x = \frac{5}{2}$ and $y = \frac{5}{4}$ the two direct partial derivatives of the second

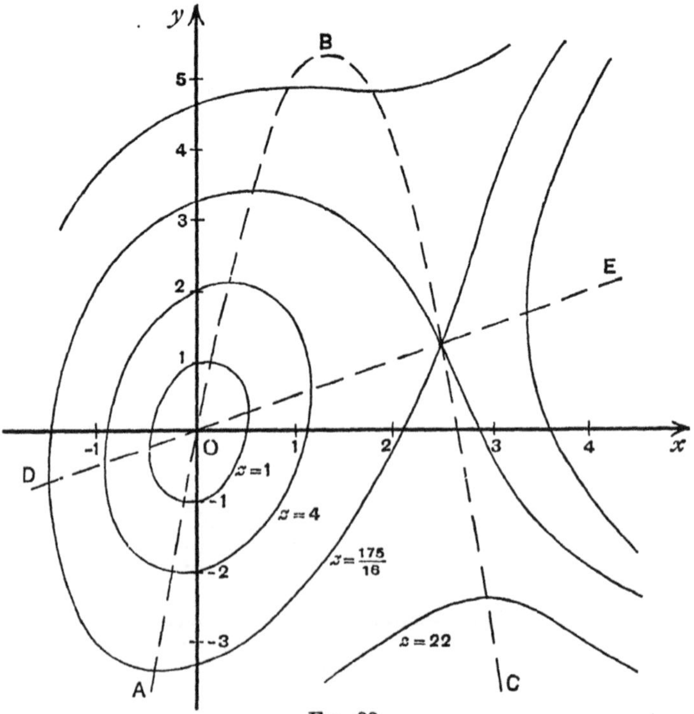

FIG. 88.

order are opposite in sign (-7 and 2) and the stationary value here occurs at a *saddle point*. Fig. 88 shows certain curves of the contour map of the function. The curve *ABC* (the parabola $y = 8x - 3x^2$) represents the ridge and trough lines of the surface in the W.E. direction. The part

$AB\left(x<\frac{4}{3}, \frac{\partial z}{\partial x}=0 \text{ and } \frac{\partial^2 z}{\partial x^2}>0\right)$ is the trough line and the part $BC\left(x>\frac{4}{3}, \frac{\partial z}{\partial x}=0 \text{ and } \frac{\partial^2 z}{\partial x^2}<0\right)$ is the ridge line. The line DE $(x=2y)$ is the trough line in the S.N. direction $\left(\frac{\partial z}{\partial y}=0 \text{ and } \frac{\partial^2 z}{\partial y^2}>0\right)$. The lie of the contours about the two points of intersection of the line and the parabola show that there is a minimum value of z at the origin and a saddle point at $x=\frac{5}{2}$, $y=\frac{5}{4}$.

Ex. 4. $x^2+y^2+3z^2-2x+2z=0.$

Taking z as a function of x and y, we find

$$\frac{\partial z}{\partial x}=-\frac{x-1}{3z+1} \quad \text{and} \quad \frac{\partial z}{\partial y}=-\frac{y}{3z+1}.$$

The only stationary values of z, for variation of x and y, occur at $x=1$, $y=0$. On substituting in the original equation, two values of z are found, $z=-1$ and $z=\frac{1}{3}$. Further, we find

$$\frac{\partial^2 z}{\partial x^2}=-\frac{1}{(3z+1)^2}\left\{(3z+1)-3(x-1)\frac{\partial z}{\partial x}\right\}$$

$$=-\frac{1}{(3z+1)^3}\{(3z+1)^2+3(x-1)^2\}=-\frac{1}{3z+1},$$

when $x=1$, $y=0$. Similarly, at the same point,

$$\frac{\partial^2 z}{\partial x\,\partial y}=0 \quad \text{and} \quad \frac{\partial^2 z}{\partial y^2}=-\frac{1}{3z+1}.$$

So, when $z=-1$, the direct second-order partial derivatives are positive and the cross derivative is zero. By the sufficient conditions, $z=-1$ is a *minimum* value of z. In the same way, $z=\frac{1}{3}$ is found to be a *maximum* value of z. The solution can be illustrated by a contour map of this double-valued function (see Examples XIV, 9).

14.4 Monopoly and joint production.

A monopolist produces amounts x_1 and x_2 of two goods X_1 and X_2 at a total cost of $\Pi=F(x_1, x_2)$. The two goods are related in consumption and the demands of the market are $x_1=\phi_1(p_1, p_2)$ and $x_2=\phi_2(p_1, p_2)$, where p_1 and p_2 are the prices charged. The net revenue of the monopolist is

$$y=x_1p_1+x_2p_2-\Pi.$$

Making use of the demand relations, this is a function of the two prices. It is taken that the monopolist fixes prices to maximise net revenue. The necessary condition for this is

$$\frac{\partial y}{\partial p_1}=\frac{\partial y}{\partial p_2}=0,$$

i.e. $$x_1+\left(p_1-\frac{\partial \Pi}{\partial x_1}\right)\frac{\partial x_1}{\partial p_1}+\left(p_2-\frac{\partial \Pi}{\partial x_2}\right)\frac{\partial x_2}{\partial p_1}=0$$

and $$x_2+\left(p_1-\frac{\partial \Pi}{\partial x_1}\right)\frac{\partial x_1}{\partial p_2}+\left(p_2-\frac{\partial \Pi}{\partial x_2}\right)\frac{\partial x_2}{\partial p_2}=0.$$

There are two relations to determine the prices. Subject to the prices obtained satisfying the conditions for a maximum (rather than a minimum) value of y, the problem is solved.

To illustrate the solution, a simple case can be considered in which the cost and demand functions are linear :

$$\Pi=\alpha_1 x_1+\alpha_2 x_2\,;\quad x_1=a_1-a_{11}p_1-a_{12}p_2 \quad\text{and}\quad x_2=a_2-a_{12}p_1-a_{22}p_2.$$

The average costs of producing the goods can be taken as constants, α_1 and α_2 respectively. Denote

$$x_{10}=a_1-a_{11}\alpha_1-a_{12}\alpha_2 \quad\text{and}\quad x_{20}=a_2-a_{12}\alpha_1-a_{22}\alpha_2,$$

i.e. the demands when the " competitive " prices (equal to average costs) are charged. Then

$$x_1=x_{10}-a_{11}(p_1-\alpha_1)-a_{12}(p_2-\alpha_2)$$

and $$x_2=x_{20}-a_{12}(p_1-\alpha_1)-a_{22}(p_2-\alpha_2)$$

and the monopolist's net revenue is

$$y=x_1p_1+x_2p_2-\Pi=x_1(p_1-\alpha_1)+x_2(p_2-\alpha_2).$$

So $$\frac{\partial y}{\partial p_1}=x_1+(p_1-\alpha_1)\frac{\partial x_1}{\partial p_1}+(p_2-\alpha_2)\frac{\partial x_2}{\partial p_1}$$

$$=x_{10}-2a_{11}(p_1-\alpha_1)-2a_{12}(p_2-\alpha_2)$$

and $$\frac{\partial y}{\partial p_2}=x_{20}-2a_{12}(p_1-\alpha_1)-2a_{22}(p_2-\alpha_2).$$

Putting these expressions equal to zero for maximum net revenue and solving, we find the monopoly prices

$$p_1=\alpha_1+\frac{a_{22}x_{10}-a_{12}x_{20}}{2(a_{11}a_{22}-a_{12}^2)} \quad\text{and}\quad p_2=\alpha_2+\frac{a_{11}x_{20}-a_{12}x_{10}}{2(a_{11}a_{22}-a_{12}^2)}.$$

Further, $\dfrac{\partial^2 y}{\partial p_1^2}=-2a_{11}$; $\dfrac{\partial^2 y}{\partial p_1\,\partial p_2}=-2a_{12}$ and $\dfrac{\partial^2 y}{\partial p_2^2}=-2a_{22}.$

If a_{11} and a_{22} are positive and such that $a_{12}^2<a_{11}a_{22}$, then the prices

give a maximum value of net revenue. These limitations on the values of the constants are assumed below.

Suppose, now, that two independent monopolists produce X_1 and X_2 with costs $\Pi_1 = \alpha_1 x_1$ and $\Pi_2 = \alpha_2 x_2$, each assuming that the other will not change his price and maximising net revenue. The net revenue of the first monopolist is

$$y_1 = x_1 p_1 - \Pi_1 = x_1 (p_1 - \alpha_1)$$

where the demand gives x_1 as before. For a given value of p_2, the maximum net revenue is obtained when $\frac{dy_1}{dp_1} = 0$, i.e. when

$$x_{10} - 2a_{11}(p_1 - \alpha_1) - a_{12}(p_2 - \alpha_2) = 0.$$

Similarly, for the second monopolist,

$$x_{20} - a_{12}(p_1 - \alpha_1) - 2a_{22}(p_2 - \alpha_2) = 0.$$

Solving these two equations, the prices charged by the independent monopolists are found to be

$$p_1 = \alpha_1 + \frac{a_{22}x_{10} - \frac{1}{2}a_{12}x_{20}}{2(a_{11}a_{22} - \frac{1}{4}a_{12}^2)} \quad \text{and} \quad p_2 = \alpha_2 + \frac{a_{11}x_{20} - \frac{1}{2}a_{12}x_{10}}{2(a_{11}a_{22} - \frac{1}{4}a_{12}^2)}.$$

Since $\frac{d^2y_1}{dp_1^2} = -2a_{11} < 0$ and similarly for $\frac{d^2y_2}{dp_2^2}$, the prices correspond to positions of maximum net revenue.

If a_{12} is negative, so that the goods are (in the broad sense) competitive in consumption, it is seen that the joint monopoly prices are greater than the separate monopoly prices ; the effect of joint monopolistic production is to restrict output and raise price. If a_{12} is positive and the goods are complementary, there are two cases : *

(1) $$a_{12} < a_{11}\frac{x_{20}}{x_{10}} \quad \text{and} \quad a_{12} < a_{22}\frac{x_{10}}{x_{20}}.$$

Here, for the joint monopoly prices, $(p_1 - \alpha_1)$ and $(p_2 - \alpha_2)$ are positive and both prices are above cost as represented by α_1 and α_2. It may or may not be that both prices are above those of separate monopoly. This case holds when, as in the majority of situations,

* Notice that the case where $a_{12} > a_{11}\frac{x_{20}}{x_{10}}$ and $a_{12} < a_{22}\frac{x_{10}}{x_{20}}$ is exactly parallel with and adds nothing to case (2) and that it is impossible, since $a_{12}^2 < a_{11}a_{22}$, for a_{12} to be greater than both $a_{11}\frac{x_{20}}{x_{10}}$ and $a_{22}\frac{x_{10}}{x_{20}}$.

the relation between the goods (as shown by a_{12}) is weak as compared with the direct price effects on demand (as shown by a_{11} and a_{22}).

$$(2) \qquad a_{12} < a_{11}\frac{x_{20}}{x_{10}} \quad \text{and} \quad a_{12} > a_{22}\frac{x_{10}}{x_{20}}.$$

Here, $(p_1 - \alpha_1)$ is negative while $(p_2 - \alpha_2)$ is positive, i.e. the good X_1 is sold by the joint monopolist at a price below cost α_1, and certainly below the separate monopoly price. It pays the monopolist now to use one good as a " loss leader " in order to promote the sales of the second good. The conditions for the case are (roughly) that the goods are strongly complementary (a_{12} large) and that the demand for the " loss leader " is sensitive to changes in its price (a_{11} large). For example, a monopolist manufacturer of razors and blades may gain by marking down the price of razors (for which the demand is elastic) below cost to promote the sales of blades (for which the demand is less elastic and strongly complementary with that for razors).*

14.5 Production, capital and interest.

To generalise the problem of 10.3 above, suppose that a variable time elapses between the beginning of a production process and the final output of a good X produced with two factors A and B. The amount of output x depends, according to technical conditions, on the amounts a and b of the factors used and on the length t of the period of production :

$$x = f(a, b, t).$$

It is assumed that the prices of the factors are given as p_a and p_b and the market rate of interest as $100r$ per cent. per year compounded continuously, all taken (for convenience) in terms of the product x. Further, it is taken that the factors are paid off at the beginning of the production. So

$$y = \frac{xe^{-rt}}{ap_a + bp_b} = \phi(a, b, t) \quad \text{(say)}$$

is the product discounted to the beginning of production as a ratio of the cost then incurred. It is assumed that the entrepreneur is a

* For an analysis of the " loss leader " problem, see Roos, *Dynamic Economics* (1934), pp. 128-47. On problems of joint production, see also Stackelberg, *Grundlagen einer reinen Kostentheorie* (1932) and Hicks, *The Theory of Monopoly*, Econometrica, 1935.

monopolist arranging production to maximise y. The variables to be chosen are a, b and t and we must have

$$\frac{\partial y}{\partial a}=\frac{\partial y}{\partial b}=\frac{\partial y}{\partial t}=0.$$

Since $$\log y=\log x-rt-\log(ap_a+bp_b),$$

we have $$\frac{1}{y}\frac{\partial y}{\partial a}=\frac{1}{x}\frac{\partial x}{\partial a}-\frac{p_a}{ap_a+bp_b},$$

$$\frac{1}{y}\frac{\partial y}{\partial b}=\frac{1}{x}\frac{\partial x}{\partial b}-\frac{p_b}{ap_a+bp_b},$$

$$\frac{1}{y}\frac{\partial y}{\partial t}=\frac{1}{x}\frac{\partial x}{\partial t}-r.$$

So $$p_a=\frac{ap_a+bp_b}{x}\frac{\partial x}{\partial a},\quad p_b=\frac{ap_a+bp_b}{x}\frac{\partial x}{\partial b}\quad\text{and}\quad r=\frac{1}{x}\frac{\partial x}{\partial t}$$

are the necessary conditions for a maximum value of y. These equations, with $x=f(a, b, t)$, determine the equilibrium values of a, b, t and x as dependent on the given values of p_a, p_b and r.

Suppose now that there is pure competition among entrepreneurs so that the market rate of interest adjusts itself to make (discounted) receipts just cover cost. The given value of r is such that the maximised value of y is unity and

$$xe^{-rt}=ap_a+bp_b.$$

So $$p_a=e^{-rt}\frac{\partial x}{\partial a}\quad\text{and}\quad p_b=e^{-rt}\frac{\partial x}{\partial b}.$$

Production is extended until the discounted marginal products of the factors are equal to the fixed prices of the factors.

For example, suppose that the production function is

$$x=Aa^{\alpha}b^{1-\alpha}t^{\beta},$$

where A is constant and α and β are fixed positive fractions. For a fixed period of production and variable factors, there are constant returns to scale. The necessary conditions for monopoly equilibrium are

$$p_a=\alpha\frac{ap_a+bp_b}{a},\quad p_b=(1-\alpha)\frac{ap_a+bp_b}{b}\quad\text{and}\quad r=\frac{\beta}{t},$$

since $$\frac{1}{x}\frac{\partial x}{\partial a}=\frac{\alpha}{a},\quad \frac{1}{x}\frac{\partial x}{\partial b}=\frac{1-\alpha}{b}\quad\text{and}\quad \frac{1}{x}\frac{\partial x}{\partial t}=\frac{\beta}{t}.$$

The period of production $t=\frac{\beta}{r}$ is here determined solely by the rate of interest and not by the prices of the factors. The first two conditions are equivalent and imply that $(1-\alpha)ap_a$ equals αbp_b,

i.e. $$\frac{a}{b}=\frac{\alpha}{1-\alpha}\frac{p_b}{p_a}.$$

Only the ratio of the amounts of the factors used is determined; the actual scale of production is immaterial (constant returns to scale). The ratio, further, is fixed by the prices of the factors and not by the rate of interest. Notice also that

$$\frac{ap_a}{bp_b}=\frac{\alpha}{1-\alpha}=\text{constant},$$

i.e. the relative share of the two factors in the product is constant, unaffected by changes in their prices or in the interest rate.

Under competition, with the particular production function now assumed, we have the added condition

$$\frac{ap_a+bp_b}{x}=e^{-rt}=e^{-\beta}=\text{constant},$$

i.e. the combined share of the factors in the product is constant, independent of their prices or the interest rate. The share of capital (as represented by the entrepreneurs) is then constant also. In the competitive case, in fact, the relative shares of the factors and of capital in the total product are constant, no matter how the prices of the factors or the interest rate change.*

14.6 Relative maximum and minimum values.

A different problem, that of relative maximum and minimum values, arises when we seek the maximum or minimum values of a function $z=f(x, y)$ where the variables are not independent but related by some given relation $\phi(x, y)=0$. The given relation is often termed a *side relation*. The side relation determines one variable (say y) as a function of the other. Substituting,

$$z=f(x, y)=\psi(x) \quad \text{(say)}$$

and our problem is effectively one of maximising or minimising a function of one variable. In simple cases, the side relation gives y

* Compare the results here obtained with those given by Edelberg, *An econometric model of production and distribution*, Econometrica, 1936.

as a single-valued function of x, in which case $\psi(x)$ also appears in single-valued form and our problem is solved without difficulty.

In general, the side relation is multi-valued and a more subtle method of attack is needed. But the essential idea is still the reduction of z to dependence on one variable only. Taking x as the independent variable and assuming that all partial derivatives are finite and continuous, the function and side relation give

$$\frac{dz}{dx}=f_x+f_y\frac{dy}{dx} \quad \text{and} \quad \phi_x+\phi_y\frac{dy}{dx}=0.$$

So

$$\frac{dz}{dx}=f_x-\frac{\phi_x}{\phi_y}f_y.$$

Equating this expression to zero, we obtain :

The function $z=f(x, y)$ has maximum or minimum values relative to the side relation $\phi(x, y)=0$ only at points where

$$\frac{f_x}{\phi_x}=\frac{f_y}{\phi_y}.$$

This is a necessary condition for maximum or minimum values. Adding the side relation, there are two equations to be solved for x and y. Each solution is a possible maximum or minimum situation and must be examined further. We find

$$\frac{d^2z}{dx^2}=\frac{d}{dx}\left(f_x-\frac{\phi_x}{\phi_y}f_y\right)=\frac{\partial}{\partial x}\left(f_x-\frac{\phi_x}{\phi_y}f_y\right)+\frac{\partial}{\partial y}\left(f_x-\frac{\phi_x}{\phi_y}f_y\right)\frac{dy}{dx}$$

$$=\frac{1}{\phi_y{}^3}\{\phi_y(f_{xx}\phi_y{}^2-2f_{xy}\phi_x\phi_y+f_{yy}\phi_x{}^2)-f_y(\phi_{xx}\phi_y{}^2-2\phi_{xy}\phi_x\phi_y+\phi_{yy}\phi_x{}^2)\}$$

on reduction. The sign of this expression at a point given as a solution of the necessary condition determines whether the point corresponds to a maximum or minimum value of z.

In diagrammatic terms, instead of locating the peaks and hollows of the surface $z=f(x, y)$, we consider only a definite path on the surface, defined by the side relation $\phi(x, y)=0$, and determine where the path attains its highest or lowest height above the zero horizontal plane Oxy. Using the contour map of the surface, our path is a given curve crossing the contours of the map and we seek those points on it where the highest or lowest contours are cut. It is clear that, at these points, the given curve *touches* a contour (see Fig. 89 below). The tangent gradient of the given curve $\phi(x, y)=0$ is then equal to that of the contour $f(x, y) = \text{constant}$, i.e. $\left(-\frac{\phi_x}{\phi_y}\right)$ equals $\left(-\frac{f_x}{f_y}\right)$ which is the necessary condition above.

For functions of several variables, we seek the maximum or minimum values of $y = f(x_1, x_2, \dots x_n)$ subject to a given side relation $\phi(x_1, x_2, \dots x_n) = 0$. The side relation gives x_1 as a function of the other $(n-1)$ variables for which

$$\frac{\partial x_1}{\partial x_2} = -\frac{\phi_{x_2}}{\phi_{x_1}}; \quad \frac{\partial x_1}{\partial x_3} = -\frac{\phi_{x_3}}{\phi_{x_1}}; \quad \dots .$$

Then y is a function of $(n-1)$ independent variables and

$$\frac{\partial y}{\partial x_2} = f_{x_1}\frac{\partial x_1}{\partial x_2} + f_{x_2} = f_{x_2} - \frac{\phi_{x_2}}{\phi_{x_1}} f_{x_1}$$

and similar expressions for the other partial derivatives. At a maximum or minimum position, all these partial derivatives must be zero. It follows that

$$\frac{f_{x_1}}{\phi_{x_1}} = \frac{f_{x_2}}{\phi_{x_2}} = \dots = \frac{f_{x_n}}{\phi_{x_n}}$$

is the necessary condition for a maximum or minimum value of y. There are $(n-1)$ equations here and these, with the side relation, determine certain sets of values of $x_1, x_2, \dots x_n$ amongst which are all those giving maximum or minimum values of y. Sufficient conditions for distinguishing maximum and minimum values are given later (19.2).

In a more general problem, the maximum or minimum values of y are required subject to r given side relations between the variables (where $r < n$). An alternative method of approach can now be used to advantage.* In the case of two side relations

$$\phi(x_1, x_2, \dots x_n) = 0 \quad \text{and} \quad \psi(x_1, x_2, \dots x_n) = 0,$$

write $\quad z = f(x_1, x_2, \dots x_n) - \lambda\phi(x_1, x_2, \dots x_n) - \mu\psi(x_1, x_2, \dots x_n).$

Then z must have a maximum or minimum value for any values of λ and μ whatever (since the coefficients of λ and μ are zero). But

$$\frac{\partial z}{\partial x_1} = f_{x_1} - \lambda\phi_{x_1} - \mu\psi_{x_1}; \quad \frac{\partial z}{\partial x_2} = f_{x_2} - \lambda\phi_{x_2} - \mu\psi_{x_2}; \quad \dots .$$

* The method, which also applies in the simple case ($r = 1$) already analysed, is known as Lagrange's method of solution by "undetermined" multipliers. See, for example, Osgood, *Advanced Calculus* (1925), pp. 180 *et seq.* and de la Vallée Poussin, *Cours d'analyse infinitésimal*, Vol. I (5th Ed. 1925), pp. 147-9.

Take the two side relations as giving x_1 and x_2 as functions of the other $(n-2)$ variables which are independent. Then we can *choose* λ and μ so that the expressions for $\frac{\partial z}{\partial x_1}$ and $\frac{\partial z}{\partial x_2}$ are zero. For a maximum or minimum value of z as a function of the $(n-2)$ independent variables, all the partial derivatives $\frac{\partial z}{\partial x_3}, \frac{\partial z}{\partial x_4}, \ldots$ must be zero. In this way, we have altogether n equations:

$$f_{x_r} - \lambda\phi_{x_r} - \mu\psi_{x_r} = 0 \quad (r = 1, 2, \ldots n).$$

These, together with the two side relations, determine certain values of λ and μ and of the n variables $x_1, x_2, \ldots x_n$. Amongst these values are those which give the relative maximum or minimum values of y required. Sufficient conditions cannot be written here.

14.7 Examples of relative maximum and minimum values.

The following examples illustrate the practical solution of relative maximum or minimum problems. It is important to note, in each case, which variables are taken as independent.

Ex. 1. To find extreme values of $z = x^2 + y^2$ relative to

$$x^2 + y^2 - 4x - 2y + 4 = 0.$$

From $f(x, y) = x^2 + y^2$, we have $f_x = 2x$ and $f_y = 2y$. Again, from $\phi(x, y) = x^2 + y^2 - 4x - 2y + 4$, we have $\phi_x = 2(x-2)$ and $\phi_y = 2(y-1)$. At a relative maximum or minimum value of z, we must have

$$\frac{x}{x-2} = \frac{y}{y-1}.$$

Hence, $x = 2y$. Substituting in the side relation,

$$5y^2 - 10y + 4 = 0.$$

So $$y = 1 \pm \frac{1}{\sqrt{5}} \quad \text{and} \quad x = 2\left(1 \pm \frac{1}{\sqrt{5}}\right).$$

The corresponding values of z are

$$z = x^2 + y^2 = 5\left(1 \pm \frac{1}{\sqrt{5}}\right)^2 = 2(3 \pm \sqrt{5}).$$

By examining the sign of the second derivative of z as a function of x, it can be shown that the plus sign above corresponds to a relative maximum and the minus sign to a relative minimum value of z. Or, Fig. 89 shows certain contours of the surface $z = x^2 + y^2$ (i.e. circles centred at 0) and the curve representing the side relation (a circle centred at $x = 2$,

$y=1$). It is seen that this latter circle touches two of the contours. For one point of contact the value of z is a relative minimum

$$z=6-2\sqrt{5}=1{\cdot}53 \text{ approximately,}$$

which is the square of the radius of the smaller contour touched. For the other point of contact

$$z=6+2\sqrt{5}=10{\cdot}47 \text{ approximately,}$$

the square of the radius of the larger circle touched.

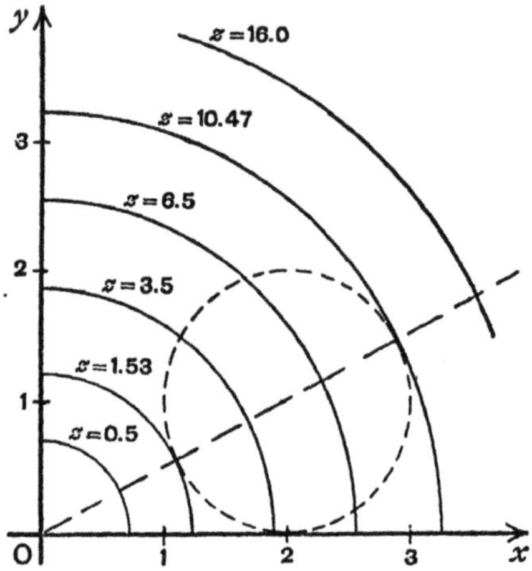

FIG. 89.

Ex. 2. To find the dimensions of the rectangular block of maximum volume which can be cut from a sphere of radius a inches.

Let the sides of the block be $2x$, $2y$ and $2z$ inches. Simple geometrical considerations show that, for maximum volume, the corners of the block must lie on the sphere's surface with diagonals as diameters of the sphere. Hence, $x^2+y^2+z^2=a^2$. We require the maximum volume $V=8xyz$ subject to this relation.

From $f(x, y, z)=8xyz$, we have $f_x=8yz$; $f_y=8xz$; $f_z=8xy$.

From $\phi(x, y, z)=x^2+y^2+z^2-a^2$, we have $\phi_x=2x$; $\phi_y=2y$; $\phi_z=2z$.

Hence, for the relative maximum of V,

$$\frac{4yz}{x}=\frac{4xz}{y}=\frac{4xy}{z}.$$

It follows that $$x^2=y^2=z^2.$$

Since $x^2+y^2+z^2=a^2$ and x, y and z are positive, we find

$$x=y=z=\frac{a}{\sqrt{3}}.$$

These values clearly correspond to a (relative) maximum volume

$$V=\frac{8a^3}{3\sqrt{3}}.$$

The block of maximum volume is a cube of side $\frac{a}{\sqrt{3}}$ inches.

Ex. 3. The function $z=f(x, y)$ has a stationary value at (a, b), so that $f_x=f_y=0$ at this point. The relation

$$y-b=m(x-a),$$

where m is a fixed value, indicates a particular direction of variation of x and y from (a, b). To find the nature of the stationary value of z in this definite direction:

The side relation gives y as a function of the independent variable x and we find

$$\frac{dy}{dx}=m, \quad \text{(from the side relation)}$$

$$\frac{dz}{dx}=f_x+f_y\frac{dy}{dx}=f_x+mf_y=0, \quad \text{at } (a, b)$$

and

$$\frac{d^2z}{dx^2}=\left(f_{xx}+f_{xy}\frac{dy}{dx}\right)+m\left(f_{xy}+f_{yy}\frac{dy}{dx}\right)$$

$$=f_{xx}+2mf_{xy}+m^2f_{yy}$$

$$=f_{yy}\left(m+\frac{f_{xy}}{f_{yy}}\right)^2+\frac{f_{xx}f_{yy}-f_{xy}^2}{f_{yy}}$$

(from the original function).

The sign of this last expression, for the given value of m, determines whether z is a maximum or minimum.

If f_{xx} and f_{yy} are both negative and such that

$$f_{xx}f_{yy}>f_{xy}^2,$$

the expression shows that $\frac{d^2z}{dx^2}$ is negative whatever the value of m, i.e. z has a maximum value in all directions from (a, b). These are sufficient conditions that $f(a, b)$ is an (unrestricted) maximum value of the function. If f_{xx} and f_{yy} are both positive subject to the same condition, it follows similarly that $f(a, b)$ is a minimum value of the function.

14.8 The demand for factors of production.

The output of a good X when amounts a and b of two factors A and B are used is given by the production function $x=f(a, b)$. It is assumed that the prices of the factors are fixed at p_a and p_b and that a given output x is required at minimum cost. We choose a and b to minimise $\Pi = ap_a + bp_b$ subject to $f(a, b) = x$. Taking a as the independent variable, the side relation gives

$$\frac{db}{da} = -\frac{f_a}{f_b}.$$

So
$$\frac{d\Pi}{da} = p_a + p_b\frac{db}{da} = p_a - \frac{f_a}{f_b}p_b = 0$$

is the necessary condition for minimum cost,

i.e.
$$\frac{p_a}{f_a} = \frac{p_b}{f_b}.$$

The factors are employed in amounts such that their marginal products are proportional to their given prices. Further,

$$\frac{d^2\Pi}{da^2} = \frac{d}{da}\left(\frac{d\Pi}{da}\right) = p_b\frac{d^2b}{da^2}.$$

So, if $\frac{d^2b}{da^2} > 0$, then the position concerned corresponds to a minimum cost. The condition implies that the constant product curve at the point concerned is convex to the origin. The equilibrium position is thus stable at all points where the constant product curves are convex to the origin.

The necessary condition with the side relation determines the demands a and b for the factors in terms of the given values of p_a, p_b and x. The (minimised) cost is then determined in the same way. As the given output x is changed, the prices of the factors remaining constant, the demands for the factors vary and the cost of production varies to give the ordinary total cost function.

In diagrammatic terms, the contour map of the function

$$\Pi = ap_a + bp_b$$

consists of a set of parallel straight lines in the plane Oab. The contour for fixed cost Π_1 is a line with gradient equal to $\left(-\frac{p_a}{p_b}\right)$ and cutting the axes at distances $\frac{\Pi_1}{p_a}$ and $\frac{\Pi_1}{p_b}$ from O. As Π_1 is increased the line moves away from O parallel to itself. If a given product x is to be obtained, then all possible combinations (a, b) of factors lie on the corresponding constant product curve. We seek that combination with minimum cost, i.e. lying on the " cost line " of the parallel set which is nearest O. The combination is given by the co-ordinates of the point P where the constant product curve touches one of the cost lines. If the constant product curve is of normal form (downward sloping and convex to O), there is only one such point P and it corresponds to minimum cost. At P, the gradient of the constant product curve $\left(-\frac{f_a}{f_b}\right)$ and that of the cost line $\left(-\frac{p_a}{p_b}\right)$ are equal. This is the necessary condition above.

As the value of x is changed, we obtain a series of different constant product curves and a series of points P_1, P_2, ... where they touch cost lines

$$A_1B_1,\ A_2B_2,\ \ldots .$$

This is shown in Fig. 90 where the values of x (x_1, x_2, ...) increase. As x changes continuously, the points describe a curve in the plane Oab. The varying co-ordinates of a point on the curve show the variation of demand for the factors as the output x varies. Further, the varying cost corresponding to the cost lines touched defines the total cost function for varying output x.

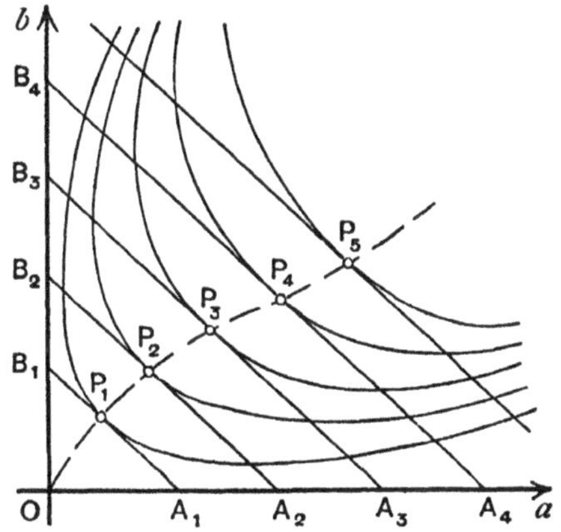

FIG. 90.

If the production function is linear and homogeneous (constant returns to scale), then Euler's Theorem gives

$$af_a + bf_b = x.$$

So
$$\frac{\Pi}{x} = \frac{ap_a + bp_b}{x} = \lambda\frac{af_a + bf_b}{x} = \lambda,$$

where λ denotes the common value of $p_a : f_a$ and $p_b : f_b$. Now

$$\frac{d\Pi}{dx} = p_a\frac{da}{dx} + p_b\frac{db}{dx} = \lambda\left(f_a\frac{da}{dx} + f_b\frac{db}{dx}\right) = \lambda,$$

since
$$f_a\frac{da}{dx} + f_b\frac{db}{dx} = 1 \quad \text{from } f(a, b) = x.$$

We consider here a variable equilibrium position with output x

changing and p_a and p_b fixed. Hence, average and marginal cost are equal at all outputs. Average cost must thus be constant irrespective of output (see Examples XVI, 22, below), a natural result in this case of constant returns to scale. In Fig. 90, the constant product curves are now of the same shape (radial projections of each other) and the locus of P is a straight line. The factors are used in the same proportion and their amounts and total cost increase in proportion as output increases.

If the good X, produced under constant returns to scale, is sold on a competitive market at a price p equal to the constant average cost, then

$$\frac{p_a}{f_a}=\frac{p_b}{f_b}=\frac{\Pi}{x}=p.$$

So $$p_a=pf_a \quad \text{and} \quad p_b=pf_b.$$

This is the law of " marginal productivity " ; the price of a factor equals the marginal product of the factor valued at the selling price of the product. Suppose, further, that the demand law for X is $x=\phi(p)$ with elasticity $\eta=-\frac{p}{x}\frac{dx}{dp}$. To determine the amounts (a and b) of the factors and the selling price (p) in terms of given factor prices (p_a and p_b), we have now three conditions :

$$(1)\ f(a, b)=\phi(p)\ ; \quad (2)\ p_a=pf_a\ ; \quad (3)\ p_b=pf_b.$$

The output $x=f(a, b)=\phi(p)$ and total cost $\Pi=px=ap_a+bp_b$ are also determined. Full competitive equilibrium, with constant returns to scale, is thus determinate.

We can trace the variation of the demands for the factors of production, given above in terms of p_a and p_b, as one of the prices (say p_a) changes while the other (p_b) remains constant. For the linear homogeneous function $x=f(a, b)$, we have

$$f_{aa}=-\frac{b}{a}f_{ab} \quad \text{and} \quad f_{bb}=-\frac{a}{b}f_{ab} \quad (12.8, (4), \text{above}).$$

The elasticity of substitution between the factors is $\sigma=\frac{f_a f_b}{x f_{ab}}$.

So $$(4)\ f_{aa}=-\frac{b}{a}\frac{f_a f_b}{x\sigma}; \quad f_{bb}=-\frac{a}{b}\frac{f_a f_b}{x\sigma} \quad \text{and} \quad f_{ab}=\frac{f_a f_b}{x\sigma}.$$

The equations (1), (2) and (3) hold for any values of p_a and p_b and can be differentiated partially with respect to p_a:

$$f_a \frac{\partial a}{\partial p_a} + f_b \frac{\partial b}{\partial p_a} = \phi'(p) \frac{\partial p}{\partial p_a} = -\eta \frac{x}{p} \frac{\partial p}{\partial p_a},$$

$$1 = f_a \frac{\partial p}{\partial p_a} + p \left(f_{aa} \frac{\partial a}{\partial p_a} + f_{ab} \frac{\partial b}{\partial p_a} \right),$$

$$0 = f_b \frac{\partial p}{\partial p_a} + p \left(f_{ab} \frac{\partial a}{\partial p_a} + f_{bb} \frac{\partial b}{\partial p_a} \right).$$

Making use of (2), (3) and (4), we have:

$$x\eta \frac{\partial p}{\partial p_a} + p_a \frac{\partial a}{\partial p_a} + p_b \frac{\partial b}{\partial p_a} = 0,$$

$$x\sigma \frac{\partial p}{\partial p_a} - \frac{b}{a} p_b \frac{\partial a}{\partial p_a} + p_b \frac{\partial b}{\partial p_a} = \frac{xp}{p_a} \sigma,$$

$$x\sigma \frac{\partial p}{\partial p_a} + p_a \frac{\partial a}{\partial p_a} - \frac{a}{b} p_a \frac{\partial b}{\partial p_a} = 0.$$

There are thus three linear equations in the "unknowns", $\frac{\partial p}{\partial p_a}$, $\frac{\partial a}{\partial p_a}$ and $\frac{\partial b}{\partial p_a}$. By simple algebraic methods,* the values of these partial derivatives can be found and written:

$$\frac{\partial a}{\partial p_a} = -\frac{a}{p_a} \left(\frac{ap_a}{xp} \eta + \frac{bp_b}{xp} \sigma \right)$$

and

$$\frac{\partial b}{\partial p_a} = -\frac{ab}{xp} (\eta - \sigma).$$

Write $\frac{Ea}{Ep_a} = \frac{p_a}{a} \frac{\partial a}{\partial p_a}$ and $\frac{Eb}{Ep_a} = \frac{p_a}{b} \frac{\partial b}{\partial p_a}$ for the elasticities of demand for the factors with respect to p_a and $\kappa_a = \frac{ap_a}{xp}$ and $\kappa_b = \frac{bp_b}{xp}$ $(\kappa_a + \kappa_b = 1)$ for the proportions of total receipts going to the factors. Then

$$\frac{Ea}{Ep_a} = -(\kappa_b \sigma + \kappa_a \eta)$$

and

$$\frac{Eb}{Ep_a} = \kappa_a (\sigma - \eta).$$

The interpretation of these results is clear. If the price of the

* For example, by eliminating $\frac{\partial p}{\partial p_a}$, we obtain two equations in $\frac{\partial a}{\partial p_a}$ and $\frac{\partial b}{\partial p_a}$ which are easily solved. Or, the method of solution by "determinants" can be utilised (see 18.6, Ex. 2, below).

factor A rises, the demands for *both* factors are affected, and in two ways. In the first place, the cost of production is increased, the product is dearer and (for a decreasing demand law with η positive) less of the product is bought. There is then a proportional decrease in the demand for both factors. This is shown by the negative term $(-\kappa_a\eta)$ in the expressions for both elasticities of demand. Secondly, the factor B is now cheaper *relatively* to the factor A and it pays to substitute B for A in production as far as possible. The demand for B thus increases at the expense of that for A, as is shown by the positive term $(\kappa_a\sigma)$ in the expression for one elasticity and the negative term $(-\kappa_b\sigma)$ in the other. The total effect on demand is found by addition. The demand for A falls in any case, but that for B may rise or fall according as the substitution effect is stronger or weaker than the effect through the demand for the product.

14.9 The demand for consumers' goods and for loans.

A similar analysis applies to the demand of a consumer for two goods X and Y obtainable at given market prices p_x and p_y. The consumer has a fixed income μ and makes his purchases to attain the highest position on his preference scale, as represented by an index-function of utility

$$u=\phi(x, y).$$

We have to find x and y, the individual's purchases, so that u is a maximum subject to the relation expressing the balance of the individual's budget:

$$xp_x+yp_y=\mu.$$

This expression of the problem is independent of which particular form of the index-function of utility is selected; all forms of the index-function increase and decrease together and so attain their maximum values together, at the same values of x and y.

The side relation gives y as a function of x for which

$$y=\frac{\mu-xp_x}{p_y}; \quad \frac{dy}{dx}=-\frac{p_x}{p_y} \quad \text{and} \quad \frac{d^2y}{dx^2}=0.$$

So
$$\frac{du}{dx}=\phi_x+\phi_y\frac{dy}{dx}=\phi_x-\frac{p_x}{p_y}\phi_y$$

and
$$\frac{d^2u}{dx^2}=\left(\phi_{xx}+\phi_{xy}\frac{dy}{dx}\right)+\left(\phi_{xy}+\phi_{yy}\frac{dy}{dx}\right)\frac{dy}{dx}+\phi_y\frac{d^2y}{dx^2}$$

$$=\frac{1}{p_y{}^2}(\phi_{yy}p_x{}^2-2\phi_{xy}p_xp_y+\phi_{xx}p_y{}^2).$$

The necessary condition for a relative maximum value of u is thus

$$\frac{\phi_x}{p_x}=\frac{\phi_y}{p_y},$$

i.e. the marginal rate of substitution between the goods in consumption ($R=\phi_x : \phi_y$) is equal to the ratio of the given market prices. This condition, together with the side relation, is sufficient to determine the purchases of the individual in terms of the given values of p_x, p_y and μ, i.e. to determine the demand functions of the individual (see 11.7 above). A sufficient condition for a relative maximum value of u is

$$\phi_{yy}p_x^2-2\phi_{xy}p_xp_y+\phi_{xx}p_y^2<0.$$

Using the necessary condition, this implies that

$$\phi_{yy}\phi_x^2-2\phi_{xy}\phi_x\phi_y+\phi_{xx}\phi_y^2<0.$$

In diagrammatic terms, the purchases of the individual are given by the co-ordinates of the point P in the Oxy plane where the fixed line with equation $xp_x+yp_y=\mu$, expressing the budget balance, touches one of the indifference curves of the system

$$\phi(x, y)=\text{constant}.$$

The point P corresponds to a maximum value of u if the indifference curve at P is convex to the origin. The condition of tangency is equivalent to the necessary condition and the condition of convexity to the sufficient condition written above. So, in the "normal" case where the indifference curves are downward sloping and convex to the origin, a unique and maximal pair of individual purchases are defined for any given market prices or given income. The "normal" case, in fact, corresponds to the case of stable demand functions. The variation of the individual's demands for the two goods as one of the market prices or the income varies is shown by the changing position of the point P as the line $xp_x+yp_y=\mu$ moves in conformity with price or income variation.*

We can consider, as a second problem, the conditions under which an individual regulates the flow of his income, and hence of his consumption, over time. This is a problem of importance in the theory

* For more complete accounts of the problem of demand for consumers' goods, see Hicks and Allen, *A Reconsideration of the Theory of Value*, Economica, 1934, and Allen and Bowley, *Family Expenditure* (1935). Expressions are here found for the elasticities of demand with respect to price or income. The methods used are extensions of those first introduced by Pareto.

of interest.* It is assumed that the individual has given resources which can produce various incomes £x this year and various incomes £y next year subject always to a given transformation relation

$$F(x, y)=0.$$

Interest is reckoned at the fixed rate of $100r$ per cent. compounded yearly. We determine, first, those incomes which maximise the present value

$$z=x+\frac{y}{1+r}$$

subject to the transformation relation. We have

$$\frac{dz}{dx}=1+\frac{1}{1+r}\frac{dy}{dx} \quad \text{and} \quad \frac{d^2z}{dx^2}=\frac{1}{1+r}\frac{d^2y}{dx^2}$$

where the derivatives of y with respect to x are given by $F(x, y)=0$. Hence, for maximum present value, we must have

$$r=\left(-\frac{dy}{dx}\right)-1,$$

i.e. the incomes in the two years are such that the marginal rate of return over cost equals the given rate of interest. The sufficient condition for a maximum present value $\left(\frac{d^2y}{dx^2}<0\right)$ is satisfied in the "normal" case where the transformation curve in the plane Oxy is concave to the origin (see 9.7 above).

The necessary condition above and the transformation relation determine the incomes x_0 and y_0 with maximum present value, both incomes depending on the interest rate r. It is now assumed that the individual can modify his incomes by borrowing or lending money this year to be repaid next year with interest at the fixed market rate. This process leaves the present value of income unchanged, i.e.

$$x+\frac{y}{1+r}=x_0+\frac{y_0}{1+r}$$

where x and y are any incomes as modified from x_0 and y_0 by loans. For example, if the individual borrows $(x-x_0)$ this year, then

$$y_0-y=(1+r)(x-x_0)$$

is the amount he must repay next year, and this is equivalent to the

* The analysis here given is based on that given by Fisher, *The Theory of Interest* (1930).

relation written above. It is assumed, further, that the loans made by the individual are such that the utility function

$$u=\psi(x, y)$$

of the individual for incomes in two years is maximised subject to the relation above. The particular form of the utility function taken is quite immaterial. The relation gives

$$\frac{dy}{dx}=-(1+r).$$

So
$$\frac{du}{dx}=\psi_x+\psi_y\frac{dy}{dx}=\psi_x-(1+r)\psi_y.$$

The necessary condition for relative maximum utility is thus

$$r=\frac{\psi_x}{\psi_y}-1,$$

i.e. loans must be made so that the marginal rate of time-preference of the individual becomes equal to the market rate of interest. This condition, together with the constant present value relation, determines the incomes x and y in terms of the given interest rate. In particular, $(x-x_0)$ is determined as a function of r, the demand function of the individual for loans. The value of $(x-x_0)$ may be positive or negative according to the interest rate, i.e. the individual may borrow or lend this year to reach his "optimum" income stream. As in the case of the consumption goods problem, if the indifference curves, $\psi(x, y)=\text{constant}$, are convex to the origin at all points, then the positions obtained are stable and correspond always to maximum, rather than to minimum, values of utility.

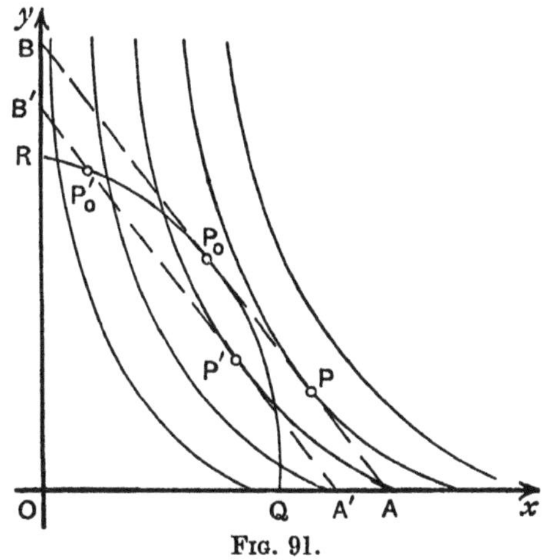

FIG. 91.

The problem can be put in a slightly different way capable of simple diagrammatic representation. Suppose that the individual, before borrowing or lending, has incomes x_0' and y_0' in the two years. Represent these incomes by a point P_0' in the plane Oxy (Fig. 91).

By borrowing or lending, the individual can modify his incomes to x and y where the point (x, y) must lie on the line $A'B'$ with equation

$$x+\frac{y}{1+r}=x_0'+\frac{y_0'}{1+r},$$

i.e. the line through P_0' with fixed gradient (to Ox) which is negative and numerically equal to $(1+r)$. The incomes selected by the individual are given by the co-ordinates of the point P' where $A'B'$ touches an indifference curve of the system

$$\psi(x, y)=\text{constant}.$$

If the indifference curves are everywhere convex to the origin, then P' is uniquely determined and corresponds to a maximum of utility. The initial incomes x_0' and y_0' can be varied according to the given transformation relation $F(x, y)=0$ and the point P_0' can take any position on the corresponding transformation curve QR. As P_0' varies, so does the point P' and we require that position of P' giving the largest possible utility level. The position required is given by P where an indifference curve is touched by the line AB (parallel to $A'B'$) which touches the transformation curve at some point P_0. In other words, we seek a line with fixed gradient $-(1+r)$ which touches the transformation curve at P_0 and an indifference curve at P. The incomes first derived by the individual are given by the co-ordinates of P_0 and the final incomes, as modified by loans, are given by the co-ordinates of P. In this position, the numerical gradient of AB $(1+r)$ is equal to that of the transformation curve $(F_x : F_y)$ and to that of the indifference curve $(\psi_x : \psi_y)$, i.e. r equals the marginal rate of return over cost and the marginal rate of time-preference. Notice that, for a line with fixed negative gradient $-(1+r)$, the present value of the incomes represented is constant. The maximum value of this present value attainable is given by the particular line AB. Hence, as in the previous analysis, we can first locate P_0 for maximum present value of incomes on the transformation curve, and then locate P for maximum utility of incomes obtainable by borrowing or lending from those at P_0.

EXAMPLES XIV

General maximum and minimum problems

1. Indicate the contour, ridge and trough lines of each of the functions $z=x^2+y^2-1$; $z=1-x^2-y^2$; $z=x^2-y^2+1$. Show that $x=y=0$ gives a stationary value of z in each case. Determine the nature of the stationary value in the various cases.

2. Show that $z=ax^2+by^2+c$ always has a single stationary value which is a minimum, a maximum or a saddle point according as a and b are positive, negative or of opposite signs.

3. Show that $z=x^2+xy+y^2$ has a minimum value and $z=x^2+xy-y^2$ a saddle point at $x=y=0$.

4. Show that $z=ax^2+2hxy+by^2+2gx+2fy+c$ has two straight ridge or trough lines and deduce that the function has only one stationary value.

5. If $z=(x-y)e^{x+y}$, show that there are ridge and trough lines which are straight and parallel and that z has no stationary value.

6. Show that $z=(x^2+y^2)e^{x^2-y^2}$ has a minimum value at $x=y=0$ and saddle points where $x=0$, $y=\pm 1$.

7. Indicate the form of the contour map of $z=x^3+y^3-3xy$ and show that the ridge and trough lines are arcs of parabolas. Deduce that z has a minimum value at $x=y=1$ and a saddle point at $x=y=0$.

8. Locate the stationary values of $z=2(x-y)^2-x^4-y^4$.

9. If z is a function of x and y defined by $x^2+y^2+3z^2-2x+2z=0$, show that the contours are circles centred at $x=1$, $y=0$ which expand as z increases $(-1<z<-\frac{1}{3})$ and then contract as z increases further $(-\frac{1}{3}<z<\frac{1}{3})$. Deduce that $z=-1$ is the minimum and $z=\frac{1}{3}$ the maximum value of z. (See 14.3, Ex. 4 above.)

10. Show that the only stationary points on the surface $z=x^2e^y$ form a line of minima lying along the axis Oy. Verify by examining the nature of the sections of the surface by planes perpendicular to Oy.

11. If $\frac{\partial z}{\partial x}=(ax+by+c)\phi(x, y)$ and $\frac{\partial z}{\partial y}=(ax+by+c)\psi(x, y)$, show that z has a line of stationary values corresponding to $ax+by+c=0$ on the contour map in the plane Oxy. Illustrate with $z=x^2e^y$.

12. If $u=f(z)$ where $z=\phi(x, y)$, show that the stationary values of u and z occur at the same values of x and y. If $f'(z)>0$ always, deduce that corresponding stationary values of u and z are also of the same type.

13. The variable ρ is defined as a function of four variables a, b, t and τ by the relation $ale^{\rho t}+bre^{\rho\tau}=f(a, b, t, \tau)$, where l and r are constants and f is a given function. Find four equations to determine a, b, t and τ for maximum ρ. (See Wicksell, *Lectures on Political Economy*, Vol. I (English Ed. 1934), p. 181.)

14. By eliminating y, show that $z=e^{x^2+y^2}$ has a minimum value e^2 relative to $x-y+2=0$. Check by means of the results of 14.6.

15. If x and y are positive, show that the maximum value of xy subject to $x^2+y^2=a^2$ occurs where $x=y=\frac{a}{\sqrt{2}}$. Illustrate by drawing the contour map of $z=xy$ and the circle $x^2+y^2=4$.

16. The variables x and y are related $\frac{x^2}{a^2}+\frac{y^2}{b^2}=1$, where $a>b$. Show that (x^2+y^2) has then a maximum value a and a minimum value b. Illustrate diagrammatically when $a=5$ and $b=3$.

17. Establish the necessary condition for an extreme value of $z=f(x, y)$ relative to $\phi(x, y)=0$ by Lagrange's method, using an "undetermined" multiplier λ.

18. Find the stationary values of $z=x^3+y^3-3xy$ subject to $y=mx$ where m is a constant. For what values of m is z (*a*) a maximum and (*b*) a minimum at $x=y=0$? Deduce that $x=y=0$ is a saddle point of the surface $z=x^3+y^3-3xy$ (see Example 7 above).

19. A rectangular brick has a given volume; show that its surface area is least when it is a cube. Show that the same result holds if the brick has a given surface area and maximum volume.

20. An open box of given volume 4 cubic feet has a rectangular base and vertical sides. It is made of wood costing 1 shilling per sq. foot. Show that, for least cost of construction, the base of the box is a square of side double the depth. What is the least cost?

21. An "ellipsoidal" block of wood has a surface with equation

$$\frac{x^2}{a^2}+\frac{y^2}{b^2}+\frac{z^2}{c^2}=1$$

referred to axes $Oxyz$ in space. Show that $\dfrac{8abc}{3\sqrt{3}}$ is the volume of the largest rectangular brick that can be cut from the wood.

22. If axes $Oxyz$ are fixed in space, what is the distance of the point (x, y, z) from O? Express the distance from O to any point on the plane $3x+4y+5z=10$ in terms of the x and y co-ordinates of the point. Deduce that $\sqrt{2}$ is the shortest and perpendicular distance from O to the plane.

23. Given $z=mf(x)+n\phi(y)$ where $mx+ny=a$ and $m+n=b$, find z as a function of x and y for given values of a and b. Deduce that z has extreme values for variation of x and y where

$$f'(x)=\phi'(y)=\frac{f(x)-\phi(y)}{x-y}.$$

(See Wicksell, *op. cit.*, p. 140.)

Economic maximum and minimum problems

24. A monopolist firm produces two chocolate lines X_1 and X_2 at constant average costs of 2s. 6d. and 3s. per lb. respectively. If p_1 and p_2 are the prices charged (in shillings per lb.), the market demands are

$$x_1=5(p_2-p_1) \quad \text{and} \quad x_2=32+5p_1-10p_2$$

thousand lbs. per week of the two lines. For maximum joint monopoly revenue, show that the prices are fixed at nearly 4s. 6d. and 4s. 9d. per lb. respectively.

25. In the problem of the previous example, find the prices of the two lines fixed by independent monopolists and show that they are less than the joint monopoly prices.

26. A monopolist produces cheap razors and blades at a constant average cost of 2s. per razor and 1s. per dozen blades. The demand of the market per week is $x_1=\dfrac{10}{p_1p_2}$ thousand razors and $x_2=\dfrac{20}{p_1p_2}$ thousand dozen blades when the prices are p_1 (shillings per razor) and p_2 (shillings per dozen blades). Show that the monopoly prices, fixed jointly, are 4s. per razor and 2s. per dozen blades.

27. If the demands of the market in the previous example are

$$x_1 = 11 - 2p_1 - 2p_2 \quad \text{and} \quad x_2 = \tfrac{67}{4} - 2p_1 - \tfrac{11}{4}p_2,$$

show that the joint monopolist would give the razors away and sell the blades at a high price.

28. The market demand law for a good X is $p = \beta - \alpha x$. A monopolist produces X at average cost $(ax + b)$ for output x and sells to a merchant at a price π which maximises his profits. The merchant is a monopolist with constant distributive costs and maximises his profits by selling on the market at price p. Show that the amount of X produced and sold is $x = \dfrac{\beta - b}{2(a + 2\alpha)}$ and that $\pi = \beta - 2\alpha x$ and $p = \beta - \alpha x$. Find the output if the producer monopolist sold direct to the market and show that "bilateral monopoly" here restricts output and raises price.

29. Generalise the problem of the previous example by taking a market demand law $p = \phi(x)$ and producer's average cost $f(x)$. Show that

$$\phi(x) + 3x\phi'(x) + x^2\phi''(x) = f(x) + xf'(x)$$

gives the output x and that $\pi = \phi(x) + x\phi'(x)$. (See Bowley, *Bilateral Monopoly*, Economic Journal, 1928.)

30. In the problem of 14.5, assume that the product after t years represents the original cost accumulated at some continuous interest rate of 100ρ per cent. Given the factor prices, express ρ as a function of a, b and t. The monopolist is assumed to arrange production to maximise ρ; find equations for a, b and t. If the maximum equals the market rate of interest r, show that the result is equivalent to that of the competitive case of 14.5.

31. If the production function of 14.5 is $x = \sqrt{ab}\, e^{\frac{1}{2}\sqrt{t}}$, show that the equilibrium period of production is $t = \dfrac{1}{16r^2}$ where r is the market rate of interest. In the competitive case, show that r is given by

$$\frac{1}{r} = 16 \log 2\sqrt{p_a p_b}.$$

32. If $x = A\sqrt{ab}$ is the production function, find the amounts of the factors used at given prices p_a and p_b to produce an output x at smallest cost. In the case of pure competition on the market for X with the demand law $x = \beta - \alpha p$, show that the demands for the factors are

$$a = \frac{\beta}{A}\sqrt{\frac{p_b}{p_a}} - \frac{2\alpha p_b}{A^2} \quad \text{and} \quad b = \frac{\beta}{A}\sqrt{\frac{p_a}{p_b}} - \frac{2\alpha p_a}{A^2}.$$

33. For any production function, if each output x is obtained at minimum cost at given factor prices, show that the common equilibrium value of $\dfrac{p_a}{f_a}$ and $\dfrac{p_b}{f_b}$ is equal to the marginal cost of output x. If the firm fixes output to maximise profits at a given selling price p, show that the marginal productivity law $p_a = pf_a$ and $p_b = pf_b$ holds. Why need not total cost equal total receipts in this case?

34. A good X is produced with two "substitutional" factors A and B and a "limitational" factor C according to the relations $x = f(a, b) = \phi(c)$. If each output x is produced at minimum cost at given factor prices p_a, p_b and p_c, show that the amounts of the factors used and the total cost are determined as functions of x and deduce that

$$\frac{p_a}{f_a}=\frac{p_b}{f_b}=\frac{d\Pi}{dx}-\frac{p_c}{\phi'(c)}.$$

The selling price p is fixed and output chosen for maximum profits. Show that

$$p_a=\left\{p-\frac{p_c}{\phi'(c)}\right\}f_a \quad \text{and} \quad p_b=\left\{p-\frac{p_c}{\phi'(c)}\right\}f_b.$$

In what sense can the factor prices be said to be equal to the "marginal revenues" of the factors? (See Georgescu, *Fixed coefficients of production and marginal productivity theory*, Review of Economic Studies, 1935.)

35. If $u=x^\alpha y^\beta$ is an individual's utility function for two goods, show that his demands for the goods are $x=\frac{\alpha}{\alpha+\beta}\frac{\mu}{p_x}$ and $y=\frac{\beta}{\alpha+\beta}\frac{\mu}{p_y}$ where p_x and p_y are the fixed prices and μ the individual's fixed income. Deduce that the elasticity of demand for either good with respect to income or to its price is equal to unity.

36. An individual's utility function is $u=2gx+2fy-ax^2-2hxy-by^2$. Show that his demands for the goods are linear in the income μ, the coefficients depending on the prices of the goods.

37. Differentiate the equilibrium conditions giving the demands for two goods X and Y for an individual with utility function $u=\phi(x, y)$ in the case where $\phi_{xy}=0$. Show that

$$\frac{\partial x}{\partial p_x}=-\frac{1}{T\phi_{xx}}\left(xp_x-\frac{\phi_y}{\phi_{yy}}p_y\right) \quad \text{where} \quad T=\frac{p_x^2}{\phi_{xx}}+\frac{p_y^2}{\phi_{yy}}$$

and deduce that the demand for a good decreases as its price increases provided that ϕ_{xx} and ϕ_{yy} are negative. Why is this statement of the problem not independent of the particular utility function selected? (See Pareto, *Manuel d'économie politique*, 2nd Ed. 1927, pp. 579-84.)

38. The incomes of an individual in two years are x_0 and y_0 and his utility function for incomes is $u=x^\alpha y^\beta$. Show that the demand $(x-x_0)$ for loans this year decreases as the given market rate of interest $100r$ per cent. increases. Deduce that the individual will not borrow this year at any (positive) rate of interest if $y_0<\frac{\beta}{\alpha}x_0$.

39. In a self-contained community, an individual can produce, with his given resources, amounts of goods X, Y, Z, ... according to the transformation function $F(x, y, z, ...)=0$ and he can then buy or sell on a market at fixed prices $p_x, p_y, p_z, \ldots$. His utility function is $u=\phi(x, y, z, ...)$. How are the amounts he demands or offers determined? Illustrate diagrammatically in the case of two goods and show that the solution is exactly analogous to that of the problem of the demand for loans (14.9 above).

40. Values (x_t) of a variable quantity X are given for a series of $(2n+1)$ years

$$t=-n,\ -(n-1),\ \ldots -1,\ 0,\ 1,\ \ldots (n-1),\ n$$

taking $t=0$ at the centre of the period. Write $x_t=at+b+v_t$ where a and b

are to be determined so that the sum of all observed values of v_t^2 is a minimum. Show that

$$a = \frac{\Sigma t x_t}{\Sigma t^2} \quad \text{and} \quad b = \frac{\Sigma x_t}{2n+1} = \bar{x}$$

where Σ implies summation over the $(2n+1)$ observed values. In what sense does $x = at + b$ represent the " trend " of the series of X?

41. The Board of Trade Wholesale Price Index-number $x_t (1924 = 100)$ at certain dates is given by the following table:

Date	t	x_t
July 1929 - - -	-3	82·7
Jan. 1930 - - -	-2	78·8
July 1930 - - -	-1	71·7
Jan. 1931 - - -	0	64·3
July 1931 - - -	1	61·5
Jan. 1932 - - -	2	63·4
July 1932 - - -	3	58·8

Find the trend $x = at + b$ by the method of the previous example. Compute the trend values for each date and compare with the actual value of the index. Represent the original values and the trend values graphically.

CHAPTER XV

INTEGRALS OF FUNCTIONS OF ONE VARIABLE

15.1 The definition of a definite integral.

THE concept of an "integral" has two very different characteristics and two correspondingly distinct applications. From one point of view, an integral is the limiting value of a certain summation expression which is of frequent appearance in mathematical analysis and which corresponds, in diagrammatic terms, to an area enclosed by a plane curve or curves. The integral, viewed in this way, is called a "definite" integral. From another point of view, an integral is the result of reversing the process of differentiation. The derivative of a function of one variable is itself a function of the same variable. The inverse problem is to obtain, from a given function, a second function which has the first as its derivative. The second function, if it can be found, is called the "indefinite" integral of the first.

It is essential, in the following development, to distinguish carefully between the two aspects of an integral, the "definite" and the "indefinite" aspects. The actual definition of an integral can be framed in one of two ways and, once one aspect is adopted as the basis of the definition, it is then our business to establish the property implicit in the second aspect. It is found that the sum-area concept of an integral provides the better starting-point. The definition given below, therefore, makes use of this concept and it is only at a later stage that the integral is connected with the inverse differentiation process. This is merely a matter of formal presentation; the sum-area and inverse-differentiation properties of integrals are equally important in application.

It is assumed that $y=f(x)$ is a single-valued function which is continuous for all values of x in the given interval from $x=a$ to

$x=b$.* The interval of x, of length $(b-a)$, is divided in any way we please into n sections by means of the dividing points :

$$a=x_1,\ x_2,\ x_3,\ \ldots,\ x_{n-1},\ x_n,\ x_{n+1}=b.$$

We form the sum :

$$f(x_1)(x_2-x_1)+f(x_2)(x_3-x_2)+\ldots+f(x_{n-1})(x_n-x_{n-1})+f(x_n)(x_{n+1}-x_n),$$

where each term is obtained from a different section and consists of the length of the section times the value of the function at the lower (or left-hand) point of the section. For convenience, a special notation is adopted in which we write the sum as

$$\sum_{r=1}^{n} f(x_r)(x_{r+1}-x_r),$$

where the term written is the typical (rth) term of the sum and the symbol $\sum_{r=1}^{n}$ indicates that all such terms, from the first ($r=1$) to the nth ($r=n$), are to be added together.† The number n of the sections into which the given interval is divided is now increased in any way such that each section becomes smaller. It can be shown that the sum above then increases and approaches a definite limiting value. No formal proof of this basic result is offered here but a diagrammatic indication of its correctness is given below. The limiting value, approached as n tends to infinity, is called the *definite integral* of the function between the lower limit a and the upper limit b and written in the symbolic form $\int_a^b f(x)\,dx$. So :

DEFINITION : $$\int_a^b f(x)\,dx=\lim_{n\to\infty}\sum_{r=1}^{n} f(x_r)(x_{r+1}-x_r).$$

The process of finding the integral of a function is called *integration*. From the definition, it is clear that the value of a definite integral is simply a number which can be positive or negative and which depends only on the form of the function and on the values of the

* The rigid definition of a definite integral does not assume that the function is continuous. The full development, following Riemann (after whom the integral is sometimes named), is somewhat involved—see Goursat, *Cours d'analyse mathématique*, Vol. I (4th Ed. 1924), pp. 169 *et seq.* The essentials of the definition, however, are clearly indicated here.

† This Σ notation is found to be of great convenience in dealing with sums of all kinds. The Greek capital Σ, an alternative to the English S, is an obvious letter to indicate a sum.

limits (a and b) taken. The variable x is not involved; it has been "integrated out".

It can be shown, further, that sums other than that written above have exactly the same limiting value, the definite integral, as n tends to infinity. Such sums are formed by taking, for each section, the length of the section times the value of the function at the upper (or right-hand) point of the section, or indeed at any point within the section. This fact, again not formally proved, is supported by the diagrammatic illustrations below. Hence:

$$\int_a^b f(x)\,dx = \underset{n\to\infty}{\text{Lim}} \sum_{r=1}^{n} f(x_r')(x_{r+1}-x_r),$$

where x_r' can take any value from x_r to x_{r+1} inclusive.

Put more broadly, the definition is as follows. The interval from $x=a$ to $x=b$ is divided into a number of sections of which Δx is a typical one with x as one point contained in it. Then

$$\int_a^b f(x)\,dx = \text{Lim}\, \Sigma f(x)\Delta x,$$

where each section length Δx is multiplied by the value of the function $f(x)$ at a point of the section, the whole set of products is summed and the limit found as the number of sections into which the given range (a, b) is divided is increased indefinitely.*

The following properties of definite integrals are derived at once from the definition. If $f(x)$ and $\phi(x)$ are single-valued functions continuous over the relevant intervals, then:

(1) $\displaystyle\int_a^b f(x)\,dx = -\int_b^a f(x)\,dx.$

(2) $\displaystyle\int_a^b \{-f(x)\}\,dx = -\int_a^b f(x)\,dx.$

(3) $\displaystyle\int_a^b kf(x)\,dx = k\int_a^b f(x)\,dx$ (k being a constant).

(4) $\displaystyle\int_a^b f(x)\,dx = \int_a^c f(x)\,dx + \int_c^b f(x)\,dx.$

* The way in which the integral *notation* has arisen is clear from this rough statement of the definition. In the limit, the sum $\Sigma f(x)\Delta x$ is written $Sf(x)dx$, S being an alternative symbol for Σ. The S then becomes elongated into the conventional sign $\int$.

(5) $$\int_a^b \{f(x)+\phi(x)\}\,dx=\int_a^b f(x)\,dx+\int_a^b \phi(x)\,dx.$$

The last of these properties can be considered as an illustration of the formal method of proof. We have

$$\sum_{r=1}^{n}\{f(x_r)+\phi(x_r)\}(x_{r+1}-x_r)=\sum_{r=1}^{n}\{f(x_r)(x_{r+1}-x_r)+\phi(x_r)(x_{r+1}-x_r)\}$$
$$=\sum_{r=1}^{n} f(x_r)(x_{r+1}-x_r)+\sum_{r=1}^{n}\phi(x_r)(x_{r+1}-x_r),$$

i.e. $$\int_a^b \{f(x)+\phi(x)\}\,dx=\int_a^b f(x)\,dx+\int_a^b \phi(x)\,dx$$

on taking the limit as n tends to infinity and using the property (4.5) that the limit of a sum is the sum of the separate limits. It can be noticed that this result extends to the integral of a difference and, in an obvious way, to the integral of sums or differences of any number of separate functions.

15.2 Definite integrals as areas.

The diagrammatic representation of a definite integral which follows makes the concept much more clear and, at the same time, leads to important applications. Suppose, first, that $b>a$ and that

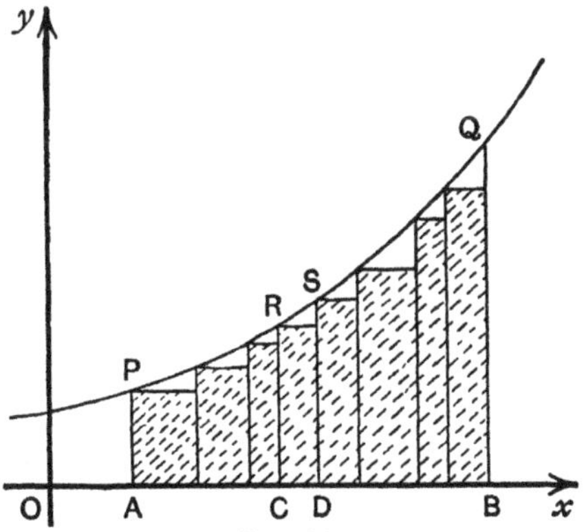

FIG. 92.

the continuous function $f(x)$ assumes only positive values between $x=a$ and $x=b$. Further, for simplicity of exposition, take the function as monotonic increasing. The function is then represented (Fig. 92) as a continuous curve lying above the horizontal axis Ox and rising from left to right between the ordinates AP (at $x=a$) and

BQ (at $x=b$). Let CD be a typical member (x_r, x_{r+1}) of the n sections into which we divide the range AB ($n=8$ in the figure). The contribution of this section to the sum $\sum_{r=1}^{n} f(x_r)(x_{r+1}-x_r)$ is CD times CR, the ordinate of the curve at C, i.e. it is the area of the rectangle with CD as base and CR as height. The whole sum is thus the sum of the rectangular areas shaded in Fig. 92, an irregular area lying entirely under the curve between P and Q. As the number of sections is increased, each section becoming smaller, the shaded area, comprising an increasing number of rectangles becoming steadily thinner, must increase and approach a limiting value which we can identify as the area under the curve, above Ox and between the ordinates AP and BQ. Hence, the sum written above has this area as limit and

$$\int_a^b f(x)\,dx = \text{area } ABQP \text{ under the curve } y=f(x).$$

It is also clear that the same result follows if we start from one of the other sums used in the definition of the integral. The sum $\sum_{r=1}^{n} f(x_{r+1})(x_{r+1}-x_r)$ is represented by a sum of rectangle areas, a typical rectangle having CD as base and DS as height. We have now an irregular area enclosing the curve and, as the number of sections is increased, the area decreases and approaches a limiting value which is the area $ABQP$ under the curve. Further, the sum $\sum_{r=1}^{n} f(x_r')(x_{r+1}-x_r)$, where x_r' is any value between x_r and x_{r+1}, is shown as the sum of rectangle areas, a typical rectangle having base CD and height equal to some ordinate of the curve above CD. This area again approaches the area $ABQP$ under the curve as the number of sections increases. It does not matter which of the sums is taken ; the limit is always the definite integral shown by the area $ABQP$ under the curve.*

* It can be noticed that we have actually given here a *definition* of the area under a curve. The only areas known to elementary geometry are those of rectangles. But, if we fit to a given curvilinear area a set of rectangles in one of the ways indicated, and if the sum of the rectangle areas tends to a limit as the number of the rectangles increases, then the limit can be defined as the curvilinear area. Compare this definition with the graphical method of estimating a curvilinear area in practice, by counting the number of small squares of the graph paper covered by the area.

A similar argument applies when the function, though positive, is not monotonic increasing between $x=a$ and $x=b$. The integral $\int_a^b f(x)dx$ can always be represented as the area under the curve $y=f(x)$, above the axis Ox and between the ordinates at $x=a$ and $x=b$. This is illustrated in Fig. 93.

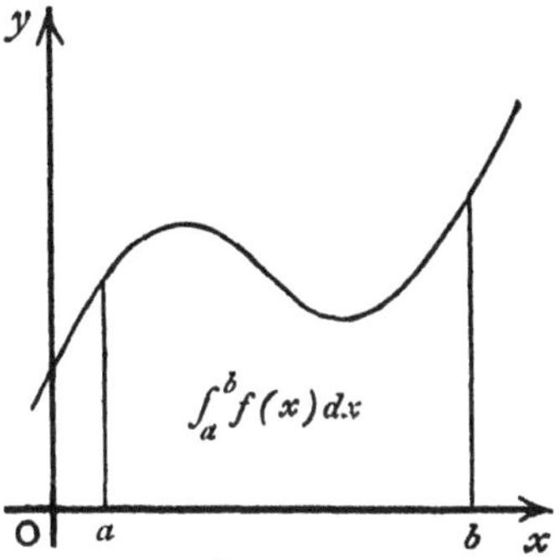

FIG. 93.

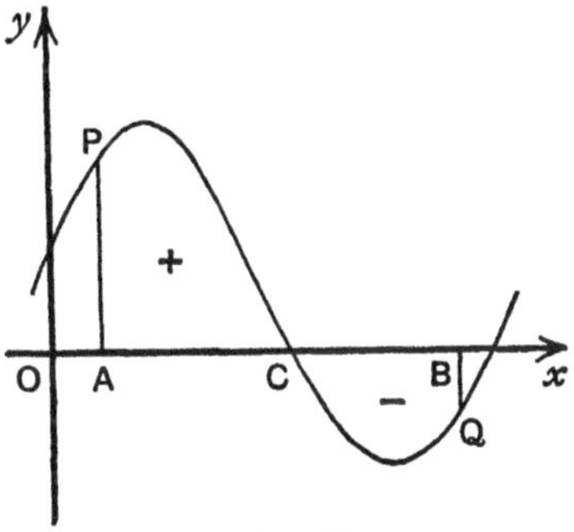

FIG. 94.

When the function $f(x)$ assumes negative values between $x=a$ and $x=b\,(b>a)$, complications arise which can only be overcome by making *conventional* distinctions between positive and negative areas. If $f(x_r)$ is negative, then the contribution of the section (x_r, x_{r+1}) to the sum $\sum_{r=1}^{n} f(x_r)(x_{r+1}-x_r)$ is negative and shown by a rectangle *below* the axis Ox. It follows, when the limit is taken, that the integral $\int_a^b f(x)dx$ is still represented by an area between the curve $y=f(x)$, the axis Ox and the ordinates at $x=a$ and $x=b$, provided that any parts of the area above the axis Ox are taken as positive and any parts below the axis Ox as negative (see Fig. 94). The integral and area as a whole may, of course, be positive or negative.*

So, if the function $y=f(x)$ is positive in the interval (a, b) and the curve above Ox, then the integral $\int_a^b f(x)dx$ is positive and measured

* It must be emphasised that the integral $\int_a^b f(x)dx$ is here not the sum, but the difference, of the numerical areas PAC and QBC of Fig. 94. If the numerical sum is required, the areas PAC and QBC must be obtained separately as $\int_a^c f(x)dx$ and $-\int_c^b f(x)\,dx$, where $OC=c$.

by an area under the curve. If the function is negative in the interval and the curve below Ox, then the integral is negative and numerically equal to the area below Ox and above the curve. If the function changes sign in the interval and the curve crosses the axis Ox, then the integral may be positive or negative and is represented by the algebraic sum of a positive area (above Ox) and a negative area (below Ox).

A further convention regarding the sign of an area is needed to interpret the integral $\int_a^b f(x)\,dx$ as an area in the cases where $a>b$. Each of the terms $(x_{r+1}-x_r)$ appearing in any of the sums defining the integral is now negative instead of positive and the corresponding rectangle areas are described from right to left instead of from left to right. We simply need the convention, therefore, that an area described from right to left is numerically equal, but opposite in sign, to the similar area described from left to right.

In all cases, the integral $\int_a^b f(x)\,dx$ is measured by the area between the curve $y=f(x)$, the axis Ox and the ordinates of the curve at $x=a$ and $x=b$. Various parts of the area must be considered as taking signs according to the conventions here indicated. The area as a whole is the algebraic (and not the numerical) sum of the separate parts and it may be positive or it may be negative.

15.3 Indefinite integrals and inverse differentiation.

In introducing the second aspect of integrals (as inverse to derivatives), we can quote a simple example of the kind of problem involved. The variable y is known to be a single-valued function of x such that, for all values of x, we have

$$\frac{dy}{dx}=1+x^2.$$

This is a "differential equation" involving the derivative $\frac{dy}{dx}$ of a function so far unknown. Is it possible to deduce the actual form of this unknown function from the equation? We know that

$$\frac{d}{dx}(x)=1 \quad \text{and} \quad \frac{d}{dx}(\tfrac{1}{3}x^3)=x^2.$$

So $$\frac{d}{dx}(x+\tfrac{1}{3}x^3)=1+x^2.$$

One possible form of the function we seek is thus

$$y=x+\tfrac{1}{3}x^3.$$

This is obviously not unique since the addition of any constant to y leaves the derivative unaltered. So, if c is any constant whatever,

$$y=x+\tfrac{1}{3}x^3+c$$

is a form of the function required. It is also fairly clear that we can expect no other form of y to have the derivative $(1+x^2)$ at all points. In a particular case such as this, it is possible to reverse the differentiation process, to solve a " differential equation " of the above type. Notice that the additive constant, which disappears on differentiation, reappears in the reverse process. We have now to generalise this step and connect it with integration.

The definite integral $\int_a^b f(x)\,dx$ depends for its value on the form of the function $f(x)$ and on the values of the limits a and b. If the form of the function and the lower limit a are given, take a variable number x as the upper limit and write the integral $\int_a^x f(x)\,dx$. The value of the integral now depends on the value allotted to x, i.e. the integral is itself a function of x. It is called the *indefinite integral* of the function $f(x)$ and written simply $\int f(x)\,dx$. So :

DEFINITION : $\int f(x)\,dx=\int_a^x f(x)\,dx=$ a function of x.

By indefinite integration, we obtain, from a given function of x, a second function of the same variable. In diagrammatic terms, the indefinite integral is a variable area, i.e. the area between the curve $y=f(x)$, the axis Ox, the fixed ordinate at $x=a$ and a variable ordinate at x.

The fundamental result connecting differentiation and integration can now be stated :

If $f(x)$ is a continuous function, then the derivative of the indefinite integral of $f(x)$ is the function $f(x)$ itself :

$$\frac{d}{dx}\left\{\int f(x)\,dx\right\}=f(x).$$

Without giving a completely rigorous proof, the correctness of this result can be seen in diagrammatic terms. Let $F(x)$ denote the indefinite integral of the continuous function $f(x)$. Then

$$F(x+h)-F(x)=\int_a^{x+h} f(x)\,dx-\int_a^x f(x)\,dx=\int_x^{x+h} f(x)\,dx$$

by property (4) of 15.1. The last integral written is shown by the area between the curve $y=f(x)$, the axis Ox and ordinates at x and $(x+h)$. If h is small, the area differs by a small amount from that of the rectangle with base h and height $f(x)$, the ordinate of the curve at x. So, if h is small,

$$F(x+h)-F(x)=hf(x) \text{ approximately,}$$

i.e.
$$\frac{F(x+h)-F(x)}{h}=f(x) \quad \text{approximately.}$$

It follows that, as $h\rightarrow 0$, the ratio on the left-hand side tends to $f(x)$, i.e. the derivative of $F(x)$ is $f(x)$. Q.E.D.

The fundamental result can be put in a more practical form. If a function $\phi(x)$ can be found with $f(x)$ as its derivative, then

$$\frac{d}{dx}\left\{\int f(x)\,dx\right\}=f(x)=\frac{d}{dx}\{\phi(x)\}$$

and $\int f(x)\,dx$ and $\phi(x)$ must be identical except for an additive and arbitrary constant:

$$\int f(x)\,dx=\phi(x)+\text{constant.}$$

The integral here is between limits a and x. Putting $x=a$,

$$\int_a^a f(x)\,dx=\phi(a)+\text{constant.}$$

The integral now is zero, representing an area on a base of zero length. It follows that

$$\text{constant}=-\phi(a),$$

i.e. the arbitrary constant of the indefinite integral can be written in terms of the arbitrary and constant lower limit of integration. Put $x=b$ in the indefinite integral and substitute for the constant. So

$$\int_a^b f(x)\,dx=\phi(b)+\text{constant}=\phi(b)-\phi(a).$$

The practical rule for integration is thus :

If $f(x)$ is a continuous function and if $\frac{d}{dx}\{\phi(x)\}=f(x)$, then

$$\int f(x)\,dx=\phi(x)+\text{constant}$$

and
$$\int_a^b f(x)\,dx=\phi(b)-\phi(a).$$

To integrate a given function $f(x)$, it is only necessary to find, by some means or other, a function $\phi(x)$ with $f(x)$ as its derivative.

The fundamental result can be put in yet a third way. If it is given that the variable y is a function of x such that $\frac{dy}{dx}=f(x)$, where $f(x)$ is a given continuous function, then

$$\frac{dy}{dx}=f(x)=\frac{d}{dx}\left\{\int f(x)\,dx\right\}$$

and y and $\int f(x)\,dx$ can differ only by a constant. Hence :

The solution of the differential equation $\frac{dy}{dx}=f(x)$ is

$$y=\int f(x)\,dx+\text{constant}.$$

We have thus established the fact that integration is inverse differentiation. If the function $f(x)$ is obtained from $\phi(x)$ by derivation, then the function $\phi(x)$ is obtained from $f(x)$ by (indefinite) integration. The only difficulty is that an arbitrary constant reappears in the integration process. The same fact is expressed by saying that the integral of $f(x)$ provides the solution of the differential equation $\frac{dy}{dx}=f(x)$. The concept of a differential equation will be greatly extended in the following chapter and the idea of integration as inverse differentiation will be generalised to the idea of the integration of differential equations of all kinds.

15.4 The technique of integration.

The practical method of evaluating a definite or indefinite integral depends, as we have just seen, on finding a function with the function to be integrated as its derivative. If $\frac{d}{dx}\{\phi(x)\}=f(x)$, then we can write at once

$$\int f(x)\,dx=\phi(x)+\text{constant}; \quad \int_a^b f(x)\,dx=\phi(b)-\phi(a).$$

In attempting to systematise the practical method on lines similar to those that proved successful in the case of derivation, we first obtain and table some standard forms. From the derivative results, we have

$$\frac{d}{dx}\left(\frac{x^{n+1}}{n+1}\right)=x^n \quad (n\neq-1); \quad \frac{d}{dx}(\log x)=\frac{1}{x}; \quad \frac{d}{dx}(e^x)=e^x.$$

Reversing these results, we have the three main *standard forms for integrals* :

$$(1) \quad \int x^n\,dx=\frac{x^{n+1}}{n+1}+\text{constant} \quad (n\neq-1).$$

$$(2) \quad \int \frac{1}{x}\,dx=\log x+\text{constant}.$$

$$(3) \quad \int e^x\,dx=e^x+\text{constant}.$$

Further, if $f(x)$ is any continuous function of x, then

$$\frac{d}{dx}\left[\frac{\{f(x)\}^{n+1}}{n+1}\right]=f'(x)\{f(x)\}^n \quad (n\neq-1);$$

$$\frac{d}{dx}\{\log f(x)\}=\frac{f'(x)}{f(x)}; \quad \frac{d}{dx}\{e^{f(x)}\}=f'(x)\,e^{f(x)}.$$

The standard forms for integrals can thus be generalised :

$$(1') \quad \int f'(x)\{f(x)\}^n\,dx=\frac{\{f(x)\}^{n+1}}{n+1}+\text{constant} \quad (n\neq-1).$$

$$(2') \quad \int \frac{f'(x)}{f(x)}\,dx=\log f(x)+\text{constant}.$$

$$(3') \quad \int f'(x)\,e^{f(x)}\,dx=e^{f(x)}+\text{constant}.$$

The second step is to frame, if possible, a set of rules for the integration of combinations of functions, functions of functions and so on. Two such rules can be written down at once from properties (3) and (5) of 15.1. For indefinite integrals, we have

$$\int kf(x)\,dx = k\int f(x)\,dx$$

and $$\int \{f_1(x)+f_2(x)\}\,dx = \int f_1(x)\,dx + \int f_2(x)\,dx.$$

By an obvious extension and combination of these results, we have

$$\int \{a_1 f_1(x) + a_2 f_2(x) + \dots\}\,dx = a_1\int f_1(x)\,dx + a_2\int f_2(x)\,dx + \dots .$$

There is no difficulty in integrating an expression consisting of sums or differences of simple functions. But we can proceed no further; any attempt to frame other rules of general applicability, for the integration of products, quotients or functions of functions, is doomed to failure. The practical process of integration can be made neither as systematic nor as complete as in the case of derivation. In integrating, we are driven back to tentative "trial and error" methods, aided by a few more general devices (such as that of "integration by parts"). In each case, the most important point to remember is that any suggested form $\phi(x)$ of the integral of $f(x)$ can always be checked by derivation: $\frac{d}{dx}\{\phi(x)\} = f(x)$. In the following illustrative examples, the additive constant that appears in all indefinite integrals is omitted for convenience of writing.

Ex. 1. $\int \sqrt{x}\,dx = \int x^{\frac{1}{2}}\,dx = \frac{2}{3}x^{\frac{3}{2}} = \frac{2}{3}\sqrt{x^3}$, by (1) above.

Ex. 2. $\int (x^2+3x+2)\,dx = \frac{1}{3}x^3 + \frac{3}{2}x^2 + 2x$

and $\int (ax^2+bx+c)\,dx = \frac{1}{3}ax^3 + \frac{1}{2}bx^2 + cx$ (in general).

Ex. 3. $\int \frac{dx}{\sqrt{2x+1}} = \frac{1}{2}\int 2(2x+1)^{-\frac{1}{2}}\,dx = \frac{1}{2}\left\{\frac{(2x+1)^{\frac{1}{2}}}{\frac{1}{2}}\right\} = \sqrt{2x+1}$, by (1′),

and $\int (ax+b)^n\,dx = \frac{1}{a}\int a(ax+b)^n\,dx = \frac{1}{a}\frac{(ax+b)^{n+1}}{n+1}$ (in general).

Ex. 4. $\int \frac{dx}{1+x} = \log(1+x)$, by (2′),

and $\int \frac{dx}{ax+b} = \frac{1}{a}\int \frac{a\,dx}{ax+b} = \frac{1}{a}\log(ax+b)$ (in general).

Ex. 5. $\int \frac{x\,dx}{1+x^2}=\frac{1}{2}\int \frac{2x\,dx}{1+x^2}=\frac{1}{2}\log(1+x^2)$, by (2′),

and $\int \frac{(2ax+b)\,dx}{ax^2+bx+c}=\log(ax^2+bx+c)$ (in general).

Ex. 6. $\int \frac{x^2+3x-2}{x}\,dx=\int\left(x+3-\frac{2}{x}\right)dx=\frac{1}{2}x^2+3x-2\log x.$

Ex. 7. $\int e^{2x}\,dx=\frac{1}{2}\int 2e^{2x}\,dx=\frac{1}{2}e^{2x}$, by (3′),

and $\int e^{ax+b}\,dx=\frac{1}{a}\int ae^{ax+b}\,dx=\frac{1}{a}e^{ax+b}$ (in general).

Ex. 8. $\int xe^{-x^2}\,dx=-\frac{1}{2}\int(-2x)e^{-x^2}\,dx=-\frac{1}{2}e^{-x^2}$, by (3′).

Ex. 9. $\int x^2e^{-x^2}\,dx.$

Since $$\frac{d}{dx}(xe^{-x^2})=e^{-x^2}+x(-2xe^{-x^2})=(1-2x^2)e^{-x^2},$$

so $$xe^{-x^2}=\int(1-2x^2)e^{-x^2}\,dx=\int e^{-x^2}\,dx-2\int x^2e^{-x^2}\,dx,$$

i.e. $$\int x^2e^{-x^2}\,dx=\frac{1}{2}\int e^{-x^2}\,dx-\frac{1}{2}xe^{-x^2}.$$

The integral $\int e^{-x^2}\,dx$, though it is known to exist, cannot be evaluated in terms of ordinary functions. We have, therefore, reduced the given integral to as simple a form as possible.

15.5 Definite integrals and approximate integration.

The integral of a function $f(x)$ between given limits is evaluated once the indefinite integral $\phi(x)$ is found as a function of x. The formula $\int_a^b f(x)\,dx=\phi(b)-\phi(a)$ suffices. For example:

$$\int_0^1 \frac{dx}{1+x}=\Big[\log(1+x)\Big]_{x=1}-\Big[\log(1+x)\Big]_{x=0}=\log 2.$$

Areas under plane curves can then be evaluated at once. The parabola with equation $y=x^2+3x+2$ lies above Ox for positive values of x. The area under the curve between ordinates at $x=0$ and $x=3$ is

$$\int_0^3(x^2+3x+2)\,dx=\left[\frac{1}{3}x^3+\frac{3}{2}x^2+2x\right]_{x=3}-\left[\frac{1}{3}x^3+\frac{3}{2}x^2+2x\right]_{x=0}=\frac{57}{2}=28{\cdot}5.$$

We have noted, however, that the indefinite integrals of some functions, even of quite simple form, are often extremely difficult, if not impossible, to evaluate.* If the definite integral of such a function is required, for example, to give the area under the corresponding curve, it is necessary to adopt methods other than that described. A number of formulae giving the *approximate* values of definite integrals have been devised and are of service in the present problem. All these formulae depend on replacing the area under the actual curve considered by that under a simpler curve which approximates to the actual curve.

Suppose that the function $f(x)$ is continuous and positive in the interval (a, b) and that we require the value of $\int_a^b f(x)\,dx$. The interval (a, b) is divided into some even number $(2n)$ of parts of equal length $h=\dfrac{b-a}{2n}$. This is shown, in Fig. 95, by the points A_1, A_2,

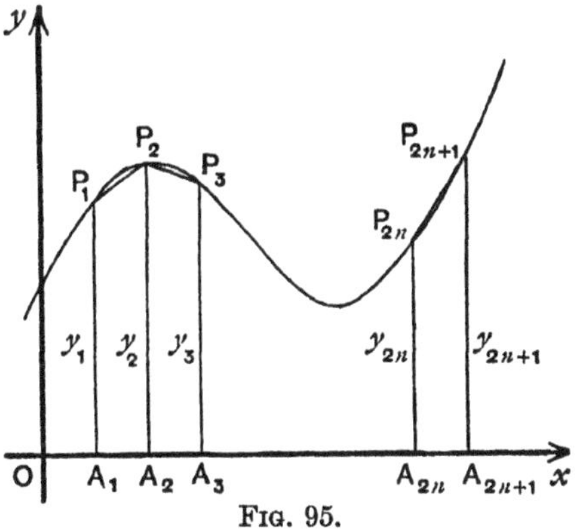

FIG. 95.

A_3, ... A_{2n}, A_{2n+1} on the axis Ox, where $OA_1=a$ and $OA_{2n+1}=b$ and $A_1A_2=A_2A_3=\ldots=h$. The ordinates $y_1=A_1P_1$, $y_2=A_2P_2$, ... can be read off the curve $y=f(x)$. The integral $\int_a^b f(x)\,dx$ is the total area

* It is not possible, for example, to find $\int e^{-x^2}\,dx$ in terms of ordinary functions. The integral exists and must be a function of x. But the function is not of algebraic form; it is, in fact, a " transcendental " function which can only be defined as taking its values from $\int_a^x e^{-x^2}\,dx$ for various values of x. It is a frequent device, in mathematical analysis, to define new functions as the integrals of other functions.

under the curve and on the base A_1A_{2n+1}, i.e. the sum of the areas under the curve and on the bases A_1A_2, A_2A_3, ...

As a first approximation, the area under the curve on the base A_1A_2 can be taken as the area of the trapezium $A_1P_1P_2A_2$:

$$\tfrac{1}{2}h(y_1+y_2).$$

Similarly, the approximate areas under the curve on the other bases are

$$\tfrac{1}{2}h(y_2+y_3),\quad \tfrac{1}{2}h(y_3+y_4),\quad \dots .$$

The approximate value of the total area, i.e. of the integral required, is found by addition. Hence,

$$\int_a^b f(x)\,dx=\tfrac{1}{2}h\{(y_1+y_{2n+1})+2(y_2+y_3+\dots+y_{2n})\}$$

approximately, the *trapezoidal rule* for approximate integration.

A closer approximation to the area under the curve on the base $A_1A_2A_3$ is given by the area under the arc of the parabola (with axis vertical) passing through the points P_1, P_2 and P_3. To calculate this latter area, let the equation of the parabola be

$$y=\alpha x^2+\beta x+\gamma$$

when the origin for x is taken, for convenience, at A_2. (This step does not affect the value of the area required.) The curve passes through $P_1(-h, y_1)$, $P_2(0, y_2)$ and $P_3(h, y_3)$. Hence,

$$y_1=\alpha h^2-\beta h+\gamma,\quad y_2=\gamma,\quad y_3=\alpha h^2+\beta h+\gamma.$$

Solving for α, β and γ, we find

$$\alpha=\frac{y_1-2y_2+y_3}{2h^2},\quad \beta=\frac{y_3-y_1}{2h}\quad \text{and}\quad \gamma=y_2.$$

The area under the parabola on the base $A_1A_2A_3$ is

$$\int_{-h}^{h}(\alpha x^2+\beta x+\gamma)\,dx=\left[\frac{1}{3}\alpha x^3+\frac{1}{2}\beta x^2+\gamma x\right]_{x=h}-\left[\frac{1}{3}\alpha x^3+\frac{1}{2}\beta x^2+\gamma x\right]_{x=-h}$$

$$=\frac{2}{3}\alpha h^3+2\gamma h=\frac{h}{3}(y_1+4y_2+y_3).$$

The approximate areas under the curve $y=f(x)$ on the bases $A_3A_4A_5$, $A_5A_6A_7$, ... are obtained similarly as

$$\frac{h}{3}(y_3+4y_4+y_5),\quad \frac{h}{3}(y_5+4y_6+y_7),\quad \dots .$$

By addition, we have the approximate area under the curve on the total base, i.e. the approximate value of the integral required. Hence,

$$\int_a^b f(x)\,dx = \frac{h}{3}\{(y_1 + y_{2n+1}) + 2(y_3 + y_5 + \ldots + y_{2n-1}) + 4(y_2 + y_4 + \ldots + y_{2n})\}$$

approximately, *Simpson's rule* for approximate integration.

Two examples illustrate the rules obtained :

Ex. 1. To find an approximate value of $\log 2 = \int_0^1 \frac{dx}{1+x}$, we divide the range (0, 1) into four equal parts by the points

$$x_1 = 0, \quad x_2 = \tfrac{1}{4}, \quad x_3 = \tfrac{1}{2}, \quad x_4 = \tfrac{3}{4} \quad \text{and} \quad x_5 = 1.$$

The ordinates of $y = \frac{1}{1+x}$ at the points of division are

$$y_1 = 1, \quad y_2 = \tfrac{4}{5}, \quad y_3 = \tfrac{2}{3}, \quad y_4 = \tfrac{4}{7} \quad \text{and} \quad y_5 = \tfrac{1}{2}.$$

The trapezoidal rule ($h = \frac{1}{4}$) gives the approximate value

$$\log 2 = \int_0^1 \frac{dx}{1+x} = \tfrac{1}{8}\{(1 + \tfrac{1}{2}) + 2(\tfrac{4}{5} + \tfrac{2}{3} + \tfrac{4}{7})\} = \tfrac{1171}{1680} = 0{\cdot}6970.$$

Simpson's rule gives the approximate value

$$\log 2 = \int_0^1 \frac{dx}{1+x} = \tfrac{1}{12}\{(1 + \tfrac{1}{2}) + 2(\tfrac{2}{3}) + 4(\tfrac{4}{5} + \tfrac{4}{7})\} = \tfrac{1747}{2520} = 0{\cdot}6933.$$

Five-figure tables give the correct value of log 2 as 0·69315. The nature of the approximation of the two rules is thus seen. The rules give closer approximations, of course, if the range (0, 1) is divided into more than four parts.

Ex. 2. The area under the curve $y = e^{-x^2}$, above Ox and between the ordinates $x = 0$ and $x = 2$ is given by the integral

$$\int_0^2 e^{-x^2}\,dx.$$

Dividing the range (0, 2) into ten equal parts ($h = \frac{1}{5}$), the ordinates at the eleven points of division are found, from tables, as

$$y_1 = 1{\cdot}0000, \quad y_2 = 0{\cdot}9608, \quad y_3 = 0{\cdot}8521, \quad y_4 = 0{\cdot}6977,$$
$$y_5 = 0{\cdot}5272, \quad y_6 = 0{\cdot}3679, \quad y_7 = 0{\cdot}2369, \quad y_8 = 0{\cdot}1409,$$
$$y_9 = 0{\cdot}0773, \quad y_{10} = 0{\cdot}0392 \quad \text{and} \quad y_{11} = 0{\cdot}0183.$$

The trapezoidal rule gives the approximate value

$$\int_0^2 e^{-x^2}\,dx = \tfrac{1}{10}\{1{\cdot}0183 + 2(3{\cdot}9000)\} = 0{\cdot}88183,$$

and Simpson's rule gives the approximate value

$$\int_0^2 e^{-x^2}\,dx = \tfrac{1}{15}\{1{\cdot}0183 + 2(1{\cdot}6935) + 4(2{\cdot}2065)\} = 0{\cdot}88209.$$

The required area, to three decimal places, is thus 0·882.

15.6 The relation between average and marginal concepts.

The demand of a market for a good X can be represented by three functions and curves showing total, average and marginal revenue respectively for various outputs of X. The relations between these functions and curves can now be examined in more detail. In Fig. 96, P and Q are points on the average and marginal revenue curves

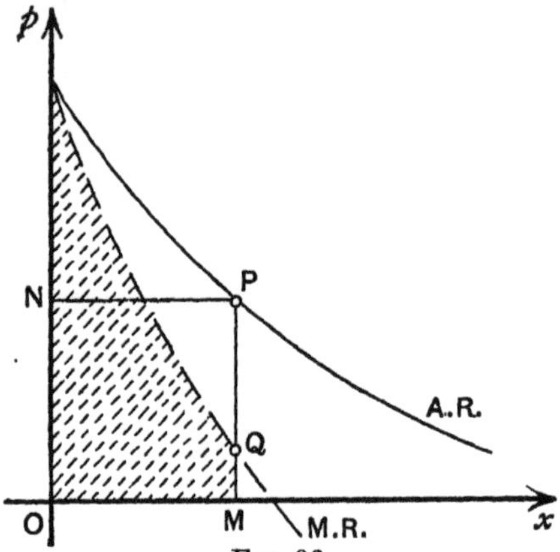

FIG. 96.

corresponding to a given output $x = OM$. Denote average revenue (price) by $p = MP$ and marginal revenue by $p_m = MQ$. Then the total revenue from the output x is

$$R = px = OM \,.\, MP,$$

i.e. the area of the rectangle $OMPN$. Further,

$$\frac{dR}{dx} = p_m,$$

i.e.
$$R = \int_0^x p_m \, dx + \text{constant}.$$

The arbitrary lower limit of the integral is here fixed at zero output. Put $x = 0$; then the constant must be zero since R is zero at zero output. Hence,

$$R = \int_0^x p_m \, dx,$$

i.e. the area under the marginal revenue curve, above Ox and between the ordinates at O and M. As long as the marginal revenue curve is above Ox, this area is positive and increases as x increases. When the curve falls below Ox, part of the area must be taken as negative and the total area decreases as x increases.

The total revenue from any output can be read off as an area either from the average or from the marginal revenue curve. We have a relation between the two curves at any output. No matter how the position of M varies, the area of the rectangle $OMPN$ is always equal to the area under the marginal revenue curve on the base OM.

A similar result holds for any pair of average and marginal curves. For example, if π and π_m denote the average and marginal cost of an output x of a good, π and π_m being functions of x, and if the cost of zero output is zero,* then total cost is

$$\Pi = x\pi = \int_0^x \pi_m \, dx.$$

Total cost is read off the average cost curve as a rectangular area and off the marginal cost curve as an area under the curve.

15.7 Capital values.

It has been shown that the present or capital value of an income stream (e.g. from land or a machine) consisting of £a_0 this year and £a_1, £a_2, ... £a_m in m following years is

$$y = \sum_{t=0}^{m} \frac{a_t}{(1+r)^t},$$

where interest is reckoned yearly at $100r$ per cent. (see 9.7 above). This calculation can be generalised to allow for a continuous income stream and for interest computed continuously.

It is assumed that income is obtained continuously over time, the rate at any time t years from the present being £$f(t)$ per year. This implies that, in the small interval of time between t and $(t+\Delta t)$ years from the present, the income obtained is approximately £$f(t)\,\Delta t$. Interest is computed continuously at the rate of $100r$ per cent. per year, where r can vary over time as a function of t. The present value of the income in the small time-interval $(t, t+\Delta t)$ is £$f(t)\,e^{-rt}\,\Delta t$ approximately. If the income stream "dries up" x years from the present, the approximate present value of the stream is

$$\Sigma f(t)\,e^{-rt}\,\Delta t,$$

where the summation extends over all small intervals Δt from $t=0$

* If the cost of zero output is Π_0, it is found that the constant in the integral expression for Π is Π_0. So $\Pi - \Pi_0 = \int_0^x \pi_m \, dx$, i.e. the area under the marginal revenue curve, is the total *variable* cost of the output x.

to $t=x$. As the number of time-intervals is increased, each interval becoming shorter, this approximation becomes closer. The exact present value is defined as the limit of the sum as the number of intervals increases indefinitely, i.e. as

$$y=\int_0^x f(t)\,e^{-rt}\,dt.$$

This is the *capital value* of the stream estimated at the present time ($t=0$). Thus, in passing from a discontinuous to a continuous income stream, the capital value changes from the sum of a finite number of terms to the limiting case of a sum as represented by an integral. We have an immediate application of the sum-area aspect of an integral.

The result obtained holds for an income stream and a rate of interest varying in any way over time. In the particular case where income is derived at a constant rate of £a per year and where the rate of interest is fixed at $100r$ per cent. per year over time, we have

$$y=a\int_0^x e^{-rt}\,dt=a\left[-\frac{1}{r}e^{-rt}\right]_{t=x}-a\left[-\frac{1}{r}e^{-rt}\right]_{t=0}$$

$$=a\left(-\frac{1}{r}e^{-rx}\right)-a\left(-\frac{1}{r}\right),$$

i.e.

$$y=\frac{a}{r}(1-e^{-rx}).$$

The capital value thus depends, in a simple way, on the size of the income stream, the number of years it flows and the rate of interest.

Problems of capital accumulation can be treated in a similar way. If £a_0, £a_1, £a_2, ... £a_m are invested in the present and in succeeding years at $100r$ per cent. compounded yearly, the total sum accumulated at the end of the period is £y where

$$y=\sum_{t=0}^{m} a_t(1+r)^{m-t}.$$

For a continuous investment which, t years from now, is made at the rate of £$f(t)$ per year, the total sum accumulated at a time x years from now is found, by an argument similar to that above, to be

$$y=\int_0^x f(t)\,e^{r(x-t)}\,dt,$$

where interest is added continuously at $100r$ per cent. per year. In the particular case where the investment £a and the interest rate $100r$ per cent. are both fixed over time, we find

$$y = a\int_0^x e^{r(x-t)}\,dt = a\left[-\frac{1}{r}e^{r(x-t)}\right]_{t=x} - a\left[-\frac{1}{r}e^{r(x-t)}\right]_{t=0},$$

i.e. $$y = \frac{a}{r}(e^{rx} - 1).$$

This expression is simply e^{rx} times the previous expression giving the present value (at $t=0$) of the same constant income stream, i.e. the accumulated value of the constant income stream after x years is the present value itself accumulated at the fixed rate of interest. This result does not hold when the investment or the interest rate varies over time.

As an application of the results above, suppose that the production process of 14.5 above is carried on continuously over time. A constant expenditure at the rate of $(ap_a + bp_b)$ per unit of time is made continuously and the product x of each investment is realised after the production period of t years. The total accumulated value of all investments in a period of t years is

$$k = \frac{ap_a + bp_b}{r}(e^{rt} - 1)$$

where interest is reckoned at the fixed continuous rate of $100r$ per cent. per year. The sum k can be interpreted either as the accumulated amount of loans needed to start the production process (between the original investment and the appearance of the first product) or as the total value of all " intermediate " product existing at any time. In any case, k represents the amount of " working " capital needed to keep the production process running continuously. In the competitive case, where the market rate of interest is such that

$$x = (ap_a + bp_b)e^{rt},$$

we have $$k = \frac{x - (ap_a + bp_b)}{r}.$$

If $x = Aa^{\alpha}b^{1-\alpha}t^{\beta}$ is the production function, then (from 14.5)

$$k = \frac{x}{r}(1 - e^{-\beta})$$

and the capital needed varies inversely with the competitive level of the market rate of interest.

15.8 A problem of durable capital goods.

The general problem of durable capital goods is complex but can be simplified, at a first approach, by making the following assumptions.* A capital good (e.g. a machine) is produced at a cost of £$f(t)$ dependent, in a known way, on the length t years of its active and continuous life. An entrepreneur, buying the good at its competitive (cost) price, obtains from its use a constant and continuous income stream of £a per year. At the time of purchase, the discounted (capital) value of all future product obtained from the good during its life of t years is

$$a\int_0^t e^{-rt}\,dt=\frac{a}{r}(1-e^{-rt}),$$

where $100r$ per cent. per year is the constant and continuous market rate of interest. It is assumed that the entrepreneur chooses the length of life of the good to maximise the capital value as a proportion of the cost:

$$y=\frac{a(1-e^{-rt})}{rf(t)}.$$

Here $$\frac{dy}{dt}=\frac{a}{r\{f(t)\}^2}\{re^{-rt}f(t)-(1-e^{-rt})f'(t)\}$$

$$=\frac{af'(t)e^{-rt}}{r\{f(t)\}^2}\left\{1-e^{rt}+r\frac{f(t)}{f'(t)}\right\}.$$

The necessary condition for the optimum length of life is thus

$$e^{rt}=1+r\frac{f(t)}{f'(t)},$$

a relation giving t in terms of the given rate of interest.

In the normal case, the cost $f(t)$ increases, but less than proportionately, with the durability (t years) of the good. As an actual example of a cost function of this type, take $f(t)=b\sqrt{t}$, where b is a constant. Here $f'(t)=\frac{b}{2\sqrt{t}}$ and the equation giving the optimum t in terms of r is

$$e^{rt}=1+2rt.$$

* The analysis is based on the work of Wicksell following Gustaf Åkerman. See Wicksell, *Lectures on Political Economy*, Vol. I (English Ed. 1934), pp. 274 *et seq.*

Write $x=rt$, and we have

$$e^x=1+2x.$$

Solving by graphical methods (2.8 above), it is found that x has one value (other than zero) which is approximately $x=1{\cdot}27$. So

$$t=\frac{1{\cdot}27}{r} \quad \text{approximately.}$$

The length of life of the capital good, as selected by the entrepreneur, varies inversely with the rate of interest. If the market rate of interest is 3 per cent., the optimum length of life is over 42 years, but at 4 per cent. the length of life is not quite 32 years.

Suppose now that production is carried on continuously, the capital goods being brought into use continuously and uniformly at the rate of n goods per year. After an initial period, it follows that the goods are scrapped at the same rate, each having served its optimum life of t years. At any moment, there are nt goods in use with ages varying uniformly from 0 to t years. The number of goods in use with between x and $x+\Delta x$ years of life to run is approximately $n\,\Delta x$. The value of the future product of one such good, discounted to the moment considered, is

$$\frac{a}{r}(1-e^{-rx}).$$

The total present value of all goods in use is approximately

$$\Sigma\frac{a}{r}(1-e^{-rx})\,n\,\Delta x,$$

the sum extending over all intervals Δx from 0 to t. Taking the limit for the continuous range of ages of the goods, the exact present value of all capital goods in use at the given moment is

$$k=n\frac{a}{r}\int_0^t(1-e^{-rx})\,dx=n\frac{a}{r}\left\{\left[x+\frac{1}{r}e^{-rx}\right]_{x=t}-\left[x+\frac{1}{r}e^{-rx}\right]_{x=0}\right\}$$

$$=\frac{na}{r}\left(t+\frac{1}{r}e^{-rt}-\frac{1}{r}\right)=\frac{na}{r^2}(e^{-rt}+rt-1).$$

This capital value, like the optimum life of each good, is dependent on the given market rate of interest.

15.9 Average and dispersion of a frequency distribution.

We are given N items showing values of a certain attribute according to the frequency distribution :

Value of attribute -	x_1 x_2 x_3 ... x_n	Total
Number of items -	y_1 y_2 y_3 ... y_n	$N=\sum_{r=1}^{n} y_r$

For example, the families in a certain town may be distributed according to the number of rooms occupied. The attribute, number of rooms occupied, takes the values $x_1=1$, $x_2=2$, $x_3=3$, ... Then y_1 represents the number of families occupying one room and so on. The *average* value of the attribute is defined as

$$\bar{x}=\frac{1}{N}\sum_{r=1}^{n} y_r x_r$$

and the *standard deviation* of the attribute as σ_x where

$$\sigma_x{}^2=\frac{1}{N}\sum_{r=1}^{n} y_r(x_r-\bar{x})^2.$$

These are familiar statistical figures, the average indicating a central value and the standard deviation indicating the " spread " or " dispersion " of the attribute in the group concerned.

Most of the variable characteristics considered statistically are not of this discontinuous nature. They are so finely divisible as to be measured by values which can be taken as continuous variables. The ages of a set of school children, the rents paid by a group of families and the death-rates in a number of districts are diverse examples of continuously variable characters. In practice, a number of items distributed according to such a characteristic is shown in definite groups or grades of the variables. Families, for example, may be arranged in grades of rent between 2 and 3 shillings, between 3 and 4 shillings, and so on. This is, however, a rough method and the results obtained are affected by the nature of the grading. Something more precise is required for a theoretical treatment. It is assumed, therefore, that the frequency distribution of a number of items is represented by a continuous function $y=f(x)$, implying that the number of items with value of the characteristic between x and $(x+\Delta x)$ is approximately $f(x)\Delta x$, the approximation being closer as the value of Δx is made smaller. The total number of items is

$$N=\Sigma f(x)\Delta x$$

approximately. Taking the limit, we have exactly

$$N=\int_a^b f(x)\,dx,$$

where a and b are the lowest and highest values of x found.

The approximate values of the average and standard deviation of the variable in the distribution are $\bar{x}$ and σ_x where

$$\bar{x}=\frac{1}{N}\Sigma xf(x)\Delta x \quad \text{and} \quad \sigma_x{}^2=\frac{1}{N}\Sigma(x-\bar{x})^2 f(x)\Delta x.$$

Taking the limit for an exact representation, we have

$$\bar{x}=\frac{1}{N}\int_a^b xf(x)\,dx=\frac{\int_a^b xf(x)\,dx}{\int_a^b f(x)\,dx},$$

and

$$\sigma_x{}^2=\frac{1}{N}\int_a^b (x-\bar{x})^2 f(x)\,dx=\frac{\int_a^b (x-\bar{x})^2 f(x)\,dx}{\int_a^b f(x)\,dx}.$$

We have here another direct application of the sum definition of an integral.

For example, Pareto's income law asserts that

$$y=\frac{A}{x^{\alpha+1}} \quad (A \text{ and } \alpha \text{ positive constants})$$

represents the number of persons with an income of £x, a continuous frequency distribution of persons according to income. The number of persons with income between £a and £b is

$$N=\int_a^b \frac{A}{x^{\alpha+1}}\,dx=\left[-\frac{A}{\alpha}\frac{1}{x^\alpha}\right]_{x=b}-\left[-\frac{A}{\alpha}\frac{1}{x^\alpha}\right]_{x=a}=\frac{A}{\alpha}\left(\frac{1}{a^\alpha}-\frac{1}{b^\alpha}\right).$$

The average income of these persons is

$$\bar{x}=\frac{1}{N}\int_a^b x\frac{A}{x^{\alpha+1}}\,dx=\frac{1}{N}\int_a^b \frac{A}{x^\alpha}\,dx=\frac{1}{N}\,\frac{A}{\alpha-1}\left(\frac{1}{a^{\alpha-1}}-\frac{1}{b^{\alpha-1}}\right).$$

Take $a=x$ and let b tend to infinity. The number of persons with income over £x is thus

$$N=\frac{A}{\alpha}\frac{1}{x^{\alpha}}$$

and the average income over £x is

$$\bar{x}=\frac{1}{N}\frac{A}{\alpha-1}\frac{1}{x^{\alpha-1}}=\frac{\alpha}{\alpha-1}x.$$

The law has the property that the average income over £x is a constant multiple $\left(\frac{\alpha}{\alpha-1}\right)$ of £x. In practice it is found that $\alpha=1{\cdot}5$ approximately. Hence, the average income over £x is approximately £$3x$, a result which agrees well with observed data of income distribution, at least for large incomes.

EXAMPLES XV

Integration

1. From the sum definition of an integral, show that $\int_a^b 1\,dx=b-a$. What is the indefinite integral of 1? Hence check the result.

2. If $f(x)$ is continuous, the sum $\sum_{r=1}^{n} f\left(\frac{x_r+x_{r+1}}{2}\right)(x_{r+1}-x_r)$ has $\int_a^b f(x)\,dx$ as its limiting value as n increases indefinitely. Deduce that $\int_a^b x\,dx=\frac{1}{2}(b^2-a^2)$ and check from the indefinite integral of x.

3. If $f(x)=x^n$, check that $\int_a^b f(x)\,dx=\int_a^c f(x)\,dx+\int_c^b f(x)\,dx$ for any values of a, b and c.

4. Show that $\frac{d}{dx}\left\{\int_x^b f(x)\,dx\right\}=-f(x)$. (Use property (1) of 15.1.)

5. Evaluate the indefinite integrals of

$$x^3,\quad \frac{1}{\sqrt{x}},\quad 1-x^2+x^4,\quad ax+b+\frac{c}{x},\quad \frac{ax+b}{\sqrt{x}}.$$

6. Use the results (1′)-(3′) of 15.4 to evaluate the integrals of

$$\sqrt{1+x},\quad \frac{1}{1-x},\quad \frac{1}{(3x+2)^2},\quad \frac{2x+1}{x^2+x-1},\quad e^{-3x},\quad xe^{x^2},\quad \frac{1}{\sqrt{x}}e^{\sqrt{x}}.$$

7. Show that $\int\frac{e^x}{1+e^x}\,dx=\log(1+e^x)$ and $\int\frac{e^x+e^{-x}}{e^x-e^{-x}}\,dx=\log(e^x-e^{-x})$.

8. From the integral of the general expression $(ax+b)^n$, show that $\int(1+x)^2\,dx=\frac{1}{3}(1+x)^3$. Expand $(1+x)^2$ in powers of x and show that $\int(1+x)^2\,dx=x+x^2+\frac{1}{3}x^3$. Why are the two results apparently different?

9. If $\dfrac{dy}{dx} = \log x$, check that $y = x(\log x - 1) + \text{constant}$.

10. Show that $\displaystyle\int (x-b)e^{-\frac{1}{2}\frac{(x-b)^2}{a^2}}\,dx = -a^2 e^{-\frac{1}{2}\frac{(x-b)^2}{a^2}}$.

11. Verify that $\dfrac{1}{x^2-1} = \dfrac{1}{2}\left(\dfrac{1}{x-1} - \dfrac{1}{x+1}\right)$ and deduce that

$$\int \frac{dx}{x^2-1} = \frac{1}{2}\log\frac{x-1}{x+1}.$$

12. Express a^x as a power of e and deduce that $\displaystyle\int a^x\,dx = \frac{a^x}{\log a}$.

13. Find $\displaystyle\int_0^2 2^x\,dx$. Check by drawing an accurate graph of $y = 2^x$ and estimating (by counting squares of the graph paper) the area under the curve, above Ox and between the ordinates at $x = 0$ and $x = 2$.

14. Find, by integration, the area under the straight line $y = x + 1$, above Ox and between the ordinates $x = 0$ and $x = 1$. Represent graphically and check by calculating the area of a trapezium.

15. Show that $\displaystyle\int_0^1 x^3\,dx = \frac{1}{4}$, $\displaystyle\int_1^4 \sqrt{x}\,dx = \frac{14}{3}$, $\displaystyle\int_0^{e-1} \frac{dx}{1+x} = 1$.

16. Find the area under the parabola $y = x^2$, above Ox and between the ordinates at $x = -3$ and $x = 3$. Verify that it is twice the area between the ordinates at $x = 0$ and $x = 3$.

17. Show that the area between the rectangular hyperbola $xy = \alpha^2$ and the axis Ox and between ordinates at $x = a$ and $x = b$ is $\alpha^2 \log\dfrac{b}{a}$, if a and b have the same sign. Why does the result not hold when a and b have opposite signs?

18. Evaluate $\displaystyle\int_0^2 (x^2 - 4x + 3)\,dx$ and check that it equals the difference between the area between the parabola $y = x^2 - 4x + 3$ and Ox from $x = 0$ to $x = 1$ and the area between Ox and the parabola from $x = 1$ to $x = 2$. Illustrate graphically.

19. By taking the difference between two integrals, show that the area enclosed between the parabolas $y = x^2$ and $y = \sqrt{x}$ is $\frac{1}{3}$.

20. Show that $\displaystyle\int_{-a}^{a} xe^{-x^2}\,dx = 0$ and $\displaystyle\int_{-a}^{a} x^2e^{-x^2}\,dx = \frac{1}{2}\int_{-a}^{a} e^{-x^2}\,dx$ for any value of a (see 15.4, Ex. 9, above).

21. Calculate approximate values of $\displaystyle\int_0^1 \frac{dx}{1+x^2}$ by the trapezoidal and Simpson's rules, dividing the range (0, 1) into ten equal parts in each case. The value of this integral is known to be $\frac{1}{4}\pi$. Hence estimate the closeness of the two approximations.

22. If the velocity of a body at time t is $v = f(t)$, show that $x = \displaystyle\int_0^t f(t)\,dt$ is the distance travelled from zero time to time t. Find the distance when $v = a - bt$ and show that the body comes to rest after travelling a distance $\dfrac{a^2}{2b}$.

23. The acceleration of a moving body is a constant a over time. Show that, after time t, the velocity is $v = u + at$ and the distance travelled $x = ut + \frac{1}{2}at^2$ where u is the velocity at zero time.

Integrals in economic problems

24. If the marginal revenue of output x is $p_m = a - bx$, find the total revenue function by integration and deduce the demand function.

25. If the marginal revenue function is $p_m = \frac{ab}{(x+b)^2} - c$, show that $p = \frac{a}{x+b} - c$ is the demand law.

26. Find the total cost function if it is known that the cost of zero output is c and that marginal cost of output x is $\pi_m = ax + b$.

27. If $\pi_m = \frac{a}{\sqrt{ax+b}}$ is the marginal cost function and if the cost of zero output is zero, find total cost as a function of x.

28. If interest is reckoned continuously at 4 per cent. per year, show that the capital value of uniform income stream of £100 per year for 10 years is approximately £824 and for 100 years approximately £2,454. ($e = 2{\cdot}71828$.)

29. Write down the capital value of a uniform income stream of £a per year for x years, reckoning interest continuously at $100r$ per cent. per year. By letting $x \to \infty$, deduce that the capital value of such an income stream going on for ever is £$\frac{a}{r}$.

30. An income stream decreases continuously over time for x years, the rate being £$a\,e^{-bt}$ per year at t years from now. Find the capital value at $100r$ per cent. continuous compound interest. Show that this equals the capital value of a uniform income stream of £a per year for x years if the rate of interest is raised to $100(r+b)$ per cent. per year.

31. An income stream decreases continuously for x years. At time t years from now, the income obtained is at the rate of £$a(1-2\alpha t)$ per year and the continuous rate of interest is $100r(1-\alpha t)$ per cent. per year. Show that the capital value of the income stream equals that of a steady and continuous income of £a per year for x years at the constant rate of interest, computed continuously, of $100r(1-\alpha x)$ per cent. per year.

32. In the problem of Examples XIV, 31, show that the value of all intermediate products at any time is

$$k = \frac{2}{\lambda}\left(1 - \frac{1}{\lambda}\right)\sqrt{p_a p_b}\,\frac{x}{r}$$

where $\frac{1}{r} = 16 \log \lambda$. In the competitive case (r being such that $ap_a + bp_b = xe^{-rt}$) show that

$$k = 16x\left(1 - \frac{1}{2\sqrt{p_a p_b}}\right)\log(2\sqrt{p_a p_b}).$$

33. In the problem of 15.8, ρ is defined so that $f(t) = \frac{a}{\rho}(1 - e^{-\rho t})$, i.e. the product of the capital good represents a net rate of return of 100ρ per cent. (computed continuously) on its cost $f(t)$. Find an equation for the length of life of the capital good so that ρ is a maximum.

34. In the competitive case of the problem of 15.8, the market rate of interest is such that the present value of the product of the capital good equals its cost. Show that the optimum length of life of the good as determined in 15.8 equals that determined in the problem of the previous example.

35. If $f(t)=3a\sqrt{t}$ and the competitive case of the previous example holds, show that the market rate of interest is approximately $4\frac{1}{2}$ per cent. per year and that the capital good has a life of approximately 28 years.

36. In the problem of 15.8, if $f(t)=b\sqrt{t}$ and the market rate of interest is $100r$ per cent. per year, show that the present value of all capital goods in use at any moment is approximately $k=0{\cdot}55\,\frac{na}{r^2}$. If the rate of interest is 5 per cent. per year, show that the present value is approximately 220 times the value of the yearly product of the capital goods.

37. In the problem of 15.8, the cost of a capital good of life t years is £$f(t)$, where $f(t)=b\left(1-\frac{1}{1+t}\right)$. Show that the optimum value of t is given by

$$e^{rt}=1+rt+rt^2.$$

When the interest rate is 4 per cent. per year, find by graphical methods the approximate value of t.

38. In a sample of 898 working-class families in Bolton, 1924, the families were classified according to the number of rooms occupied :

No. of rooms -	2	3	4	5	6	7	8	Total
No. of families -	15	477	227	169	7	2	1	898

Show that 3·65 is the average number of rooms occupied per family. Find the standard deviation of the number of rooms occupied per family in this group of families.

39. By differentiating both sides of the equation, show that

$$\int x^r e^{-x}\,dx = r\int x^{r-1}e^{-x}\,dx - x^r e^{-x}$$

for any value of r. If I_r denotes $\int_0^\infty x^r e^{-x}\,dx = \operatorname*{Lim}_{b\to\infty}\int_0^b x^r e^{-x}\,dx$, deduce that $I_r = rI_{r-1}$ and that $I_0=1$. Hence show that the average and standard deviation of the frequency distribution $y=xe^{-x}$ for the range from $x=0$ to infinity are $\bar{x}=2$ and $\sigma_x=\sqrt{2}$.

CHAPTER XVI

DIFFERENTIAL EQUATIONS

16.1 The nature of the problem.

DATA provided by scientific observation usually relate to small movements or changes in the quantities concerned. Information in " bulk ", covering a wide range of events, is scarcely to be looked for in practice. Translating the data into forms suitable for mathematical expression, we obtain relations between increments in variable quantities and, in the limit, the relations involve the derivatives or differentials of the variables. Mathematical theories in the sciences, if they are to be " realistic ", must thus be built on the basis of *differential equations*, relations between the derivatives or differentials of varying quantities. It is then the business of the theories to deduce the *functional equations* between the variables which lie behind the differential equations, i.e. to express the " general laws " whose variations correspond to the given data. In physics, for example, much of the data is expressed in terms of variations in " energy ", of one form or another, and leads to certain fundamental differential equations from which the laws of " conservation of energy " are to be derived. Differential equations thus occupy a fundamental position in most highly developed mathematical sciences.

The nature of a differential equation can be examined first in the case where only first-order variations in two variables x and y are involved. The simplest case can be written

$$\frac{dy}{dx}=f(x),$$

where $f(x)$ is some given function. We have seen (15.3 above) that this case is easily dealt with by means of the concept of an integral.

A more general form of the case is

$$\frac{dy}{dx}=f(x, y)$$

where the given function $f(x, y)$ involves both x and y. Given the values of x and y, then the value of the derivative of the relation that must exist between x and y is also known. From this information, we attempt to deduce the actual form of the functional relation, to find that particular function which gives a derivative satisfying the above relation. This is the problem of the solution, or "integration", of a differential equation, a problem which is clearly an extension of that already solved by simple integration. A still more general form of the type of differential equation here considered is a relation involving the derivative in a non-linear way, a relation including, for example, such terms as $\left(\frac{dy}{dx}\right)^2$ or $\log\left(\frac{dy}{dx}\right)$. The development of the following sections is concerned almost entirely with the simpler linear differential equation, but it must be remembered that more complicated forms can be considered and solved.

The meaning of a differential equation of the form $\frac{dy}{dx}=f(x, y)$ is most clearly seen in diagrammatic terms. At each point (x, y) of the plane Oxy, the equation provides a definite value of the derivative $\frac{dy}{dx}$. This is the gradient of the direction in which the variables are allowed to vary from (x, y), i.e. the gradient of the tangent to the (at present unknown) curve through (x, y) expressing the relation between x and y implied by the differential equation. Our problem is thus that of constructing complete curves from knowledge only of their tangent gradients. An example illustrates. The equation

$$\frac{dy}{dx}=-\frac{x^2-y^2-1}{2xy}$$

gives a tangent gradient at each point of the plane Oxy. At the point (1, 1), for example, the tangent gradient is $\frac{dy}{dx}=\frac{1}{2}$, a line sloping upwards with gradient $\frac{1}{2}$ through the point (1, 1). A convenient set of points, covering the plane Oxy fairly uniformly, is selected and the tangent gradient evaluated at each point. When lines with the appropriate gradients are drawn through the points, a graph of the kind shown in Fig. 97 is obtained. It now appears that a system of curves, one through each point of the plane, can be drawn to touch the lines of the graph. This curve system, some members of

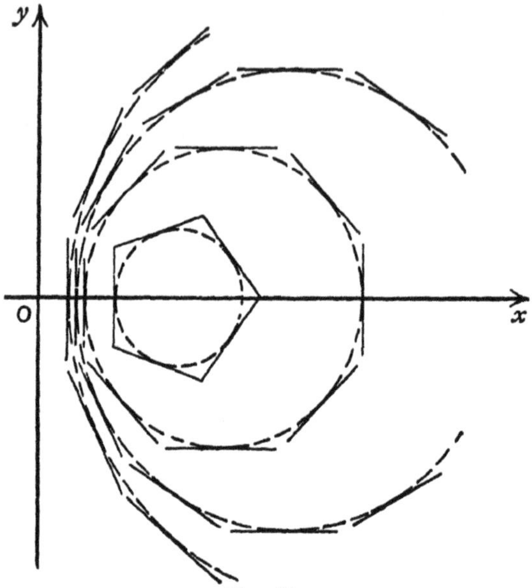

FIG. 97.

which are shown in Fig. 97, corresponds to the solution of the given differential equation. In this particular case, the curve system consists of circles with equation $x^2+y^2-2ax+1=0$ where a is a parameter. This equation is the "integral" of the differential equation.*

From a slightly different point of view, we can start from any point of the plane Oxy and describe a curve through it by moving always in the direction indicated by the differential equation. Having obtained this curve, we can select a point not on it and describe a second curve passing through this point. The process can be continued until a whole system of curves is obtained to represent the solution of the given differential equation.

* The use of the term "integral" to describe the solution of a differential equation is clear. The solution corresponds to a curve system which is built up, or "integrated", step by step, from the tangent gradients given by the differential equation.

An important point now arises. The integral of a differential equation of the type considered is shown by a curve system and so it must involve a single parameter or *arbitrary constant.* This is illustrated by the integral, $x^2+y^2-2ax+1=0$, of the equation above. The point is worth emphasising by putting the converse problem. The following examples show that a single arbitrary constant can be eliminated from a functional relation by differentiation :

Ex. 1. From the relation $x^2-e^y=a$,

we obtain $$2x\,dx-e^y\,dy=0,$$

i.e. the differential equation $$\frac{dy}{dx}=2xe^{-y}.$$

Ex. 2. From $$x^2+y^2-2ax+1=0,$$

we obtain $$2x\,dx+2y\,dy-2a\,dx=0,$$

i.e. $$a=\frac{(x\,dx+y\,dy)}{dx}.$$

Substituting in the original equation, we find

$$x^2+y^2-2x\frac{x\,dx+y\,dy}{dx}+1=0,$$

i.e. $$(x^2-y^2-1)dx+2xy\,dy=0,$$

i.e. $$\frac{dy}{dx}=-\frac{x^2-y^2-1}{2xy},$$

a differential equation in which the parameter a does not appear.

Ex. 3. Differentiating $y^2-2ax+a^2=0$,

we have $$2y\,dy-2a\,dx=0.$$

Substituting $a=y\dfrac{dy}{dx}$ in the original equation we find a differential equation which is not linear in the derivative $\dfrac{dy}{dx}$:

$$y\left(\frac{dy}{dx}\right)^2-2x\frac{dy}{dx}+y=0.$$

The integration of a differential equation reverses the procedure of these examples, taking us from a relation in the derivative to the complete relation between x and y. So the integral of $\dfrac{dy}{dx}=2xe^{-y}$ is $x^2-e^y=a$, and similarly for the others. In each case the integral involves one arbitrary constant :

The integral of a differential equation in the derivative of y with respect to x is a relation between x and y which includes a single arbitrary constant.

The result stated, though not formally proved, has been illustrated sufficiently for our purpose.*

To extend to cases of more than two variables, suppose that we have a relation between the first-order variations in three variables x, y and z. In its simple form the relation can be written

$$f(x, y, z)\,dx + g(x, y, z)\,dy - dz = 0,$$

giving the values of the partial derivatives

$$\frac{\partial z}{\partial x} = f(x, y, z) \quad \text{and} \quad \frac{\partial z}{\partial y} = g(x, y, z).$$

In diagrammatic terms, we are given the tangent plane (by its gradients in the two fundamental directions) at each point of $Oxyz$ space and require to " integrate " the planes into a system of surfaces. The integral of the differential equation thus appears as a relation between x, y and z, including one arbitrary constant, and is shown by a system of surfaces. An example makes this clear :

Ex. The relation $\quad x^2 + y^2 = az \quad$ (a being a parameter)

gives $\quad 2x\,dx + 2y\,dy = a\,dz.$

Substituting $a = 2\dfrac{x\,dx + y\,dy}{dz}$ in the original relation, we find

$$x^2 + y^2 = 2z\frac{x\,dx + y\,dy}{dz},$$

i.e.

$$2xz\,dx + 2yz\,dy - (x^2 + y^2)\,dz = 0,$$

or

$$\frac{\partial z}{\partial x} = \frac{2xz}{x^2 + y^2} \quad \text{and} \quad \frac{\partial z}{\partial y} = \frac{2yz}{x^2 + y^2}.$$

We have a differential equation in which the parameter a does not appear. Conversely, the integral of the differential equation shown is the relation $x^2 + y^2 = az$ including the arbitrary constant a.

More involved differential equations in three or more variables can be written and some examples will appear later. The simple type here considered suffices for the moment.

* It is to be noticed that the indefinite integral which is the solution of $\dfrac{dy}{dx} = f(x)$ contains an additive constant, a particular case of the present general result.

16.2 Linear differential equations and their integration.

Of the first-order differential equations in two variables x and y, the most useful is the *linear differential equation* in which the derivative or differentials are related linearly, squares, cubes and other expressions in the derivative or differentials being excluded. We can write the equation either as

$$P\,dx+Q\,dy=0,$$

where P and Q are given functions of x and y, or as

$$\frac{dy}{dx}=f(x,\,y),$$

where $f(x, y)$ is some function of x and y. To pass from one form to the other, it is only necessary to associate $f(x, y)$ with the ratio $\left(-\frac{P}{Q}\right)$. It is to be noticed that, though the derivative appears linearly, there is no such restriction on the variables x and y.

As in the case of simple integration discussed in the previous chapter, there is no comprehensive practical scheme for integrating differential equations, even of the linear type. Some particular methods for certain cases are given below but there remain many instances in which the integral can be obtained only after much trial and error or which cannot be integrated by any practical method.

The following are the four main cases for which definite practical methods of integration can be given :

(1) The form $\frac{dy}{dx}=f(x)$.

By simple integration, the solution of this form is

$$y=\int f(x)\,dx+a,$$

where a is an arbitrary constant.

(2) The form $\frac{1}{y}\frac{dy}{dx}=f(x)$.

This form can also be solved by simple integration. Since

$$\frac{d}{dx}(\log y)=\frac{1}{y}\frac{dy}{dx},$$

the equation is
$$\frac{d}{dx}(\log y)=f(x).$$

Hence, $$\log y = \int f(x)\,dx + a,$$

where a is an arbitrary constant. The variable y is found explicitly in terms of x by taking the exponential of the right-hand side.

Ex. 1. $y(1-x)\,dx - x\,dy = 0.$

Here $$\frac{1}{y}\frac{dy}{dx} = \frac{1-x}{x} = \frac{1}{x} - 1,$$

and $$\log y = \int\left(\frac{1}{x} - 1\right)dx + \text{constant} = \log x - x + b.$$

So $$\log\left(\frac{y}{x}\right) = -(x-b),$$

i.e. $$y = xe^{-(x-b)} = e^b(xe^{-x}).$$

Writing $a = e^b$, where a is thus an arbitrary constant, we have

$$y = axe^{-x}.$$

Ex. 2. To find y as a function of x when it is given that the elasticity of y with respect to x is a constant α. We have

$$\frac{x}{y}\frac{dy}{dx} = \alpha, \quad \text{i.e.} \quad \frac{1}{y}\frac{dy}{dx} = \frac{\alpha}{x}.$$

So $$\log y = \alpha\int\frac{dx}{x} + \text{constant} = \alpha \log x + \text{constant},$$

i.e. $$\log\left(\frac{y}{x^\alpha}\right) = \text{constant},$$

i.e. $$\frac{y}{x^\alpha} = \text{constant} = a,$$

i.e. $$y = ax^\alpha.$$

Ex. 3. As x increases, the variable y increases at a rate always equal to a multiple ρ of its current value : $\frac{dy}{dx} = \rho y$. To find y as a function of x, we write :

$$\frac{1}{y}\frac{dy}{dx} = \rho.$$

So $$\log y = \rho x + b,$$

and $$y = e^{\rho x + b} = e^b e^{\rho x} = ae^{\rho x},$$

where a is an arbitrary constant. Compound interest growth at the continuous rate of 100ρ per cent. per year is a particular case of this result.

(3) The form $\phi(x)\,dx + \psi(y)\,dy = 0.$

This case, in which the variables are said to be "separate", holds whenever the differential equation can be so arranged that the coefficient of dx depends on x only and the coefficient of dy on y only. The integration is then effected at once:

$$\int \phi(x)\,dx + \int \psi(y)\,dy = \text{constant.}$$

It is only necessary, therefore, to evaluate two simple integrals.

Ex. 1. $e^y \dfrac{dy}{dx} = 2x.$

Here $\qquad 2x\,dx - e^y\,dy = 0, \quad \text{i.e. } 2\int x\,dx - \int e^y\,dy = \text{constant.}$

So $$x^2 - e^y = a, \quad \text{i.e. } e^y = x^2 - a,$$

or $$y = \log(x^2 - a),$$

where a is an arbitrary constant.

Ex. 2. $(y+k)\,dx + (x+h)\,dy = 0$, where h and k are fixed constants.

Here $$\frac{dx}{x+h} + \frac{dy}{y+k} = 0, \quad \text{i.e. } \int \frac{dx}{x+h} + \int \frac{dy}{y+k} = \text{constant.}$$

So $$\log(x+h) + \log(y+k) = b, \quad \text{i.e. } \log(x+h)(y+k) = b.$$

Hence, writing $a = e^b$, we have

$$(x+h)(y+k) = a,$$

where a is an arbitrary constant. The integral is represented by a system of rectangular hyperbolas (see 3.7 above).

Ex. 3. The variable y increases as x increases at a rate given by

$$\frac{dy}{dx} = \rho y\left(1 - \frac{y}{L}\right),$$

where L and ρ are given constants.* To find y as a function of x:

Here $$\frac{L\,dy}{y(L-y)} = \rho\,dx, \quad \text{i.e. } \int \frac{L\,dy}{y(L-y)} = \rho x + b.$$

But $$\int \frac{L\,dy}{y(L-y)} = \int \left(\frac{1}{y} + \frac{1}{L-y}\right) dy = \log y - \log(L-y) = -\log \frac{L-y}{y}.$$

So $$\log \frac{L-y}{y} = -\rho x - b, \quad \text{i.e. } \frac{L-y}{y} = e^{-\rho x} e^{-b} = ae^{-\rho x},$$

where $a = e^{-b}$ is an arbitrary constant. After rearrangement,

$$y = \frac{L}{1 + ae^{-\rho x}}.$$

* This is a modified version of growth at compound interest. The curve representing y as a function of x is called a *logistic curve* and it is sometimes used to express the growth of a population over time.

(4) The " exact " form.

It may happen that the differential equation can be arranged, after multiplying through by a suitable factor, in the form

$$P\,dx+Q\,dy=0,$$

where the left-hand side is the differential of some function $\phi(x, y)$, i.e. $P\,dx+Q\,dy=d\phi(x, y)$. The equation is then said to be " exact " and its integral is clearly

$$\phi(x, y)=\text{constant}.$$

The following examples illustrate how equations can be arranged so that they appear in the " exact " form.

Ex. 1. $y\,dx+x\,dy=0.$

The equation is exact as it stands and can be written $d(xy)=0$. The integral is $xy=a$, where a is an arbitrary constant.

Ex. 2. $(x^2-y^2-1)\,dx+2xy\,dy=0.$

Now $$d\left(\frac{x^2+y^2+1}{2x}\right)=\frac{2(x\,dx+y\,dy)x-(x^2+y^2+1)\,dx}{2x^2}$$
$$=\frac{(x^2-y^2-1)\,dx+2xy\,dy}{2x^2}.$$

Hence, on multiplying through by $\frac{1}{2x^2}$, the given equation becomes

$$d\left\{\frac{(x^2+y^2+1)}{2x}\right\}=0,$$

and the integral is $\frac{x^2+y^2+1}{2x}=a$, where a is an arbitrary constant.

So $$x^2+y^2-2ax+1=0,$$

as represented by the system of circles of Fig. 97.

Ex. 3. $\frac{dy}{dx}=-2\sqrt{y+k}\,\frac{h-\sqrt{y+k}}{x+h}$, where h and k are fixed constants.

We can write the equation in the form

$$\frac{1}{2}\frac{x+h}{\sqrt{y+k}}\frac{dy}{dx}+(h-\sqrt{y+k})=0,$$

i.e. $$(x+h)^2\frac{d}{dx}\left(\frac{h-\sqrt{y+k}}{x+h}\right)=0.$$

The integral is thus

$$\frac{h-\sqrt{y+k}}{x+h}=a, \text{ where } a \text{ is an arbitrary constant.}$$

This integral is shown by a system of curves, each being the arc of a parabola (see 3.7 above).

The differential equation of the linear type in three variables x, y and z appears in the form

$$P\,dx+Q\,dy+R\,dz=0,$$

where P, Q and R are three given functions of x, y and z. This is often termed the *total differential equation.* The integral is a relation between the variables including a single arbitrary constant and reproducing, on differentiation, the given differential equation. Since we can take z as a function of x and y, we have

$$\frac{\partial z}{\partial x}=\left(\frac{dz}{dx}\right)_{y\text{ constant}}=-\frac{P}{R}\quad\text{and}\quad\frac{\partial z}{\partial y}=\left(\frac{dz}{dy}\right)_{x\text{ constant}}=-\frac{Q}{R}.$$

An alternative expression of the differential equation is thus

$$\frac{\partial z}{\partial x}=f(x,\,y,\,z)\quad\text{and}\quad\frac{\partial z}{\partial y}=g(x,\,y,\,z),$$

where the functions f and g are to be associated with $\left(-\frac{P}{R}\right)$ and $\left(-\frac{Q}{R}\right)$ respectively. The cases where there are more than three variables are exactly similar. Practical methods of the kinds indicated above serve to integrate particular forms of the total differential equation.* The following examples illustrate the procedure :

Ex. 1. The partial elasticities of z as a function of x and y are constants α and β respectively. To find z in terms of x and y :

$$\frac{x}{z}\frac{\partial z}{\partial x}=\alpha\quad\text{and}\quad\frac{y}{z}\frac{\partial z}{\partial y}=\beta,$$

i.e. $$\alpha\frac{dx}{x}+\beta\frac{dy}{y}-\frac{dz}{z}=0.$$

So $$\alpha\int\frac{dx}{x}+\beta\int\frac{dy}{y}-\int\frac{dz}{z}=b\quad(b\text{ an arbitrary constant})$$

and $$\alpha\log x+\beta\log y-\log z=b,\quad\text{i.e. }\log\left(\frac{x^{\alpha}y^{\beta}}{z}\right)=b.$$

Writing $a=e^{-b}$, the integral of the differential equation is

$$z=ax^{\alpha}y^{\beta}.$$

Ex. 2. $x\,dx+y\,dy+z\,dz=0.$

So $$\int x\,dx+\int y\,dy+\int z\,dz=\text{constant},$$

i.e. $$\tfrac{1}{2}x^2+\tfrac{1}{2}y^2+\tfrac{1}{2}z^2=\text{constant}.$$

* For a method which serves in most ordinary cases, see Piaggio, *Differential Equations* (1920), p. 139.

The integral required is thus represented by the system of spheres

$$x^2+y^2+z^2=a,$$

where a is an arbitrary constant.

Ex. 3. $2xz\,dx+2yz\,dy-(x^2+y^2)\,dz=0.$

Multiplying through by $\frac{1}{z^2}$, the equation is thrown into "exact" form

$$d\left(\frac{x^2+y^2}{z}\right)=0.$$

So
$$\frac{x^2+y^2}{z}=a,$$

and the integral of the equation is $x^2+y^2=az$, where a is an arbitrary constant. This integral is represented by a system of surfaces known as "paraboloids" (see 11.2-3 above).

16.3 The general integral of a linear differential equation.

Two points arise in considering linear differential equations from a general and theoretical, as opposed to a practical, angle. One concerns the question whether an integral of a given equation exists at all, and the other whether the integral, if it exists, is uniquely determined or not.

The fundamental "existence" results, which answer the first question, can be stated, but not formally proved, here.* The linear differential equation in two variables,

$$P\,dx+Q\,dy=0 \quad \text{or} \quad \frac{dy}{dx}=f(x, y),$$

can be shown to possess an integral in all cases where certain conditions are satisfied by the functions P and Q, or by the function f. These conditions are satisfied in all ordinary cases, e.g. when f is a function with continuous partial derivatives. In the two-variable case, therefore, we can assume that all the usual linear differential equations are integrable.

A complication arises when we pass to cases of three or more variables. It can be shown that the linear differential equation $P\,dx+Q\,dy+R\,dz=0$ fails to possess an integral even in quite

* See de la Vallée Poussin, *Cours d'analyse infinitésimale*, Vol. II (5th Ed. 1925), pp. 133 *et seq.*, pp. 292 *et seq.*; Piaggio, *Differential Equations* (1920), pp. 139 *et seq.*

ordinary cases. The integral exists only if the functions P, Q and R, having continuous partial derivatives, satisfy a certain relation

$$P\left(\frac{\partial Q}{\partial z}-\frac{\partial R}{\partial y}\right)+Q\left(\frac{\partial R}{\partial x}-\frac{\partial P}{\partial z}\right)+R\left(\frac{\partial P}{\partial y}-\frac{\partial Q}{\partial x}\right)=0,$$

which is called the *integrability condition*. We cannot assume that the equation has an integral at all without first imposing or verifying the integrability condition. For example, the equation

$$2xz\,dx+2yz\,dy-(x^2+y^2)\,dz=0$$

is found to satisfy the integrability condition; its integral has, in fact, been found above. On the other hand, the equation

$$dx+z\,dy+2y\,dz=0,$$

though quite simple in form, fails to satisfy the integrability condition and possesses no integral.

In diagrammatic terms, these important results imply that a set of tangent lines in the plane Oxy with gradients given by $\frac{dy}{dx}=f(x, y)$, where f is a given function with continuous partial derivatives, can always be built up, or "integrated", into a system of curves. But this is not true of the corresponding case of a set of tangent planes in space $Oxyz$. It may not be possible, even in quite simple cases, to build up the planes into a system of surfaces, i.e. to find a set of surfaces to fit the given tangent planes. It is not easy to see why the extra dimension spoils the simplicity of the previous result.

Turning to the question of the uniqueness of integration, the integral of a linear differential equation in two variables is an equation in the variables including one arbitrary constant. This constant must be involved linearly. If, for example, squares or cubes of the constant appear, then the process of eliminating it must lead to an equation involving squares or cubes of the derivative $\frac{dy}{dx}$, i.e. to a non-linear differential equation. Example 3 of 16.1 illustrates this point. It follows that the constant can be separated off linearly from the variables of the integral,* i.e.

* In fact, the curve system representing the integral of the differential equation must be such that one, and only one, curve of the system passes through each point of the relevant part of the plane Oxy (see 3.7 above).

The integral of the linear differential equation $P\,dx + Q\,dy = 0$ can be written

$$\phi(x, y) = \text{constant},$$

where ϕ is a function with partial derivatives proportional to P and Q.

Ex. The equation $(x^2 - y^2 - 1)\,dx + 2xy\,dy = 0$ has an integral

$$x^2 + y^2 - 2ax + 1 = 0,$$

which can be arranged in the form

$$\frac{x^2 + y^2 + 1}{x} = \text{constant}.$$

Here
$$\phi(x, y) = \frac{x^2 + y^2 + 1}{x},$$

and
$$\frac{\partial \phi}{\partial x} = \frac{x^2 - y^2 - 1}{x^2}, \quad \frac{\partial \phi}{\partial y} = \frac{2xy}{x^2},$$

which are proportional to the coefficients of dx and dy in the original differential equation.

It is easily seen that the form of the function $\phi(x, y)$ is by no means unique. In fact, if $\phi(x, y) = \text{constant}$ is one form of the integral of the given equation and if F is any function, then

$$F\{\phi(x, y)\} = \text{constant}$$

is also a form of the integral. For, if $\phi(x, y)$ has a constant value for certain values of x and y, then $F\{\phi(x, y)\}$ also has a constant value for these x and y. Hence :

If $\phi(x, y) = \text{constant}$ is one integral of $P\,dx + Q\,dy = 0$, then the general integral is $F\{\phi(x, y)\} = \text{constant}$, where F is an arbitrary function.

This lack of uniqueness only affects the *form* of the analytical expression of the integral ; it does not correspond to any lack of uniqueness in the integral itself. The curve system which represents the integral is quite unique, being obtained equally well from $F\{\phi(x, y)\} = \text{constant}$ as from $\phi(x, y) = \text{constant}$. The contour map of the function $z = F\{\phi(x, y)\}$ does not depend on which form of F is selected and always gives the integral of our differential equation. For example, the integral of $(x^2 - y^2 - 1)\,dx + 2xy\,dy = 0$ is shown by the perfectly definite system of circles (Fig. 97). The analytical form of the integral can, however, appear in various disguises, e.g.

$$\frac{x^2 + y^2 + 1}{x} = \text{const.}; \quad \frac{(x^2 + y^2 + 1)^2}{x^2} = \text{const.}; \quad \log \frac{x^2 + y^2 + 1}{x} = \text{const.}$$

Exactly similar results hold for the total differential equation in three variables $P\,dx + Q\,dy + R\,dz = 0$, *provided* that the integrability condition is satisfied. The form of the integral appears

$$F\{\phi(x, y, z)\} = \text{constant},$$

where $\phi(x, y, z) = \text{constant}$ is one integral and F is an arbitrary function. The partial derivatives of ϕ are proportional to P, Q and R. But the integral itself, as represented by a system of surfaces in space, is quite definite. The result extends, of course, to cases of more than three variables.

16.4 Simultaneous linear differential equations.

Two linear differential equations in three variables must give rise to two functional relations between the variables, each including one arbitrary constant, and represented by two systems of surfaces in space. An actual example will illustrate the position.

Ex. By differentiating and eliminating the constants in

$$x + y + z = a \quad \text{and} \quad x^2 + y^2 = bz,$$

we have $dx + dy + dz = 0$ and $2xz\,dx + 2yz\,dy - (x^2 + y^2)\,dz = 0$.

The two functional relations provide the solution of this pair of linear differential equations. The latter can always be arranged to give the ratios of dx, dy and dz in terms of x, y and z. Here, on solving the differential equations for the differentials, we find

$$\frac{dy}{dx} = -\frac{x^2 + y^2 + 2xz}{x^2 + y^2 + 2yz} \quad \text{and} \quad \frac{dz}{dx} = 2z\frac{x - y}{x^2 + y^2 + 2yz}$$

or

$$\frac{dx}{x^2 + y^2 + 2yz} = \frac{-dy}{x^2 + y^2 + 2xz} = \frac{dz}{2z(x - y)}.$$

Further, the relations giving the differential equations can be arranged to give any two of the variables (say, y and z) as functions of the third (x). In the present case, we find

$$y = \pm\tfrac{1}{2}\sqrt{(b^2 + 4ab) - 4bx - 4x^2} - \tfrac{1}{2}b$$

and

$$z = \mp\tfrac{1}{2}\sqrt{(b^2 + 4ab) - 4bx - 4x^2} - x + a + \tfrac{1}{2}b.$$

These functions involve the two arbitrary constants a and b. We can now say that the derivatives of these two functions must correspond to the values of $\frac{dy}{dx}$ and $\frac{dz}{dx}$ given from the differential equations. In diagrammatic terms, the solutions of our differential equations is shown by a pair of systems of surfaces in space. But the intersection of each pair of surfaces (one from each system) is a curve in space. The solution is thus shown by a whole system of curves, not of the plane kind, but of the spatial kind. In the present case, the curves are parallel plane sections (by $x+y+z=a$) of paraboloid surfaces ($x^2+y^2=bz$). These curves are in space ; they do not lie in one and the same plane.

In general, two given linear differential equations in three variables can be arranged in the form

$$\frac{dx}{P}=\frac{dy}{Q}=\frac{dz}{R},$$

where P, Q and R are some functions of x, y and z. So, if we consider y and z as functions of x, we can write

$$\frac{dy}{dx}=f(x, y, z) \quad \text{and} \quad \frac{dz}{dx}=g(x, y, z),$$

where the functions f and g are equivalent to the ratios $\frac{Q}{P}$ and $\frac{R}{P}$. It can be shown that the integral of these equations exists in all cases subject to the usual conditions concerning the continuity of the functions P, Q and R, or the functions f and g.* The integral appears in the form of two relations between the variables, each involving an arbitrary constant which can be separated off linearly :

$$\phi(x, y, z)=a \quad \text{and} \quad \psi(x, y, z)=b.$$

As before, the analytical forms of ϕ and ψ are not unique. In fact, the integral can be expressed in general as

$$F\{\phi(x, y, z)\}=\text{constant} \quad \text{and} \quad G\{\psi(x, y, z)\}=\text{constant},$$

where F and G are any functions. Alternatively, the integral can be written, on solving the two relations, as giving (say) y and z as functions of x. Each function involves two arbitrary constants and its analytical form is not unique.

The integral described can be shown, in diagrammatic terms, as a pair of systems of surfaces in space, i.e. as a system of curves in space. This curve system is perfectly definite and not affected by the lack of uniqueness of the analytical form of the integral. The nature of this diagrammatic representation can be examined further.

* See de la Vallée Poussin, *Cours d'analyse infinitésimale*, Vol. II (5th Ed. 1925), pp. 141 *et seq.*

Let P be any point on a curve in space, and draw the tangent line to the curve at P (Fig. 98). The " projection " of the curve and its tangent on to the plane Oxy gives a plane curve passing through a point Q with a definite tangent, Q being the foot of the perpendicular from P to Oxy. The plane curve can be regarded as the " shadow " of the spatial curve on the plane Oxy (taken as horizontal). The curve in space is represented by two functions, y and z as functions of x. The shadow plane curve is then represented by the first of these functions, y as a function of x, and the gradient of its tangent at Q is indicated by the derivative $\frac{dy}{dx}$. In the same way, the space curve has a "shadow" on the plane Oxz, a curve representing z as a function of x and with tangent gradient given by $\frac{dz}{dx}$. The curve in space and its tangent at P can thus be investigated by means of two " shadow " curves and their tangents.

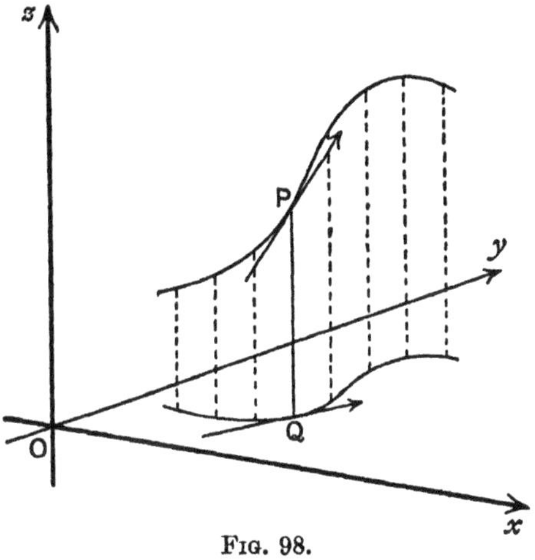

FIG. 98.

Now, the given differential equations

$$\frac{dy}{dx}=f(x, y, z) \quad \text{and} \quad \frac{dz}{dx}=g(x, y, z)$$

give values of $\frac{dy}{dx}$ and $\frac{dz}{dx}$ at each point $P(x, y, z)$ of space, and so determine two " shadow " tangents, one in the plane Oxy and the other in Oxz. These " shadows " can be combined to give a definite tangent line in space at P. The integral of the differential equations is then obtained by building up the tangent lines into a system of curves in space, one curve passing through each point of space.

The practical process of integrating a pair of differential equations of linear form is often extremely difficult.* The following examples illustrate the practical integration.

* See Piaggio, *Differential Equations* (1920), pp. 133 *et seq.*

Ex. 1. $\frac{dx}{x}=\frac{dy}{y}=\frac{dz}{z}$.

One of these equations can be written $\frac{dy}{dx}=\frac{y}{x}$ or $\frac{1}{y}\frac{dy}{dx}=\frac{1}{x}$.

On integrating this by ordinary methods (16.2 above), we have

$$y=ax \quad (a \text{ being an arbitrary constant}).$$

Similarly, from the other equation $\frac{dz}{dx}=\frac{z}{x}$, we find

$$z=bx \quad (b \text{ being an arbitrary constant}).$$

The integral of the pair of differential equations is thus

$$y=ax \quad \text{and} \quad z=bx,$$

shown by a system of straight lines in space, all passing through O.

Ex. 2. $\frac{dx}{2xz}=\frac{dy}{2yz}=\frac{-dz}{x^2+y^2}$.

It can be verified that these differential equations can be arranged in the form (amongst many others)

$$\frac{dx}{x}-\frac{dy}{y}=0 \quad \text{and} \quad x\,dx+y\,dy+2z\,dz=0.$$

We can now integrate at once :

$$\int\frac{dx}{x}-\int\frac{dy}{y}=\text{constant} \quad \text{and} \quad \int x\,dx+\int y\,dy+2\int z\,dz=\text{constant},$$

i.e. $\log x-\log y=\text{constant}$ and $\frac{1}{2}x^2+\frac{1}{2}y^2+z^2=\text{constant}$.

The integral of the equations can then be written

$$y=ax \quad \text{and} \quad x^2+y^2+2z^2=b,$$

where a and b are arbitrary constants. The two surface systems are now a system of planes ($y=ax$) and a system of surfaces known as "ellipsoids" ($x^2+y^2+2z^2=b$). The integral is shown by a set of plane sections of a system of "ellipsoids".

Ex. 3. $\frac{dx}{x^2+y^2+2yz}=\frac{-dy}{x^2+y^2+2xz}=\frac{dz}{2z(x-y)}$.

It can be verified, in this case, that the two relations

$$dx+dy+dz=0 \quad \text{and} \quad 2xz\,dx+2yz\,dy-(x^2+y^2)\,dz=0$$

hold. The integral of the first relation is found at once as

$$x+y+z=a$$

and that of the second (16.2 above) as

$$x^2+y^2=bz,$$

where a and b are arbitrary constants. The complete integral is thus obtained in the form already considered above.

16.5 Orthogonal curve and surface systems.

The linear differential equations in two variables

$$P\,dx+Q\,dy=0 \quad \text{and} \quad \frac{dx}{P}=\frac{dy}{Q}$$

give two tangent lines at each point (x, y) of the plane Oxy, one with gradient (to Ox) equal to $\left(-\frac{P}{Q}\right)$ and the other with gradient equal to $\frac{Q}{P}$. Since the product of these gradients is -1, the tangent lines are always perpendicular. Hence, the two curve systems representing the integrals of the equations must be such that, wherever

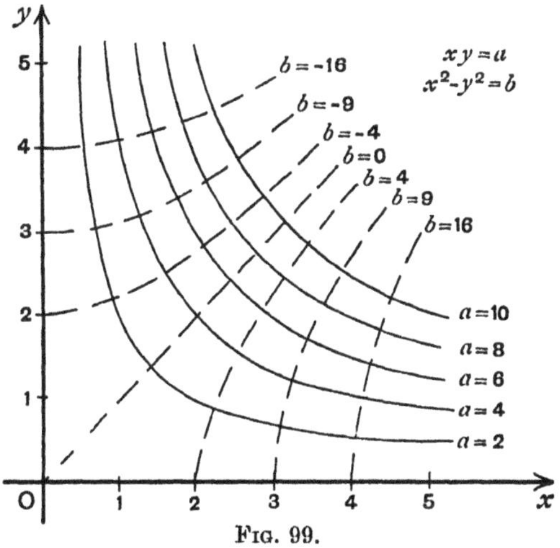

FIG. 99.

a curve of one system cuts a curve of the other system, the two curves are at right angles. Such curve systems are described as *orthogonal* systems. An example of differential equations of the orthogonal type is given by $y\,dx+x\,dy=0$ and $x\,dx-y\,dy=0$. It is easily seen that the integrals are respectively $xy=a$ and $x^2-y^2=b$, where a and b are arbitrary constants. The curve system, in each case, is a system of rectangular hyperbolas, one system being obtained from the other by rotation through 45°. Fig. 99 shows certain curves of the systems in the positive quadrant of Oxy and clearly indicates the orthogonal property of the systems.

The total differential equation in three variables

$$P\,dx+Q\,dy+R\,dz=0$$

gives a tangent plane at each point (x, y, z) of space, and its gradients are determined by the ratios of the functions P, Q and R. The pair of linear differential equations

$$\frac{dx}{P}=\frac{dy}{Q}=\frac{dz}{R}$$

gives a tangent line at each point of space and its direction is determined by the functions P, Q and R. It is a well-known result that this tangent line is perpendicular to the previous tangent plane at the same point. This can be verified by taking projections on the co-ordinate planes. The tangent plane contains a horizontal line (parallel to Oxy) with gradient referred to Ox equal to $\frac{dy}{dx}=-\frac{P}{Q}$. This projects directly on to the plane Oxy. The tangent line, as we have seen, has a projection or " shadow " on the Oxy plane with gradient to Ox equal to $\frac{dy}{dx}=\frac{Q}{P}$. These two lines in Oxy are perpendicular. A similar result holds for the projections on to one of the other co-ordinate planes, and it is then clear that the tangent line in space must be perpendicular to the corresponding tangent plane.

The total differential equation (if it is integrable) is shown by a system of surfaces in space and the pair of linear differential equations by a system of curves in space. These systems must here possess the property that, wherever one of the surfaces is cut by one of the curves, the intersection is at right angles. The surfaces and curves are said to form *orthogonal* systems. For example, the total differential equation $x\,dx+y\,dy+z\,dz=0$ has an integral which is the system of concentric spheres (centre at O) $x^2+y^2+z^2=c$. The pair of linear equations $\frac{dx}{x}=\frac{dy}{y}=\frac{dz}{z}$ has been found to integrate to $y=ax$ and $z=bx$, a system of straight lines all passing through O. The lines are the diameters of the spheres and the two systems are clearly orthogonal.

16.6 Other differential equations.

There are many types of differential equations more complicated than the simple linear equations so far considered. For example, $y\left(\frac{dy}{dx}\right)^2-2x\frac{dy}{dx}+y=0$ is a non-linear equation involving only a first-order derivative and its integral (see 16.1) is $y^2-2ax+a^2=0$, where a is an arbitrary constant. Again, by differentiating the relation $y=ax+\frac{b}{x}$ twice and eliminating the parameters a and b, we obtain the differential equation $x^2\frac{d^2y}{dx^2}+x\frac{dy}{dx}-y=0$ which involves derivatives of the first and second orders. The integral of the equation is the relation above and includes two arbitrary constants.* The practical methods of integrating such differential equations are not easy. The following general problem can be taken as an illustration.†

Ex. *The "dog and his master" problem.*

A man stands at a point O and his dog at a point A in an open field. The man begins to walk at a uniform pace along a path at right angles to OA, while the dog runs at a uniform pace (greater than that of the man) always in the direction pointing to the position of the man. When does the dog catch up with the man?

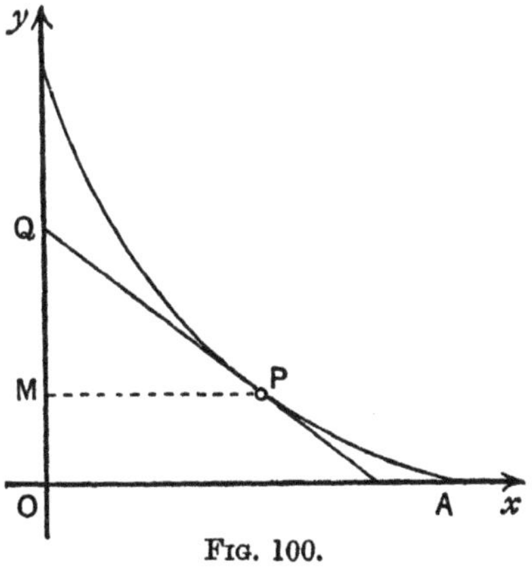

FIG. 100.

Take co-ordinate axes as shown in Fig. 100, Oy being the direction of the path taken by the man. Choose the unit of distance as OA, so that A is $(1, 0)$. Let the ratio of the man's speed to that of the dog be u, where $u<1$. After a certain time, suppose that the dog is at the point $P(x, y)$ and the man at Q on Oy. If the path taken by the dog is as shown with equation $y=f(x)$, then PQ is the tangent to the path at P with gradient

$$\frac{dy}{dx}=-\frac{MQ}{MP}=-\frac{OQ-y}{x}.$$

So

$$OQ=y-x\frac{dy}{dx}.$$

* This suggests an extension of the result of 16.1 above. It can, in fact, be shown that, if the nth derivative is that of the highest order in a differential equation in two variables, then the integral of the equation is a relation between the variables which includes n arbitrary constants.

† For an account of this and similar problems, see Osgood, *Advanced Calculus* (1925), p. 332 *et seq.*

Let $s = \text{arc } AP$ of the dog's path

$$= \frac{1}{u} \text{ (distance travelled by man)} = \frac{OQ}{u} = \frac{1}{u}\left(y - x\frac{dy}{dx}\right).$$

Now s is a decreasing function of x with negative derivative

$$\frac{ds}{dx} = \frac{1}{u}\left(\frac{dy}{dx} - x\frac{d^2y}{dx^2} - \frac{dy}{dx}\right) = -\frac{x}{u}\frac{d^2y}{dx^2}.$$

The increment ds, being compounded of perpendicular increments dx and dy along the dog's path, is given by $ds^2 = dx^2 + dy^2$.

So $$\left(\frac{ds}{dx}\right)^2 = 1 + \left(\frac{dy}{dx}\right)^2, \quad \text{i.e. } \frac{ds}{dx} = -\sqrt{1 + \left(\frac{dy}{dx}\right)^2},$$

taking the negative root since the derivative is negative. Hence,

$$-\sqrt{1 + \left(\frac{dy}{dx}\right)^2} = -\frac{x}{u}\frac{d^2y}{dx^2}.$$

The equation, $y = f(x)$, of the dog's path is thus the integral of

$$x\frac{d^2y}{dx^2} - u\sqrt{1 + \left(\frac{dy}{dx}\right)^2} = 0,$$

a differential equation of the second order. To integrate, write $p = \frac{dy}{dx}$ so that $\frac{dp}{dx} = \frac{d^2y}{dx^2}$. Then

$$x\frac{dp}{dx} - u\sqrt{1 + p^2} = 0,$$

i.e. $$\int \frac{dp}{\sqrt{1 + p^2}} = u\int \frac{dx}{x} + \text{constant}.$$

So $$\log(p + \sqrt{1 + p^2}) = u \log x + \text{const.}, \quad \text{i.e. } p + \sqrt{1 + p^2} = ax^u.$$

Since $p = \frac{dy}{dx} = $ (gradient of path of dog at A) $= 0$ when $x = 1$, we find, on substitution, that the constant $a = 1$. Hence,

$$x^u = \sqrt{1 + p^2} + p.$$

So $$x^{-u} = \frac{1}{\sqrt{1 + p^2} + p} = \frac{\sqrt{1 + p^2} - p}{(\sqrt{1 + p^2})^2 - p^2} = \sqrt{1 + p^2} - p$$

and $$x^u - x^{-u} = 2p.$$

Hence, $$\frac{dy}{dx} = \frac{1}{2}(x^u - x^{-u}),$$

and $$y = \frac{1}{2}\left(\int x^u\,dx - \int x^{-u}\,dx\right) + \text{constant} = \frac{1}{2}\left(\frac{x^{1+u}}{1+u} - \frac{x^{1-u}}{1-u}\right) + b.$$

At A, $y=0$ when $x=1$, and so we find $b=\frac{u}{1-u^2}$. The equation of the dog's path is thus

$$y=\frac{x^{1+u}}{2(1+u)}-\frac{x^{1-u}}{2(1-u)}+\frac{u}{1-u^2}.$$

The dog catches up with the man when $x=0$, in which case

$$OQ=y=\frac{u}{1-u^2}.$$

The man walks a certain multiple $\left(\frac{u}{1-u^2}\right)$ of the distance OA before the dog catches up with him. For example, if the dog's speed is twice that of the man ($u=\frac{1}{2}$), the equation of the dog's path is $y=\frac{1}{3}(\sqrt{x^3}-3\sqrt{x}+2)$ and the man walks only two-thirds of the distance OA before the dog reaches him.

Similar complications arise in equations which involve the differentials of three or more variables. Further, when the number of variables exceeds two, we can define equations of a rather different type involving partial derivatives instead of differentials. The simplest form of this type of equation occurs when the partial derivatives are related linearly as in the following example :

Ex. $x\frac{\partial z}{\partial x}+y\frac{\partial z}{\partial y}=z.$

The equation is integrated when we have expressed z explicitly in terms of x and y. The following trick is needed in the particular case considered. Write $u=\frac{z}{x}$ so that

$$du=\frac{1}{x^2}(x\,dz-z\,dx)=\frac{1}{x^2}\left\{x\left(\frac{\partial z}{\partial x}dx+\frac{\partial z}{\partial y}dy\right)-z\,dx\right\}$$

$$=\frac{1}{x^2}\left\{x\frac{\partial z}{\partial y}dy-\left(z-x\frac{\partial z}{\partial x}\right)dx\right\}.$$

From the original differential equation, we have $z-x\frac{\partial z}{\partial x}=y\frac{\partial z}{\partial y}$.

So $$du=\frac{\partial z}{\partial y}\frac{x\,dy-y\,dx}{x^2}=\frac{\partial z}{\partial y}d\left(\frac{y}{x}\right).$$

Hence the variation of u as a function of x and y depends only on the variation of $\frac{y}{x}$, and not on the separate variations of x and y. It follows that u is some function of $\frac{y}{x}$ only, and so

$$z=x\phi\left(\frac{y}{x}\right),$$

where ϕ is any function whatever. This is the integral of the given differential equation. It is to be noticed that the arbitrary element in the integral consists, not of an arbitrary *constant*, but of an arbitrary *function*. This is a special feature of equations in partial derivatives.

The form of the integral shows that z is a linear homogeneous function of x and y (see 12.8 above). The integral, in fact, provides the converse of Euler's Theorem :

If $x\frac{\partial z}{\partial x}+y\frac{\partial z}{\partial y}=z$ at all points (x, y), then z is some linear and homogeneous function of x and y.

As an application of the result, suppose that a good X is produced by using two factors A and B according to the production function $x=f(a, b)$ and that the product is always divided (without excess or deficit) according to the marginal productivity law. Hence, each unit of the factor A receives a return equal to the marginal product $\frac{\partial x}{\partial a}$ and similarly for the factor B. It follows that

$$a\frac{\partial x}{\partial a}+b\frac{\partial x}{\partial b}=x$$

for all combinations of the factors. This is only possible, as our present result shows, if the production function is linear and homogeneous.

16.7 Dynamic forms of demand and supply functions.

A market for a good X is made up of a competitive group of buyers and a competitive group of sellers. Under static conditions, assuming that the prices of other goods are fixed, the demand of the buyers can be represented by a function $x=\phi(p)$ and the supply of the seller by a function $x=f(p)$, where p denotes the uniform price of the good. A position of static equilibrium is said to be determined at a price which equates demand and supply : $\phi(p)=f(p)$. This is an equation in one variable with a definite number of solutions, each corresponding to a possible equilibrium price.

The static problem implicitly assumes that there is no "speculative" demand or supply and that there is no "time-lag" in production on the supply side. Our problem can be extended by dropping these assumptions in turn. The market price of X is now taken as varying over time, the current price being a function of t, the time (in years or some other unit) since a fixed base period. We have to find, under various assumed conditions, the form of the price function $p(t)$. The static problem gives a definite price and equilibrium position; the extended "dynamic" problem must give the course of prices over time and the path of a moving equilibrium.

The demand of the market (as a rate per unit of time) at the current time t is taken as depending on the price $p(t)$ and on certain speculative elements, including the buyers' view of the future course of prices. We consider here only the simple case in which the buyers' "foresight" is confined to an allowance for the rate at which the price is changing at the time concerned, demand depending on whether prices are rising or falling and at what rate they are changing. We now have demand as dependent on $p(t)$ and its derivative $p'(t)$:

$$x = \phi\{p(t), p'(t)\}, \quad \text{the } \textit{dynamic demand function.}$$

In the same way, assuming the same kind of speculation (but no time-lag) on the supply side, we can write

$$x = f\{p(t), p'(t)\}, \quad \text{the } \textit{dynamic supply function.}$$

For equilibrium over time, the price function $p(t)$ must be such that demand and supply are equal at all times t:

$$\phi\{p(t), p'(t)\} = f\{p(t), p'(t)\}.$$

This is a differential equation including the first derivative of p with respect to t, and its integral gives p as a function of t involving one arbitrary constant. The latter can be determined in terms of the price p_0 ruling in the base period $(t=0)$. Hence, given the initial price, the course of prices over time is determined.

The problem can be illustrated by taking the particular case where the demand and supply functions are linear. Under static conditions, we write demand as $x = ap + b$ and supply as $x = \alpha p + \beta$. In the "normal" case, the constants a and β are negative and b and α are positive. For equilibrium, demand equals supply, i.e.

$$ap + b = \alpha p + \beta,$$

giving one equilibrium price $\bar{p} = \dfrac{b - \beta}{\alpha - a}$.

In the simple speculative case considered, we write:

Demand $\qquad x = ap(t) + b + cp'(t).$

Supply $\qquad x = \alpha p(t) + \beta + \gamma p'(t),$

where a, b, α and β are constants with the signs indicated above and where c and γ are new constants which can be taken as positive. This implies that a rising price, other things being equal, stimulates both demand and production. It is possible, however, to take γ as negative if a rising price causes sellers to hold back their supplies. For a moving equilibrium, the price function $p(t)$ must be such that

$$ap(t)+b+cp'(t)=\alpha p(t)+\beta+\gamma p'(t),$$

i.e. $$p'(t)=\frac{\alpha-a}{c-\gamma}\{p(t)-\bar{p}\} \quad \text{where} \quad \bar{p}=\frac{b-\beta}{\alpha-a}.$$

Write $$q(t)=p(t)-\bar{p} \quad \text{and} \quad \lambda=\frac{\alpha-a}{c-\gamma}.$$

Then $$q'(t)=\lambda q(t), \quad \text{i.e.} \quad \frac{1}{q}\frac{dq}{dt}=\lambda.$$

This familiar differential equation has integral $q=Ae^{\lambda t}$.

So $$p(t)=\bar{p}+Ae^{\lambda t}.$$

Putting $t=0$ and $p=p_0$ (the initial price), we find $A=p_0-\bar{p}$. The course of prices over time is thus uniquely given by

$$p(t)=\bar{p}+(p_0-\bar{p})e^{\lambda t},$$

where $\bar{p}=\dfrac{b-\beta}{\alpha-a}$, $\lambda=\dfrac{\alpha-a}{c-\gamma}$ and p_0 is the initial price.

The important term in the expression for $p(t)$ is $e^{\lambda t}$. This term increases and tends to infinity, or decreases and tends to zero, according as λ is positive or negative. Since we have taken $(\alpha-a)$ as positive, the sign of λ is governed by that of $(c-\gamma)$. Hence,

(1) If $\gamma<c$, λ is positive and the price steadily diverges from the static equilibrium value $\bar{p}$ as time goes on.

(2) If $\gamma>c$, λ is negative and the price steadily approaches the static equilibrium value $\bar{p}$ as time goes on.

A stable course of prices is thus only possible if $\gamma>c$, i.e. if the speculative element in supply is stronger than that in demand.

If, in addition to the speculative element, there is a time-lag of constant length θ years in production, then the supply of X is dependent on the price and the price changes at the earlier time $(t-\theta)$ rather than at the current time t. The current supply was arranged for θ years ago at the beginning of the production process and was subject to the price then ruling. The general condition for moving equilibrium is now

$$\phi\{p(t),\ p'(t)\} = f\{p(t-\theta),\ p'(t-\theta)\}$$

and with linear demand and supply functions

$$ap(t) + b + cp'(t) = \alpha p(t-\theta) + \beta + \gamma p'(t-\theta).$$

Such differential equations can be integrated only with the aid of "complex numbers", entities designed for the description of variables which oscillate in value. But the general solution is found to be very similar to that of one special case which we can easily solve.*

When there is a production time-lag but no speculation, the equation for $p(t)$ ceases to be differential and takes the form

$$ap(t) + b = \alpha p(t-\theta) + \beta,$$

i.e. $$p(t) - \bar{p} = \frac{\alpha}{a}\{p(t-\theta) - \bar{p}\}.$$

Here, linear demand and supply functions are assumed and $\bar{p}$ has the same value as before. If the course of prices over the initial period from $t=0$ to $t=\theta$ is given, then the equation above gives the price $p(t)$ at any subsequent time t. In fact, measuring all prices from the level $\bar{p}$, the current price is just a constant multiple of the price θ years ago. The multiple is the negative number $\frac{\alpha}{a}$. There are three cases to consider. Suppose, for simplicity, that the prices in the period $(0, \theta)$ oscillate about $\bar{p}$, being as much below $\bar{p}$ at $t=\theta$ as above $\bar{p}$ at $t=0$. Then:

(1) If a and α are numerically equal, then the price variation in the period $(\theta, 2\theta)$ completes the oscillation of the period $(0, \theta)$, the price at $t=2\theta$ being the same as that at $t=0$. In successive periods of 2θ years, the same cycle of prices is repeated. The course of price over time is illustrated in diagram A of Fig. 101.

(2) If $a > \alpha$ numerically, then the ratio $\frac{\alpha}{a}$ is negative and numerically less than unity. The price again oscillates, describing a cycle

* On the problems considered here, see Tinbergen, *Utilisation des équations fonctionelles*, Econometrica, 1933, and Theis, *A quantitative theory of industrial fluctuations*, Journal of Political Economy, 1933. The first work in this field is due to Roos and Evans.

in every period of 2θ years, but now the extent of the oscillation diminishes as time goes on. The oscillations are "damped" as shown in diagram B of Fig. 101.

(3) If $a < \alpha$ numerically, the case is similar to the previous one

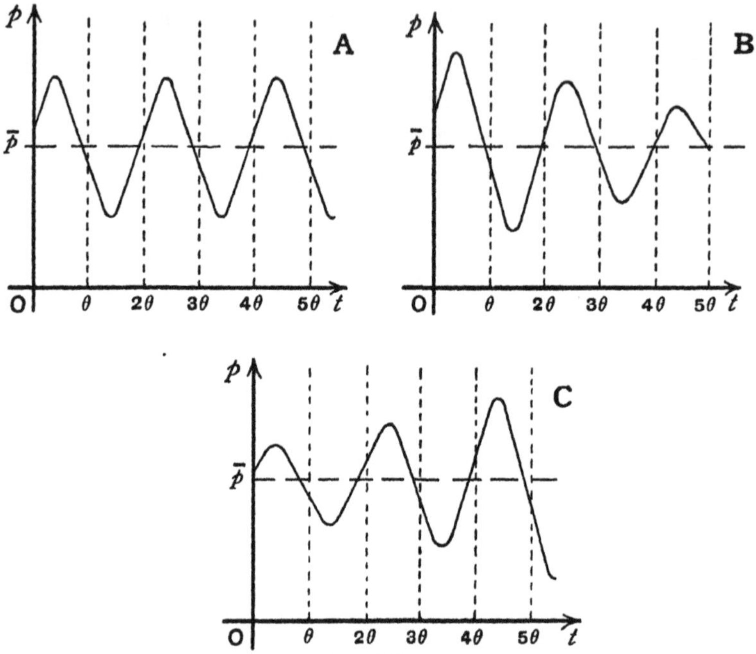

Fig. 101.

except that the oscillations increase in amplitude as time goes on. This case of "explosive" oscillations in price is illustrated by diagram C of Fig. 101.

16.8 The general theory of consumers' choice.

The analysis of the demand of an individual for consumers' goods has proceeded, so far, on the assumption that the individual's scale of preferences can be represented by an indifference map, i.e. by a utility function (measurable or not). This assumption can now be scrutinised.

Suppose that x and y are the individual's current purchases of two goods X and Y. It is now assumed simply that the individual has a definite scale of preferences for *small* changes in these purchases. He can distinguish that small increase Δy in his purchase of

Y which just compensates (i.e. leaves him indifferent) for a given small decrease $(-\Delta x)$ in his purchase of X. Larger increases than Δy are preferred by the individual whereas smaller increases leave him in a position worse than before. Hence, for an indifferent change, Δy is a definite multiple of $(-\Delta x)$. Taking the limit for smaller and smaller changes, the assumption defines the differential dy as a definite multiple of $(-dx)$. The multiple (which is positive) is defined for each initial set of purchases (x, y) we care to select and varies as these purchases vary. The multiple is, in fact, a function of the variables x and y and we can write it as $R(x, y)$. Our assumption thus gives the differential equation for indifferent changes :

$$-\frac{dy}{dx}=R(x, y) \quad \text{or} \quad R(x, y)\,dx+dy=0.$$

The function R takes only positive values. It is further assumed that the scale of preferences is continuous in the sense that R has continuous partial derivatives.

The differential equation above can be integrated in all cases (16.3 above) to give the general integral

$$F\{\phi(x, y)\}=\text{constant}.$$

Here F is any function and $\phi(x, y)$ is a function such that

$$\frac{\phi_x}{\phi_y}=-\frac{dy}{dx}=R(x, y).$$

In diagrammatic terms, our assumption corresponds to the definition of a definite " indifference direction " given by $-\frac{dy}{dx}=R(x, y)$ at each point (x, y) of the plane Oxy. The indifference directions can then be integrated to give a whole system of indifference curves defined by $F\{\phi(x, y)\}=\text{constant}$. The situation is exactly as before but based on a different assumption. We do not assume a complete scale of preferences but a scale for small changes from any given set of purchases. Finally, the utility function index

$$u=F\{\phi(x, y)\},$$

with a constant value along any one indifference curve, follows from our new assumption.

From 13.8 above, the function $R(x, y)$ here defined is to be interpreted as the marginal rate of substitution of Y for X. Previously, we deduced this concept from an assumed utility function. The position is now reversed; the marginal rate of substitution is the fundamental concept and the indifference map and utility function are deduced from it. This is an important change of approach.

The general case where any number of goods appear in the consumer's budget is sufficiently represented by the case of three goods X, Y and Z. The assumption to be adopted asserts that the individual has a scale of preferences for small changes from a given set of purchases (x, y, z) so that he can distinguish that small increase Δx in the purchase of X which compensates for given small decreases $(-\Delta y)$ and $(-\Delta z)$ in Y and Z. The individual prefers a larger increase than Δx, but rejects a smaller increase. Taking smaller and smaller changes and writing differentials for the increments, we have dx as a multiple of $(-dy)$ plus a multiple of $(-dz)$. The multiples are positive and functions of the purchases (x, y, z). So

$$dx = R_1(x, y, z)(-dy) + R_2(x, y, z)(-dz)$$

or

$$dx + R_1(x, y, z)dy + R_2(x, y, z)dz = 0$$

is the differential equation for indifferent changes in purchases. The functions R_1 and R_2 take positive values and are assumed to have continuous partial derivatives.

The interpretation of R_1 and R_2 is not difficult. The ratio of compensating increments in the purchases of X and Y when the purchase of Z remains unaltered can be written

$$\left(-\frac{dx}{dy}\right)_{z \text{ constant}} = R_1(x, y, z).$$

The function R_1 thus represents, at the purchases (x, y, z), the *marginal rate of substitution* of X for Y. In the same way R_2 represents the marginal rate of substitution of X for Z. Our assumption is equivalent to the definition of two marginal rates of substitution as describing the individual's scale of preferences. In diagrammatic terms, we have, at each point of $Oxyz$ space, an indifference plane given by $dx + R_1 dy + R_2 dz = 0$ and with gradients $(-R_1)$ and $(-R_2)$ in the Oy and Oz directions. The plane includes all the indifferent directions of change from the purchases (x, y, z).

The differential equation $dx + R_1 dy + R_2 dz = 0$ need not possess an integral (16.3 above). In general, therefore, we cannot integrate the set of indifference planes into a complete set of indifference surfaces, and we cannot assume that any utility function exists. The assumption of a scale of preferences for small changes of purchases does *not* imply that a complete scale of preferences exists. The consumer can discriminate between small changes from his established purchases but need not be able to discriminate between widely different sets of purchases. The assumption of the marginal rates of substitution as fundamental has shown up a new possibility.

The differential equation is integrable only in the special case where R_1 and R_2 satisfy the integrability condition (16.3 above):

$$\left(R_1\frac{\partial R_2}{\partial x}-R_2\frac{\partial R_1}{\partial x}\right)+\left(\frac{\partial R_1}{\partial z}-\frac{\partial R_2}{\partial y}\right)=0.$$

In this case we can write an integral in the form

$$F\{\phi(x\,.\,y\,.\,z)\}=\text{constant},$$

where F is any function and $\phi(x, y, z)$ a function such that

$$\frac{\phi_y}{\phi_x}=R_1(x, y, z) \quad \text{and} \quad \frac{\phi_z}{\phi_x}=R_2(x, y, z).$$

The integral is shown by a system of indifference surfaces touching the given indifference planes and the function $u=F\{\phi(x, y, z)\}$ can be taken as a function index of utility. The given marginal rates of substitution are equal to the ratios of the partial derivatives of u in any one of its forms. It must be remembered now that the indifference map and utility function are deduced from the marginal rates of substitution and not conversely.

In developing the theory of individual demand for three or more goods, therefore, we proceed from the assumption of marginal rates of substitution. As a special case, we can then add the integrability assumption, take an indifference map and a function index of utility and deduce certain results which hold only in this case. We have a general theory and a special (integrability case) theory of demand, a distinction which will be made clear later (19.7-8 below).

The results of 16.5 above are of direct application to the present problem. The differential equations $\frac{dx}{1}=\frac{dy}{R_1}=\frac{dz}{R_2}$ define a tangent direction at the point (x, y, z) in space, a direction perpendicular to the indifference plane given by $dx+R_1\,dy+R_2\,dz=0$. If the individual increases his purchases from the given set (x, y, z) in the way indicated by this direction, these are the increases he prefers above all others. The direction can thus be termed the individual's *preference direction* at (x, y, z).* The values of R_1 and R_2 at any point thus indicate, not only the marginal rates of substitution for *indifferent* changes, but also the ratios of the increases in purchases which are *most preferred* by the individual.

Since R_1 and R_2 have continuous partial derivatives, the differential equations for the preference direction can be integrated to give a system of curves in space, the *lines of preference* of the individual (16.5 above). One such curve passes through each point of space and indicates the most preferred direction of change from the point. The complete lines of preference exist even when there is no indifference map. But, in the integrability case where indifference surfaces exist, it follows that the lines of preference and the indifference surfaces form orthogonal systems. A line of preference cuts an indifference surface at right angles and indicates the "quickest" way from one indifference surface to the next, i.e. the direction in which the individual's utility level is increased most rapidly.

EXAMPLES XVI

Differential equations

1. Draw a graph showing a sufficient number of the tangent directions $\frac{dy}{dx}=\frac{x}{y}$ to indicate the curve system which is the integral of the equation. Draw a similar graph of $\frac{dy}{dx}=-\frac{y}{x}$. Check by integrating the equations by the "variables separate" method (see 16.5 above).

2. Illustrate graphically the differential equation $\frac{dy}{dx}=\frac{y(1-x)}{x}$ and its integral. From the equation, prove that each curve of the integral system has a maximum point at the same value of x.

3. If $y=1+\frac{a}{x-1}$, show that $(y-1)dx+(x-1)dy=0$. Conversely, show that the differential equation has the integral shown (using the "variables separate" method). What are the curves represented?

* Preference directions were introduced by Edgeworth, *Mathematical Psychics* (1881, reprinted 1932) and by Irving Fisher, *Mathematical investigations in the theory of value and prices* (1892, reprinted 1925). In the case of two goods, indifference curves are defined by the relation $\frac{dy}{dx}=-R(x, y)$ and preference lines by the relation $\frac{dy}{dx}=\frac{1}{R(x, y)}$. We have, in all cases, orthogonal systems of indifference curves and lines of preference. Either system defines the individual's scale of preferences.

4. Show that $y = a(x+a)$ is a certain system of straight lines and eliminate a to obtain the differential equation of the system.

5. Obtain a differential equation of the second order, representing the curve system with equation $ax^2 + by^2 = 1$ (a and b parameters).

6. The equation $y = ax^2 + b$ represents all parabolas with axis along Oy. Find a differential equation to represent the curve system. Of what order is the equation? Consider, in the same way, the equation $y = ax^2 + bx + c$, representing all parabolas with axis parallel to Oy.

7. Derive linear differential equations by eliminating a from the relation $x + y = ae^x$ and from the relation $z = e^{a\frac{x}{y}}$.

8. Find an integral of each of the differential equations:

$$x^2\,dx + y^2\,dy = 0\ ;\quad 2y\,dx + x\,dy = 0\ ;\quad y\,dx - (x^2 - 1)\,dy = 0\ ;\quad 2\frac{dy}{dx} + 3\sqrt[3]{y} = 0.$$

9. If α is a given constant, obtain an integral of

$$\frac{dy}{dx} = \frac{\alpha - x}{x}y.$$

Show that the curve system obtained can be represented in either of the forms $ye^x x^{-\alpha} = \text{constant}$ and $x - \alpha \log x + \log y = \text{constant}$.

10. Show that $y = ae^{x^n}$ is an integral of $\dfrac{dy}{dx} = nyx^{n-1}$.

11. Show that the integral of $(y^2 - 1)dx - 2dy = 0$ is $y = \dfrac{1 + ae^x}{1 - ae^x}$.

12. Integrate $(2x + y)\,dx + (x + 2y)\,dy = 0$ by showing that the left-hand side of the equation is an exact differential.

13. Show that $y\,dx + (2x + y)\,dy = 0$ becomes exact on multiplying through by $3y$. Hence integrate the equation.

14. Find the partial derivatives of $\dfrac{x^3 + y^3}{3xy}$ and hence solve the differential equation

$$y(2x^3 - y^3)\,dx - x(x^3 - 2y^3)\,dy = 0.$$

15. Integrate $\dfrac{dx}{x} + \dfrac{dy}{y} + \dfrac{dz}{z} = 0.$

16. Show that a differential equation of the form

$$\phi(x)\,dx + \psi(y)\,dy + \chi(z)\,dz = 0$$

is always integrable. How is the integral obtained?

17. Which of the differential equations

$$yz\,dx + xz\,dy + xy\,dz = 0\ ;\quad dx - z\,dy - 2y\,dz = 0\ ;$$

$$z\,dx + (x + y)\,dy + dz = 0 \text{ and } y(y+z)\,dx + x(x - z)\,dy + x(x + y)\,dz = 0$$

are integrable and which not?

18. Integrate $2x\,dx + z\,dy + y\,dz = 0$ by showing that the left-hand side is an exact differential.

19. Find the partial derivatives of $\dfrac{x(y+z)}{x+y+z}$ with respect to x, y and z. Hence integrate $(y + z)^2\,dx + x^2\,dy + x^2\,dz = 0$.

20. Find the integral of the simultaneous differential equations

$$x\,dx = y\,dy = z\,dz.$$

Examine this integral and that of example 15 above in diagrammatic terms and show that the curve and surface systems concerned are orthogonal.

Economic applications of differential equations

21. It is known that the elasticity of a demand law $x = \phi(p)$ is of the form $(a - bp)$ where a and b are given constants. By integrating a differential equation, show that the law is $x = p^a e^{-b(p+c)}$, c being an arbitrary constant.

22. If Π is the total cost of an output x, it is known that marginal cost $\left(\frac{d\Pi}{dx}\right)$ always equals average cost $\left(\frac{\Pi}{x}\right)$. Integrate and show that Π is a fixed multiple of x, i.e. that average cost is constant.

23. If y is the number of incomes of £x, it is found that y decreases as x increases according to the law $\frac{dy}{dx} = -m\frac{y}{x}$ where m is a given constant. Integrate and show that the dependence of y on x is $y = \frac{a}{x^m}$.

24. The price of tea, initially 3s. per lb., is $p(t)$ pence per lb. after t weeks. The demand is $x = 120 - 2p + 5\frac{dp}{dt}$ and the supply $x = 3p - 30 + 50\frac{dp}{dt}$ thousand lbs. per week. Show that, for demand always to equal supply, the price of tea must vary over time according to the law $p = 30 + 6e^{-\frac{1}{9}t}$. What are the prices after 10 and after 50 weeks? Draw a graph to show the approach of the price to the equilibrium value of 2s. 6d. per lb.

25. The demand (per unit of time) for a good is $x = ap + b$ and the supply $\xi = \alpha p + \beta$ where p is the price. If there is (e.g.) an excess of demand over supply in any period, it is assumed that the price changes to decrease the excess at a rate proportional to the excess.

If $p(t)$ is the price at time t, show that the assumption implies that

$$\frac{d}{dt}(x - \xi) = -k(x - \xi)$$

where k is some positive constant. Deduce that $p(t)$ is given by $\frac{dp}{dt} + k(p - \bar{p}) = 0$ where $\bar{p} = \frac{b - \beta}{\alpha - a}$. If p_0 is the initial price ($t = 0$), show that $p(t) = \bar{p} + (p_0 - \bar{p})e^{-kt}$ and that the price tends to the equilibrium value p. See Evans, *Mathematical Introduction to Economics* (1930), p. 48.

26. An individual's preference scale for two goods X and Y is defined by the marginal rate of substitution of Y for X, $R = \frac{x - a}{y - b}$. Show that $u = (x - a)^2 + (y - b)^2$ is one form of the utility function.

27. If $R = \frac{\alpha}{\beta}\frac{y + b}{x + a}$ is the marginal rate of substitution of Y for X, show that one form of the individual's utility function is $u = (x + a)^\alpha (y + b)^\beta$ where a, b, α and β are given constants.

28. The marginal rate of substitution of Y for X is $R=\dfrac{ax+by+c}{\alpha x+\beta y+\gamma}$ where the coefficients are given constants. If $\alpha=b$, show that one form of the utility function of the individual is quadratic in x and y. If $\alpha\neq b$, show that the utility function need not be of this form. Illustrate by considering the particular cases of the two previous examples.

29. Three goods X, Y and Z are said to be "independent" in consumption if the ratios of the individual's marginal rates of substitution $1 : R_1 : R_2$ are of the form $\phi(x) : \psi(y) : \chi(z)$, where ϕ, ψ and χ are functions of the single variables named. Show that the integrability condition is always satisfied and indicate the form of the utility function. Illustrate when

$$1 : R_1 : R_2=(x-a) : (y-b) : (z-c).$$

30. The marginal rates of substitution of X for Y and Z are known to be ratios of linear expressions in x, y and z, so that

$$(a_1x+b_1y+c_1z+d_1)dx+(a_2x+b_2y+c_2z+d_2)dy+(a_3x+b_3y+c_3z+d_3)dz=0$$

is the differential equation of the indifference plane. What is the condition for the existence of indifference surfaces? Show that the condition is satisfied if $b_1=a_2$, $c_2=b_3$ and $a_3=c_1$. In this case, show that the utility function can be written as a quadratic in x, y and z.

CHAPTER XVII

EXPANSIONS, TAYLOR'S SERIES AND HIGHER ORDER DIFFERENTIALS

17.1 Limits and infinite series.

WE have seen (4.1 above) that there are two broad classes of number sequences. A sequence either tends or does not tend to a finite limit, the second case including sequences tending to infinity and sequences which oscillate. For example,

$$\tfrac{1}{2},\ \tfrac{3}{4},\ \tfrac{7}{8},\ \tfrac{15}{16},\ \ldots \to 1,$$

$$1,\ 4,\ 9,\ 16,\ \ldots \to \infty .$$

The nth members of these sequences are $\left(1-\frac{1}{2^n}\right)$ and n^2 respectively and the results can be written in the alternative forms

$$1-\frac{1}{2^n} \to 1 \text{ as } n \to \infty \quad \text{and} \quad n^2 \to \infty \text{ as } n \to \infty .$$

Here n represents any positive integer.

A number sequence can be treated in a slightly different way by forming from it a second sequence, a sequence of differences. The nth member of the new sequence is the difference between the nth and the $(n-1)$th members of the old sequence. The new sequence is written down with plus signs between successive members (to indicate our intention to add them together), continued indefinitely and called an *infinite series*. For example, the two sequences above give rise to the infinite series

$$\frac{1}{2}+\frac{1}{4}+\frac{1}{8}+\ldots+\frac{1}{2^n}+\ldots ,$$

$$1+3+5+\ldots+(2n-1)+\ldots ,$$

where the expression in n represents the nth term of the series in each case (n being a positive integer). From the definition, the *sum of n terms* of the infinite series equals the nth member of the original sequence. So, for any positive integral value of n,

$$\frac{1}{2}+\frac{1}{4}+\frac{1}{8}+\ldots+\frac{1}{2^n}=1-\frac{1}{2^n},$$

$$1+3+5+\ldots+(2n-1)=n^2.$$

In general, if $f(n)$ is the nth member of a given sequence, then the infinite series of differences has sum of n terms equal to $f(n)$.

If the original sequence has a finite limit, then the sum of n terms of the derived series has the same limit. The series is said to be *convergent* and the limit of the sum is the *sum to infinity*. If the original sequence has no limit, neither has the sum of n terms of the series. The series is said to be *divergent* and there is no sum to infinity. For example, the series first written above is convergent with unity as the sum to infinity; the second series is divergent, the sum of n terms tending to infinity with n.

The limit of any sequence (if it exists) can thus be written as the sum to infinity of a convergent infinite series. Any member of the sequence and the sum of any number of terms of the series can then serve as an *approximate* value of the limit, a fact which provides one of the main uses of a convergent series. If λ is the sum to infinity of a certain series, then λ is given by the limit of the sum of n terms and found approximately by adding together a sufficient number of terms of the series. This may be the simplest way of getting an approximate value of λ if the latter is a complicated expression. Again, the sum to infinity provides an approximate value of the sum of a large number of terms of the series. The sum of (say) 100 terms of the series $(\frac{1}{2}+\frac{1}{4}+\frac{1}{8}+\ldots)$ is not easily calculated. But we know that the sum to infinity is 1 and this can be taken as the approximate value of the sum of the 100 terms.

Conversely, if we are given a certain infinite series, we can form a sequence of sums, the nth member of the sequence being the sum of n terms of the series. By definition, the given series is convergent if the derived sequence tends to a finite limit.

Two important cases of series can be considered in detail. A *geometric series* consists of terms which increase (or decrease) from one term to the next by a given factor, called the *common ratio.* In general, if the first term is a and the common ratio r, the series is

$$a+ar+ar^2+ar^3+\ldots+ar^{n-1}+\ldots.$$

It is shown, in elementary algebra, that the sum of n terms is

$$S_n=a\,\frac{1-r^n}{1-r}\quad(n=1,\ 2,\ 3,\ \ldots).$$

The derived sequence is thus

$$a,\quad a\frac{1-r^2}{1-r},\quad a\frac{1-r^3}{1-r},\quad \ldots\quad a\frac{1-r^n}{1-r},\quad \ldots.$$

If r is numerically less than unity, the expression r^n becomes smaller and tends to zero as n increases indefinitely. Hence,

$$S_n=\frac{a}{1-r}-\frac{a}{1-r}r^n\to\frac{a}{1-r}\quad\text{as } n\to\infty.$$

The geometric series is convergent with sum to infinity $\frac{a}{1-r}$, *provided that r is numerically less than one.* In other cases, the expression r^n does not tend to zero as n increases and the geometric series is not convergent. The sum of n terms either tends to infinity or oscillates.

Ex. 1. $\frac{1}{2}+\frac{1}{4}+\frac{1}{8}+\ldots+\frac{1}{2^n}+\ldots.$

Here $a=\frac{1}{2}$, $r=\frac{1}{2}$ and $S_n=\frac{1}{2}\frac{1-(\frac{1}{2})^n}{1-\frac{1}{2}}=1-\frac{1}{2^n}\to 1$ as $n\to\infty$.

The series is convergent with sum to infinity 1 (as obtained above).

Ex. 2. The recurring decimal $0\cdot\dot{1}$ is a short way of writing the sum to infinity of the series $\frac{1}{10}+\frac{1}{100}+\frac{1}{1000}+\ldots.$

This is a convergent geometric series with $a=r=\frac{1}{10}$. The sum to infinity is $\frac{a}{1-r}=\frac{1}{9}$. Hence, the decimal form of the fraction $\frac{1}{9}$ is $0\cdot\dot{1}$.

Ex. 3. An income stream of £a starts next year and continues for n years. Reckoning interest yearly at $100r$ per cent. per year, the present value of the stream (see 9·7 above) is

$$\frac{a}{1+r}+\frac{a}{(1+r)^2}+\frac{a}{(1+r)^3}+\ldots+\frac{a}{(1+r)^n}$$

$$=\frac{a}{1+r}\,\frac{1-\left(\dfrac{1}{1+r}\right)^n}{1-\dfrac{1}{1+r}}=\frac{a}{r}\left\{1-\left(\frac{1}{1+r}\right)^n\right\}.$$

If the income stream goes on for ever, the present value is obtained by letting $n \to \infty$, i.e. it is the sum to infinity of the convergent geometric series above. Hence, the present value of £a per year for ever, at interest at $100r$ per cent. reckoned yearly, is £$\frac{a}{r}$. For example, the present value of £3 per year for ever is £100 if interest is reckoned yearly at 3 per cent. per year.

An *arithmetic series* consists of terms which increase (or decrease) from one term to the next by a given amount, called the *common difference*. If the first term is a and the common difference d, the series appears

$$a + (a+d) + (a+2d) + \dots + (a + \overline{n-1}\,d) + \dots .$$

It is easily shown that the sum of n terms is

$$S_n = \tfrac{1}{2}n\{2a + (n-1)d\}.$$

Since $S_n \to \pm\infty$ as $n \to \infty$, the arithmetic series is not convergent and never possesses a sum to infinity.

Ex. 1. $1 + 2 + 3 + \dots + n + \dots .$

Here $S_n = \frac{1}{2}n(n+1)$ the sum of the first n positive integers.

Ex. 2. $1 + 3 + 5 + \dots + (2n-1) + \dots .$

Here $S_n = \frac{1}{2}n\{2 + (n-1)2\} = n^2$ the sum of the first n odd positive integers.

17.2 The expansion of a function of one variable (Taylor's series).

By a process of long division, we find

$$\frac{1}{1-x} = 1 + x + x^2 + \dots + x^n + \frac{x^{n+1}}{1-x}.$$

If x has a value numerically less than one, the term containing x^{n+1} decreases and tends to zero as n increases indefinitely. Hence,

$$\frac{1}{1-x} = 1 + x + x^2 + \dots + x^n + \dots ,$$

if x is numerically less than one. We have thus expressed $\frac{1}{1-x}$ as the sum to infinity of a convergent infinite series. The result can be checked in this case. The series written is a geometric series with common ratio x (<1); the series is convergent with sum to infinity equal to $\frac{1}{1-x}$ (by the formula of 17.1).

The advantage of the series expression for $\frac{1}{1-x}$ is clear. If x is given a small value, then, instead of evaluating $\frac{1}{1-x}$ by long division, we find its approximate value (to any required degree of accuracy) by adding a sufficient number of terms of the series. For example, if $x=\frac{1}{30}$,

$$\frac{30}{29}=\frac{1}{1-x}=1+x+x^2+\ldots=1+0{\cdot}03333+0{\cdot}00111+0{\cdot}00004+\ldots$$
$$=1{\cdot}03448.$$

Hence to four decimal places, $\frac{30}{29}=1{\cdot}0345$, as can be verified by long division. The fifth term of the series is too small to affect the approximation and is not included. Again, if x is a small number, then to a degree of accuracy represented by x^3

$$\frac{1}{1-x}=1+x+x^2,$$

i.e. for approximate purposes the expression $\frac{1}{1-x}$ can be replaced by the simpler quadratic expression $(1+x+x^2)$.

The problem can be generalised. Given a single-valued function $f(x)$, it is desired to find a convergent infinite series, the successive terms of which involve increasing powers of x, with $f(x)$ as its sum to infinity. In symbols, we wish to write

$$f(x)=a_0+a_1x+a_2x^2+a_3x^3+\ldots \text{ (to infinity)},$$

where the a's are certain constants. The problem is a little complex since we have first to see whether the process indicated can be carried out at all, then to determine for what values of x it is valid and finally to allot the appropriate values to the constant coefficients.

If the process is possible, the series obtained is called the *expansion* of $f(x)$ as a power series, i.e. as a series in ascending powers of x. The particular case above indicates the uses of the expansion when obtained. For a definite value of x, the *accurate* value of $f(x)$ is obtainable only by substituting directly in $f(x)$ or by finding the limit of the sum of n terms of the series. But, if *approximate* results suffice, we need only add a limited number of terms of the expansion. It is possible, for example, to replace $f(x)$ approximately by a quadratic or cubic in x when x is small, a great simplification if $f(x)$ is at all complicated.*

It is assumed, in the following, that $f(x)$ has continuous derivatives up to any desired order at the points concerned. We start with a simple case where the problem is really solved in advance but which will indicate the more general method. Suppose that

$$f(x)=a_0+a_1x+a_2x^2+a_3x^3+\ldots+a_nx^n,$$

a given polynomial of the nth degree in which the coefficients have known values. By successive derivation,

$$f'(x)=a_1+2a_2x+3a_3x^2+\ldots+na_nx^{n-1},$$
$$f''(x)=2a_2+6a_3x+\ldots+n(n-1)a_nx^{n-2},$$
$$f'''(x)=6a_3+\ldots+n(n-1)(n-2)a_nx^{n-3},$$
$$\ldots\ldots\ldots\ldots\ldots\ldots\ldots\ldots\ldots\ldots\ldots\ldots$$
$$f^{(n)}(x)=n(n-1)(n-2)\ldots3\,.\,2\,.\,1a_n.$$

Putting $x=0$,

$$f(0)=a_0,\quad f'(0)=a_1,\quad f''(0)=2a_2,\quad f'''(0)=6a_3,\ \ldots,$$

and
$$f^{(n)}(0)=n(n-1)(n-2)\ldots3\,.\,2\,.\,1a_n.$$

A convenient algebraic notation can be introduced here. The product of the positive integers from 1 to n inclusive is called n *factorial* and written $\lfloor\underline{n}$. Hence, from the results above,

$$a_0=f(0),\quad a_1=f'(0),\quad a_2=\frac{1}{\lfloor\underline{2}}f''(0),\quad a_3=\frac{1}{\lfloor\underline{3}}f'''(0),\ \ldots\ a_n=\frac{1}{\lfloor\underline{n}}f^{(n)}(0).$$

The original polynomial can thus be written in the form

$$f(x)=f(0)+f'(0)x+f''(0)\frac{x^2}{\lfloor\underline{2}}+f'''(0)\frac{x^3}{\lfloor\underline{3}}+\ldots+f^{(n)}(0)\frac{x^n}{\lfloor\underline{n}}.$$

A function which is not a polynomial cannot be expressed exactly in this way. But we can always find the difference between a given function $f(x)$ and the series written above and determine whether the difference is large or small. To be quite general, let n be a fixed positive integer and $x=a$ a fixed value of the variable x. Then, for the given function $f(x)$, write

$$f(a+x)=f(a)+f'(a)x+f''(a)\frac{x^2}{\lfloor\underline{2}}+\ldots+f^{(n)}(a)\frac{x^n}{\lfloor\underline{n}}+R_n(x),$$

* This is an extension of the method of approximations to the value of a function by a linear expression (see 6.4 above).

where $R_n(x)$ is the difference, or "remainder", between $f(a+x)$ and the series written. The fundamental result is that

$$R_n(x) = f^{(n+1)}(a+\theta x)\frac{x^{n+1}}{\underline{|n+1}},$$

where θ is a positive fraction. In other words, $R_n(x)$ is the term we expect to follow $f^{(n)}(a)\frac{x^n}{\underline{|n}}$ except that the value of the $(n+1)$th derivative is taken, not at a, but at some point (not otherwise specified) between a and $(a+x)$. Without giving a complete proof of the result,* we can illustrate (in diagrammatic terms) the simple but important case where n is zero. The curve $y=f(x)$ is continuous with a tangent at all points (Fig. 102). Let A and P be points on the curve with abscissae a and $(a+x)$ and ordinates $f(a)$ and $f(a+x)$. The chord AP has gradient $\frac{f(a+x)-f(a)}{x}$ referred to Ox. Since the curve passes continuously from A to P, it is clear that there must be a point Q on the curve between A and P where the tangent is parallel to AP. Let ξ be the abscissa of Q so that ξ is between a and $(a+x)$, i.e. $\xi = a+\theta x$ for some fractional value of θ. The tangent gradient at Q is $f'(\xi)=f'(a+\theta x)$. So

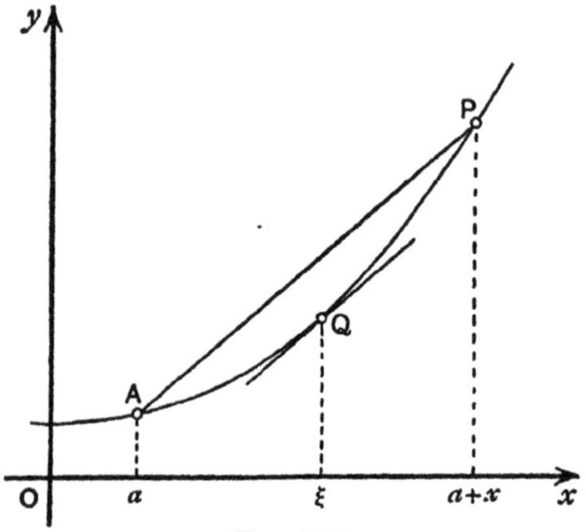

FIG. 102.

$$\frac{f(a+x)-f(a)}{x} = f'(a+\theta x),$$

i.e. $\qquad f(a+x)=f(a)+R_0(x)$ where $R_0(x)=f'(a+\theta x)x$

and our result for $n=0$ is obtained.†

* The rigid proof of this result is complex. See Hardy, *Pure Mathematics* (3rd Ed. 1921), pp. 262 *et seq.*

† The result when $n=0$ is known as the Mean Value Theorem and its meaning can be expressed as follows. The mean increment in the function $f(x)$ for the increment in x from a to $(a+x)$ is $\frac{f(a+x)-f(a)}{x}$. As $x \to a$, this ratio tends to $f'(a)$. Our result shows that the ratio equals the derivative of $f(x)$ at *some* point between a and $(a+x)$ even when x has not a small value.

Consider, now, the result of letting n take larger and larger integral values. The number of terms in the series for $f(x)$ then increases, i.e. we are deriving an infinite series. All now turns on the limiting tendency of the "remainder" $R_n(x)$. If, for a certain value of x, $R_n(x)$ tends to zero as n tends to infinity, then the infinite series we obtain is convergent and we have

$$f(a+x)=f(a)+f'(a)x+f''(a)\frac{x^2}{\underline{|2}}+f'''(a)\frac{x^3}{\underline{|3}}+\dots \text{ (to } \infty \text{)}.$$

On the other hand, if $R_n(x)$ does not tend to zero, then the infinite we obtain is not convergent and cannot have $f(a+x)$ as a sum to infinity. The broad outline of the solution of the problem of expanding $f(a+x)$ as a power series is now clear. The possibility of the expansion turns on the behaviour of $R_n(x)$ as $n \to \infty$. $R_n(x)$ may tend to zero for some values of x and not for others. The former values of x are those for which expansion is possible. Once this has been determined, the actual expansion is the series written above.

Collecting our results :

The function $f(x)$ has finite and continuous derivatives of all orders at $x=a$. For a given positive integral value of n,

$$f(a+x)=f(a)+f'(a)x+f''(a)\frac{x^2}{\underline{|2}}+\dots+f^{(n)}(a)\frac{x^n}{\underline{|n}}+f^{(n+1)}(a+\theta x)\frac{x^{n+1}}{\underline{|n+1}},$$

where θ is some positive fraction. Further,

$$f(a+x)=f(a)+f'(a)x+f''(a)\frac{x^2}{\underline{|2}}+\dots+f^{(n)}(a)\frac{x^n}{\underline{|n}}+\dots,$$

for any value of x such that $f^{(n+1)}(a+\theta x)\dfrac{x^{n+1}}{\underline{|n+1}} \to 0$ as $n \to \infty$.

The expansion of $f(a+x)$ shown is known as *Taylor's series.*

To summarise the position, the values of $f(x)$ and its derivatives are known *at the point* $x=a$. It is then possible, subject to conditions named, to express the value of the function at any point $(a+x)$ *in the neighbourhood of* $x=a$ as a series in ascending powers of x, the coefficients involving only values at $x=a$. In short, the values of $f(x)$ and its derivatives at $x=a$ provide also the value of $f(x)$ at any neighbouring point. The particular point $x=a$, in the neighbourhood of which the expansion holds, can be chosen at will. By selecting different points we get different expansions, e.g.

$$f(x)=f(0)+f'(0)x+f''(0)\frac{x^2}{\lfloor 2}+\dots \quad (a=0),$$

$$f(1+x)=f(1)+f'(1)x+f''(1)\frac{x^2}{\lfloor 2}+\dots \quad (a=1).$$

Finally, it is to be stressed that each expansion holds only for certain values of x. As extreme cases, it may be that $f(a+x)$ can be expanded for no values of x other than $x=0$ (the most unfavourable case) or that $f(a+x)$ can be expanded for all finite values of x (the most favourable case).

17.3 Examples of the expansion of functions.

The most useful and frequently quoted cases of expansions are considered in some detail in the following examples :

Ex. 1. If $f(x)=x^r$ (r any real number), then

$$f'(x)=rx^{r-1}, \quad f''(x)=r(r-1)x^{r-2}, \quad \dots .$$

In general, $\quad f^{(n)}(x)=r(r-1)(r-2)\dots(r-n+1)x^{r-n}.$

Taylor's series ($a=1$, $n=1$) gives

$$(1+x)^r=1+rx+\lambda x^2,$$

where $\lambda=\frac{1}{2}r(r-1)(1+\theta x)^{r-2}$ is finite even when $x=0$. This gives a convenient approximate expression for $(1+x)^r$ when x is small. More generally,

$$(1+x)^r=1+rx+\frac{r(r-1)}{\lfloor 2}x^2+\dots+\frac{r(r-1)(r-2)\dots(r-n+1)}{\lfloor n}x^n+\dots$$

provided that

$$R_n(x)=\frac{r(r-1)(r-2)\dots(r-n)}{\lfloor n+1}(1+\theta x)^{r-n-1}x^{n+1}$$

$$=r(r-1)(\tfrac{1}{2}r-1)\dots\left(\frac{r}{n}-1\right)\frac{x^{n+1}}{(n+1)(1+\theta x)^{n+1}}(1+\theta x)^r$$

tends to zero as n tends to infinity. If x is positive and less than unity, then $x^{n+1}\to 0$ and $(n+1)(1+\theta x)^{n+1}\to\infty$ as $n\to\infty$. Hence, $R_n(x)$ as written above must tend to zero. The same result holds, but is more difficult to prove, if x is negative and numerically less than unity. We thus have the *binomial expansion*,

$$(1+x)^r = 1 + rx + \frac{r(r-1)}{\lfloor 2}x^2 + \dots + \frac{r(r-1)(r-2)\dots(r-n+1)}{\lfloor n}x^n + \dots,$$

for any value of x which is numerically less than one. E.g.

$$\sqrt{1+x} = 1 + \tfrac{1}{2}x - \tfrac{1}{8}x^2 + \tfrac{1}{16}x^3 + \dots.$$

To obtain an approximate value of such a binomial expression, we have only to add a sufficient number of terms of the series. So,

$$\begin{aligned}\sqrt{1\cdot 1} = \sqrt{1+\tfrac{1}{10}} &= 1 + \tfrac{1}{2}\tfrac{1}{10} - \tfrac{1}{8}(\tfrac{1}{10})^2 + \tfrac{1}{16}(\tfrac{1}{10})^3 + \dots \\ &= 1 + 0\cdot 05 - 0\cdot 00125 + 0\cdot 0000625 + \dots \\ &= 1\cdot 0488 \text{ to four decimal places.}\end{aligned}$$

The fifth term of the series is not needed here.

Ex. 2. All the derivatives of e^x are equal to e^x. Hence, by Taylor's series with $a=0$,

$$e^x = 1 + x + \frac{x^2}{\lfloor 2} + \frac{x^3}{\lfloor 3} + \dots + \frac{x^n}{\lfloor n} + \dots,$$

provided that $$R_n(x) = \frac{x^{n+1}}{\lfloor n+1}e^{\theta x} \to 0 \text{ as } n \to \infty.$$

The latter is seen to hold for any finite value of x whatever since

$$\lfloor n+1 = n(n-1)(n-2)\dots 3 \,.\, 2 \,.\, 1$$

increases more rapidly than any power x^{n+1} as n tends to infinity. The series written above thus holds for any finite value of x and it is called the *exponential series*.

If x is small, the exponential e^x can be replaced approximately by the quadratic $(1+x+\frac{1}{2}x^2)$ or by the cubic $(1+x+\frac{1}{2}x^2+\frac{1}{6}x^3)$. The latter gives e^x to a degree of approximation represented by x^4. E.g., if $x = 0\cdot 1$, the cubic expression certainly gives e^x correct to three places of decimals. We must include more terms when the value of x is larger or when greater accuracy is required.

Taking $x=1$, it is found that the series

$$e = 1 + 1 + \frac{1}{\lfloor 2} + \frac{1}{\lfloor 3} + \dots$$

gives e correct to five decimal places when ten terms are added. Noting that each term can be obtained, as a decimal, by dividing the previous term by 2, 3, 4, ... , the ten terms are easily evaluated each to six decimal places as follows:

$$
\begin{aligned}
e &= 2{\cdot}5 \\
&\quad +0{\cdot}166667 \\
&\quad +0{\cdot}041667 \\
&\quad +0{\cdot}008333 \\
&\quad +0{\cdot}001389 \\
&\quad +0{\cdot}000198 \\
&\quad +0{\cdot}000025 \\
&\quad +0{\cdot}000003 \\
&= 2{\cdot}718282
\end{aligned}
$$

i.e. $e = 2{\cdot}71828$ correct to five decimal places.

Ex. 3. If $f(x) = \log x$, then

$$f'(x) = \frac{1}{x}, \quad f''(x) = -\frac{1}{x^2}, \quad f'''(x) = \frac{2}{x^3}, \quad \ldots .$$

In general, $$f^{(n)}(x) = (-1)^{n-1} \frac{\lfloor n-1}{x^n}.$$

Taylor's series with $a = 1$ gives

$$\log(1+x) = x - \frac{x^2}{2} + \frac{x^3}{3} - \frac{x^4}{4} + \ldots + (-1)^{n-1} \frac{x^n}{n} + \ldots ,$$

provided that $R_n(x) = (-1)^n \frac{\lfloor n}{(1+\theta x)^{n+1}} \frac{x^{n+1}}{\lfloor n+1} = (-1)^n \frac{x^{n+1}}{(n+1)(1+\theta x)^{n+1}}$

tends to zero as $n \to \infty$. This condition (exactly as in Ex. 1) is satisfied if x is numerically less than one. The series written above, the *logarithmic series*, is valid for such values of x.

Here again we can replace $\log(1+x)$ by an approximate quadratic or cubic expression if x is small. The series gives a method of finding approximate values of particular (natural) logarithms. E.g.

$$
\begin{aligned}
\log 1{\cdot}1 = \log(1 + \tfrac{1}{10}) &= \tfrac{1}{10} - \tfrac{1}{2}(\tfrac{1}{10})^2 + \tfrac{1}{3}(\tfrac{1}{10})^3 - \tfrac{1}{4}(\tfrac{1}{10})^4 + \ldots \\
&= 0{\cdot}1 - 0{\cdot}005 + 0{\cdot}00033 - 0{\cdot}000025 + \ldots \\
&= 0{\cdot}0953 \text{ to four decimal places.}
\end{aligned}
$$

17.4 The expansion of a function of two or more variables.

The function $z = f(x, y)$ is assumed to have continuous and finite partial derivatives up to any desired order at (a, b).

Write $$\phi(t) = f(a + xt, b + yt) = f(u, v) \quad (u = a + xt, \ v = b + yt).$$

Then $$\phi'(t) = \frac{\partial f}{\partial u}\frac{du}{dt} + \frac{\partial f}{\partial v}\frac{dv}{dt} = x\frac{\partial f}{\partial u} + y\frac{\partial f}{\partial v} \quad \text{(13.4 above)}$$

and $$\phi''(t)=\frac{d}{dt}\left(x\frac{\partial f}{\partial u}+y\frac{\partial f}{\partial v}\right)=x\frac{d}{dt}\left(\frac{\partial f}{\partial u}\right)+y\frac{d}{dt}\left(\frac{df}{\partial v}\right)$$
$$=x\left(\frac{\partial^2 f}{\partial u^2}\frac{du}{dt}+\frac{\partial^2 f}{\partial u\,\partial v}\frac{dv}{dt}\right)+y\left(\frac{\partial^2 f}{\partial u\,\partial v}\frac{du}{dt}+\frac{\partial^2 f}{\partial v^2}\frac{dv}{dt}\right)$$
$$=x^2\frac{d^2 f}{\partial u^2}+2xy\frac{\partial^2 f}{\partial u\,\partial v}+y^2\frac{\partial^2 f}{\partial v^2},$$

and so on. In general, we find

$$\phi^{(n)}(t)=x^n\frac{\partial^n f}{\partial u^n}+nx^{n-1}y\frac{\partial^n f}{\partial u^{n-1}\partial v}+\frac{n(n-1)}{\lfloor 2}x^{n-2}y^2\frac{\partial^n f}{\partial u^{n-2}\partial v^2}+\dots$$
$$+nxy^{n-1}\frac{\partial^n f}{\partial x\,\partial v^{n-1}}+y^n\frac{\partial^n f}{\partial v^n},$$

where the numerical coefficients follow the "binomial" form indicated above (17.3, Ex. 1). By Taylor's series for $\phi(t)$,

$$\phi(t)=\phi(0)+\phi'(0)t+\phi''(0)\frac{t^2}{\lfloor 2}+\dots+\phi^{(n)}(0)\frac{t^n}{\lfloor n}+\phi^{(n+1)}(\theta t)\frac{t^{n+1}}{\lfloor n+1}.$$

Putting $t=1$ and substituting the values of $\phi(t)$ and its derivatives (all evaluated at $t=0$, so that $u=a$ and $v=b$), we have

$$f(a+x,\,b+y)=f(a,\,b)+\left(x\frac{\partial f}{\partial x}+y\frac{\partial f}{\partial y}\right)_{a,\,b}$$
$$+\frac{1}{\lfloor 2}\left(x^2\frac{\partial^2 f}{\partial x^2}+2xy\frac{\partial^2 f}{\partial x\,\partial y}+y^2\frac{\partial^2 f}{\partial y^2}\right)_{a,\,b}+\dots$$
$$+\frac{1}{\lfloor n}\left(x^n\frac{\partial^n f}{\partial x^n}+nx^{n-1}y\frac{\partial^n f}{\partial x^{n-1}\partial y}+\dots+y^n\frac{\partial^n f}{\partial y^n}\right)_{a,\,b}+R_n(x,\,y),$$

where

$$R_n(x,\,y)=\frac{1}{\lfloor n+1}\left(x^{n+1}\frac{\partial^{n+1}f}{\partial x^{n+1}}+\overline{n+1}\,x^n y\frac{\partial^{n+1}f}{\partial x^n\,\partial y}+\dots+y^{n+1}\frac{\partial^{n+1}f}{\partial y^{n+1}}\right)_{a+\theta x,\,b+\theta y},$$

θ being some positive fraction. The notation adopted here needs a little explanation. The values a and b appearing as suffixes of successive terms are to be inserted for x and y in the partial derivatives of $f(x,\,y)$ but not in the powers of x and y included in the terms. Thus

$$\left(x\frac{\partial f}{\partial x}+y\frac{\partial f}{\partial y}\right)_{a,\,b}=xf_x'(a,\,b)+yf_y'(a,\,b),$$

and similarly for the other terms. Again, in the expression for $R_n(x,\,y)$, the values $a+\theta x$ and $b+\theta x$ are to be substituted for x and y in the $(n+1)$th order partial derivatives concerned.

We have now obtained an extended form of Taylor's series and

$$f(a+x,\, b+y)=f(a,\, b)+\left(x\frac{\partial f}{\partial x}+y\frac{\partial f}{\partial y}\right)_{a,\,b}$$

$$+\frac{1}{\lfloor 2}\left(x^2\frac{\partial^2 f}{\partial x^2}+2xy\frac{\partial^2 f}{\partial x\,\partial y}+y^2\frac{\partial^2 f}{\partial y^2}\right)_{a,\,b}+\dots \quad \text{(to infinity)},$$

provided that $\qquad R_n(x,\, y)\to 0$ as $n\to\infty$.

This is the expansion of $f(a+x,\, b+y)$ as a series in ascending powers of x and y, the coefficients being the values of the function and its partial derivatives at the point $(a,\, b)$. The expansion holds only for those values of x and y which make $R_n(x,\, y)\to 0$ as $n\to\infty$. It is to be noticed that the terms of the expansion are grouped so that each contains all powers and products of x and y of a certain degree, the first term being linear in x and y, the second term quadratic and so on. Successive terms are variations of higher and higher orders.

The use of Taylor's series in expanding and finding approximate values of given functions is exactly as before. If x and y are small, then $f(a+x,\, b+y)$ can be replaced approximately by a quadratic or cubic expression in x and y by retaining only the first few terms of the series and the complex later terms are not needed.

Ex. All the partial derivatives of e^{x+y} are equal to e^{x+y}.

Fixing the point (0, 0) and applying Taylor's series,

$$e^{x+y}=1+(x+y)+\frac{1}{\lfloor 2}(x^2+2xy+y^2)+\frac{1}{\lfloor 3}(x^3+3x^2y+3xy^2+y^3)+\dots,$$

provided that $\qquad R_n(x,\, y)=\dfrac{(x+y)^{n+1}}{\lfloor n+1}e^{\theta(x+y)}\to 0$ as $n\to\infty$.

The condition holds for all values of x and y (as in 17.3, Ex. 2). So

$$e^{x+y}=1+(x+y)+\frac{1}{\lfloor 2}(x+y)^2+\frac{1}{\lfloor 3}(x+y)^3+\dots,$$

which agrees with the ordinary exponential expansion on substituting $(x+y)$ for the index.

Taylor's series can be extended to cases where more than two variables appear. As the number of the variables increases, each term of the series involves more and more powers and products.

In general, if $f(x_1, x_2, \ldots x_n)$ is a function of n variables and if $(a_1, a_2, \ldots a_n)$ is a fixed point, then

$$f(a_1+x_1, a_2+x_2, \ldots a_n+x_n)$$
$$=f(a_1, a_2, \ldots a_n)+\left(x_1\frac{\partial f}{\partial x_1}+x_2\frac{\partial f}{\partial x_2}+\ldots+x_n\frac{\partial f}{\partial x_n}\right)_{a_1 a_2 \ldots a_n}$$
$$+\frac{1}{\lfloor 2}\left(x_1^2\frac{\partial^2 f}{\partial x_1^2}+x_2^2\frac{\partial^2 f}{\partial x_2^2}+\ldots+x_n^2\frac{\partial^2 f}{\partial x_n^2}+2x_1x_2\frac{\partial^2 f}{\partial x_1\,\partial x_2}\right.$$
$$\left.+2x_1x_3\frac{\partial^2 f}{\partial x_1\,\partial x_3}+\ldots+2x_{n-1}x_n\frac{\partial^2 f}{\partial x_{n-1}\,\partial x_n}\right)_{a_1 a_2 \ldots a_n}+\ldots,$$

subject to conditions similar to those already given. Our results are thus perfectly general and the only difficulty is the labour involved in writing the appropriate series.

17.5 A complete criterion for maximum and minimum values.

In addition to its practical use in the expansion of functions, Taylor's series is a powerful theoretical tool of service in such problems as that of maximum and minimum values. We can, first, complete the tests for extreme values of a function of one variable (8.2, 8.3 and 8.6 above) and then, in a later chapter, use similar methods to develop tests for functions of several variables.

A convenient form of the definition of extreme values is:

The function $f(x)$ has a maximum (minimum) value at $x=a$ if $f(a+x)-f(a)$ is negative (positive) for all small values of x.

It is again assumed that the function has finite and continuous derivatives up to any desired order at $x=a$. Taylor's series gives

$$f(a+x)-f(a)=f'(a)x+f''(a)\frac{x^2}{\lfloor 2}+\ldots+f^{(n)}(a)\frac{x^n}{\lfloor n}+f^{(n+1)}(a+\theta x)\frac{x^{n+1}}{\lfloor n+1}$$

for any fixed positive integral value of n. This form of the series is clearly of direct application to our problem.

(1) Suppose $f'(a)\neq 0$.

By the continuity of the derivative, $f'(a+\theta x)$ has the same sign as $f'(a)$ for a certain range of small (positive and negative) values of x. Now

$$f(a+x)-f(a)=f'(a+\theta x)x \quad (n=0)$$

and so $f(a+x)-f(a)$ must change sign as x changes sign. The point $x=a$ cannot, therefore, give an extreme value of $f(x)$. Hence, if $x=a$ gives an extreme value of $f(x)$, we must have $f'(a)=0$.

(2) Suppose $f'(a)=0$, $f''(a)\neq 0$.

As before, $f''(a+\theta x)$ has the sign of $f''(a)$ for a certain range of positive and negative values of x. But

$$f(a+x)-f(a)=f'(a)x+f''(a+\theta x)\frac{x^2}{\lfloor 2}=\frac{1}{2}f''(a+\theta x)x^2 \quad (n=1),$$

i.e. $f(a+x)-f(a)$ has the sign of $f''(a+\theta x)$, i.e. of $f''(a)$. Hence,

If $f'(a)=0$ and $f''(a)<0$, $x=a$ gives a maximum value of $f(x)$.
If $f'(a)=0$ and $f''(a)>0$, $x=a$ gives a minimum value of $f(x)$.

(3) Suppose $f'(a)=f''(a)=0$, $f'''(a)\neq 0$.

Again $f'''(a+\theta x)$ has the sign of $f'''(a)$ for sufficiently small positive and negative values of x. So

$$f(a+x)-f(a)=f'(a)x+f''(a)\frac{x^2}{\lfloor 2}+f'''(a+\theta x)\frac{x^3}{\lfloor 3}$$
$$=\tfrac{1}{6}f'''(a+\theta x)x^3 \quad (n=2)$$

shows that $f(a+x)-f(a)$ changes sign as x changes sign. The point $x=a$, which is a stationary point, is thus a point of inflexion and does not give an extreme value of $f(x)$.

(4) Suppose $f'(a)=f''(a)=\ldots=f^{(n-1)}(a)=0$, $f^{(n)}(a)\neq 0$.

Taylor's series can now be written

$$f(a+x)-f(a)=f^{(n)}(a+\theta x)\frac{x^n}{\lfloor n}.$$

Of the two terms on the right-hand side, the derivative has a constant sign, that of $f^{(n)}(a)$, for sufficiently small (positive or negative) values of x. The other term $\left(\frac{x^n}{\lfloor n}\right)$ changes sign with x if n is an odd integer and is always positive if n is an even integer. In the former case, $f(a+x)-f(a)$ changes sign as x changes sign and $x=a$ must be a point of inflexion. In the latter case, $f(a+x)-f(a)$ has a constant sign and $x=a$ gives a maximum or minimum value of $f(x)$ according as $f^{(n)}(a)$ is negative or positive. Hence,

COMPLETE CRITERION FOR MAXIMUM AND MINIMUM VALUES

(1) If $f(x)$ has an extreme value at $x=a$, then $f'(a)=0$.

(2) If $f'(a)=f''(a)=\ldots=f^{(n-1)}(a)=0,\ f^{(n)}(a)\neq 0,$ then $f(x)$ has a stationary value at $x=a$ which is an inflexional value if n is odd, a maximum value if n is even and $f^{(n)}(a)<0$ and a minimum value if n is even and $f^{(n)}(a)>0$.

Subject to the condition that the derivatives involved are finite and continuous, the criterion is complete and so both necessary and sufficient. There is no case of failure; unless the function is a constant (and so without maxima or minima) there must always be some derivative which is not zero. The order and sign of this derivative determines the nature of the point considered. The practical method of finding maximum and minimum values (8.4 above) is scarcely affected; a few doubtful cases (such as that of the following example) are cleared up and that is all.

Ex. $y=(x-1)^4$.

So $\dfrac{dy}{dx}=4(x-1)^3,\quad \dfrac{d^2y}{dx^2}=12(x-1)^2,\quad \dfrac{d^3y}{dx^3}=24(x-1)$ and $\dfrac{d^4y}{dx^4}=24.$

There is only one stationary value, $y=0$, occurring at $x=1$. At this point, the first three derivatives are zero and the fourth derivative is positive. Hence, $y=0$ is a *minimum* value of the function. This can be checked since y is positive for all values of x except that it is zero when $x=1$. A graph of the function shows a curve very similar in shape to the parabola and with vertex at $(1, 0)$ on the axis Ox.

17.6 Second and higher order differentials.

The differential of a function of several variables describes the "first-order" variation of the function. But, just as derivatives of various orders are needed to describe completely the variation of a function of one variable, so something more than the ordinary differential is required in the case of a function of two or more variables. The need is met by the definition of differentials of the second and higher orders. In the following, the definitions and results are given only for a function, $z=f(x, y)$, of two variables, but they are easily extended to more general functions.

The function $z=f(x, y)$ is assumed to have finite and continuous partial derivatives of all orders required. The first-order variation is described by $dz = \frac{\partial z}{\partial x} dx + \frac{\partial z}{\partial y} dy$ whether the variables x and y are independent or not. It is very important, however, to distinguish carefully between the two cases. If x and y are independent, then $\frac{\partial z}{\partial x}$ and $\frac{\partial z}{\partial y}$ are functions of these variables but dx and dy are arbitrary and *constant* increments in the variables. But, if x and y are dependent on other variables, then $\frac{\partial z}{\partial x}$ and $\frac{\partial z}{\partial y}$ are also functions of the genuine independent variables and dx and dy are expressed in terms of the same variables plus their arbitrary and constant increments. In either case, dz involves functions of the independent variables (whether they are x and y or others) and certain constants, the arbitrary increments in the independent variables. The differential of dz, as a function of certain independent variables, can now be defined in the ordinary way and is called the *second differential*:

$$d^2z = d(dz).$$

The process can be extended and, in general, the nth differential of z is defined as the differential of the $(n-1)$th differential:

$$d^nz = d(d^{n-1}z).$$

As in the case of derivatives, however, it is seldom that we need differentials of higher order than the second.

Our problem now is to express higher order differentials in terms of successive partial derivatives of the function. The importance of distinguishing the independent variables becomes evident here. Unlike the expression for dz, which is the same in all cases, our results for the higher order differentials are different and simpler when x and y are the independent variables than they are in other cases. The reason for this is not far to seek. In differentiating dz to obtain d^2z, we must know how to treat the dx and dy appearing in dz. If x and y are independent variables, dx and dy are constants and do not worry us. In other cases, this is not so and due account must be taken of the variation of dx and dy in the differentiation process. Before beginning to differentiate, therefore, it is essential to know whether x and y are independent or dependent variables.

17.7 Differentials of a function of two independent variables.

Differentiating the first differential of a function of two independent variables, we have

$$d^2z = d\left(\frac{\partial z}{\partial x}dx + \frac{\partial z}{\partial y}dy\right) = d\left(\frac{\partial z}{\partial x}\right)dx + d\left(\frac{\partial z}{\partial y}\right)dy$$

$$= \left\{\frac{\partial}{\partial x}\left(\frac{\partial z}{\partial x}\right)dx + \frac{\partial}{\partial y}\left(\frac{\partial z}{\partial x}\right)dy\right\}dx + \left\{\frac{\partial}{\partial x}\left(\frac{\partial z}{\partial y}\right)dx + \frac{\partial}{\partial y}\left(\frac{\partial z}{\partial y}\right)dy\right\}dy,$$

i.e.
$$d^2z = \frac{\partial^2 z}{\partial x^2}dx^2 + 2\frac{\partial^2 z}{\partial x\,\partial y}dx\,dy + \frac{\partial^2 z}{\partial y^2}dy^2.$$

The essential point is that dx and dy are treated throughout as constant increments ; otherwise the ordinary rules of differentiation (13.3 above) are used. It appears, therefore, that d^2z is a quadratic expression in dx and dy, the coefficients being given by the second-order partial derivatives of z at the point (x, y) in question.

In finding second differentials in practice, we can *either* evaluate the second-order partial derivatives and use the above result, *or* we can differentiate twice by rule in the particular case considered. The latter method usually involves less labour and is adopted in the following examples. In each case, however, the results can be checked by means of the partial derivatives.

Ex. 1. $z = x^2 + 2xy - y^2$.

Here $\quad dz = 2(x+y)dx + 2(x-y)dy$

and $\quad d^2z = 2d(x+y)dx + 2d(x-y)dy = 2\{(dx+dy)dx + (dx-dy)dy\}$

$$= 2(dx^2 + 2dx\,dy - dy^2).$$

Ex. 2. $z = \dfrac{x^2}{x-y+1}$.

Here
$$dz = \frac{x(x-2y+2)dx + x^2\,dy}{(x-y+1)^2}$$

and
$$d^2z = \frac{\{d(x^2-2xy+2x)dx + d(x^2)dy\}(x-y+1)^2 - \{(x^2-2xy+2x)dx + x^2\,dy\}\,d(x-y+1)^2}{(x-y+1)^4}$$

$$= \frac{2(x-y+1)^2\,dx^2 - 2\{(x^2-2xy+2x)dx + x^2\,dy\}(dx-dy)}{(x-y+1)^3}$$

$$= 2\,\frac{\{(y-1)dx - x\,dy\}^2}{(x-y+1)^3}.$$

Ex. 3. $z = \log(x^2 + y^2)$.

Here $$dz = 2\frac{x\,dx + y\,dy}{x^2 + y^2}$$

and $$d^2z = 2\frac{d(x\,dx + y\,dy)(x^2 + y^2) - d(x^2 + y^2)(x\,dx + y\,dy)}{(x^2 + y^2)^2}$$
$$= 2\frac{(dx^2 + dy^2)(x^2 + y^2) - 2(x\,dx + y\,dy)^2}{(x^2 + y^2)^2}$$
$$= -2\frac{(x^2 - y^2)\,dx^2 + 4xy\,dx\,dy - (x^2 - y^2)\,dy^2}{(x^2 + y^2)^2}.$$

The process of differentiation can be repeated to give

$$d^3z = d\left(\frac{\partial^2 z}{\partial x^2}dx^2 + 2\frac{\partial^2 z}{\partial x\,\partial y}dx\,dy + \frac{\partial^2 z}{\partial y^2}dy^2\right)$$
$$= \left(\frac{\partial^3 z}{\partial x^3}dx + \frac{\partial^3 z}{\partial x^2\,\partial y}dy\right)dx^2 + 2\left(\frac{\partial^3 z}{\partial x^2\,\partial y}dx + \frac{\partial^3 z}{\partial x\,\partial y^2}dy\right)dx\,dy$$
$$+ \left(\frac{\partial^3 z}{\partial x\,\partial y^2}dx + \frac{\partial^3 z}{\partial y^3}dy\right)dy^2,$$

i.e. $$d^3z = \frac{\partial^3 z}{\partial x^3}dx^3 + 3\frac{\partial^3 z}{\partial x^2\,\partial y}dx^2\,dy + 3\frac{\partial^3 z}{\partial x\,\partial y^2}dx\,dy^2 + \frac{\partial^3 z}{\partial y^3}dy^3.$$

This is a cubic expression in dx and dy with coefficients given by the third-order partial derivatives of z. Higher order differentials are obtained in exactly the same way and it is to be noticed that the numerical coefficients of successive terms follow the "binomial" law (17, 3, Ex. 1, above).

We can note here a connection between successive differentials of a function of independent variables and the corresponding terms of the Taylor expansion of the function. Assuming that Taylor's series is valid, it can now be written in the form

$$\Delta z = f(x+h,\ y+k) - f(x,\ y) = dz + \frac{d^2z}{\underline{|2}} + \frac{d^3z}{\underline{|3}} + \dots,$$

where the differentials of z are to be taken at (x, y) and with respect to arbitrary increments $dx = h$ and $dy = k$ in the variables. Hence, the increment Δz in the function for increments h and k in the variables from (x, y) is compounded of the successive differentials, $dz, d^2z, d^3z, \dots$, of the function at (x, y). The complete variation of the function is described only by using differentials of all orders. But, the successive differentials involve higher powers of the increments h and k, i.e. they are of higher order of "smallness" if h and k are small. For approximate results, only a few differentials (i.e. a few terms of Taylor's series) are needed to describe the variation of the function. As a first approximation $\Delta z = dz$; as a second approximation $\Delta z = dz + \frac{1}{2}d^2z$; and so on.

The higher order differentials of a function of a single variable, $y = f(x)$, appear as a special case of the general process of differentiation described above:

$$dy = f'(x)\,dx, \quad d^2y = f''(x)\,dx^2, \quad d^3y = f'''(x)\,dx^3, \quad \ldots .$$

In general, $$d^ny = f^{(n)}(x)\,dx^n.$$

The first derivative, as we have seen, can be interpreted as the ratio of the differential dy to the differential dx. This result now extends. The second derivative is the ratio of d^2y to dx^2, i.e. the ratio of the second differential of the dependent variable y to the square of the differential (or arbitrary increment) of the independent variable x. The notation $\frac{d^2y}{dx^2}$ for the second derivative is thus justified. Similar results hold for higher order derivatives.

17.8 Differentials of a function of two dependent variables.

In the function $z = f(x, y)$, the variables x and y are dependent on a set of independent variables $u, v, w, \ldots$. The process of obtaining the second differential d^2z now proceeds:

$$d^2z = d\left(\frac{\partial z}{\partial x}dx + \frac{\partial z}{\partial y}dy\right) = d\left(\frac{\partial z}{\partial x}dx\right) + d\left(\frac{\partial z}{\partial y}dy\right)$$

$$= d\left(\frac{\partial z}{\partial x}\right)dx + \frac{\partial z}{\partial x}d(dx) + d\left(\frac{\partial z}{\partial y}\right)dy + \frac{\partial z}{\partial y}d(dy)$$

$$= \left(\frac{\partial^2 z}{\partial x^2}dx + \frac{\partial^2 z}{\partial x\,\partial y}dy\right)dx + \frac{\partial z}{\partial x}d^2x + \left(\frac{\partial^2 z}{\partial x\,\partial y}dx + \frac{\partial^2 z}{\partial y^2}dy\right)dy + \frac{\partial z}{\partial y}d^2y,$$

i.e. $$d^2z = \frac{\partial^2 z}{\partial x^2}dx^2 + 2\frac{\partial^2 z}{\partial x\,\partial y}dx\,dy + \frac{\partial^2 z}{\partial y^2}dy^2 + \frac{\partial z}{\partial x}d^2x + \frac{\partial z}{\partial y}d^2y.$$

The expression for d^2z now includes two additional terms, the terms in the second differentials of x and y as functions of the independent variables $u, v, w, \ldots$. If x and y happen to be independent variables, these terms are zero (since dx and dy are constants) and the expression for d^2z reduces to that previously obtained. In general, however, d^2x and d^2y must be retained and interpreted in the light of the particular problem considered.

Two simple applications of the result can be noted. If $z=f(x, y)$ where x and y are functions of a single variable t, the partial derivatives and derivatives of the functions being known, then from $dz=\frac{\partial z}{\partial x}dx+\frac{\partial z}{\partial y}dy$, we derive (13.4 above):

$$\frac{dz}{dt}=\frac{\partial z}{\partial x}\frac{dx}{dt}+\frac{\partial z}{\partial y}\frac{dy}{dt}.$$

From $$d^2z=\frac{\partial^2 z}{\partial x^2}dx^2+2\frac{\partial^2 z}{\partial x\,\partial y}dx\,dy+\frac{\partial^2 z}{\partial y^2}dy^2+\frac{\partial z}{\partial x}d^2x+\frac{\partial z}{\partial y}d^2y,$$
we derive

$$\frac{d^2z}{dt^2}=\frac{\partial^2 z}{\partial x^2}\left(\frac{dx}{dt}\right)^2+2\frac{\partial^2 z}{\partial x\,\partial y}\frac{dx}{dt}\frac{dy}{dt}+\frac{\partial^2 z}{\partial y^2}\left(\frac{dy}{dt}\right)^2+\frac{\partial z}{\partial x}\frac{d^2x}{dt^2}+\frac{\partial z}{\partial y}\frac{d^2y}{dt^2}.$$

The ratio of d^2z to dt^2 is to be interpreted as the second derivative of z as a function of t (defined by means of x and y). We have thus a simple formula for this derivative in terms of the partial derivatives of z and the derivatives of x and y. The same result is obtained by taking $\frac{dz}{dt}$ as a function of t and by finding its derivative directly. In particular, if $z=f(x, y)$ where y is a function of x,

$$\frac{dz}{dx}=\frac{\partial z}{\partial x}+\frac{\partial z}{\partial y}\frac{dy}{dx}$$

and $$\frac{d^2z}{dx^2}=\frac{\partial^2 z}{\partial x^2}+2\frac{\partial^2 z}{\partial x\,\partial y}\frac{dy}{dx}+\frac{\partial^2 z}{\partial y^2}\left(\frac{dy}{dx}\right)^2+\frac{\partial z}{\partial y}\frac{d^2y}{dx^2}.$$

In this case, d^2x does not appear in d^2z since x is the independent variable.

If y is an implicit function of x defined by the relation $f(x, y)=0$, we have seen (13.5 above) that

$$f_x\,dx+f_y\,dy=0,$$

which gives the derivative $\frac{dy}{dx}=-\frac{f_x}{f_y}$. Differentiating again,

$$d(f_x\,dx+f_y\,dy)=0,$$

i.e. $$f_{xx}\,dx^2+2f_{xy}\,dx\,dy+f_{yy}\,dy^2+f_y\,d^2y=0,$$

noting that d^2y, but not d^2x, appears since y is taken as a function of x. Dividing through by dx^2,

$$f_{xx}+2f_{xy}\frac{dy}{dx}+f_{yy}\left(\frac{dy}{dx}\right)^2+f_y\frac{d^2y}{dx^2}=0,$$

i.e. $$\frac{d^2y}{dx^2}=-\frac{1}{f_y}\left\{f_{xx}+2f_{xy}\frac{dy}{dx}+f_{yy}\left(\frac{dy}{dx}\right)^2\right\}.$$

This gives the second derivative of y as a function of x. If the value of $\frac{dy}{dx}$ previously obtained is substituted, the result already written (13.5 above) is again obtained.

If the same relation is regarded as giving x as a function of y, then $f_x\,dx + f_y\,dy = 0$ gives the first derivative $\frac{dx}{dy} = -\frac{f_y}{f_x}$ and from $d(f_x\,dx + f_y\,dy) = 0$, we find

$$f_{xx}\,dx^2 + 2f_{xy}\,dx\,dy + f_{yy}\,dy^2 + f_x\,d^2x = 0,$$

which gives the second derivative,

$$\frac{d^2x}{dy^2} = -\frac{1}{f_x}\left\{f_{yy} + 2f_{xy}\frac{dx}{dy} + f_{xx}\left(\frac{dx}{dy}\right)^2\right\}.$$

This can be expanded, as before, on substituting the value of $\frac{dx}{dy}$. It is to be noted that, though the first derivatives $\frac{dy}{dx}$ and $\frac{dx}{dy}$ are reciprocal, the same relation does not hold between the second derivatives.

In practice, the derivatives of implicit functions are best obtained by differentiating the relation between the variables as it stands, always remembering which of the variables is dependent :

Ex. 1. $x^2 + y^2 = 16$.

Taking y as a double-valued function of x, we have

$$2x + 2y\frac{dy}{dx} = 0, \quad \text{i.e. } \frac{dy}{dx} = -\frac{x}{y}.$$

Differentiating again,

$$2 + 2\left(\frac{dy}{dx}\right)^2 + 2y\frac{d^2y}{dx^2} = 0,$$

i.e.
$$\frac{d^2y}{dx^2} = -\frac{1}{y}\left\{1 + \left(\frac{dy}{dx}\right)^2\right\} = -\frac{1}{y}\frac{x^2+y^2}{y^2} = -\frac{16}{y^3}.$$

But, taking x as a double-valued function of y,

$$2x\frac{dx}{dy} + 2y = 0, \quad \text{i.e. } \frac{dx}{dy} = -\frac{y}{x}$$

and
$$2\left(\frac{dx}{dy}\right)^2 + 2x\frac{d^2x}{dy^2} + 2 = 0,$$

i.e.
$$\frac{d^2x}{dy^2} = -\frac{1}{x}\left\{1 + \left(\frac{dx}{dy}\right)^2\right\} = -\frac{1}{x}\frac{x^2+y^2}{x^2} = -\frac{16}{x^3}.$$

Ex. 2. $x^3+y^3-3xy=0$.

Regarding y as a triple-valued function of x,

$$3x^2+3y^2\frac{dy}{dx}-3\left(x\frac{dy}{dx}+y\right)=0.$$

So
$$(x^2-y)+(y^2-x)\frac{dy}{dx}=0, \quad \text{i.e.} \quad \frac{dy}{dx}=-\frac{x^2-y}{y^2-x}.$$

Differentiating again,

$$2x-\frac{dy}{dx}+\left(2y\frac{dy}{dx}-1\right)\frac{dy}{dx}+(y^2-x)\frac{d^2y}{dx^2}=0,$$

giving
$$\frac{d^2y}{dx^2}=-\frac{2}{y^2-x}\left\{y\left(\frac{dy}{dx}\right)^2-\frac{dy}{dx}+x\right\}$$

$$=-\frac{2}{(y^2-x)^3}\{y(x^2-y)^2+(x^2-y)(y^2-x)+x(y^2-x)^2\}$$

$$=-\frac{2xy}{(y^2-x)^3}(x^3+y^3-3xy+1)=-\frac{2xy}{(y^2-x)^3}.$$

Similarly, if x is regarded as a function of y, we find

$$\frac{dx}{dy}=-\frac{y^2-x}{x^2-y} \quad \text{and} \quad \frac{d^2x}{dy^2}=-\frac{2xy}{(x^2-y)^3}.$$

Returning to the general case where $z=f(x, y)$ and x and y are functions of certain independent variables, the third differential d^3z can be obtained by differentiating the second differential d^2z :

$$d^3z=d\left(\frac{\partial^2 z}{\partial x^2}dx^2+2\frac{\partial^2 z}{\partial x\,\partial y}dx\,dy+\frac{\partial^2 z}{\partial y^2}dy^2+\frac{\partial z}{\partial x}d^2x+\frac{\partial z}{\partial y}d^2y\right)$$

$$=\left(\frac{\partial^3 z}{\partial x^3}dx^3+3\frac{\partial^3 z}{\partial x^2\,\partial y}dx^2\,dy+3\frac{\partial^3 z}{\partial x\,\partial y^2}dx\,dy^2+\frac{\partial^3 z}{\partial y^3}dy^3\right)$$

$$+3\left(\frac{\partial^2 z}{\partial x^2}dx+\frac{\partial^2 z}{\partial x\,\partial y}dy\right)d^2x+3\left(\frac{\partial^2 z}{\partial x\,\partial y}dx+\frac{\partial^2 z}{\partial y^2}dy\right)d^2y$$

$$+\frac{\partial z}{\partial x}d^3x+\frac{\partial z}{\partial y}d^3y$$

on reduction. The expression of the differential has now become very cumbersome. Higher order differentials are obtained by further differentiation and are exceedingly involved.

As a particular case of the above results, let $y=f(x)$ where x is dependent on other variables. Then

$$dy=f'(x)\,dx, \quad d^2y=f''(x)\,dx^2+f'(x)\,d^2x, \quad \dots .$$

For example, if x is a function of a single variable t,

$$\frac{dy}{dt}=f'(x)\frac{dx}{dt}, \quad \frac{d^2y}{dt^2}=f''(x)\left(\frac{dx}{dt}\right)^2+f'(x)\frac{d^2x}{dt^2}, \quad \ldots .$$

This is the ordinary function of a function rule for functions of one variable, extended to the second and higher order derivatives.

EXAMPLES XVII

Infinite series

1. Write down the sequence of numbers with nth member $\left(\frac{1}{2}-\frac{n}{|\underline{n+1}}\right)$ and obtain the infinite series of which the sum of n terms is given by this expression. Show that the series is convergent to $\frac{1}{2}$.

2. Find the infinite series with sum of n terms given by (*a*) $n^2(n+1)$ and (*b*) $\frac{1}{6}n(n+1)(2n+1)$. Show that neither series is convergent.

3. Write the series with sum of n terms $\dfrac{n}{n+1}$ and show that it is convergent.

4. It is known that π is the sum to infinity of the series

$$4(1-\tfrac{1}{3}+\tfrac{1}{5}-\tfrac{1}{7}+\tfrac{1}{9}-\tfrac{1}{11}+\ldots).$$

Find the value of π to five decimal places.

5. Show that $\sqrt{2}$ is the sum to infinity of the binomial series

$$\tfrac{3}{2}\{1-\tfrac{1}{2}\tfrac{1}{9}-\tfrac{1}{8}(\tfrac{1}{9})^2-\tfrac{1}{16}(\tfrac{1}{9})^3+\ldots\}, \quad \text{(cf. 17, 3, Ex. 1.)}$$

and deduce the value of $\sqrt{2}$ correct to four decimal places.

6. Express the recurring decimals $0\cdot\dot{1}\dot{3}$ and $0\cdot 2\dot{7}$ as fractions.

7. Write down the sum to infinity of each of the convergent series

$$1-\frac{1}{2}+\frac{1}{4}-\frac{1}{8}+\frac{1}{16}-\frac{1}{32}+\ldots; \quad 2+\sqrt{2}+1+\frac{1}{\sqrt{2}}+\frac{1}{2}+\frac{1}{2\sqrt{2}}+\ldots .$$

Illustrate the approximation to the sum to infinity by finding the sum of ten terms in each case.

8. A ball is dropped from height h feet on a hard floor, bounces, falls, bounces again, and so on. Each bounce is to height e times the previous fall, e being a positive fraction. Show that the distance travelled by the ball before coming to rest can be represented as the sum to infinity of an infinite geometric series. If the distance is $2h$, show that $e=\frac{1}{3}$.

9. An income stream, starting with £a next year, is such that each year's income is half that of the previous year. Adding interest yearly at $100r$ per cent. per year, find the present value of the stream flowing for n years, and of the stream flowing for ever.

10. An income stream starts next year with £a and flows for ever so that each year's income is a fixed percentage ($100s$ per cent.) less than the previous year's income. Find the present value of the stream, adding interest yearly at $100r$ per cent. per year.

Expansions

11. If θ is some positive fraction, use Taylor's series $(n=0)$ to show that $\log(1+x)=\frac{x}{1+\theta x}$ and deduce that

$$\frac{x}{1+x}<\log(1+x)<x \qquad (x>0).$$

Then, from Taylor's series $(n=1)$, show that

$$\frac{x^2}{2(1+x^2)}<x-\log(1+x)<\tfrac{1}{2}x^2 \qquad (x>0).$$

12. For values of x numerically less than one, show that

$$\frac{1}{(1-x)^2}=1+2x+3x^2+4x^3+\ldots .$$

13. Write $\sqrt{6}=\frac{1}{2}\sqrt{25-1}$ and use a binomial series to calculate its value correct to five decimal places.

14. Find a quadratic expression which approximates to $\frac{1}{\sqrt{1+x}}$ when x is small. Put $x=\frac{1}{80}$ and find $\sqrt{5}$ correct to four decimal places.

15. If $x<1$ numerically, write expansions for $\sqrt{1+x}$, $\sqrt{1-x}$ and $\sqrt{1-x^2}$. By algebraic multiplication of infinite series, verify that the expansion of a function which is the product of two parts is the product of the expansions of the separate parts.

16. Form an infinite series each term of which is the integral of the corresponding term of the expansion of $(1+x)^{-1}$ in ascending powers of x. What is the series? Show that its sum to infinity is the integral of $(1+x)^{-1}$. What general rule does this result suggest?

17. By taking sufficient terms of the appropriate series, find the values of $\sqrt{e}$ and $\frac{1}{e}$ correct to four decimal places.

18. Show that $\quad \log(n+1)=\log n+\left(\frac{1}{n}-\frac{1}{2}\frac{1}{n^2}+\frac{1}{3}\frac{1}{n^3}-\frac{1}{4}\frac{1}{n^4}+\ldots\right).$

Given $\log 2=0{\cdot}6931$, find values of the natural logarithms of 3, 4 and 5 correct to three decimal places. (*Note*: log 4 can be found without using the infinite series.)

19. If interest at $100r$ per cent. compounded yearly is equivalent to interest at $100s$ per cent. compounded n times a year (cf. Examples IX, 25), show that r exceeds s by approximately $\frac{n-1}{2n}s^2$.

20. If interest at $100r$ per cent. compounded yearly and at 100ρ per cent. compounded continuously are equivalent, show that $r-\rho=\frac{1}{2}r^2$ approximately when r is small (cf. Examples IX, 26).

21. Find the amount of £1 after n years at $100r$ per cent. interest which is (*a*) simple, (*b*) compounded yearly, and (*c*) compounded continuously. If $(rn)^3$ is small enough to be neglected, show that the last amount exceeds the first by $\frac{1}{2}(rn)^2$ and the second by $\frac{1}{2}r^2n$ approximately.

22. A man's income is £x and his (measurable) utility is then $\phi(x)$ where $\phi(x)$ increases at a decreasing rate as x increases. The man makes a fair bet on an event with chance p, laying £pa against £$(1-p)a$ that the event happens. Show that his expectation of utility is now £$\{p\phi(x+\overline{1-p}a)+(1-p)\phi(x-pa)\}$. If the bet is small, use Taylor's series to show that the expectation gives less utility than the original income £x. (See Marshall, *Principles of Economics*, 8th Ed. 1927, p. 843.)

23. By evaluating derivatives and using Taylor's series, show that

$$\log \tfrac{1}{2}(1+e^x)=\tfrac{1}{2}x+\tfrac{1}{8}x^2-\tfrac{1}{192}x^4+\dots,$$

assuming that the series is convergent.

24. Expand, by Taylor's series for a function of two variables, the expression $\dfrac{1}{1+x-y}$ as a series in ascending powers of x and y. Deduce linear and quadratic approximations to this expression when x and y are both small. (Cf. Examples XIII, 7.)

Higher order differentials

25. Show that neither $y=(x-1)^3$ nor $y=(x-1)^5$ has a maximum or minimum value and that each has an inflexional point at (1, 0).

26. Find the partial derivatives, and hence the first two differentials, of each of $z=x^2+2xy-y^2$, $z=\dfrac{x^2}{x-y+1}$ and $z=\log(x^2+y^2)$.

27. Find the second differential of each of $z=x^3+y^3-3xy$, $z=\sqrt{x^2+y^2}$, $z=e^{x-y}$ and $z=\log\dfrac{x}{x+y}$ where x and y are independent variables.

28. If x, y and z are independent variables, find the second differential of $u=(x^2+2xy-y^2)e^z$.

29. If $z=\dfrac{1}{x+y}$ where $x=e^t$ and $y=e^{-t}$, evaluate $\dfrac{d^2z}{dt^2}$.

30. Given that $x^3+y^3-3xy=0$, use the derivatives of y as a function of x and of x as a function of y to show that y has a single maximum value for variation of x and conversely. Illustrate diagrammatically. (See Example XIII, 22.)

31. The relation $f(x, y, z)=0$ gives z as a function of x and y. Differentiate the relation and obtain the first and second order partial derivatives of z in terms of those of f. Illustrate by taking the relation $x^3+y^3+z^3-3xy=0$. (See Examples XIII, 23.)

32. If $z=f(x, y)$ where x and y are given functions of the independent variables u and v, show that

$$\frac{\partial^2 z}{\partial u^2}=\frac{\partial^2 z}{\partial x^2}\left(\frac{\partial x}{\partial u}\right)^2+2\frac{\partial^2 z}{\partial x\,\partial y}\frac{\partial x}{\partial u}\frac{\partial y}{\partial u}+\frac{\partial^2 z}{\partial y^2}\left(\frac{\partial y}{\partial u}\right)^2+\frac{\partial z}{\partial x}\frac{\partial^2 x}{\partial u^2}+\frac{\partial z}{\partial y}\frac{\partial^2 y}{\partial u^2}$$

and similar results for the other two partial derivatives of the second order.

33. From the results of the previous example, find the second-order partial derivatives of $z=\log(x-y)$ where $x=\dfrac{u}{v}$ and $y=\dfrac{v}{u}$. Express z explicitly in terms of u and v and find the partial derivatives directly.

CHAPTER XVIII

DETERMINANTS, LINEAR EQUATIONS AND QUADRATIC FORMS

18.1 The general notion of a determinant.

DESPITE its somewhat terrifying name, a "determinant" is a mathematical tool of a very ordinary kind and involves no new ideas of any description. Briefly, a determinant is a *notation* that is found convenient in handling certain involved, but essentially commonplace, algebraic processes. Certain expressions of a common form appear in algebraic problems such as that of the solution of linear equations, expressions consisting of sums or differences of a number of terms each of which is the product of a number of quantities. The expressions $(ab-h^2)$ and $(abc-af^2-bg^2-ch^2+2fgh)$ are cases in point. Quite apart from other considerations, the labour of writing out the more complicated of these expressions is severe and there is every reason to welcome a compact and general notation for them. The determinant notation is justified on these grounds alone.

There is, however, more in the determinant notation than this. Once the notation is introduced, the expressions denoted by determinants are seen to obey quite simple rules and the algebraic processes in which they appear become simpler and more uniform than before. As a consequence, results can be established which would almost certainly be missed were it not for the new notation. It is for such reasons that determinants have become of general use; no notation can be retained unless it saves labour and enables us to carry out processes more easily and with greater generality than before.

The foundation of a determinant is a square "block" of quantities written down in rows and columns in "crossword" form. There is no restriction on the quantities except that each must be capable of taking a single numerical value; they can be numbers, constants, variables, functions, derivatives of functions, and so on. The number of the quantities must be a perfect square; four quantities can be written down in a square form of two rows and two columns, nine quantities in a square form of three rows and three columns, and so on. Examples are provided by the following arrangements:

$$\begin{matrix} 2 & -1 \\ 0 & 3 \end{matrix} \quad ; \quad \begin{matrix} \frac{\partial u}{\partial x} & \frac{\partial u}{\partial y} \\ \frac{\partial v}{\partial x} & \frac{\partial v}{\partial y} \end{matrix} \quad \text{and} \quad \begin{matrix} x^2 & xy & xz \\ xy & y^2 & yz \\ xz & yz & z^2 \end{matrix}$$

A determinant of two rows and columns is called a determinant of the second order, of three rows and columns a determinant of the third order, and so on. In general, a determinant of the nth order has n^2 quantities arranged in n rows and n columns. The quantities themselves are called the *elements* of the determinant and we can speak of the elements of the first row or column, of the second row or column, and so on. Each element is allotted, of course, to one row and to one column.

A general method of denoting the block of quantities making up a determinant is required. From the point of view of stressing the arrangement in rows and columns, the best notation is

$$\begin{matrix} a_{11} & a_{12} & a_{13} & \cdots & a_{1n} \\ a_{21} & a_{22} & a_{23} & \cdots & a_{2n} \\ \cdots & \cdots & \cdots & \cdots & \cdots \\ a_{n1} & a_{n2} & a_{n3} & \cdots & a_{nn} \end{matrix}$$

where the integer n denotes the order of the determinant. The two suffixes of an element denote the row and the column into which the element is to be placed. Thus, a_{53} is the element to be inserted in the " cell " at the intersection of the fifth row and third column. It must be noted, however, that the notation is designed for dealing with determinants in general; in any particular case the elements appear as actual numbers or quantities of one kind or another.

18.2 The definition of determinants of various orders.

A good working definition of determinants can be given in successive stages as follows.* A *determinant of the second order* is

* For a more strict and general definition, see Netto, *Die Determinanten* (2nd Ed. 1925), pp. 8-13.

defined as taking its value from the "cross multiplication" of the four elements which compose it and we write

$$\begin{vmatrix} a_{11} & a_{12} \\ a_{21} & a_{22} \end{vmatrix} = a_{11}a_{22} - a_{12}a_{21}.$$

The determinant on the left is simply another way of writing the algebraic expression on the right.

Ex. 1. $\begin{vmatrix} 2 & -1 \\ 0 & 3 \end{vmatrix} = 2 \times 3 - (-1) \times 0 = 6.$

Ex. 2. $\begin{vmatrix} a & h \\ h & b \end{vmatrix} = a \times b - h \times h = ab - h^2.$

Ex. 3. $\begin{vmatrix} x^2 & 2xy \\ 2xy & y^2 \end{vmatrix} = x^2 \times y^2 - 2xy \times 2xy = -3x^2y^2.$

Ex. 4. $\begin{vmatrix} \frac{\partial u}{\partial x} & \frac{\partial u}{\partial y} \\ \frac{\partial v}{\partial x} & \frac{\partial v}{\partial y} \end{vmatrix} = \frac{\partial u}{\partial x}\frac{\partial v}{\partial y} - \frac{\partial u}{\partial y}\frac{\partial v}{\partial x},$

where u and v are two given functions of x and y. This determinant, which is often written

$$\frac{\partial (u, v)}{\partial (x, y)},$$

is termed the *Jacobian* of u and v with respect to x and y.

A *determinant of the third order* is defined in terms of those of the second order by the rule:

$$\begin{vmatrix} a_{11} & a_{12} & a_{13} \\ a_{21} & a_{22} & a_{23} \\ a_{31} & a_{32} & a_{33} \end{vmatrix} = a_{11}\begin{vmatrix} a_{22} & a_{23} \\ a_{32} & a_{33} \end{vmatrix} - a_{12}\begin{vmatrix} a_{21} & a_{23} \\ a_{31} & a_{33} \end{vmatrix} + a_{13}\begin{vmatrix} a_{21} & a_{22} \\ a_{31} & a_{32} \end{vmatrix}$$

$$= a_{11}(a_{22}a_{33} - a_{23}a_{32}) - a_{12}(a_{21}a_{33} - a_{23}a_{31}) + a_{13}(a_{21}a_{32} - a_{22}a_{31})$$

$$= a_{11}a_{22}a_{33} + a_{12}a_{23}a_{31} + a_{13}a_{21}a_{32} - a_{11}a_{23}a_{32} - a_{12}a_{21}a_{33} - a_{13}a_{22}a_{31}.$$

The second-order determinant multiplying a_{11} is derived from the original determinant by omitting the first row and the first column (the row and column intersecting in a_{11}), and similarly for the other determinants. The third-order determinant is thus seen to be a short way of writing an algebraic sum of six terms, each term being the product of three elements of the determinant so chosen that one element comes from each row and one from each column. Further, half the terms are added and half subtracted to form the algebraic sum.

Ex. 1. $$\begin{vmatrix} 1 & 3 & -2 \\ -1 & 0 & 3 \\ 0 & -2 & 1 \end{vmatrix} = \begin{vmatrix} 0 & 3 \\ -2 & 1 \end{vmatrix} - 3\begin{vmatrix} -1 & 3 \\ 0 & 1 \end{vmatrix} - 2\begin{vmatrix} -1 & 0 \\ 0 & -2 \end{vmatrix}$$
$$=(0+6)-3(-1-0)-2(2-0)=6+3-4=5.$$

Ex. 2. $$\begin{vmatrix} 0 & \alpha & \beta \\ \alpha & a & h \\ \beta & h & b \end{vmatrix} = -\alpha\begin{vmatrix} \alpha & h \\ \beta & b \end{vmatrix} + \beta\begin{vmatrix} \alpha & a \\ \beta & h \end{vmatrix} = 2h\alpha\beta - b\alpha^2 - a\beta^2.$$

Ex. 3. $$\begin{vmatrix} a & h & g \\ h & b & f \\ g & f & c \end{vmatrix} = a\begin{vmatrix} b & f \\ f & c \end{vmatrix} - h\begin{vmatrix} h & f \\ g & c \end{vmatrix} + g\begin{vmatrix} h & b \\ g & f \end{vmatrix}$$
$$= abc - af^2 - bg^2 - ch^2 + 2fgh.$$

Ex. 4. $$\begin{vmatrix} \frac{\partial u}{\partial x} & \frac{\partial u}{\partial y} & \frac{\partial u}{\partial z} \\ \frac{\partial v}{\partial x} & \frac{\partial v}{\partial y} & \frac{\partial v}{\partial z} \\ \frac{\partial w}{\partial x} & \frac{\partial w}{\partial y} & \frac{\partial w}{\partial z} \end{vmatrix} = \frac{\partial u}{\partial x}\frac{\partial v}{\partial y}\frac{\partial w}{\partial z} + \frac{\partial u}{\partial y}\frac{\partial v}{\partial z}\frac{\partial w}{\partial x} + \frac{\partial u}{\partial z}\frac{\partial v}{\partial x}\frac{\partial w}{\partial y} - \frac{\partial u}{\partial x}\frac{\partial v}{\partial z}\frac{\partial w}{\partial y} - \frac{\partial u}{\partial y}\frac{\partial v}{\partial x}\frac{\partial w}{\partial z} - \frac{\partial u}{\partial z}\frac{\partial v}{\partial y}\frac{\partial w}{\partial x}$$

This is the Jacobian, $\frac{\partial(u, v, w)}{\partial(x, y, z)}$, of u, v and w as functions of x, y and z.

A *determinant of the fourth order* is then defined :

$$\begin{vmatrix} a_{11} & a_{12} & a_{13} & a_{14} \\ a_{21} & a_{22} & a_{23} & a_{24} \\ a_{31} & a_{32} & a_{33} & a_{34} \\ a_{41} & a_{42} & a_{43} & a_{44} \end{vmatrix} = a_{11}\begin{vmatrix} a_{22} & a_{23} & a_{24} \\ a_{32} & a_{33} & a_{34} \\ a_{42} & a_{43} & a_{44} \end{vmatrix} - a_{12}\begin{vmatrix} a_{21} & a_{23} & a_{24} \\ a_{31} & a_{33} & a_{34} \\ a_{41} & a_{43} & a_{44} \end{vmatrix}$$
$$+ a_{13}\begin{vmatrix} a_{21} & a_{22} & a_{24} \\ a_{31} & a_{32} & a_{34} \\ a_{41} & a_{42} & a_{44} \end{vmatrix} - a_{14}\begin{vmatrix} a_{21} & a_{22} & a_{23} \\ a_{31} & a_{32} & a_{33} \\ a_{41} & a_{42} & a_{43} \end{vmatrix}$$

This is an obvious extension of the rule for third-order determinants and it remains to evaluate each of the determinants shown on the right-hand side and collect terms. It is then found that the determinant denotes the algebraic sum of 24 terms, each term consisting of four of the total of sixteen elements so chosen that one comes from each row and one from each column. Again, half the terms are positive and half negative in the same kind of way as before. The determinant notation represents more and more lengthy expressions, and so becomes more and more convenient, as we proceed.

In general, a *determinant of the nth order* is written down in terms of determinants of the $(n-1)$th order by the rule:

$$\begin{vmatrix} a_{11} & a_{12} & a_{13} & \cdots & a_{1n} \\ a_{21} & a_{22} & a_{23} & \cdots & a_{2n} \\ \cdots & \cdots & \cdots & \cdots & \cdots \\ a_{n1} & a_{n2} & a_{n3} & \cdots & a_{nn} \end{vmatrix} = a_{11} \begin{vmatrix} a_{22} & a_{23} & \cdots & a_{2n} \\ \cdots & \cdots & \cdots & \cdots \\ a_{n2} & a_{n3} & \cdots & a_{nn} \end{vmatrix} - a_{12} \begin{vmatrix} a_{21} & a_{23} & \cdots & a_{2n} \\ \cdots & \cdots & \cdots & \cdots \\ a_{n1} & a_{n3} & \cdots & a_{nn} \end{vmatrix} + \dots .$$

The signs of the successive entries on the right-hand side are alternatively positive and negative until all the n elements of the first row are taken. The $(n-1)$th order determinant multiplying a_{1r} is obtained from the original determinant by the omission of the first row and rth column. The determinant is now to be evaluated by going, step by step, back to determinants of the second order and so to the elements themselves. It is then found that we have the algebraic sum of terms each of which is the product of n of the n^2 elements selected so that one element comes from each row and one from each column. Half the terms have a positive and half a negative sign. The number of terms in the sum is

$$\lfloor n = n(n-1)(n-2) \dots 3 \cdot 2 \cdot 1,$$

which is very large when the order of the determinant is high.

The determinant notation as written above can be sometimes abbreviated to

$$| a_{rs} | \quad (r \text{ and } s = 1, 2, 3, \dots n),$$

where a_{rs} is a typical element (in the rth row and sth column).

A determinant, therefore, is a notation expressing an algebraic sum of terms which are products of the elements of the determinant and its value is of the same nature (e.g. numerical or a function of certain variables) as the elements which compose it. The practical way of finding the value of a given determinant is indicated by the definition, i.e. by successive reduction to determinants of lower and lower order until second-order determinants are obtained and evaluated by cross multiplication.

18.3 Properties of determinants.

The following are simple properties of the determinant notation :

(1) A determinant is unchanged in value if its rows and columns are transposed :

$$\begin{vmatrix} a_{11} & a_{12} & a_{13} & \cdots & a_{1n} \\ a_{21} & a_{22} & a_{23} & \cdots & a_{2n} \\ \multicolumn{5}{c}{\dotfill} \\ a_{n1} & a_{n2} & a_{n3} & \cdots & a_{nn} \end{vmatrix} = \begin{vmatrix} a_{11} & a_{21} & a_{31} & \cdots & a_{n1} \\ a_{12} & a_{22} & a_{32} & \cdots & a_{n2} \\ \multicolumn{5}{c}{\dotfill} \\ a_{1n} & a_{2n} & a_{3n} & \cdots & a_{nn} \end{vmatrix}$$

(2) The value of a determinant is unaltered numerically but changed in sign if two rows (or two columns) are interchanged :

e.g.
$$\begin{vmatrix} a_{12} & a_{11} & a_{13} & \cdots & a_{1n} \\ a_{22} & a_{21} & a_{23} & \cdots & a_{2n} \\ \multicolumn{5}{c}{\dotfill} \\ a_{n2} & a_{n1} & a_{n3} & \cdots & a_{nn} \end{vmatrix} = - \begin{vmatrix} a_{11} & a_{12} & a_{13} & \cdots & a_{1n} \\ a_{21} & a_{22} & a_{23} & \cdots & a_{2n} \\ \multicolumn{5}{c}{\dotfill} \\ a_{n1} & a_{n2} & a_{n3} & \cdots & a_{nn} \end{vmatrix}$$

(3) The value of a determinant is increased k-fold when each element in one row (or column) is increased in this way :

e.g.
$$\begin{vmatrix} ka_{11} & ka_{12} & ka_{13} & \cdots & ka_{1n} \\ a_{21} & a_{22} & a_{23} & \cdots & a_{2n} \\ \multicolumn{5}{c}{\dotfill} \\ a_{n1} & a_{n2} & a_{n3} & \cdots & a_{nn} \end{vmatrix} = k \begin{vmatrix} a_{11} & a_{12} & a_{13} & \cdots & a_{1n} \\ a_{21} & a_{22} & a_{23} & \cdots & a_{2n} \\ \multicolumn{5}{c}{\dotfill} \\ a_{n1} & a_{n2} & a_{n3} & \cdots & a_{nn} \end{vmatrix}$$

(4) A determinant has zero value if the elements of one row (or of one column) are equal or proportional to the corresponding elements of a second row (or of a second column).

(5) If the elements of one row (or of one column) of a determinant appear each as the sum of two parts, then an additive rule applies :

e.g.
$$\begin{vmatrix} a_{11}+a_{11}' & a_{12}+a_{12}' & \cdots & a_{1n}+a_{1n}' \\ a_{21} & a_{22} & \cdots & a_{2n} \\ \multicolumn{4}{c}{\dotfill} \\ a_{n1} & a_{2n} & \cdots & a_{nn} \end{vmatrix} = \begin{vmatrix} a_{11} & a_{12} & \cdots & a_{1n} \\ a_{21} & a_{22} & \cdots & a_{2n} \\ \multicolumn{4}{c}{\dotfill} \\ a_{n1} & a_{n2} & \cdots & a_{nn} \end{vmatrix}$$

$$+ \begin{vmatrix} a_{11}' & a_{12}' & \cdots & a_{1n}' \\ a_{21} & a_{22} & \cdots & a_{2n} \\ \multicolumn{4}{c}{\dotfill} \\ a_{n1} & a_{n2} & \cdots & a_{nn} \end{vmatrix}$$

(6) The value of a determinant is unchanged when a multiple of

the elements of one row (or of one column) is added to the corresponding elements of a second row (or of a second column):

e.g. $$\begin{vmatrix} a_{11}+ka_{21} & a_{12}+ka_{22} & \cdots & a_{1n}+ka_{2n} \\ a_{21} & a_{22} & \cdots & a_{2n} \\ \cdots & \cdots & \cdots & \cdots \\ a_{n1} & a_{n2} & \cdots & a_{nn} \end{vmatrix} = \begin{vmatrix} a_{11} & a_{12} & \cdots & a_{1n} \\ a_{21} & a_{22} & \cdots & a_{2n} \\ \cdots & \cdots & \cdots & \cdots \\ a_{n1} & a_{n2} & \cdots & a_{nn} \end{vmatrix}$$

To make clear the meaning of these properties, it is a useful exercise to verify that they hold in the particular cases of second and third-order determinants. It is also a relatively easy matter to give general proofs, using properties of the values and signs of the terms in the expanded form of a determinant (18.2 above).* The practical evaluation of determinants is often simplified by using the above properties, e.g. a combination of (6) and (4) sometimes shows that a determinant has zero value. The following examples illustrate:

Ex. 1. Show that $\begin{vmatrix} 1 & 2 & 3 \\ 4 & 5 & 6 \\ 7 & 8 & 9 \end{vmatrix} = 0.$

Take the elements of the first row from those of the second and third rows, using property (6) with $k = -1$. Then

$$\begin{vmatrix} 1 & 2 & 3 \\ 4 & 5 & 6 \\ 7 & 8 & 9 \end{vmatrix} = \begin{vmatrix} 1 & 2 & 3 \\ 3 & 3 & 3 \\ 6 & 6 & 6 \end{vmatrix} = 0 \text{ by property (4).}$$

The same result holds if the elements of the determinant (reading by rows from left to right) form any set of integers in Arithmetic Progression.

Ex. 2. Express $\begin{vmatrix} 1 & a & a^2 \\ 1 & b & b^2 \\ 1 & c & c^2 \end{vmatrix}$ in factorial form.

If $a = b$, the first two rows of the determinant become identical and it has zero value. Hence, $(a-b)$ must be a factor of the value of the determinant. Similarly, $(b-c)$ and $(c-a)$ are factors. Since the value of the determinant is of the third degree in a, b and c, it must be a numerical multiple of these factors. The leading term in the expanded value is $1 \times b \times c^2$ (from the "diagonal" of the determinant) and the multiple can only be unity. The value of the determinant is thus $(a-b)(b-c)(c-a)$.

18.4 Minors and co-factors of determinants.

Selecting an element of a given determinant, we delete the row and column intersecting in the element and obtain a determinant of

* See Netto, *op. cit.*, pp. 13-9, and Courant, *Differential and Integral Calculus*, Vol. II (English Ed. 1936), pp. 20-3.

order one less than that of the original determinant. The derived determinant is called the *minor* of the selected element in the given determinant. There are n^2 minors of a determinant of the nth order and each is a determinant of the $(n-1)$th order. In

$$\Delta = | a_{rs} | \quad (r \text{ and } s = 1, 2, 3, \dots n)$$

the minor of the typical element a_{rs} is denoted by Δ_{rs}, as obtained from Δ by omitting the rth row and sth column.

The *co-factor* of a selected element in a given determinant is the minor of the element with a sign attached. The rule of signs is quite simple. If the numbers of the row and column containing the element add to an *even* number, then a *plus* sign is given to the co-factor ; if they add to an *odd* number, then a *minus* sign is allotted. The co-factor of a_{rs} in Δ is denoted by A_{rs} where

$$\begin{aligned} A_{rs} &= \Delta_{rs} \qquad \text{if } (r+s) \text{ is even} \\ &= -\Delta_{rs} \qquad \text{if } (r+s) \text{ is odd.} \end{aligned}$$

It is important to distinguish between the closely related concepts of minors and co-factors. The *minors* of various elements in

$$\begin{vmatrix} a_{11} & a_{12} & a_{13} \\ a_{21} & a_{22} & a_{23} \\ a_{31} & a_{32} & a_{33} \end{vmatrix}$$

are $\Delta_{11} = \begin{vmatrix} a_{22} & a_{23} \\ a_{32} & a_{33} \end{vmatrix}$; $\Delta_{12} = \begin{vmatrix} a_{21} & a_{23} \\ a_{31} & a_{33} \end{vmatrix}$; $\Delta_{21} = \begin{vmatrix} a_{12} & a_{13} \\ a_{32} & a_{33} \end{vmatrix}$; ....

On the other hand, the *co-factors* of the elements are

$$A_{11} = \begin{vmatrix} a_{22} & a_{23} \\ a_{32} & a_{33} \end{vmatrix} ; \; A_{12} = -\begin{vmatrix} a_{21} & a_{23} \\ a_{31} & a_{33} \end{vmatrix} ; \; A_{21} = -\begin{vmatrix} a_{12} & a_{13} \\ a_{32} & a_{33} \end{vmatrix} ; \dots .$$

Our definition of determinants can be re-framed in terms of minors or co-factors. We have

$$\Delta = | a_{rs} | = a_{11}\Delta_{11} - a_{12}\Delta_{12} + a_{13}\Delta_{13} - \dots + (-1)^{n+1} a_{1n}\Delta_{1n},$$

or $$= a_{11}A_{11} + a_{12}A_{12} + a_{13}A_{13} + \dots + a_{1n}A_{1n},$$

i.e. the value of Δ equals the sum of the elements of the first row each multiplied by the corresponding co-factor. A similar result can be shown, by property (2) of 18.3, to hold for the elements of any row. For example, taking the second row, we have

$$\begin{aligned} &a_{21}A_{21} + a_{22}A_{22} + a_{23}A_{23} + \dots + a_{2n}A_{2n} \\ &\qquad = -(a_{21}\Delta_{21} - a_{22}\Delta_{22} + a_{23}\Delta_{23} - \dots \pm a_{2n}\Delta_{2n}). \end{aligned}$$

The bracket on the right-hand side is, by the definition, the value of the determinant obtained by interchanging the first two rows of Δ. The bracket thus equals $(-\Delta)$ and so

$$\Delta = a_{21}A_{21} + a_{22}A_{22} + a_{23}A_{23} + \ldots + a_{2n}A_{2n}.$$

Again, take the sum of the elements of one row of the determinant each multiplied, not by the corresponding co-factors, but by the co-factors of the elements of another row, e.g.

$$a_{21}A_{11} + a_{22}A_{12} + a_{23}A_{13} + \ldots + a_{2n}A_{1n}.$$

This sum must, by the definition, equal Δ except that the first row in Δ is replaced by the row of elements $(a_{21}, a_{22}, a_{23}, \ldots a_{2n})$, i.e. the sum represents a determinant with the first two rows consisting of identical elements. By property (4) of 18.3, the sum is zero.

The following results are now established :

The sum of the elements of any row of $\Delta = |a_{rs}|$ each multiplied by the corresponding co-factor is equal to the value of the determinant. The sum of the elements each multiplied by the co-factor of the corresponding element of another row is zero. In symbols, for unequal values of r and s,

$$\left.\begin{aligned} a_{r1}A_{r1} + a_{r2}A_{r2} + a_{r3}A_{r3} + \ldots + a_{rn}A_{rn} &= \Delta \\ a_{r1}A_{s1} + a_{r2}A_{s2} + a_{r3}A_{s3} + \ldots + a_{rn}A_{sn} &= 0 \end{aligned}\right\}.$$

The first result, an extension of the definition, is often called the *expansion rule* and its use in evaluating determinants is evident. By property (1) of 18.3, the results are true also for the elements of any column of the determinant.

Minors of higher order than those discussed above can also be defined. Selecting two elements not in one row or column and omitting the two rows and two columns containing them, a second-order minor is derived as a determinant of order two less than that of the original determinant. The process can be continued by selecting more and more elements and omitting the relevant rows and columns. In particular, the *principal minors* of $\Delta = |a_{rs}|$ are the minors of various orders obtained by selecting the last, the last two, ... elements of the "principal diagonal" $(a_{11}, a_{22}, a_{33}, \ldots a_{nn})$ and then carrying out the deletion process indicated.

18.5 Linear and homogeneous functions of several variables.

The determinant notation serves to express precisely the properties of linear and homogeneous functions of more than two variables. Let $y=f(x_1, x_2, x_3, \dots x_n)$ be a linear and homogeneous function with continuous partial derivatives of the first two orders:

$$f_r=\frac{\partial y}{\partial x_r} \quad \text{and} \quad f_{rs}=\frac{\partial^2 y}{\partial x_r\,\partial x_s} \quad (r \text{ and } s=1, 2, 3, \dots n).$$

Then, Euler's Theorem (12.8 above) can be extended to give:

$$x_1 f_1 + x_2 f_2 + x_3 f_3 + \dots + x_n f_n = y \quad \text{.................(1)}$$

at any point $(x_1, x_2, x_3, \dots x_n)$. Since the result is identically true, it can be differentiated with respect to any variable x_r:

$$f_r + x_1 f_{1r} + x_2 f_{2r} + x_3 f_{3r} + \dots + x_n f_{rn} = f_r,$$

i.e.

$$x_1 f_{1r} + x_2 f_{2r} + x_3 f_{3r} + \dots + x_n f_{rn} = 0, \quad \text{.................(2)}$$

$$(r=1, 2, 3, \dots n).$$

The n relations (2) are extensions of the results (4) of 12.8 and give each direct second-order partial derivative in terms of the cross-partial derivatives.

We denote by F the determinant of the $(n+1)$th order formed from the block of second-order partial derivatives of f, "bordered" with the first-order partial derivatives:

$$F=\begin{vmatrix} 0 & f_1 & f_2 & f_3 & \dots & f_n \\ f_1 & f_{11} & f_{12} & f_{13} & \dots & f_{1n} \\ f_2 & f_{12} & f_{22} & f_{23} & \dots & f_{2n} \\ \dots & \dots & \dots & \dots & \dots & \dots \\ f_n & f_{1n} & f_{2n} & f_{3n} & \dots & f_{nn} \end{vmatrix}$$

Let F_0 denote the co-factor of the element 0 in F and F_{0rs} the co-factor of the element f_{rs} in F_0 (r and $s=1, 2, 3, \dots n$). Then

$$F_0 = 0 \quad \text{.................................(3)}$$

and

$$F_{0rs} = -\frac{x_r x_s}{y^2} F. \quad \text{...........................(4)}$$

The proofs of (3) and (4) are interesting exercises in the manipulation of determinants:

Multiply the relation (2) by F_{01r} $(r=1, 2, 3, \dots n)$ and add together the n equations obtained. Then

$$x_1(f_{11}F_{011} + f_{12}F_{012} + \dots + f_{1n}F_{01n}) \\ + x_2(f_{21}F_{011} + f_{22}F_{012} + \dots + f_{2n}F_{01n}) + \dots = 0.$$

By the results of 18.4, the coefficient of x_1 is F_0 and the coefficients of $x_2, x_3, \ldots x_n$ are all zero. Hence,

$$F_0 x_1 = 0$$

for any value of x_1, i.e. $F_0 = 0$. This proves (3).

The determinant F can be evaluated as follows:

$$F = \frac{1}{x_1}\begin{vmatrix} 0 & f_1 & f_2 & f_3 & \cdots & f_n \\ x_1 f_1 & x_1 f_{11} & x_1 f_{12} & x_1 f_{13} & \cdots & x_1 f_{1n} \\ f_2 & f_{12} & f_{22} & f_{23} & \cdots & f_{2n} \\ \cdots & \cdots & \cdots & \cdots & \cdots & \cdots \\ f_n & f_{1n} & f_{2n} & f_{3n} & \cdots & f_{nn} \end{vmatrix} \quad \text{by 18.3, (3)}$$

$$= \frac{1}{x_1}\begin{vmatrix} 0 & f_1 & f_2 & f_3 & \cdots & f_n \\ y & 0 & 0 & 0 & \cdots & 0 \\ f_2 & f_{12} & f_{22} & f_{23} & \cdots & f_{2n} \\ \cdots & \cdots & \cdots & \cdots & \cdots & \cdots \\ f_n & f_{1n} & f_{2n} & f_{3n} & \cdots & f_{nn} \end{vmatrix} \quad \text{by 18.3, (6)}$$

adding x_2 times the elements of the third row, x_3 times the elements of the fourth row, ... to the elements of the second row and using (1) and (2) above. Expanding in terms of the elements of the second row and applying a similar manipulative process all over again (with columns instead of rows), we find

$$F = -\frac{y}{x_1}\begin{vmatrix} f_1 & f_2 & f_3 & \cdots & f_n \\ f_{12} & f_{22} & f_{23} & \cdots & f_{2n} \\ \cdots & \cdots & \cdots & \cdots & \cdots \\ f_{1n} & f_{2n} & f_{3n} & \cdots & f_{nn} \end{vmatrix} = -\frac{y}{x_1^2}\begin{vmatrix} x_1 f_1 & f_2 & f_3 & \cdots & f_n \\ x_1 f_{12} & f_{22} & f_{23} & \cdots & f_{2n} \\ \cdots & \cdots & \cdots & \cdots & \cdots \\ x_1 f_{1n} & f_{2n} & f_{3n} & \cdots & f_{nn} \end{vmatrix}$$

$$= -\frac{y}{x_1^2}\begin{vmatrix} y & f_2 & f_3 & \cdots & f_n \\ 0 & f_{22} & f_{23} & \cdots & f_{2n} \\ \cdots & \cdots & \cdots & \cdots & \cdots \\ 0 & f_{2n} & f_{3n} & \cdots & f_{nn} \end{vmatrix} = -\frac{y^2}{x_1^2} F_{011}.$$

Hence, $F_{011} = -\dfrac{x_1^2}{y^2} F$, which is one of the results (4). The other results follow by similar reductions of F.

18.6 The solution of linear equations.

The determinant notation is of particular use in solving linear equations, operating with quadratic forms and making linear substitutions. We propose to give here some account of the use of the notation in the first two problems.*

* For applications of the determinant notation to linear substitutions, see Netto, *op. cit.*, pp. 77 *et seq.* and pp. 100 *et seq.*

As we have seen (11.6 above), a system of exactly n equations in n variables is sufficient to determine the values of the variables provided that the equations are consistent and independent. But this tells us nothing about the actual values obtained in the solution, or even whether one or more alternative solutions exist. These points can be cleared up at least in the simple case where all the equations are of linear form. The solution of the pair of linear equations

$$a_1x+b_1y=c_1 \quad \text{and} \quad a_2x+b_2y=c_2$$

in two variables is (by the method of 2.9 above) given by

$$x=\frac{c_1b_2-c_2b_1}{a_1b_2-a_2b_1} \quad \text{and} \quad y=\frac{a_1c_2-a_2c_1}{a_1b_2-a_2b_1}.$$

Using the determinant notation, the solution can be written

$$\frac{x}{\begin{vmatrix} c_1 & b_1 \\ c_2 & b_2 \end{vmatrix}}=\frac{y}{\begin{vmatrix} a_1 & c_1 \\ a_2 & c_2 \end{vmatrix}}=\frac{1}{\begin{vmatrix} a_1 & b_1 \\ a_2 & b_2 \end{vmatrix}}.$$

The solution is unique and appears in terms of second-order determinants involving the coefficients of the equations.

To generalise the result, consider a system of n linear equations in n variables, $x_1, x_2, x_3, \ldots x_n$:

$$\begin{aligned} a_{11}x_1+a_{12}x_2+a_{13}x_3+\ldots+a_{1n}x_n&=a_1 \\ a_{21}x_1+a_{22}x_2+a_{23}x_3+\ldots+a_{2n}x_n&=a_2 \\ \ldots\ldots\ldots\ldots\ldots\ldots\ldots\ldots\ldots\ldots\ldots\ldots \\ a_{n1}x_1+a_{n2}x_2+a_{n3}x_3+\ldots+a_{nn}x_n&=a_n. \end{aligned}$$

The given coefficients in the equations form a square block of elements giving the determinant $\Delta=|\,a_{rs}$ (r and $s=1, 2, \ldots n$). Here, a_{rs} is the coefficient of the sth variable (x_s) in the rth equation. The constant terms ($a_1, a_2, \ldots a_n$) on the right-hand sides of the equations are kept separate. Assuming that $\Delta \neq 0$, the equations are solved by the following device. Multiply the equations respectively by $A_{11}, A_{21}, \ldots A_{n1}$, the co-factors of the elements of the first column of Δ, and add. Then

$$(a_{11}A_{11}+a_{21}A_{21}+\ldots+a_{n1}A_{n1})x_1+(a_{12}A_{11}+a_{22}A_{21}+\ldots+a_{n2}A_{n1})x_2 \\ +\ldots+(a_{1n}A_{11}+a_{2n}A_{21}+\ldots+a_{nn}A_{n1})x_n=a_1A_{11}+a_2A_{21}+\ldots+a_nA_{n1}.$$

By the results of 18.3, the coefficient of x_1 here is Δ and the coefficients of the other variables are all zero. Further, let Δ_1

denote the value of Δ when the first column of elements is replaced by the column $(a_1, a_2, \ldots a_n)$, so that

$$\Delta_1 = a_1 A_{11} + a_2 A_{21} + \ldots + a_n A_{n1}.$$

The equation then becomes $\Delta x_1 = \Delta_1$, i.e.

$$x_1 = \frac{\Delta_1}{\Delta}.$$

Similar processes give the values of the other variables and so:

The solution of the n linear equations written above is unique and can be expressed in the form

$$\frac{x_1}{\Delta_1} = \frac{x_2}{\Delta_2} = \frac{x_3}{\Delta_3} = \ldots = \frac{x_n}{\Delta_n} = \frac{1}{\Delta},$$

where Δ is the nth order determinant $|a_{rs}|$ and $\Delta_1, \Delta_2, \Delta_3, \ldots \Delta_n$ are the determinants obtained by substituting the column $(a_1, a_2, \ldots a_n)$ for the first, second, third, ... and nth columns of Δ respectively.

The result for two variables given above can be verified as a particular case of this general solution.*

Ex. 1. The equations

$$3x + y - 6z = 0, \quad -x + 3y + 4z = 5, \quad x + y + 2z = 4$$

give

$$\frac{x}{\begin{vmatrix} 0 & 1 & -6 \\ 5 & 3 & 4 \\ 4 & 1 & 2 \end{vmatrix}} = \frac{y}{\begin{vmatrix} 3 & 0 & -6 \\ -1 & 5 & 4 \\ 1 & 4 & 2 \end{vmatrix}} = \frac{z}{\begin{vmatrix} 3 & 1 & 0 \\ -1 & 3 & 5 \\ 1 & 1 & 4 \end{vmatrix}} = \frac{1}{\begin{vmatrix} 3 & 1 & -6 \\ -1 & 3 & 4 \\ 1 & 1 & 2 \end{vmatrix}}$$

i.e.

$$\frac{x}{48} = \frac{y}{36} = \frac{z}{30} = \frac{1}{36},$$

i.e.

$$x = \tfrac{4}{3}, \quad y = 1, \quad z = \tfrac{5}{6}.$$

Ex. 2. In considering the variation of demand for factors of production (14.8 above), we had to solve the equations

$$\eta u + p_a v + p_b w = 0, \quad \sigma u - \frac{bp_b}{a} v + p_b w = \frac{xp}{p_a} \sigma, \quad \sigma u + p_a v - \frac{ap_a}{b} w = 0$$

for the three variables

$$u = x \frac{\partial p}{\partial p_a}, \quad v = \frac{\partial a}{\partial p_a} \quad \text{and} \quad w = \frac{\partial b}{\partial p_a}.$$

* We can now see, at least roughly, why $\Delta = 0$ is a case of failure of our result. If $\Delta = 0$ and $\Delta_1, \Delta_2, \Delta_3, \ldots \Delta_n \neq 0$, only infinite values of the variables can satisfy the equations, and the latter are thus inconsistent for all finite values of the variables. If some of $\Delta_1, \Delta_2, \Delta_3, \ldots \Delta_n$ are also zero, then at least one variable becomes indeterminate (zero divided by zero) and the equations are not independent.

The solution can be written in determinant form

$$\frac{u}{\begin{vmatrix} 0 & p_a & p_b \\ \frac{xp}{p_a}\sigma & -\frac{bp_b}{a} & p_b \\ 0 & p_a & -\frac{ap_a}{b} \end{vmatrix}} = \frac{v}{\begin{vmatrix} \eta & 0 & p_b \\ \sigma & \frac{xp}{p_a}\sigma & p_b \\ \sigma & 0 & -\frac{ap_a}{b} \end{vmatrix}}$$

$$= \frac{w}{\begin{vmatrix} \eta & p_a & 0 \\ \sigma & -\frac{bp_b}{a} & \frac{xp}{p_a}\sigma \\ \sigma & p_a & 0 \end{vmatrix}} = \frac{1}{\begin{vmatrix} \eta & p_a & p_b \\ \sigma & -\frac{bp_b}{a} & p_b \\ \sigma & p_a & -\frac{ap_a}{b} \end{vmatrix}}$$

The first three determinants are easily expanded since one column in each case contains two zeros. The other determinant, expanding in terms of the first column, equals

$$\eta(p_a p_b - p_a p_b) - \sigma\left(-\frac{ap_a^2}{b} - p_a p_b\right) + \sigma\left(p_a p_b + \frac{bp_b^2}{a}\right)$$

$$= \frac{\sigma}{ab}(ap_a + bp_b)^2 = \frac{\sigma}{ab} x^2 p^2,$$

since $xp = ap_a + bp_b$. Hence,

$$\frac{u}{\frac{xp}{b}\sigma(ap_a + bp_b)} = \frac{v}{-\frac{xp}{bp_a}\sigma(ap_a\eta + bp_b\sigma)} = \frac{w}{-xp\sigma(\eta - \sigma)} = \frac{1}{\frac{\sigma}{ab}x^2p^2},$$

i.e. $$(v=)\frac{\partial a}{\partial p_a} = -\frac{a}{p_a}\left(\frac{ap_a}{xp}\eta + \frac{bp_b}{xp}\sigma\right)$$

and $$(w=)\frac{\partial b}{\partial p_a} = -\frac{ab}{xp}(\eta - \sigma).$$

18.7 Quadratic forms in two and three variables.

An expression which is the sum of a number of terms each of the same (given) degree in certain variables can be termed a *form*. Classifying forms according to the degree of the terms, we distinguish linear forms, quadratic forms, cubic forms, and so on. The following analysis is concerned with forms no more complicated than the quadratic and makes their nature clearer than any general remarks we can offer.

The general expression of a linear form is $(ax + by)$, $(ax + by + cz)$, ... when there are two, three, ... variables. Here $a, b, c, \ldots$ are constant coefficients. It is clear that linear forms present no algebraic difficulties and no special theory need be developed.

The general *quadratic form* is

$$(ax^2+by^2+2hxy), \quad (ax^2+by^2+cz^2+2fyz+2gxz+2hxy), \ \ldots$$

according to the number of variables. There is now more algebraic difficulty in dealing with such forms and the main problem that presents itself is to determine whether a given form has the same sign (i.e. positive always or negative always) for values of the variables which are not all zero. The solution of many problems in mathematical analysis depends on the exact conditions under which this property holds.

With two variables, we can write the quadratic form

$$F(x, y)=ax^2+by^2+2hxy=a\left(x+\frac{h}{a}y\right)^2+\frac{ab-h^2}{a}y^2,$$

adopting a process (familiar in elementary algebra) known as "completing the square" in the variable x. It follows that $F(x, y)$ is only positive for all values of x and y not both zero, if the two squares in the above expression have positive coefficients, *i.e.* if a and $(ab-h^2)$ are both positive. Similarly, $F(x, y)$ is negative under the same conditions only if the coefficients are negative, i.e. only if a is negative and $(ab-h^2)$ positive. These conditions are easily seen to be both necessary and sufficient. Writing the expression $(ab-h^2)$ in determinant form, we have the result:

The quadratic form (ax^2+by^2+2hxy) is positive for all values of x and y (other than $x=y=0$) if, and only if,

$$a>0 \quad \text{and} \quad \begin{vmatrix} a & h \\ h & b \end{vmatrix}>0$$

and negative for all values of x and y (other than $x=y=0$) if, and only if, the same expressions are negative and positive respectively.

It may be objected that the condition written is unsymmetrical in the sense that the coefficient of x^2, rather than that of y^2, is selected for determination of sign. In fact, if we complete the square in y instead of in x, we find that $F(x, y)$ is always positive if b and $(ab-h^2)$ are both positive and always negative if b is negative and $(ab-h^2)$ positive. The essential point, however, is that these second conditions are equivalent to the first since $(ab-h^2)$ is only positive if ab is positive, i.e. if a and b have the same sign. Our condition is thus quite symmetrical. It may be objected also that the determinant notation is not necessary here. This is true but the notation is introduced to facilitate the generalisation of the result to cases of more than two variables. It is to be noticed that the determinant used is most easily written from the coefficients of the quadratic form when arranged :

$$\begin{aligned} F(x, y) = ax^2 &+ hxy \\ +hxy &+ by^2. \end{aligned}$$

The determinant is called the *discriminant* of the quadratic form.

In the case of three variables, we write

$$\begin{aligned} F(x, y, z) &= ax^2 + by^2 + cz^2 + 2fyz + 2gxz + 2hxy \\ &= a\left(x + \frac{h}{a}y + \frac{g}{a}z\right)^2 + \frac{ab-h^2}{a}y^2 + \frac{ac-g^2}{a}z^2 + 2\frac{af-gh}{a}yz \\ &= a\left(x + \frac{h}{a}y + \frac{g}{a}z\right)^2 + \frac{ab-h^2}{a}\left(y + \frac{af-gh}{ab-h^2}z\right)^2 \\ &\qquad + \frac{abc-af^2-bg^2-ch^2+2fgh}{ab-h^2}z^2, \end{aligned}$$

where we have "completed the square" twice in succession. Now, $F(x, y, z)$ is positive for any values of the variables not all zero if, and only if, the coefficients of the three squares in the above expression are all positive, i.e. if

$$a, \quad (ab-h^2) \quad \text{and} \quad (abc-af^2-bg^2-ch^2+2fgh)$$

are positive. Similarly, $F(x, y, z)$ is negative for any values of the variables not all zero if, and only if, the same coefficients are all negative, i.e. if the three expressions written above are negative, positive and negative respectively. The determinant notation is again applicable and we have the result :

The quadratic form $(ax^2 + by^2 + cz^2 + 2fyz + 2gxz + 2hxy)$ is positive for all values of x, y and z (other than $x=y=z=0$) if, and only if,

$$a > 0, \quad \begin{vmatrix} a & h \\ h & b \end{vmatrix} > 0 \quad \text{and} \quad \begin{vmatrix} a & h & g \\ h & b & f \\ g & f & c \end{vmatrix} > 0$$

and negative for all values of x, y and z (other than $x=y=z=0$) if, and only if, these expressions are negative, positive and negative respectively.

It can be shown, as before, that the conditions set out here are symmetrical and can be written in two equivalent ways by selecting the coefficient b or c instead of a. The third-order determinant written is called the *discriminant* of the quadratic form and is most easily derived when the form is arranged:

$$\begin{aligned} F(x, y, z) = ax^2 &+ hxy + gxz \\ &+ hxy + by^2 + fyz \\ &+ gxz + fyz + cz^2. \end{aligned}$$

A more restricted problem of the same nature is that of finding the conditions under which a quadratic form preserves a given sign, not for all values of the variables but for values *satisfying a given linear relation* in the variables. The general method of solving this problem is quite simple. The linear relation gives one variable in terms of the others and the quadratic form is then reduced to a form in one less than the original number of variables. The previous results apply at once. Suppose that the two variables of the quadratic form $(ax^2 + by^2 + 2hxy)$ are related $\alpha x + \beta y = 0$, where α and β are constants as well as a, b and c. Then $y = -\frac{\alpha}{\beta} x$ and the quadratic form becomes:

$$ax^2 + b\frac{\alpha^2}{\beta^2}x^2 - 2h\frac{\alpha}{\beta}x^2 = (b\alpha^2 + a\beta^2 - 2h\alpha\beta)\frac{x^2}{\beta^2},$$

which is positive (negative) for all values of x (and for the corresponding values of y given by the relation) if, and only if, the expression $(b\alpha^2 + a\beta^2 - 2h\alpha\beta)$ is positive (negative) From an example given above (18.2), the condition can be written in determinant form:

The quadratic form $(ax^2 + by^2 + 2hxy)$ is positive for all values of x and y (other than $x = y = 0$) which satisfy $\alpha x + \beta y = 0$ if, and only if,

$$\begin{vmatrix} 0 & \alpha & \beta \\ \alpha & a & h \\ \beta & h & b \end{vmatrix} < 0$$

and negative under the same conditions if, and only if, the determinant is positive.

Notice that the third-order determinant is obtained from the discriminant of the original quadratic form by "bordering" with the coefficients α and β from the linear relation.

In the three variables case, the variables in the quadratic form $(ax^2+by^2+cz^2+2fyz+2gxz+2hxy)$ are related $\alpha x+\beta y+\gamma z=0$. Eliminating $x=-\frac{1}{\alpha}(\beta y+\gamma z)$, the form becomes

$$\frac{a}{\alpha^2}(\beta y+\gamma z)^2+by^2+cz^2+2fyz-\frac{2g}{\alpha}(\beta y+\gamma z)z-\frac{2h}{\alpha}(\beta y+\gamma z)y$$

$$=Ay^2+Bz^2+2Hyz,$$

where $\quad A=\frac{1}{\alpha^2}(b\alpha^2+a\beta^2-2h\alpha\beta),\quad B=\frac{1}{\alpha^2}(c\alpha^2+a\gamma^2-2g\alpha\gamma)$

and $\quad H=\frac{1}{\alpha^2}(a\beta\gamma-g\alpha\beta-h\alpha\gamma+f\alpha^2).$

The form is positive for all values of y and z (x being given by the linear relation) if $A>0$ and $AB-H^2>0$, i.e. if

$$b\alpha^2+a\beta^2-2h\alpha\beta>0$$

and
$$(bc-f^2)\alpha^2+(ab-g^2)\beta^2+(ab-h^2)\gamma^2$$
$$-2(af-gh)\beta\gamma-2(bg-fh)\alpha\gamma-2(ch-fg)\alpha\beta>0.$$

The form is negative under the same conditions if the two expressions given above are negative and positive respectively. It is easily verified that the expressions are simple determinants and so :

The quadratic form $(ax^2+by^2+cz^2+2fyz+2gxz+2hxy)$ is positive for all values of x, y and z (other than $x=y=z=0$) which satisfy $\alpha x+\beta y+\gamma z=0$ if, and only if,

$$\begin{vmatrix} 0 & \alpha & \beta \\ \alpha & a & h \\ \beta & h & b \end{vmatrix}<0 \quad\text{and}\quad \begin{vmatrix} 0 & \alpha & \beta & \gamma \\ \alpha & a & h & g \\ \beta & h & b & f \\ \gamma & g & f & c \end{vmatrix}<0$$

and negative under the same conditions if, and only if, the two determinants are positive and negative respectively.

18.8 Examples of quadratic forms.

The following examples illustrate our results :

Ex. 1. $(x^2+2y^2+z^2+yz+2xy)$ is positive for all values of x, y and z which are not all zero ; for

$$1>0,\quad \begin{vmatrix} 1 & 1 \\ 1 & 2 \end{vmatrix}=1>0 \quad\text{and}\quad \begin{vmatrix} 1 & 1 & 0 \\ 1 & 2 & \frac{1}{2} \\ 0 & \frac{1}{2} & 1 \end{vmatrix}=\tfrac{3}{4}>0.$$

Ex. 2. $(x^2-y^2-7z^2+xy)$ is negative for all values of x, y and z subject to the relation $x+y+2z=0$ since

$$\begin{vmatrix} 0 & 1 & 1 \\ 1 & 1 & \frac{1}{2} \\ 1 & \frac{1}{2} & -1 \end{vmatrix} = 1 > 0 \quad \text{and} \quad \begin{vmatrix} 0 & 1 & 1 & 2 \\ 1 & 1 & \frac{1}{2} & 0 \\ 1 & \frac{1}{2} & -1 & 0 \\ 2 & 0 & 0 & -7 \end{vmatrix} = -1 < 0.$$

Or, eliminate $z=-\frac{1}{2}(x+y)$ and write the quadratic form

$$x^2-y^2-\tfrac{7}{4}(x+y)^2+xy=-\tfrac{1}{4}(3x^2+10xy+11y^2).$$

But $(3x^2+10xy+11y^2)$ is positive for all values of x and y since

$$3>0 \quad \text{and} \quad \begin{vmatrix} 3 & 5 \\ 5 & 11 \end{vmatrix} = 8 > 0.$$

The original quadratic form is thus negative for all values of x and y and for $z=-\frac{1}{2}(x+y)$.

Ex. 3. y_1, y_2 and y_3 are given functions of three variables x_1, x_2 and x_3 such that

$$\frac{\partial y_1}{\partial x_2}=\frac{\partial y_2}{\partial x_1}, \quad \frac{\partial y_2}{\partial x_3}=\frac{\partial y_3}{\partial x_2} \quad \text{and} \quad \frac{\partial y_3}{\partial x_1}=\frac{\partial y_1}{\partial x_3}.$$

It is required to find the conditions under which

$$\frac{\partial y_1}{\partial x_1}dx_1^2+\frac{\partial y_1}{\partial x_2}dx_1\,dx_2+\frac{\partial y_1}{\partial x_3}dx_1\,dx_3$$
$$+\frac{\partial y_2}{\partial x_1}dx_1\,dx_2+\frac{\partial y_2}{\partial x_2}dx_2^2+\frac{\partial y_2}{\partial x_3}dx_2\,dx_3$$
$$+\frac{\partial y_3}{\partial x_1}dx_1\,dx_3+\frac{\partial y_3}{\partial x_2}dx_2\,dx_3+\frac{\partial y_3}{\partial x_3}dx_3^2$$

is negative for all values of dx_1, dx_2 and dx_3.* The conditions are

$$\frac{\partial y_1}{\partial x_1}<0 \quad \begin{vmatrix} \frac{\partial y_1}{\partial x_1} & \frac{\partial y_1}{\partial x_2} \\ \frac{\partial y_2}{\partial x_1} & \frac{\partial y_2}{\partial x_2} \end{vmatrix} > 0 \quad \text{and} \quad \begin{vmatrix} \frac{\partial y_1}{\partial x_1} & \frac{\partial y_1}{\partial x_2} & \frac{\partial y_1}{\partial x_3} \\ \frac{\partial y_2}{\partial x_1} & \frac{\partial y_2}{\partial x_2} & \frac{\partial y_2}{\partial x_3} \\ \frac{\partial y_3}{\partial x_1} & \frac{\partial y_3}{\partial x_2} & \frac{\partial y_3}{\partial x_3} \end{vmatrix} < 0,$$

i.e. the Jacobians $\dfrac{\partial y_1}{\partial x_1}$, $\dfrac{\partial (y_1, y_2)}{\partial (x_1, x_2)}$ and $\dfrac{\partial (y_1, y_2, y_3)}{\partial (x_1, x_2, x_3)}$ are negative, positive and negative respectively.

* See Ramsey, *A Contribution to the Theory of Taxation*, Economic Journal, 1927, p. 50.

18.9 Two general results for quadratic forms.

The problems considered above can now be generalised. The general quadratic form in n variables $x_1, x_2, x_3, \dots x_n$ is

$$F(x_1, x_2, x_3, \dots x_n) = a_{11}x_1^2 + a_{22}x_2^2 + \dots \\ + 2a_{12}x_1x_2 + 2a_{13}x_1x_3 + \dots + 2a_{23}x_2x_3 + \dots ,$$

which can be written more symmetrically as

$$\begin{aligned} F(x_1, x_2, x_3, \dots x_n) = {} & a_{11}x_1^2 + a_{12}x_1x_2 + \dots + a_{1n}x_1x_n \\ & + a_{21}x_1x_2 + a_{22}x_2^2 + \dots + a_{2n}x_2x_n \\ & + \dots\dots\dots\dots\dots\dots\dots\dots \\ & + a_{n1}x_1x_n + a_{n2}x_2x_n + \dots + a_{nn}x_n^2 \end{aligned}$$

where the a's denote given coefficients such that the order of the suffixes is immaterial ($a_{21} = a_{12}, \dots$). The determinant $\Delta = |a_{rs}|$, written down from coefficients of the quadratic form, is called the *discriminant* of the form. We also need the " bordered " determinant

$$\Delta' = \begin{vmatrix} 0 & \alpha_1 & \alpha_2 & \dots & \alpha_n \\ \alpha_1 & a_{11} & a_{12} & \dots & a_{1n} \\ \alpha_2 & a_{21} & a_{22} & \dots & a_{2n} \\ \dots & \dots & \dots & \dots & \dots \\ \alpha_n & a_{n1} & a_{n2} & \dots & a_{nn} \end{vmatrix}$$

where $\alpha_1, \alpha_2, \dots \alpha_n$ are some constants.

The quadratic form $F(x_1, x_2, x_3, \dots x_n)$ is said to be *positive definite* if it takes only positive values for all permissible values of the variables which are not all zero. Similarly, the form is *negative definite* if it takes only negative values. Two general results can now be stated :

(1) The quadratic form is positive definite if, and only if, Δ and all its principal minors are positive, i.e.

$$a_{11} > 0, \quad \begin{vmatrix} a_{11} & a_{12} \\ a_{21} & a_{22} \end{vmatrix} > 0, \quad \dots \quad \begin{vmatrix} a_{11} & a_{12} & \dots & a_{1n} \\ a_{21} & a_{22} & \dots & a_{2n} \\ \dots & \dots & \dots & \dots \\ a_{n1} & a_{n2} & \dots & a_{nn} \end{vmatrix} > 0$$

and the form is negative definite if, and only if, the above expressions are alternatively negative and positive.

(2) The quadratic form is positive definite subject to the

relation $\alpha_1 x_1 + \alpha_2 x_2 + \ldots + \alpha_n x_n = 0$ between the variables if, and only if, Δ' and all its principal minors are negative, i.e.

$$\begin{vmatrix} 0 & \alpha_1 & \alpha_2 \\ \alpha_1 & a_{11} & a_{12} \\ \alpha_2 & a_{21} & a_{22} \end{vmatrix} < 0, \quad \ldots \quad \begin{vmatrix} 0 & \alpha_1 & \alpha_2 & \ldots & \alpha_n \\ \alpha_1 & a_{11} & a_{12} & \ldots & a_{1n} \\ \alpha_2 & a_{21} & a_{22} & \ldots & a_{2n} \\ \ldots & \ldots & \ldots & \ldots & \ldots \\ \alpha_n & a_{n1} & a_{n2} & \ldots & a_{nn} \end{vmatrix} < 0$$

and the form is negative definite under the same conditions if, and only if, the above expressions are alternatively positive and negative.

The proofs of these general results follow lines similar to those indicated in the particular cases already established and no further details need be given here. The proof of (1) depends on the reduction of the quadratic form to the sum of n squares, all the coefficients of which must be positive (or all negative). The proof of (2) is obtained when the quadratic form is reduced from one in n variables to one in $(n-1)$ variables by means of the given relation.

EXAMPLES XVIII

Determinants

1. Show that $\begin{vmatrix} 1 & 1 & 1 \\ 1 & 2 & 3 \\ 1 & 3 & 6 \end{vmatrix} = 1$ and $\begin{vmatrix} 1+a & 1 & 1 \\ 1 & 1+b & 1 \\ 1 & 1 & 1+c \end{vmatrix} = abc + ab + ca + bc.$

2. Evaluate $\begin{vmatrix} 0 & \alpha & \beta & \gamma \\ \alpha & a & h & g \\ \beta & h & b & f \\ \gamma & g & f & c \end{vmatrix}$

3. Use property (6) of 18.3 to show that

$$\begin{vmatrix} a & b-a & c+b \\ b & c-b & a+c \\ c & a-c & b+a \end{vmatrix} = \begin{vmatrix} a & b & c \\ b & c & a \\ c & a & b \end{vmatrix} \quad \text{and} \quad \begin{vmatrix} 0 & 1 & 1 \\ 1 & a+b & b \\ 1 & a & a+b \end{vmatrix} = \begin{vmatrix} 0 & 1 & 1 \\ 1 & a & 0 \\ 1 & 0 & b \end{vmatrix}$$

Hence evaluate the determinants.

4. Show that $\begin{vmatrix} b+c & a & a \\ b & c+a & b \\ c & c & a+b \end{vmatrix} = 4abc$

by noticing that the determinant vanishes if (e.g.) $a=0$.

5. Show that $\begin{vmatrix} x & a & a \\ a & x & a \\ a & a & x \end{vmatrix} = (x-a)^2 \begin{vmatrix} 1 & 0 & a \\ 0 & 1 & a \\ -1 & -1 & x \end{vmatrix}$ (18.3, (3) and (6)).

Deduce that the determinant equated to zero gives a cubic equation in x with two roots $x=a$ and a third root $x=-2a$.

6. Prove that $\begin{vmatrix} a & a^2 & a^3-1 \\ b & b^2 & b^3-1 \\ c & c^2 & c^3-1 \end{vmatrix} = (abc-1)\begin{vmatrix} 1 & a & a^2 \\ 1 & b & b^2 \\ 1 & c & c^2 \end{vmatrix}$ (18.3, (2) and (5)),

and hence evaluate the determinant.

7. Factorise $\begin{vmatrix} 1 & a_1 & a_1^2 & \dots & a_1^{n-1} \\ 1 & a_2 & a_2^2 & \dots & a_2^{n-1} \\ \cdots & \cdots & \cdots & \cdots & \cdots \\ 1 & a_n & a_n^2 & \dots & a_n^{n-1} \end{vmatrix}$

8. Write down the co-factors of the elements of the second and third rows of the second determinant of Example 1 above and verify the results of 18.4 in this case :

$$(1+a)A_{21}+A_{22}+A_{23}=0 \quad \text{and} \quad (1+a)A_{31}+A_{32}+A_{33}=0.$$

9. Show that $u=ax^{\alpha}y^{\beta}z^{\gamma}$ (where $\alpha+\beta+\gamma=1$) is a linear and homogeneous function for which

$$\frac{\partial u}{\partial x}=\frac{\alpha u}{x}, \ \dots, \ \frac{\partial^2 u}{\partial x^2}=\frac{\alpha(\alpha-1)u}{x^2}, \ \dots, \ \frac{\partial^2 u}{\partial x\,\partial y}=\frac{\alpha\beta u}{xy}, \ \dots .$$

Verify the determinant properties of 18.5.

Linear equations

10. Solve $2x-y=2$, $3y+2z=16$ and $3z+5x=21$.

11. If a, b and c are given constants, find the solution of

$$-x+y+z=a, \quad x-y+z=b \quad \text{and} \quad x+y-z=c.$$

12. Use the determinant of 18.3, Ex. 2, to solve the equations

$$x+y+z=1, \quad ax+by+cz=d, \quad a^2x+b^2y+c^2z=d^2,$$

where a, b and c are constants.

• **13.** From the result of 18.6 verify that the equations

$$\begin{aligned} a_1x+b_1y+c_1z+d_1&=0,\\ a_2x+b_2y+c_2z+d_2&=0,\\ a_3x+b_3y+c_3z+d_3&=0, \end{aligned}$$

have a unique solution which can be written

$$\frac{x}{\Delta_1}=\frac{y}{\Delta_2}=\frac{z}{\Delta_3}=\frac{1}{\Delta_4},$$

where Δ_1, Δ_2, Δ_3 and Δ_4 are the co-factors of λ_1, λ_2, λ_3 and λ_4 in

$$\begin{vmatrix} \lambda_1 & \lambda_2 & \lambda_3 & \lambda_4 \\ a_1 & b_1 & c_1 & d_1 \\ a_2 & b_2 & c_2 & d_2 \\ a_3 & b_3 & c_3 & d_3 \end{vmatrix}.$$

Generalise this form of the solution of linear equations.

14. Show that the solution of the three equations of the previous example satisfies a fourth equation $a_4x+b_4y+c_4z+d_4=0$ provided that

$$\begin{vmatrix} a_1 & b_1 & c_1 & d_1 \\ a_2 & b_2 & c_2 & d_2 \\ a_3 & b_3 & c_3 & d_3 \\ a_4 & b_4 & c_4 & d_4 \end{vmatrix} = 0.$$

This is the condition that four linear equations in three variables are consistent. Generalise it.

Quadratic forms

15. Show that $(4xy-2x^2-3y^2)$ is negative definite and $(x^2+y^2+z^2-yz)$ positive definite for all values of the variables (not all zero).

16. If x and y take any values subject to $2x+y=0$, show that (x^2+y^2+3xy) is negative definite. Is this quadratic form also negative definite for any values of the variables?

17. Show that $(xy+yz+xz-x^2-y^2-z^2)$ is negative definite for values of the variables which are such that $x+y+z=0$.

18. If a, b, h, α and β have *positive* values, show that the quadratic form (ax^2+by^2+2hxy) is always positive definite subject to $\alpha x-\beta y=0$.

19. $u=f(x, y, z)$ is a function of three independent variables. Obtain the second differential d^2u and find the conditions that it is positive definite for values of dx, dy and dz, (*a*) without restriction and (*b*) subject to

$$du=f_x dx+f_y dy+f_z dz=0.$$

20. Write the condition that (ax^2+by^2+2hxy) is positive definite subject to $\alpha x+\beta y=0$. Show that the quadratic form is positive definite for *any* values of α and β under conditions identical with those required for the form to be positive definite for unrestricted variation of x and y. Generalise and indicate the relation of the conditions for relative definite quadratic forms to those for unrestricted definite quadratic forms.

CHAPTER XIX

FURTHER PROBLEMS OF MAXIMUM AND MINIMUM VALUES

19.1 Maximum and minimum values of a function of several variables.

THE definition of extreme values of a function of any number of variables can be put most precisely in the form (see 17.5 above):

The function $y=f(x_1, x_2, \ldots x_n)$ of n independent variables has a maximum (minimum) value at $x_1=a_1$, $x_2=a_2$, ... $x_n=a_n$ if $f(a_1+x_1, a_2+x_2, \ldots a_n+x_n)-f(a_1, a_2, \ldots a_n)$ is negative (positive) for all small values of $x_1, x_2, \ldots x_n$.

The analysis which follows is set out, for convenience, in terms of functions of two variables. It is, however, quite general in form.

It is assumed that the function $z=f(x, y)$ has continuous partial derivatives up to any desired order at the point (a, b). Then Taylor's Theorem gives, for any integral value of n,

$$f(a+x, b+y)-f(a, b)=df+\frac{1}{\underline{|2}}d^2f+\frac{1}{\underline{|3}}d^3f+\ldots+\frac{1}{\underline{|n}}d^nf+R_n(x, y),$$

where the successive differentials of the function are taken at the point (a, b) and with arbitrary increments $dx=x$ and $dy=y$, and where

$$R_n(x, y)=\frac{1}{\underline{|n+1}}d^{n+1}f \quad \text{at the point } (a+\theta x, b+\theta y).$$

As before, θ denotes some positive fraction.

(1) Suppose that $df\neq 0$ for some variations in the variables from the point (a, b). Then by the continuity of the partial derivatives of the function, $df\neq 0$ also for some variations from the point $(a+\theta x, b+\theta y)$, provided only that x and y are sufficiently small. Further, if $df=x\frac{\partial z}{\partial x}+y\frac{\partial z}{\partial y}$ happens to be positive for certain values

of x and y, then it must be negative for values of x and y which are numerically equal but opposite in sign. Hence, df must take both positive and negative values for different variations from the point $(a+\theta x, b+\theta y)$. Taylor's Theorem $(n=0)$ gives

$$f(a+x, b+y)-f(a, b)=df \quad \text{at } (a+\theta x, b+\theta y),$$

i.e. $f(a+x, b+y)-f(a, b)$ is sometimes positive and sometimes negative according to the values allotted to x and y. It follows, from our definition, that $f(a, b)$ cannot be an extreme (maximum or minimum) value of the function. If $z=f(x, y)$ has an extreme value at (a, b), then df must be zero for all variations from (a, b).

(2) Suppose that $df=0$ for all variations while $d^2f \neq 0$ for some variations in the variables from the point (a, b). By continuity again, we can take $d^2f \neq 0$ also for some variations from $(a+\theta x, b+\theta y)$, provided that x and y are sufficiently small. Taylor's Theorem, with $n=1$, now gives

$$f(a+x, b+y)-f(a, b)=\frac{1}{\lfloor 2}d^2f \quad \text{at } (a+\theta x, b+\theta y)$$

and, using the definition of extreme values, we can distinguish three cases * :

(*a*) If $d^2f<0$ for all variations from the point (a, b), then $z=f(x, y)$ has a maximum value at this point.

(*b*) If $d^2f>0$ for all variations from the point (a, b), then $z=f(x, y)$ has a minimum value at this point.

(*c*) If d^2f is positive for some variations and negative for others, then $z=f(x, y)$ has a stationary, but not an extreme, value at the point (a, b), i.e. the point corresponds to a " saddle point " of the function.

(3) If $df=d^2f=0$ for all variations from the point (a, b), it is necessary to proceed to the terms in the Taylor series which involve the third and higher order differentials. The analysis, though possible, becomes complex and, as the cases considered are not often met with in practice, we need not examine them.

Collecting our results and extending them to the general case of a function of any number of variables, we have

* It is possible that $d^2f<0$ (or $d^2f>0$) for all variations *except* some which make $d^2f=0$. This is not included in the three cases here distinguished and further treatment is necessary to allow for such possibilities.

GENERAL CRITERION FOR MAXIMUM AND MINIMUM VALUES

(1) If $y=f(x_1, x_2, \dots x_n)$ has an extreme value at $x_1=a_1$, $x_2=a_2, \dots x_n=a_n$, then $dy=0$ for all variations of the variables from these values.

(2) If $dy=0$ for all variations and $d^2y \neq 0$ for some variations of the variables from the values $(a_1, a_2, \dots a_n)$, then y has a stationary value at this point which is a maximum value if $d^2y<0$ for all variations, a minimum value if $d^2y>0$ for all variations and a "saddle" value if d^2y takes positive and negative values for different variations.

The conditions given are respectively necessary and sufficient for extreme values. But the criterion is not complete, i.e. not necessary *and* sufficient, since no allowance is made for cases where $d^2y=0$ for all variations from the values $(a_1, a_2, \dots a_n)$.

The differential form of the criterion can be expanded to give conditions relating to the partial derivatives of the function. The necessary condition $(dy=0)$ simply implies that all the partial derivatives of y must vanish at an extreme value (see 14.2 above). The sufficient conditions relate to the second differential

$$\begin{aligned} d^2y &= f_{x_1x_1}\, dx_1^2 + f_{x_1x_2} dx_1\, dx_2 + \dots \\ &\quad + f_{x_1x_2}\, dx_1 dx_2 + f_{x_2x_2}\, dx_2^2 + \dots \\ &\quad + \dots , \end{aligned}$$

all the partial derivatives being evaluated at the point $(a_1, a_2, \dots a_n)$. A maximum (minimum) value of y is obtained if this quadratic form is negative (positive) definite for all variations $dx_1, dx_2, \dots dx_n$. The conditions for either case are given in 18.9 above.

In the case of two independent variables, a point where $f_x=f_y=0$ gives a maximum value of $z=f(x, y)$ if

$$\begin{aligned} d^2z &= f_{xx}\, dx^2 + f_{xy}\, dx\, dy \\ &\quad + f_{xy} dx\, dy + f_{yy} dy^2 \end{aligned}$$

is negative definite, i.e. if

$$f_{xx}<0 \quad \text{and} \quad \begin{vmatrix} f_{xx} & f_{xy} \\ f_{xy} & f_{yy} \end{vmatrix} > 0.$$

The point gives a minimum value of z if the first inequality is reversed. The second inequality, which is the same in both cases of extreme values, is

$$f_{xx}f_{yy} > (f_{xy})^2.$$

The conditions are those previously given (14.2 above).

If there are three independent variables, then a point where $f_x = f_y = f_z = 0$ gives a maximum value of $u = f(x, y, z)$ if

$$\begin{aligned} d^2u = f_{xx}\,dx^2 &+ f_{xy}\,dx\,dy + f_{xz}\,dx\,dz \\ + f_{xy}\,dx\,dy &+ f_{yy}\,dy^2 \quad + f_{yz}\,dy\,dz \\ + f_{xz}\,dx\,dz &+ f_{yz}\,dy\,dz \quad + f_{zz}\,dz^2 \end{aligned}$$

is negative definite, i.e. if

$$f_{xx} < 0, \quad \begin{vmatrix} f_{xx} & f_{xy} \\ f_{xy} & f_{yy} \end{vmatrix} > 0 \quad \text{and} \quad \begin{vmatrix} f_{xx} & f_{xy} & f_{xz} \\ f_{xy} & f_{yy} & f_{yz} \\ f_{xz} & f_{yz} & f_{zz} \end{vmatrix} < 0.$$

The point gives a minimum value of u if the first and third inequalities are reversed.

19.2 Relative maximum and minimum values.

The advantage of the differential form of the criterion for extreme values is that it applies, with suitable restrictions, even when the variables are not independent. A function $y = f(x_1, x_2, \dots x_n)$ has an extreme value relative to a number of given side relations, $\phi_1(x_1, x_2, \dots x_n) = 0$, $\phi_2(x_1, x_2, \dots x_n) = 0, \dots$, if the conditions previously written are satisfied, provided that the differentials of y are expressed in terms of the independent variables only, the other variables being eliminated by the side relations.

With a single side relation, we seek the extreme values of $y = f(x_1, x_2, \dots x_n)$ relative to $\phi(x_1, x_2, \dots x_n) = 0$. The necessary condition is that

$$dy = f_{x_1}\,dx_1 + f_{x_2}\,dx_2 + f_{x_3}\,dx_3 + \dots + f_{x_n}\,dx_n = 0$$

for all increments which satisfy

$$\phi_{x_1}\,dx_1 + \phi_{x_2}\,dx_2 + \phi_{x_3}\,dx_3 + \dots + \phi_{x_n}\,dx_n = 0.$$

Taking x_1 as the dependent variable given by the side relation, we can eliminate dx_1 and obtain

$$dy = \left(f_{x_2} - f_{x_1}\frac{\phi_{x_2}}{\phi_{x_1}}\right)dx_2 + \left(f_{x_3} - f_{x_1}\frac{\phi_{x_3}}{\phi_{x_1}}\right)dx_3 + \dots + \left(f_{x_n} - f_{x_1}\frac{\phi_{x_n}}{\phi_{x_1}}\right)dx_n = 0$$

for all increments in $x_2, x_3, \ldots x_n$. Hence,

$$\frac{f_{x_1}}{\phi_{x_1}} = \frac{f_{x_2}}{\phi_{x_2}} = \frac{f_{x_3}}{\phi_{x_3}} = \ldots = \frac{f_{x_n}}{\phi_{x_n}}.$$

These are the necessary conditions (see 14.6 above). To obtain sufficient conditions, we write

$$d^2y = f_{x_1}\, d^2x_1 + f_{x_1x_1}\, dx_1{}^2 + f_{x_2x_2}\, dx_2{}^2 + \ldots + 2f_{x_1x_2}\, dx_1\, dx_2 + \ldots,$$

where $$\phi_{x_1}\, dx_1 + \phi_{x_2}\, dx_2 + \ldots + \phi_{x_n}\, dx_n = 0$$

and $$\phi_{x_1}\, d^2x_1 + \phi_{x_1x_1}\, dx_1{}^2 + \phi_{x_2x_2}\, dx_2{}^2 + \ldots + 2\phi_{x_1x_2}\, dx_1\, dx_2 + \ldots = 0.$$

Eliminating d^2x_1 by means of the last relation, we have

$$\begin{aligned} d^2y = &\left(f_{x_1x_1} - \frac{f_{x_1}}{\phi_{x_1}}\phi_{x_1x_1}\right) dx_1{}^2 + \left(f_{x_1x_2} - \frac{f_{x_1}}{\phi_{x_1}}\phi_{x_1x_2}\right) dx_1\, dx_2 + \ldots \\ &+ \left(f_{x_1x_2} - \frac{f_{x_1}}{\phi_{x_1}}\phi_{x_1x_2}\right) dx_1\, dx_2 + \left(f_{x_2x_2} - \frac{f_{x_1}}{\phi_{x_1}}\phi_{x_2x_2}\right) dx_2{}^2 + \ldots \\ &+ \ldots, \end{aligned}$$

where $$\phi_{x_1}\, dx_1 + \phi_{x_2}\, dx_2 + \ldots + \phi_{x_n}\, dx_n = 0.$$

A point which satisfies the necessary conditions gives a relative maximum (minimum) value of y if d^2y is a negative (positive) definite quadratic form for all increments $dx_1, dx_2, \ldots dx_n$ subject to the relation written above. The conditions for either case, given in 18.9 above, indicate the appropriate signs for the principal minors of the determinant formed from the coefficients in d^2y, " bordered " with the partial derivatives $\phi_{x_1}, \phi_{x_2}, \ldots \phi_{x_n}$.

In the case of two variables, a point satisfying the necessary conditions $\phi(x, y) = 0$ and $\dfrac{f_x}{\phi_x} = \dfrac{f_y}{\phi_y}$ gives a maximum value of $z = f(x, y)$ relative to $\phi(x, y) = 0$ if

$$\begin{aligned} d^2z = &\left(f_{xx} - \frac{f_x}{\phi_x}\phi_{xx}\right) dx^2 + \left(f_{xy} - \frac{f_x}{\phi_x}\phi_{xy}\right) dx\, dy \\ &+ \left(f_{xy} - \frac{f_x}{\phi_x}\phi_{xy}\right) dx\, dy + \left(f_{yy} - \frac{f_x}{\phi_x}\phi_{yy}\right) dy^2 \end{aligned}$$

is negative definite subject to $\phi_x dx + \phi_y dy = 0$, i.e. if

$$\begin{vmatrix} 0 & \phi_x & \phi_y \\ \phi_x & f_{xx} - \dfrac{f_x}{\phi_x}\phi_{xx} & f_{xy} - \dfrac{f_x}{\phi_x}\phi_{xy} \\ \phi_y & f_{xy} - \dfrac{f_x}{\phi_x}\phi_{xy} & f_{yy} - \dfrac{f_x}{\phi_x}\phi_{yy} \end{vmatrix} > 0.$$

Expanding the determinant, the condition is

$$2\left(f_{xy}-\frac{f_x}{\phi_x}\phi_{xy}\right)\phi_x\phi_y-\left(f_{yy}-\frac{f_x}{\phi_x}\phi_{yy}\right)\phi_x^2-\left(f_{xx}-\frac{f_x}{\phi_x}\phi_{xx}\right)\phi_y^2>0,$$

i.e.

$$\phi_x(f_{xx}\phi_y^2-2f_{xy}\phi_x\phi_y+f_{yy}\phi_x^2)-f_x(\phi_{xx}\phi_y^2-2\phi_{xy}\phi_x\phi_y+\phi_{yy}\phi_x^2)<0.$$

This agrees with the result previously obtained (14.6 above) when we remember that $\frac{f_x}{\phi_x}=\frac{f_y}{\phi_y}$ by the necessary conditions.

19.3 Examples of maximum and minimum values.

The results of the previous two sections are illustrated by the following examples :

Ex. 1. If $u=x^2+y^2+z^2$ is a function of three independent variables, then $\quad du=2(x\,dx+y\,dy+z\,dz)\quad$ and $\quad d^2u=2(dx^2+dy^2+dz^2)$.

Extreme values of u occur only where $du=0$ for all variations, i.e. where $x=y=z=0$. Since $d^2u>0$ for all variations from any point, these values must give a minimum value of u. Hence, u has a single minimum value, $u=0$, which occurs where $x=y=z=0$. This is also clear since u is positive except when x, y and z are all zero.

Ex. 2. If x, y and z are independent variables and

$$u=x+2z+yz-x^2-y^2-z^2,$$

then

$$du=(1-2x)\,dx+(z-2y)\,dy+(2+y-2z)\,dz$$

and

$$d^2u=-2(dx^2+dy^2+dz^2)+2dy\,dz.$$

Extreme values of u can occur where

$$1-2x=z-2y=2+y-2z=0,$$

i.e.

$$x=\tfrac{1}{2},\quad y=\tfrac{2}{3},\quad z=\tfrac{4}{3},\quad \text{giving}\quad u=\tfrac{19}{12}.$$

But d^2u is negative definite at all points since

$$-2<0,\quad \begin{vmatrix}-2 & 0\\ 0 & -2\end{vmatrix}=4>0\quad\text{and}\quad\begin{vmatrix}-2 & 0 & 0\\ 0 & -2 & 1\\ 0 & 1 & -2\end{vmatrix}=-6<0.$$

Hence, u has a single maximum value $\frac{19}{12}$.

Ex. 3. To find the shortest distance of the origin of co-ordinates from the plane with equation $ax+by+cz=d$:

Let u be the square of the distance from the origin to the point with co-ordinates (x, y, z) on the plane. Then

$$u=x^2+y^2+z^2\quad\text{where}\quad ax+by+cz=d.$$

The relative maximum of u is required. The necessary condition is

$$du = 2(x\,dx + y\,dy + z\,dz) = 0,$$

subject to $$a\,dx + b\,dy + c\,dz = 0.$$

So $$\frac{x}{a} = \frac{y}{b} = \frac{z}{c}.$$

Since $ax + by + cz = d$, these equations give

$$x = \frac{ad}{a^2 + b^2 + c^2}, \quad y = \frac{bd}{a^2 + b^2 + c^2}, \quad z = \frac{cd}{a^2 + b^2 + c^2}$$

and the corresponding value of u is

$$u = \frac{d^2}{a^2 + b^2 + c^2}.$$

Taking the equation of the plane as giving x as a function of y and z, then $d^2x = 0$, and

$$d^2u = 2x\,d^2x + 2(dx^2 + dy^2 + dz^2) = 2(dx^2 + dy^2 + dz^2) > 0$$

for variations from any point. The positions obtained is thus one of relative minimum. The shortest distance required is the square root of the minimum value of u, i.e. it is

$$\frac{d}{\sqrt{a^2 + b^2 + c^2}}.$$

Ex. 4. To find the rectangular block of maximum volume we can cut from a sphere of radius a (14.7, Ex. 2), we require the maximum value of $V = 8xyz$ relative to $x^2 + y^2 + z^2 = a^2$. The necessary condition is

$$dV = 8(yz\,dx + xz\,dy + xy\,dz) = 0 \quad \text{subject to} \quad 2(x\,dx + y\,dy + z\,dz) = 0.$$

Hence, $$\frac{yz}{x} = \frac{xz}{y} = \frac{xy}{z}.$$

Using $x^2 + y^2 + z^2 = a^2$, we find $x = y = z = \dfrac{a}{\sqrt{3}}$, giving $V = \dfrac{8a^3}{3\sqrt{3}}$.

Now, $$d^2V = 8yz\,d^2x + 16(z\,dx\,dy + x\,dy\,dz + y\,dx\,dz),$$

where $$x\,dx + y\,dy + z\,dz = 0$$

and $$x\,d^2x + dx^2 + dy^2 + dz^2 = 0,$$

where the given relation determines x as a function of y and z. On eliminating d^2x, we have

$$d^2V = 8\left\{2z\,dx\,dy + 2x\,dy\,dz + 2y\,dx\,dz - \frac{yz}{x}(dx^2 + dy^2 + dz^2)\right\},$$

where $$x\,dx + y\,dy + z\,dz = 0.$$

At the point where $x = y = z = \dfrac{a}{\sqrt{3}}$,

$$d^2V = \frac{8a}{\sqrt{3}}(2dx\,dy + 2dy\,dz + 2dx\,dz - dx^2 - dy^2 - dz^2),$$

subject to $$dx + dy + dz = 0.$$

For a relative maximum of V, the quadratic form

$$\begin{array}{lll} -dx^2 & +dx\,dy & +dx\,dz \\ +dx\,dy & -dy^2 & +dy\,dz \\ +dx\,dz & +dy\,dz & -dz^2 \end{array}$$

must be negative definite subject to $dx+dy+dz=0$. The conditions for this are satisfied since

$$\begin{vmatrix} 0 & 1 & 1 \\ 1 & -1 & 1 \\ 1 & 1 & -1 \end{vmatrix} = 4 > 0 \quad \text{and} \quad \begin{vmatrix} 0 & 1 & 1 & 1 \\ 1 & -1 & 1 & 1 \\ 1 & 1 & -1 & 1 \\ 1 & 1 & 1 & -1 \end{vmatrix} = -12 < 0.$$

Hence, $V=\dfrac{8a^3}{3\sqrt{3}}$ is the relative maximum value of V.

19.4 The stability of demand for factors of production.

The demands of a firm or industry for two factors of production (14.8 above) are stable if the constant product curves, obtained from the given production function, are convex to the origin at all relevant points. The conditions for equilibrium and stability of production can now be examined in more general cases.

A good X is produced with n variable factors $A_1, A_2, A_3, \dots A_n$ according to the production function $x=f(a_1, a_2, a_3, \dots a_n)$, which is assumed to have continuous partial derivatives of the first two orders:

$$f_r=\frac{\partial x}{\partial a_r} \quad \text{and} \quad f_{rs}=\frac{\partial^2 x}{\partial a_r\,\partial a_s} \quad (r \text{ and } s=1, 2, 3, \dots n).$$

Our first problem is, given the market prices $p_1, p_2, p_3, \dots p_n$ of the factors, to find the grouping which produces a given output x at minimum cost. We have to minimise $\Pi=a_1p_1+a_2p_2+a_3p_3+\dots+a_np_n$ relative to $f(a_1, a_2, a_3, \dots a_n)=x$. The necessary condition is

$$d\Pi=p_1\,da_1+p_2\,da_2+p_3\,da_3+\dots+p_n\,da_n=0,$$

subject to $$f_1\,da_1+f_2\,da_2+f_3\,da_3+\dots+f_n\,da_n=0,$$

i.e. $$\frac{p_1}{f_1}=\frac{p_2}{f_2}=\frac{p_3}{f_3}=\dots=\frac{p_n}{f_n}.$$

There are $(n-1)$ equations which, with the given side relation, determine the equilibrium employments of the factors in terms of the given output and the given prices of the factors. We thus have the demand for each factor and we can write

$$a_r=\phi_r(x, p_1, p_2, p_3, \dots p_n) \quad (r=1, 2, 3, \dots n).$$

Taking the side relation as giving a_1 as a function of $a_2, a_3, \ldots a_n$, then, differentiating twice, we have

$$f_1\,d^2a_1 + f_{11}\,da_1^{\,2} + f_{22}\,da_2^{\,2} + \ldots + 2f_{12}\,da_1\,da_2 + \ldots = 0.$$

But $\quad d^2\Pi = d(p_1\,da_1 + p_2\,da_2 + \ldots + p_n\,da_n) = p_1\,d^2a_1$

$$= -\frac{p_1}{f_1}(f_{11}\,da_1^{\,2} + f_{22}\,da_2^{\,2} + \ldots + 2f_{12}\,da_1\,da_2 + \ldots).$$

For a minimum value of Π, we have $d^2\Pi > 0$ and

$$\begin{aligned} & f_{11}\,da_1^{\,2} + f_{12}\,da_1\,da_2 + f_{13}\,da_1\,da_3 + \ldots + f_{1n}\,da_1\,da_n \\ & + f_{12}\,da_1\,da_2 + f_{22}\,da_2^{\,2} + f_{23}\,da_2\,da_3 + \ldots + f_{2n}\,da_2\,da_n \\ & + \quad \cdots\cdots\cdots\cdots\cdots\cdots\cdots\cdots \\ & + f_{1n}\,da_1\,da_n + f_{2n}\,da_2\,da_n + f_{3n}\,da_3\,da_n + \ldots + f_{nn}\,da_n^{\,2} \end{aligned}$$

is a negative definite quadratic form subject to

$$f_1\,da_1 + f_2\,da_2 + f_3\,da_3 + \ldots + f_n\,da_n = 0.$$

From 18.9, (2), a sufficient condition is

$$\begin{vmatrix} 0 & f_1 & f_2 \\ f_1 & f_{11} & f_{12} \\ f_2 & f_{12} & f_{22} \end{vmatrix} > 0, \quad \begin{vmatrix} 0 & f_1 & f_2 & f_3 \\ f_1 & f_{11} & f_{12} & f_{13} \\ f_2 & f_{12} & f_{22} & f_{23} \\ f_3 & f_{13} & f_{23} & f_{33} \end{vmatrix} < 0, \ldots .$$

Suppose that $f_1, f_2, f_3, \ldots f_n$ are all positive and that the inequalities above are satisfied for all combinations of factors within a relevant range. Then, for *any* set of outputs and prices of the factors (within a certain range), equilibrium is possible and the position determined is stable. We have the *stability conditions* for the demand for factors of production, conditions which are given as limitations on the form of the production function. For a production function satisfying the conditions, the demand functions (ϕ_r) for the factors are uniquely determined and stable. If there are only three factors, it is easily shown that the stability conditions imply simply that the constant product surfaces in factor space are downward sloping and convex to the origin at all points, at least within a certain range.

19.5 Partial elasticities of substitution.

We require the following notations which represent certain features of the production function $x = f(a_1, a_2, a_3, \ldots a_n)$. For any combination of the factors, define

$$\kappa_r = \frac{a_r f_r}{a_1 f_1 + a_2 f_2 + \ldots + a_n f_n} \quad (r = 1, 2, 3, \ldots n).$$

Hence, $$\kappa_1+\kappa_2+\kappa_3+\ldots+\kappa_n=1.$$

At an equilibrium position, $f_1, f_2, f_3, \ldots f_n$ are proportional to the prices $p_1, p_2, p_3, \ldots p_n$ of the factors. Hence,

$$\kappa_r=\frac{a_r p_r}{a_1p_1+a_2p_2+\ldots+a_np_n}=\frac{a_rp_r}{\Pi},$$

i.e. κ_r is the proportion of total cost (Π) which is spent on the factor A_r.

Let
$$F=\begin{vmatrix} 0 & f_1 & f_2 & f_3 & \ldots & f_n \\ f_1 & f_{11} & f_{12} & f_{13} & \ldots & f_{1n} \\ f_2 & f_{12} & f_{22} & f_{23} & \ldots & f_{2n} \\ \ldots & \ldots & \ldots & \ldots & \ldots & \ldots \\ f_n & f_{1n} & f_{2n} & f_{3n} & \ldots & f_{nn} \end{vmatrix}$$

and let F_{rs} denote the co-factor of f_{rs} in F (r and $s=1, 2, 3, \ldots n$).

Write $$\sigma_{rs}=\frac{a_1f_1+a_2f_2+\ldots+a_nf_n}{a_ra_s}\frac{F_{rs}}{F} \quad (r \text{ and } s=1, 2, 3, \ldots n).$$

For any unequal values of r and s, the value of σ_{rs} is called the *partial elasticity of substitution* of the pair of factors A_r and A_s (as against all other factors). Its value depends on, and varies with, the grouping of factors employed. The interpretation of the partial elasticities of substitution will appear in the following section and it can be shown that they are related to the ordinary elasticity of substitution between two factors as defined in 13.7 above (cf. Examples XIX, 9).

A number of relations exist between the values of σ_{rs} for various values of r and s. If each element of the first row of F is multiplied by the co-factor of the corresponding element of another row, then the sum of the products is zero (18.4 above). So

$$f_1F_{r1}+f_2F_{r2}+f_3F_{r3}+\ldots+f_nF_{rn}=0 \quad (r=1, 2, 3, \ldots n).$$

From the definitions above, we have

$$\kappa_1\sigma_{r1}=\frac{f_1F_{r1}}{a_rF}, \quad \kappa_2\sigma_{r2}=\frac{f_2F_{r2}}{a_rF}, \quad \ldots \quad \kappa_n\sigma_{rn}=\frac{f_nF_{rn}}{a_rF}.$$

So $$\kappa_1\sigma_{r1}+\kappa_2\sigma_{r2}+\kappa_3\sigma_{r3}+\ldots+\kappa_n\sigma_{rn}=0 \quad (r=1, 2, 3, \ldots n).$$

We can write this relation in the form

$$\kappa_1\sigma_{r1}+\kappa_2\sigma_{r2}+\ldots+\kappa_{r-1}\sigma_{rr-1}+\kappa_{r+1}\sigma_{rr+1}+\ldots+\kappa_n\sigma_{rn}=-\kappa_r\sigma_{rr}.$$

The last two inequalities of the stability conditions (19.4) imply that F and F_{nn} are of opposite sign. The same is true of F and F_{rr} for any value of r.* But σ_{rr} is proportional to the ratio of F_{rr} to F and so

$$\sigma_{rr} < 0 \quad (r = 1, 2, 3, \ldots n).$$

Hence, for a production function satisfying the stability conditions,

$$\kappa_1 \sigma_{r1} + \kappa_2 \sigma_{r2} + \ldots + \kappa_{r-1} \sigma_{rr-1} + \kappa_{r+1} \sigma_{rr+1} + \ldots + \kappa_n \sigma_{rn} > 0$$

for any value of r ($r = 1, 2, 3, \ldots n$). In any case, there are n relations limiting the values that can be taken by the partial elasticities of substitution for any grouping of the factors.†

The values of the partial elasticities of substitution can be positive or negative. But the limitations above show that the positive values must be more numerous or important than the negative values. In particular, the $(n-1)$ partial elasticities of substitution between any one factor and the others cannot all be negative. If there are only three factors (A_1, A_2 and A_3), then

$$\kappa_2 \sigma_{12} + \kappa_3 \sigma_{13} > 0, \quad \kappa_1 \sigma_{12} + \kappa_3 \sigma_{23} > 0 \quad \text{and} \quad \kappa_1 \sigma_{13} + \kappa_2 \sigma_{23} > 0.$$

It follows *either* that all three partial elasticities of substitution (σ_{12}, σ_{13} and σ_{23}) are positive *or* that one of the partial elasticities is negative and the other two positive.

19.6 Variation of demand for factors of production.

It is now assumed that there are constant returns to scale in the production of the good X, the production function $x = f(a_1 a_2, a_3, \ldots a_n)$ being linear and homogeneous with the properties set out in 18.5 above. Given the output (x) and the prices ($p_1, p_2, p_3, \ldots p_n$) of the factors, the demands for the factors are determined by

$$\frac{p_1}{f_1} = \frac{p_2}{f_2} = \frac{p_3}{f_3} = \ldots = \frac{p_n}{f_n} = \lambda,$$

and $$f(a_1, a_2, a_3, \ldots a_n) = x.$$

The average cost of the output x is

$$\frac{\Pi}{x} = \frac{a_1 p_1 + a_2 p_2 + \ldots + a_n p_n}{x} = \lambda \frac{a_1 f_1 + a_2 f_2 + \ldots + a_n f_n}{x} = \lambda,$$

since $a_1 f_1 + a_2 f_2 + a_3 f_3 + \ldots + a_n f_n = x$ by Euler's Theorem. Further,

* The order in which the factors are enumerated is immaterial; we can take the factor A_r last just as well as the factor A_n.

† The other conditions of stability give rise to further limitations on the partial elasticities of substitution.

if the output varies while the prices of the factors remain fixed, the marginal cost at the output x is

$$\frac{d\Pi}{dx}=p_1\frac{da_1}{dx}+p_2\frac{da_2}{dx}+p_3\frac{da_3}{dx}+\ldots+p_n\frac{da_n}{dx}$$

$$=\lambda\left(f_1\frac{da_1}{dx}+f_2\frac{da_2}{dx}+f_3\frac{da_3}{dx}+\ldots+f_n\frac{da_n}{dx}\right)=\lambda,$$

since $f_1\frac{da_1}{dx}+f_2\frac{da_2}{dx}+f_3\frac{da_3}{dx}+\ldots+f_n\frac{da_n}{dx}=1$ from the production function. Average cost and marginal cost are equal and, therefore, constant for all outputs.

Extending the problem of 14.8 above, it is assumed that X is sold on a competitive market at a price p equal to the constant average cost. Then the common equilibrium value of $\frac{p_1}{f_1}, \frac{p_2}{f_2}, \frac{p_3}{f_3}, \ldots \frac{p_n}{f_n}$ is equal to p, i.e.

$$p_r=pf_r \quad (r=1, 2, 3, \ldots n),$$

which is the law of " marginal productivity ". Let the demand of the market for X be given by $x=\phi(p)$ with elasticity $\eta=-\frac{p}{x}\frac{dx}{dp}$. Then the position of competitive equilibrium is described by

$$f(a_1, a_2, a_3, \ldots a_n)=\phi(p),$$

and

$$\frac{p_1}{f_1}=\frac{p_2}{f_2}=\frac{p_3}{f_3}=\ldots=\frac{p_n}{f_n}=p.$$

These equations determine the values of $a_1, a_2, a_3, \ldots a_n$ and p in terms of the given prices of the factors. The output $x=\phi(p)$ and the total cost $\Pi=xp=a_1p_1+a_2p_2+a_3p_3+\ldots+a_np_n$ are also given in similar terms.

The demands of the competitive market for the factors are determined, by the equilibrium conditions, as functions of the prices of the factors. Let one of these prices (p_1) vary while all the other prices remain fixed. The resulting variations in the demands for the factors are then to be found by differentiating the equilibrium equations (which hold for any prices of the factors) with respect to p_1. From $f(a_1, a_2, a_3, \ldots a_n)=\phi(p)$,

$$f_1\frac{\partial a_1}{\partial p_1}+f_2\frac{\partial a_2}{\partial p_1}+\ldots+f_n\frac{\partial a_n}{\partial p_1}=-\eta\frac{x}{p}\frac{\partial p}{\partial p_1}.$$

From $pf_1 = p_1$, $pf_2 = p_2$, ... $pf_n = p_n$,

$$f_1 \frac{\partial p}{\partial p_1} + p\left(f_{11}\frac{\partial a_1}{\partial p_1} + f_{12}\frac{\partial a_2}{\partial p_1} + \ldots + f_{1n}\frac{\partial a_n}{\partial p_1}\right) = 1,$$

$$f_2 \frac{\partial p}{\partial p_1} + p\left(f_{12}\frac{\partial a_1}{\partial p_1} + f_{22}\frac{\partial a_2}{\partial p_1} + \ldots + f_{2n}\frac{\partial a_n}{\partial p_1}\right) = 0,$$

....

$$f_n \frac{\partial p}{\partial p_1} + p\left(f_{1n}\frac{\partial a_1}{\partial p_1} + f_{2n}\frac{\partial a_2}{\partial p_1} + \ldots + f_{nn}\frac{\partial a_n}{\partial p_1}\right) = 0.$$

Hence
$$x\eta\left(\frac{1}{p}\frac{\partial p}{\partial p_1}\right) + f_1\frac{\partial a_1}{\partial p_1} + f_2\frac{\partial a_2}{\partial p_1} + \ldots + f_n\frac{\partial a_n}{\partial p_1} = 0,$$

$$f_1\left(\frac{1}{p}\frac{\partial p}{\partial p_1}\right) + f_{11}\frac{\partial a_1}{\partial p_1} + f_{12}\frac{\partial a_2}{\partial p_1} + \ldots + f_{1n}\frac{\partial a_n}{\partial p_1} = \frac{1}{p},$$

$$f_2\left(\frac{1}{p}\frac{\partial p}{\partial p_1}\right) + f_{12}\frac{\partial a_1}{\partial p_1} + f_{22}\frac{\partial a_2}{\partial p_1} + \ldots + f_{2n}\frac{\partial a_n}{\partial p_1} = 0,$$

....

$$f_n\left(\frac{1}{p}\frac{\partial p}{\partial p_1}\right) + f_{1n}\frac{\partial a_1}{\partial p_1} + f_{2n}\frac{\partial a_2}{\partial p_1} + \ldots + f_{nn}\frac{\partial a_n}{\partial p_1} = 0.$$

F stands for the determinant already written (19.5 above) and F_0 is the co-factor of the element 0 in F. Further, F_{rs} and F_{0rs} denote the co-factors of the element f_{rs} in F and F_0 respectively (r and $s = 1, 2, 3, \ldots n$). Then, by 18.5 and 19.5,

$$F_0 \doteq 0, \quad F_{rs} = \frac{a_r}{f_s} F\kappa_s\sigma_{rs} = \frac{a_s}{f_r} F\kappa_r\sigma_{rs}, \quad F_{0rs} = -\frac{a_r a_s}{x^2} F,$$

where σ_{rs} is the partial elasticity of substitution of A_r and A_s and

$$\kappa_r = \frac{a_r f_r}{a_1 f_1 + a_2 f_2 + \ldots + a_n f_n} = \frac{a_r p_r}{xp}$$

is the proportion of total cost (=total receipts) spent on A_r.

Using the determinant notation (18.6 above) to solve the linear equations above for $\frac{\partial a_1}{\partial p_1}$, we find

$$\frac{\partial a_1}{\partial p_1} = \begin{vmatrix} x\eta & 0 & f_2 & \cdots & f_n \\ f_1 & \frac{1}{p} & f_{12} & \cdots & f_{1n} \\ f_2 & 0 & f_{22} & \cdots & f_{2n} \\ \cdots & \cdots & \cdots & \cdots & \cdots \\ f_n & 0 & f_{2n} & \cdots & f_{nn} \end{vmatrix} \div \begin{vmatrix} x\eta & f_1 & f_2 & \cdots & f_n \\ f_1 & f_{11} & f_{12} & \cdots & f_{1n} \\ f_2 & f_{12} & f_{22} & \cdots & f_{2n} \\ \cdots & \cdots & \cdots & \cdots & \cdots \\ f_n & f_{1n} & f_{2n} & \cdots & f_{nn} \end{vmatrix}$$

With the aid of 18.3, (5), the numerator can be written

$$\frac{1}{p}\begin{vmatrix} x\eta & f_2 & \cdots & f_n \\ f_2 & f_{22} & \cdots & f_{2n} \\ \cdots & \cdots & \cdots & \cdots \\ f_n & f_{2n} & \cdots & f_{nn} \end{vmatrix} = \frac{1}{p}\left(\begin{vmatrix} 0 & f_2 & \cdots & f_n \\ f_2 & f_{22} & \cdots & f_{2n} \\ \cdots & \cdots & \cdots & \cdots \\ f_n & f_{2n} & \cdots & f_{nn} \end{vmatrix} + \begin{vmatrix} x\eta & f_2 & \cdots & f_n \\ 0 & f_{22} & \cdots & f_{2n} \\ \cdots & \cdots & \cdots & \cdots \\ 0 & f_{2n} & \cdots & f_{nn} \end{vmatrix} \right)$$

$$= \frac{1}{p}(F_{11} + x\eta F_{011}) = \frac{a_1 F}{pf_1}\left(\kappa_1\sigma_{11} - \frac{a_1 f_1}{x}\eta\right) = \frac{a_1 F}{p_1}\kappa_1(\sigma_{11} - \eta),$$

since $pf_1 = p_1$ and $\kappa_1 = \dfrac{a_1 p_1}{xp}$. The denominator equals

$$\begin{vmatrix} 0 & f_1 & f_2 & \cdots & f_n \\ f_1 & f_{11} & f_{12} & \cdots & f_{1n} \\ f_2 & f_{12} & f_{22} & \cdots & f_{2n} \\ \cdots & \cdots & \cdots & \cdots & \cdots \\ f_n & f_{1n} & f_{2n} & \cdots & f_{nn} \end{vmatrix} + \begin{vmatrix} x\eta & f_1 & f_2 & \cdots & f_n \\ 0 & f_{11} & f_{12} & \cdots & f_{1n} \\ 0 & f_{12} & f_{22} & \cdots & f_{2n} \\ \cdots & \cdots & \cdots & \cdots & \cdots \\ 0 & f_{1n} & f_{2n} & \cdots & f_{nn} \end{vmatrix} = F + x\eta F_0 = F$$

Hence,
$$\frac{\partial a_1}{\partial p_1} = \frac{a_1}{p_1}\kappa_1(\sigma_{11} - \eta),$$

i.e.
$$\frac{Ea_1}{Ep_1} = \frac{p_1}{a_1}\frac{\partial a_1}{\partial p_1} = \kappa_1(\sigma_{11} - \eta).$$

Proceeding in exactly the same way, we find

$$\frac{Ea_2}{Ep_1} = \frac{p_1}{a_2}\frac{\partial a_2}{\partial p_1} = \kappa_1(\sigma_{12} - \eta)$$

and similar results for the demands for other factors.

The results, though expressed in terms of variations in the price p_1, are clearly quite general. We can, in fact, write one formula to express the effect of a change in the price of any factor on the demand for any factor :

$$\frac{Ea_s}{Ep_r} = \frac{p_r}{a_s}\frac{\partial a_s}{\partial p_r} = \kappa_r(\sigma_{rs} - \eta) \quad (r \text{ and } s = 1, 2, 3, \ldots n).$$

These expressions for the partial elasticities of demand for the factors are of the same form as those obtained in the case of two factors (14.8 above). Their interpretation proceeds as before but some new conclusions are now reached. If the market price of one factor A_r rises, then the demand for this or any other factor is affected in two ways. Firstly, the cost of production is now higher and the product dearer. For a decreasing demand law ($\eta > 0$), the amount of the product sold is less and there is an all-round and proportional decrease in the demand for the factors. This effect is shown by the negative term $(-\kappa_r\eta)$ which appears in each of the elasticities of demand with respect to the price of A_r. Abstracting from this effect, the factor A_r is now relatively more expensive than other factors and it pays to substitute other factors for A_r in production. The demand for A_r thus decreases on account of substitution, as shown by the negative term $(\kappa_r\sigma_{rr})$ in $\frac{Ea_r}{Ep_r}$. (By 19.5 above, σ_{rr} is negative for stable demands.) The effect on the demand for one of the other factors is more complex but the net result (on the demand for A_s) is shown by the term $(\kappa_r\sigma_{rs})$ in $\frac{Ea_s}{Ep_r}$. There are two cases:

(1) If $\sigma_{rs}>0$, then the demand for A_s increases on account of substitution; the factor A_s takes part in the replacement of A_r in production. In this case, the factor A_s is said to be *competitive* with the factor A_r at the grouping of factors considered.

(2) If $\sigma_{rs}<0$, then substitution results in a decrease in the demand for A_s; the factor A_s, like the factor A_r, has been partly replaced by other factors in production. Here, the factors A_r and A_s are said to be *complementary* at the grouping considered.

The sign of σ_{rs}, therefore, indicates whether A_r and A_s are competitive or complementary factors. From the limitations described in 19.5, we see that competition between factors is, on the whole, more general than complementarity. One factor, in any case, cannot be complementary with all other factors. Where there are only three factors, for example, *either* all factors are competitive with each other *or* one pair of the factors is complementary while the other two pairs are competitive.

19.7 The demand for consumers' goods (integrability case).

We have considered (14.9 above) the demands of a consumer with a given income when there are two goods obtainable on a market at given prices. The demands are stable if the indifference curves of the consumer are downward sloping and convex to the origin at all relevant points. In extending the analysis to the general case where there are more than two goods available, we assume first that the scale of preferences of the consumer can be represented by an indifference map and a utility function. Complications relating to "integrability" are deliberately avoided in this first approach.

If there are n goods $X_1, X_2, X_3, \dots X_n$, let one form of the utility function of the consumer be

$$u = \phi(x_1, x_2, x_3, \dots x_n)$$

with continuous partial derivatives of the first two orders:

$$\phi_r = \frac{\partial u}{\partial x_r} \quad \text{and} \quad \phi_{rs} = \frac{\partial^2 u}{\partial x_r \, \partial x_s} \quad (r \text{ and } s = 1, 2, 3, \dots n).$$

The consumer has a given income μ and can purchase the goods at given market prices, $p_1, p_2, p_3, \dots p_n$. He makes his purchases so that u has a maximum value subject to the condition expressing the fact that he must balance his budget:

$$x_1 p_1 + x_2 p_2 + x_3 p_3 + \dots + x_n p_n = \mu.$$

Since each utility function increases and decreases with any other, a maximum of u corresponds to a maximum of any form of the utility function and our results are thus independent of the fact that utility is not a measurable concept. The necessary condition for the maximum value of u we seek is

$$du = \phi_1 \, dx_1 + \phi_2 \, dx_2 + \phi_3 \, dx_3 + \dots + \phi_n \, dx_n = 0,$$

subject to $$p_1 \, dx_1 + p_2 \, dx_2 + p_3 \, dx_3 + \dots + p_n \, dx_n = 0,$$

i.e. $$\frac{\phi_1}{p_1} = \frac{\phi_2}{p_2} = \frac{\phi_3}{p_3} = \dots = \frac{\phi_n}{p_n}.$$

These $(n-1)$ equations, with the given side relation, determine the equilibrium purchases of the consumer in terms of his given income and the given prices, i.e. the consumer's demands as functions of $\mu, p_1, p_2, p_3, \dots p_n$.

The condition of "budget balance" gives x_1 as a linear function of the other purchases and so $d^2x_1 = 0$. It follows that a sufficient condition for maximum u is that

$$\begin{aligned} d^2u = {} & \phi_{11} \, dx_1^2 + \phi_{12} \, dx_1 \, dx_2 + \phi_{13} \, dx_1 \, dx_3 + \dots + \phi_{1n} \, dx_1 \, dx_n \\ & + \phi_{12} \, dx_1 \, dx_2 + \phi_{22} \, dx_2^2 + \phi_{23} \, dx_2 \, dx_3 + \dots + \phi_{2n} \, dx_2 \, dx_n \\ & + \quad \cdots\cdots\cdots\cdots\cdots\cdots\cdots\cdots \\ & + \phi_{1n} \, dx_1 \, dx_n + \phi_{2n} \, dx_2 \, dx_n + \phi_{3n} \, dx_3 \, dx_n + \dots + \phi_{nn} \, dx_n^2 \end{aligned}$$

is a negative definite quadratic form subject to

$$p_1 \, dx_1 + p_2 \, dx_2 + p_3 \, dx_3 + \dots + p_n \, dx_n = 0,$$

i.e. to $$\phi_1 \, dx_1 + \phi_2 \, dx_2 + \phi_3 \, dx_3 + \dots + \phi_n \, dx_n = 0,$$

(using the necessary conditions).

Hence, by 18.9, (2),

$$\begin{vmatrix} 0 & \phi_1 & \phi_2 \\ \phi_1 & \phi_{11} & \phi_{12} \\ \phi_2 & \phi_{12} & \phi_{22} \end{vmatrix} > 0, \quad \begin{vmatrix} 0 & \phi_1 & \phi_2 & \phi_3 \\ \phi_1 & \phi_{11} & \phi_{12} & \phi_{13} \\ \phi_2 & \phi_{12} & \phi_{22} & \phi_{23} \\ \phi_3 & \phi_{13} & \phi_{23} & \phi_{33} \end{vmatrix} < 0, \ \ldots .$$

The equilibrium position and the demand functions are stable for any income and market prices if $\phi_1, \phi_2, \phi_3, \ldots \phi_n$ are positive and if the determinant inequalities above are satisfied for all purchases of the consumer, at least within a relevant range. These stability conditions, it should be noticed, are limitations on the form of the utility function. When there are only three goods, the conditions imply that the indifference surfaces of the consumer are downward sloping and convex to the origin at all points, i.e. the "normal" form of the indifference map is sufficient for stability of demand.

The stability conditions appear in a form which involves a particular utility function ϕ, and it is not clear that they are independent of the non-measurable character of utility. We can, however, translate the conditions so that they involve only the marginal rates of substitution between the goods (16.8 above). The ratios

$$\frac{\phi_2}{\phi_1}, \ \frac{\phi_3}{\phi_1}, \ \ldots \ \frac{\phi_n}{\phi_1}$$

express the marginal rates of substitution of X_1 for $X_2, X_3, \ldots X_n$ respectively. The necessary conditions for equilibrium show that these ratios are equal to the corresponding ratios of market prices, i.e. the equilibrium purchases of the consumer are such that each marginal rate of substitution equals the ratio of the prices of the goods concerned. Now

$$\frac{\partial}{\partial x_1}\left(\frac{\phi_2}{\phi_1}\right) = \frac{\phi_{12}\phi_1 - \phi_{11}\phi_2}{\phi_1^2}, \quad \frac{\partial}{\partial x_2}\left(\frac{\phi_2}{\phi_1}\right) = \frac{\phi_{22}\phi_1 - \phi_{12}\phi_2}{\phi_1^2}, \ \ldots,$$

and similarly for the partial derivatives of the other marginal rates of substitution. The determinant inequalities of the stability conditions, using these results, can be reduced to

$$\begin{vmatrix} 1 & \dfrac{\phi_2}{\phi_1} \\ \dfrac{\partial}{\partial x_1}\left(\dfrac{\phi_2}{\phi_1}\right) & \dfrac{\partial}{\partial x_2}\left(\dfrac{\phi_2}{\phi_1}\right) \end{vmatrix} < 0, \quad \begin{vmatrix} 1 & \dfrac{\phi_2}{\phi_1} & \dfrac{\phi_3}{\phi_1} \\ \dfrac{\partial}{\partial x_1}\left(\dfrac{\phi_2}{\phi_1}\right) & \dfrac{\partial}{\partial x_2}\left(\dfrac{\phi_2}{\phi_1}\right) & \dfrac{\partial}{\partial x_3}\left(\dfrac{\phi_2}{\phi_1}\right) \\ \dfrac{\partial}{\partial x_1}\left(\dfrac{\phi_3}{\phi_1}\right) & \dfrac{\partial}{\partial x_2}\left(\dfrac{\phi_3}{\phi_1}\right) & \dfrac{\partial}{\partial x_3}\left(\dfrac{\phi_3}{\phi_1}\right) \end{vmatrix} > 0, \ \ldots .$$

The stability conditions thus involve only the marginal rates of substitution and are independent of the actual form chosen for the utility function.

Exactly as in 19.5 above, we define

$$\kappa_r = \frac{x_r \phi_r}{x_1 \phi_1 + x_2 \phi_2 + \ldots + x_n \phi_n} \quad \text{and} \quad \sigma_{rs} = \frac{x_1 \phi_1 + x_2 \phi_2 + \ldots + x_n \phi_n}{x_r x_s} \frac{\Phi_{rs}}{\Phi}$$

$$(r \text{ and } s = 1, 2, 3, \ldots n),$$

where Φ_{rs} is the co-factor of ϕ_{rs} in the determinant

$$\Phi = \begin{vmatrix} 0 & \phi_1 & \phi_2 & \phi_3 & \cdots & \phi_n \\ \phi_1 & \phi_{11} & \phi_{12} & \phi_{13} & \cdots & \phi_{1n} \\ \phi_2 & \phi_{12} & \phi_{22} & \phi_{23} & \cdots & \phi_{2n} \\ \cdots & \cdots & \cdots & \cdots & \cdots & \cdots \\ \phi_n & \phi_{1n} & \phi_{2n} & \phi_{3n} & \cdots & \phi_{nn} \end{vmatrix}$$

At the equilibrium position, κ_r represents the proportion of total expenditure ($\mu = x_1 p_1 + x_2 p_2 + \ldots + x_n p_n$) which is spent on the good X_r. The value of σ_{rs} is defined as the *partial elasticity of substitution* of the pair of goods X_r and X_s in consumption.

The demand of the consumer for each good is defined as a function of μ, p_1, p_2, p_3, ... p_n by the conditions

$$\frac{\phi_1}{p_1} = \frac{\phi_2}{p_2} = \frac{\phi_3}{p_3} = \ldots = \frac{\phi_n}{p_n} = \lambda$$

and
$$x_1 p_1 + x_2 p_2 + x_3 p_3 + \ldots + x_n p_n = \mu.$$

The parameter λ (dependent only on μ, p_1, p_2, p_3, ... p_n) is introduced for convenience and often described as the "marginal utility of money". Following the method of 19.6 above, these equations can be differentiated with respect to μ or to one of the prices to give the variations of demand for any good:

$$\frac{Ex_s}{E\mu} = \frac{\mu}{x_s} \frac{\partial x_s}{\partial \mu} \quad \text{and} \quad \frac{Ex_s}{Ep_r} = \frac{p_r}{x_s} \frac{\partial x_s}{\partial p_r} \quad (r \text{ and } s = 1, 2, 3, \ldots n).$$

It is found (see Examples XIX, 17 and 18) that

$$\frac{Ex_s}{Ep_r} = \kappa_r \left(\sigma_{rs} - \frac{Ex_s}{E\mu} \right) \quad (r \text{ and } s = 1, 2, 3, \ldots n).$$

It follows that the demand for each good is affected in two ways by an increase in the price of any one good (X_r). Since the *money* income of the consumer is fixed, the increase in the price results in a lower *real* income and causes a change (usually a decrease) in the purchase of each good. This is shown by the term $\left(-\kappa_r \frac{Ex_s}{E\mu}\right)$ in the expression for $\frac{Ex_s}{Ep_r}$. Further, the good X_r is now relatively more expensive than other goods and the consumer proceeds to substitute other goods for X_r in consumption. The demand for X_r is thus decreased, as shown by the negative term $(\kappa_r \sigma_{rr})$ in the expression for $\frac{Ex_r}{Ep_r}$. (The value of σ_{rr} is negative by the stability conditions.) The effect of substitution on the demand for one of the other goods (X_s) is represented by the term $(\kappa_r \sigma_{rs})$ in $\frac{Ex_s}{Ep_r}$. The consumption of X_s increases or decreases on account of substitution according as this term is positive or negative. Hence, X_r and X_s are *competitive* in consumption if $\sigma_{rs} > 0$ and *complementary* if $\sigma_{rs} < 0$. As in the production problem, the stability conditions impose limitations on the values of σ_{rs} (r and $s = 1, 2, 3, \ldots n$) which imply that the competitive relations between the goods outweigh, on the whole, the complementary relations.*

19.8 Demands for three consumers' goods (general case).

In generalising the results of the previous section, we must assume that the preference scale of the consumer is defined, not by a utility function and a complete indifference map, but only by the marginal rates of substitution between the various goods at different levels of consumption. To simplify the exposition, we take the case where there are three goods X, Y and Z and make use of the notation and results of 16.8 above. Each possible set of purchases of the consumer is represented by a point (x, y, z) in space referred to axes

* We have given here definitions of competitive and complementary goods which are more strict than those previously suggested (12.6 above). On the old definition, X_r and X_s are competitive goods if $\frac{Ex_s}{Ep_r}$ and $\frac{Ex_r}{Ep_s}$ are both positive. From $\frac{Ex_s}{Ep_r} = \kappa_r \left(\sigma_{rs} - \frac{Ex_s}{E\mu}\right)$, we see that the new definition $(\sigma_{rs} > 0)$ usually implies the same thing. But one or both of $\frac{Ex_s}{Ep_r}$ and $\frac{Ex_r}{Ep_s}$ can be negative if $\frac{Ex_s}{E\mu}$ or $\frac{Ex_r}{E\mu}$ is positive and greater than σ_{rs}, i.e. if the effect on demand via changes in real income is stronger than the substitution effect. Similarly, the two definitions of complementarity are roughly, but not exactly, in agreement.

Ox, Oy and Oz. The preference scale of the consumer is then described by an indifference plane $dx + R_1 dy + R_2 dz = 0$ at any given point of space. Here, R_1 and R_2 are given functions of x, y and z (the purchases of the consumer) which express the marginal rates of substitution of X for Y and Z respectively. The functions are assumed to have continuous partial derivatives. The differential equation of the indifference plane is not necessarily integrable and we are not at liberty to assume that a utility function or a complete system of indifference surfaces exists.

With given market prices, p_x, p_y and p_z, of the goods and a given income μ, the purchases of the consumer must conform to the condition of " budget balance " $(xp_x + yp_y + zp_z = \mu)$ and any variations in the purchases from an established set (x, y, z) must satisfy

$$p_x\, dx + p_y\, dy + p_z\, dz = 0.$$

If the purchases of Y and Z are increased by dy and dz respectively, the *necessary* decrease in the purchase of X is

$$(-dx)_N = \frac{p_y}{p_x} dy + \frac{p_z}{p_x} dz.$$

The indifference equation shows that the *compensating* decrease is

$$(-dx)_C = R_1\, dy + R_2\, dz.$$

If the necessary decrease is less than the compensating decrease, then the consumer tends to increase his purchases of Y and Z as indicated. Conversely, if the necessary decrease is greater than the compensating decrease, the consumer will tend to decrease his purchases of Y and Z. In fact, equilibrium of consumer's choice is only possible if the necessary and compensating decreases are equal for all values of dy and dz. Hence, the necessary conditions for equilibrium are

$$R_1 = \frac{p_y}{p_x} \quad \text{and} \quad R_2 = \frac{p_z}{p_x}$$

and the marginal rates of substitution, as before, are equal to ratios of the market prices.

To examine the stability of demand, we suppose that increases dy and dz in the purchases of Y and Z have been made from the equilibrium position, together with the necessary decrease

$\left(\frac{p_y}{p_x}dy+\frac{p_z}{p_x}dz\right)$ in the purchase of X. The new values of the marginal rates of substitution are approximately (R_1+dR_1) and (R_2+dR_2) where

$$dR_1=\frac{\partial R_1}{\partial x}dx+\frac{\partial R_1}{\partial y}dy+\frac{\partial R_1}{\partial z}dz \quad \text{and} \quad dR_2=\frac{\partial R_2}{\partial x}dx+\frac{\partial R_2}{\partial y}dy+\frac{\partial R_2}{\partial z}dz.$$

Suppose, now, that an exactly similar change from the new purchases is contemplated. The necessary decrease in the purchase of X is

$$(-dx)_N=\frac{p_y}{p_x}dy+\frac{p_z}{p_x}dz,$$

but the compensating decrease in X is now

$$(-dx)_C=(R_1+dR_1)\,dy+(R_2+dR_2)\,dz.$$

The original purchases are stable provided that this second change will *not* be made, i.e. provided that the compensating decrease in X is less than the necessary decrease :

$$(R_1+dR_1)\,dy+(R_2+dR_2)\,dz<\frac{p_y}{p_x}dy+\frac{p_z}{p_x}dz=R_1\,dy+R_2\,dz,$$

i.e. $$dR_1\,dy+dR_2\,dz<0,$$

where the variations in purchases are related

$$p_x\,dx+p_y\,dy+p_z\,dz=0 \quad \text{or} \quad dx+R_1\,dy+R_2\,dz=0,$$

using the necessary conditions. Substituting for dR_1 and dR_2,

$$\frac{\partial R_1}{\partial y}dy^2+\frac{\partial R_2}{\partial z}dz^2+\frac{\partial R_1}{\partial x}dx\,dy+\frac{\partial R_2}{\partial x}dx\,dz+\left(\frac{\partial R_2}{\partial y}+\frac{\partial R_1}{\partial z}\right)dy\,dz<0,$$

i.e. $$0\,dx^2+\frac{1}{2}\frac{\partial R_1}{\partial x}dx\,dy+\frac{1}{2}\frac{\partial R_2}{\partial x}dx\,dz$$

$$+\frac{1}{2}\frac{\partial R_1}{\partial x}dx\,dy+\frac{\partial R_1}{\partial y}dy^2+\frac{1}{2}\left(\frac{\partial R_2}{\partial y}+\frac{\partial R_1}{\partial z}\right)dy\,dz$$

$$+\frac{1}{2}\frac{\partial R_2}{\partial z}dx\,dz+\frac{1}{2}\left(\frac{\partial R_2}{\partial y}+\frac{\partial R_1}{\partial z}\right)dy\,dz+\frac{\partial R_2}{\partial z}dz^2$$

must be a negative definite quadratic form, subject to

$$dx+R_1\,dy+R_2\,dz=0.$$

Hence, from 18.9, (2),

$$D=\begin{vmatrix} 0 & 1 & R_1 \\ 1 & 0 & \frac{1}{2}\frac{\partial R_1}{\partial x} \\ R_1 & \frac{1}{2}\frac{\partial R_1}{\partial x} & \frac{\partial R_1}{\partial y} \end{vmatrix} > 0,$$

$$\Delta=\begin{vmatrix} 0 & 1 & R_1 & R_2 \\ 1 & 0 & \frac{1}{2}\frac{\partial R_1}{\partial x} & \frac{1}{2}\frac{\partial R_2}{\partial x} \\ R_1 & \frac{1}{2}\frac{\partial R_1}{\partial x} & \frac{\partial R_1}{\partial y} & \frac{1}{2}\left(\frac{\partial R_2}{\partial y}+\frac{\partial R_1}{\partial z}\right) \\ R_2 & \frac{1}{2}\frac{\partial R_2}{\partial x} & \frac{1}{2}\left(\frac{\partial R_2}{\partial y}+\frac{\partial R_1}{\partial z}\right) & \frac{\partial R_2}{\partial z} \end{vmatrix} < 0.$$

The demands of the consumer, given by the necessary conditions, are stable for all prices and incomes if R_1 and R_2 are positive and if $D>0$ and $\Delta<0$ for all purchases. These are the general stability conditions. Since an indifference map need not exist, the conditions cannot be interpreted as before.

A certain simplification can be made :

$$D=-\begin{vmatrix} 1 & R_1 \\ \frac{1}{2}\frac{\partial R_1}{\partial x} & \frac{\partial R_1}{\partial y} \end{vmatrix} + R_1\begin{vmatrix} 1 & R_1 \\ 0 & \frac{1}{2}\frac{\partial R_1}{\partial x} \end{vmatrix}$$

$$=-\frac{\partial R_1}{\partial y}+\frac{1}{2}R_1\frac{\partial R_1}{\partial x}+\frac{1}{2}R_1\frac{\partial R_1}{\partial x}=R_1\frac{\partial R_1}{\partial x}-\frac{\partial R_1}{\partial y}=-\begin{vmatrix} 1 & R_1 \\ \frac{\partial R_1}{\partial x} & \frac{\partial R_1}{\partial y} \end{vmatrix}$$

Further, write

$$\Delta'=\begin{vmatrix} 1 & R_1 & R_2 \\ \frac{\partial R_1}{\partial x} & \frac{\partial R_1}{\partial y} & \frac{\partial R_1}{\partial z} \\ \frac{\partial R_2}{\partial x} & \frac{\partial R_2}{\partial y} & \frac{\partial R_2}{\partial z} \end{vmatrix}$$

By manipulation or expansion of the determinants, it can be shown that

$$\Delta+\Delta'=\lambda^2$$

where

$$\lambda=\frac{1}{2}\left\{\left(R_1\frac{\partial R_2}{\partial x}-R_2\frac{\partial R_1}{\partial x}\right)+\left(\frac{\partial R_1}{\partial z}-\frac{\partial R_2}{\partial y}\right)\right\}.$$

The stability conditions ($D>0$ and $\Delta<0$) thus reduce to

$$\begin{vmatrix} 1 & R_1 \\ \dfrac{\partial R_1}{\partial x} & \dfrac{\partial R_1}{\partial y} \end{vmatrix} < 0 \quad \text{and} \quad \begin{vmatrix} 1 & R_1 & R_2 \\ \dfrac{\partial R_1}{\partial x} & \dfrac{\partial R_1}{\partial y} & \dfrac{\partial R_1}{\partial z} \\ \dfrac{\partial R_2}{\partial x} & \dfrac{\partial R_2}{\partial y} & \dfrac{\partial R_2}{\partial z} \end{vmatrix} > \lambda^2.$$

These conditions are very similar to those of the integrability case but the third-order determinant shown must not only be positive but also greater than a certain positive amount λ^2. The conditions of the integrability case must still hold but they are only necessary, and not sufficient, for stability. Notice that the integrability condition (16.8 above) is simply $\lambda=0$ and that the general stability conditions then reduce to those previously written.

The demand of the consumer for each of the three goods is determined, as a function of μ, p_x, p_y and p_z, by the conditions:

$$R_1=\frac{p_y}{p_x}, \quad R_2=\frac{p_z}{p_x} \quad \text{and} \quad xp_x+yp_y+zp_z=\mu.$$

This is true of all cases. In the special integrability case, the forms assumed by the marginal rates of substitution and by the stability conditions are simpler than in the general case. The results (19.7 above) relating to the variations of demand and the partial elasticities of substitution hold only in the simpler case and need modification when we pass to the general case. In particular, the distinction between goods which are competitive and goods which are complementary in consumption is found to be less clear-cut in the general than in the special case.*

EXAMPLES XIX

General maximum and minimum problems

1. Show that $u=x^2+y^2+z^2+xy$ has a minimum value at $x=y=z=0$.

2. Show that $u=x^2+y^2+xyz$ has minimum values at all points where $x=y=0$ and $z<2$ numerically.

3. If $u=x+y+xz-x^2-y^2+z^2$, show that u has a single stationary value which corresponds to a saddle point.

* An analysis of the competitive and complementary relations between goods and of the variations of demand in the general case are given by Hicks and Allen, *A Reconsideration of the Theory of Value*, Economica, 1934. The stability conditions given in this article are not fully expressed and the present development can be taken as replacing the earlier work.

4. Show that $u=\sqrt[3]{xyz}$, subject to $x+y+z=3a$, has a maximum value ($u=a$) when $x=y=z=a$. Deduce that the geometric mean between three positive quantities is always less than the arithmetic mean except in the case where the quantities are equal. Generalise the result to apply to any number of quantities.

5. Find the shortest distance from the origin of co-ordinates to any point on the surface $xyz=a$ in the positive quadrant.

6. If a is a positive constant, show that the maximum value of

$$u=a(a-x)(a-y)(a-z)$$

relative to $x+y+z=2a$ occurs where x, y and z are each equal to $\frac{2}{3}a$. Interpret this result in the light of the fact that the area of a triangle with sides x, y and z is given by $\sqrt{a(a-x)(a-y)(a-z)}$, where $2a$ is the perimeter.

Economic maximum and minimum problems

7. A monopolist produces three goods X_1, X_2 and X_3. The total cost of outputs x_1, x_2 and x_3 of the three goods is $(\alpha_1 x_1+\alpha_2 x_2+\alpha_3 x_3)$, where α_1, α_2 and α_3 are constants. The monopolist charges prices p_1, p_2 and p_3 and the demands of his market are

$$x_1=x_{10}-a_{11}(p_1-\alpha_1)-a_{12}(p_2-\alpha_2)-a_{13}(p_3-\alpha_3),$$
$$x_2=x_{20}-a_{12}(p_1-\alpha_1)-a_{22}(p_2-\alpha_2)-a_{23}(p_3-\alpha_3),$$
$$x_3=x_{30}-a_{13}(p_1-\alpha_1)-a_{23}(p_2-\alpha_2)-a_{33}(p_3-\alpha_3),$$

where x_{10}, x_{20}, x_{30} and the a's are given constants. Find the prices the monopolist must charge for maximum joint revenue and show that the following conditions must be satisfied by the a's :

$$a_{11}>0, \quad \begin{vmatrix} a_{11} & a_{12} \\ a_{12} & a_{22} \end{vmatrix}>0 \quad \text{and} \quad \begin{vmatrix} a_{11} & a_{12} & a_{13} \\ a_{12} & a_{22} & a_{23} \\ a_{13} & a_{23} & a_{33} \end{vmatrix}>0.$$

Extend the results of 14.4 above.

8. Find the conditions under which the equilibrium of the general problem of 14.5 above is stable, expressing them in terms of the partial derivatives of the production function $x=f(a, b, t)$.

9. A good X is produced with three variable factors according to the production function $x=f(a_1, a_2, a_3)$. Show that σ_{12} (as defined in 19.5 above) is a negative multiple of the ratio

$$\frac{a_2}{a_3}d\left(\frac{a_3}{a_2}\right) : \frac{f_3}{f_2}d\left(\frac{f_2}{f_3}\right)$$

where the differentials apply to a variation of the factors A_1 and A_3 (the factor A_2 being held constant) so that product is unchanged. Express this result in terms of an elasticity of substitution of the kind defined in 13.7 above.

10. Three factors are obtainable at given prices, p_1, p_2 and p_3. When amounts a_1, a_2 and a_3 of the factors are employed, the output of a good X is $x=A\sqrt[3]{a_1 a_2 a_3}$. The good is sold at the competitive price p on a market with demand law $x=Bp^{-\alpha}$. Find the amounts of the factors demanded in terms of p_1, p_2 and p_3. By differentiation, show that

$$\frac{Ea_1}{Ep_1}=-\frac{1}{3}(\alpha+2), \quad \frac{Ea_2}{Ep_1}=\frac{Ea_3}{Ep_1}=\frac{1}{3}(1-\alpha).$$

11. For the production function of the previous example, show that the demands for the factors are stable and verify that the partial elasticities of substitution are all equal to unity. Hence use the results of the previous example to check the formulae of 19.6.

12. A good X is produced with n factors of production, the production function, $x=f(a_1, a_2, a_3, \ldots a_n)$, being of any given form. The equations

$$\frac{p_1}{f_1}=\frac{p_2}{f_2}=\frac{p_3}{f_3}=\ldots=\frac{p_n}{f_n}=\lambda \quad \text{and} \quad f(a_1, a_2, a_3, \ldots a_n)=x$$

give the demands for the factors in terms of the given output (x) and the given prices ($p_1, p_2, p_3, \ldots p_n$) of the factors. By differentiating these equations, show that the variations of the demands are given by

$$\frac{\partial a_s}{\partial x}=\frac{F_s}{F} \quad (s=1, 2, 3, \ldots n)$$

when output varies (prices of the factors fixed), and by

$$\frac{Ea_s}{Ep_r}=\frac{p_r}{a_s}\frac{\partial a_s}{\partial p_r}=\kappa_r\sigma_{rs} \quad (r \text{ and } s=1, 2, 3, \ldots n)$$

when the price of one factor varies (output and prices of other factors fixed). Here, F_s is the co-factor of f_s in the determinant F and κ_r and σ_{rs} have the meanings of 19.5. Interpret the second results in terms of the competitive and complementary relations between the factors.

13. A monopolist sells a good X at a price p on a market with demand law $x=\phi(p)$ and fixes output so that net revenue is a maximum (i.e. so that marginal revenue equals marginal cost). He uses n factors of production, obtainable at given prices ($p_1, p_2, p_3, \ldots p_n$), and his production function, $x=f(a_1, a_2, a_3, \ldots a_n)$, is linear and homogeneous. Show that the price p and the amounts of the factors employed are given by

$$\frac{p_1}{f_1}=\frac{p_2}{f_2}=\frac{p_3}{f_3}=\ldots=\frac{p_n}{f_n}=p\left(1-\frac{1}{\eta}\right) \quad \text{and} \quad f(a_1, a_2, a_3, \ldots a_n)=\phi(p)$$

where η is the elasticity of the demand for the good.

14. In the problem of the previous example, the price of the factor A_1 is varied while the prices of other factors are fixed. Differentiate the equilibrium equations and show that

$$\frac{Ea_r}{Ep_1}=\frac{p_1}{a_r}\frac{\partial a_r}{\partial p_1}=\kappa_1\left\{\sigma_{1r}-\frac{\eta}{1+\dfrac{p}{\eta(\eta-1)}\dfrac{d\eta}{dp}}\right\} \quad (r=1, 2, 3, \ldots n)$$

where κ_1 and σ_{1r} have the meanings of 19.5. Deduce that, in the normal case where $\eta>1$ and $\dfrac{d\eta}{dp}>0$, a fall in the price of A_1 increases the demand for each factor less in this monopoly case than in the case of competition (19.6 above).

15. A consumer has a given income (μ) and can buy three goods at given prices (p_1, p_2 and p_3). One form of his utility function is $u=x_1^{\alpha_1}x_2^{\alpha_2}x_3^{\alpha_3}$, where α_1, α_2 and α_3 are positive constants. Find the demand of the consumer for each good in terms of μ, p_1, p_2 and p_3, and show that

$$\frac{Ex_1}{E\mu}=1, \quad \frac{Ex_1}{Ep_1}=-1, \quad \frac{Ex_1}{Ep_2}=\frac{Ex_1}{Ep_3}=0,$$

with similar results for the other two goods.

16. Show that the utility function of the previous example satisfies the stability conditions and gives partial elasticities of substitution which are all unity. Use the results of the previous example to verify, in this case, that the general formulae of 19.7 hold.

17. The equilibrium equations of 19.7 give the demands of a consumer as functions of his income (μ) and of the prices ($p_1, p_2, p_3, \ldots p_n$) of the n goods obtainable. If income varies while prices remain fixed, differentiate the equations with respect to μ and show that

$$\frac{Ex_r}{E\mu}=\frac{\mu}{x_r}\frac{\partial x_r}{\partial \mu}=\frac{x_1\phi_1+x_2\phi_2+\ldots+x_n\phi_n}{x_r}\frac{\Phi_r}{\Phi} \quad (r=1, 2, 3, \ldots n)$$

where the utility function and the notation of 19.7 is adopted.

18. In the problem of the previous example, the price of the good X_1 varies while income and the other prices remain given. By differentiating the equilibrium equations with respect to p_1, show that

$$\frac{Ex_r}{Ep_1}=\frac{p_1}{x_r}\frac{\partial x_r}{\partial p_1}=\kappa_1\left(\sigma_{1r}-\frac{Ex_r}{E\mu}\right) \quad (r=1, 2, 3, \ldots n)$$

where κ_1 and σ_{1r} have the meanings defined in 19.7.

19. The scale of preferences of a consumer for n goods is defined by the differential equation

$$(a_1+a_{11}x_1)\,dx_1+(a_2+a_{22}x_2)\,dx_2+\ldots+(a_n+a_{nn}x_n)\,dx_n=0$$

where $x_1, x_2, x_3, \ldots x_n$ represent any set of purchases made by the consumer and where the a's are constants. Show that a utility function exists and verify that the demands of the consumer are stable, provided that a_{11}, a_{22}, $a_{33}, \ldots a_{nn}$ are all negative. Find the demand for each good as a function of the given income of the consumer and of the given market prices. Show that the demand depends linearly on the income (prices fixed) and is a ratio of quadratic expressions in the prices (income fixed).

20. A firm with given resources (labour, plant, raw materials, etc.) produces n goods and its outputs of the various goods are related by a transformation function $F(x_1, x_2, x_3, \ldots x_n)=0$. The goods are sold at given market prices, $p_1, p_2, p_3, \ldots p_n$, and the firm fixes the outputs to maximise its revenue $R=x_1p_1+x_2p_2+\ldots+x_np_n$. Find equations giving the amount of each good supplied as a function of all the market prices. Under what conditions is the supply stable? How can partial elasticities of substitution between the goods in production be defined and used to describe the variations of supply of different goods as the market price of one good changes?

21. Extend the analysis of the demand of an individual for loans (14.9 above) to the case where £x_1, £x_2, £x_3, ... £x_n are incomes obtained in n successive years, related by a transformation function $F(x_1, x_2, x_3, \ldots x_n)=0$. Assume that the individual has a utility function for incomes which can be written $u=\psi(x_1, x_2, x_3, \ldots x_n)$ and that the rate of interest is $100r_1$ per cent. from the first to the second year, $100r_2$ per cent. from the second to the third year, and so on.

CHAPTER XX

SOME PROBLEMS IN THE CALCULUS OF VARIATIONS

20.1 The general theory of functionals.

We have been concerned so far with the theory of functions of a finite number of variables, with the variations of quantities depending on one or more other quantities. There remains for brief consideration an extension of the analysis, opening up many new fields of application, which takes a quantity as dependent not upon a finite set of other quantities but upon one or more *variable functions.* In the simplest case, a variable quantity u is defined as taking its value from the form assumed by a function $x=\phi(t)$. To each function $\phi(t)$ there corresponds a definite value of u, and as the *form* of the function is changed so is the *value* of u. The dependence of u upon $\phi(t)$ is called a *functional* and written $u=F\{\phi\}$. The functional symbol F is to be regarded in a manner analogous to the ordinary symbol for a function. It is essential, however, that the functional $u=F\{\phi\}$ should not be confused with the function of a function $u=F\{\phi(t)\}$. The latter assumes that $\phi(t)$ is a given function of t and hence that u is also a definite function of t. The former takes ϕ as a variable function, the functional F associating one value of u with each whole function ϕ. The variable t does not itself appear in the determination of u in the functional relation.*

In diagrammatic terms, the variable function $x=\phi(t)$ is shown by a variable curve C in the plane Oxt. As the form of the function changes (ϕ_1, ϕ_2, ϕ_3, ...), the curve C shifts and takes up different positions and shapes (C_1, C_2, C_3, ...). If u is a functional of ϕ, then its value depends on the particular position taken by the curve C,

* More generally, u may be a functional of several variable functions $x=\phi(t)$, $y=\psi(t)$, $z=\chi(t)$, ... We write $u=F\{\phi, \psi, \chi, ...\}$. Further, the functions ϕ, ψ, χ, ... may be functions of several variables instead of one variable t only. For the general theory of functionals, see Volterra, *Theory of Functionals* (English Ed. 1931).

and to the series of positions (C_1, C_2, C_3, ...) there corresponds a series of definite value of u (u_1, u_2, u_3, ...). For example, Fig. 103 shows five positions of a curve limited to pass through two fixed end-points A and B and there will be five corresponding values of u.

An example of a functional can be taken from quite simple economic theory. Each member of a competitive group of firms producing a good X with known cost functions fixes that output (x) which makes marginal cost equal to the given market price (p) of

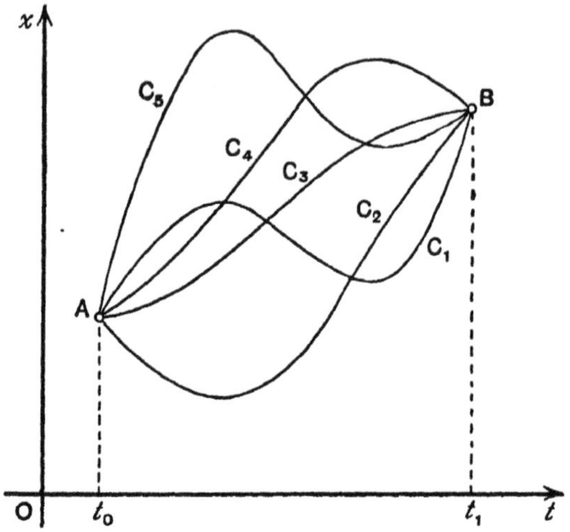

Fig. 103.

the good. Further, the output fixed by the firm varies when different market prices are given. A *supply function*, $x = f(p)$, can thus be defined for each firm. But if the firm has monopoly control and fixes output and price subject only to certain demand conditions, then the output is such that marginal cost equals marginal revenue. If the marginal revenue curve is given (given demand conditions and given prices of all other goods), there corresponds one definite monopoly output x_m. As the marginal revenue curve shifts in position (changing demand conditions or varying prices of other goods), the monopoly output is determined afresh and varies in value. The monopoly output x_m depends on the form of the marginal revenue function $\phi(x)$ and we have the *supply functional* $x_m = F\{\phi\}$.*

The step from functions to functionals can be regarded as another instance of the step from the finite to the infinite so characteristic

* See Schneider, *Theorie der Produktion* (1934), p. 74.

of mathematical analysis. If $u = f(x_1, x_2, \ldots x_n)$ is a function of a finite number (n) of variables, we can increase n indefinitely by the introduction of more and more variables. The function u then tends to involve an infinite number of variables. But this is just what is implied by saying that u depends upon the form of a variable function $x = \phi(t)$. The function, in fact, comprises an infinite number of particular values $(x_1, x_2, \ldots, x_n, \ldots)$ and u depends, in the functional notion, on the whole of this infinite set of values. The functional $u = F\{\phi\}$ implies that u depends on the infinity of values making up the function $x = \phi(t)$.

20.2 The calculus of variations.

The most important and frequent case of a functional $u = F\{\phi\}$ occurs when the form of F appears as an *integral*. Suppose that $f(t)$ is a function changing in form as $\phi(t)$ changes. Write

$$u = \int_{t_0}^{t_1} f(t)\, dt$$

where t_0 and t_1 are certain limits of integration. Then the value of u depends on what particular form we take for $\phi(t)$ and hence for $f(t)$, i.e. u is a functional of the variable function ϕ. The problem that usually arises is to determine that function ϕ which makes u a maximum or a minimum. The analysis of this problem is termed the *calculus of variations*, one branch of the much wider theory of functionals.* Many problems of importance in the varied applications of mathematical analysis are found to relate to the calculus of variations. Two examples will illustrate this fact :

Ex. 1. *The problem of the surface of revolution of minimum area.*

A problem arising early in the development of the calculus of variations is that of determining that curve of all curves joining two fixed points A and B on the same side of a given line L which forms a surface of the smallest area when revolved about the line L. Axes Oxt are fixed in any way so that Ot lies along L and so that the fixed points have co-ordinates $A(t_0, x_0)$ and $B(t_1, x_1)$ which are all positive and $t_1 > t_0$. If $x = \phi(t)$ is any curve lying above Ot and joining A and B, then it can be shown that the

* Most text-books on advanced mathematical analysis include an account of the calculus of variations. See, for example, Courant, *Differential and Integral Calculus*, Vol. II (English Ed. 1936), Chapter VII; Osgood, *Advanced Calculus* (1925), Chapter XVII; de la Vallée Poussin, *Cours d'analyse infinitésimale*, Vol. II (1925), Chapter X.

area of the surface formed by revolving the curve about Ot is proportional to

$$u = \int_{t_0}^{t_1} x \sqrt{1 + \left(\frac{dx}{dt}\right)^2}\, dt.$$

Here u is a functional of the variable function ϕ and we seek that function (subject to the limitations indicated) which makes u a minimum.

Ex. 2. *The brachistochrone problem of John Bernouilli.*

One of the first problems in the calculus of variations, posed by John Bernouilli in 1696, is to fix a curve joining two given points O and A so that the time taken by a particle to slide under gravity along the curve from O to A is least. Axes Oxt are chosen, as shown in Fig. 104, with origin at the given point O and with Ox drawn vertically downwards. Let A have co-ordinates (t_1, x_1) which are both positive. If $x = \phi(t)$ is the equation of any curve joining O to A so that $\phi(0) = 0$ and $\phi(t_1) = x_1$, then it can be shown that the time taken by a particle to slide under gravity from O to A is proportional to

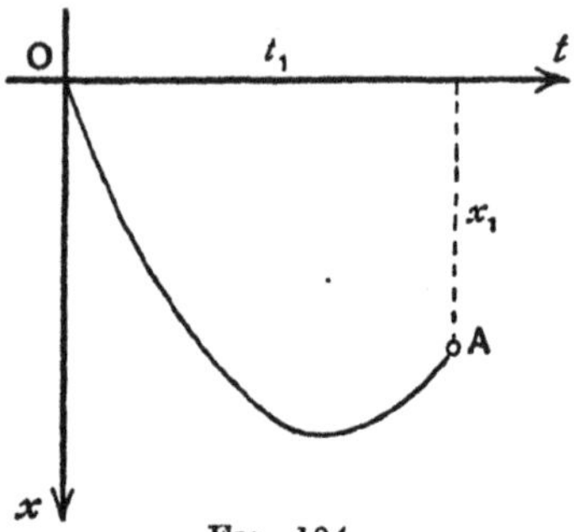

FIG. 104.

$$u = \int_0^{t_1} \sqrt{\frac{1}{x}\left\{1 + \left(\frac{dx}{dt}\right)^2\right\}}\, dt.$$

Here, again, u is a functional dependent on the form assumed by the variable function ϕ and we seek that form of ϕ which corresponds to the minimum value of u.

20.3 The method of the calculus of variations.

Two important preliminary points must be considered in devising practical methods of solving problems in the calculus of variations. The first point concerns what can be called *boundary conditions.* The problem to be solved is usually framed so that only certain arcs of the variable curve $x = \phi(t)$ in the plane Oxt are needed, definite conditions being imposed upon the points which mark the ends of the arcs. In the most frequent case, the conditions are that the arcs should start and finish at two *fixed* points A and B in the plane. The boundary conditions, in such a case, impose limitations on the field of possible variation of the curve $x = \phi(t)$ and we have to consider, in fact, only those curves which can be drawn from

A to B. This is so in the two problems instanced above and it is illustrated by Fig. 103. In analytical terms, the function $x=\phi(t)$ can only be selected provided that

$$\phi(t_0)=x_0 \quad \text{and} \quad \phi(t_1)=x_1$$

where $A(t_0, x_0)$ and $B(t_1, x_1)$ are the two fixed end-points. Other boundary conditions may be imposed in less usual cases. It may be given, for example, that the variable curve $x=\phi(t)$ should join two points P and Q which lie one on each of two given curves in the plane Oxt. Here, we have the boundary conditions $\phi(t_0)=x_0$ where (t_0, x_0) are the co-ordinates of P and satisfy some given relation $\psi(t_0, x_0)=0$; and similarly for the co-ordinates (t_1, x_1) of Q.

Even when the boundary conditions are taken into account, the field of possible variation of the function $x=\phi(t)$ is so large that analysis of the problem is practically impossible and it becomes essential to limit the field of variation by some further device. An obvious step is to take only those functions, satisfying the boundary conditions, which are continuous and possess continuous derivatives up to any desired order. Further, the field can be more severely limited by taking functions only of a particular type or curves of a particular class. We may take, for example, only functions of the quadratic type represented by parabolas with vertical axis. We know that a function type or curve class can be represented by a relation involving certain parameters $\alpha, \beta, \gamma, \ldots$; the larger the number of the parameters the more general is the function type or curve class. The result of this limitation on the field of the variable function is to replace the function $x=\phi(t)$ of variable form by

$$x=\phi(t\,;\ \alpha, \beta, \gamma, \ldots),$$

where ϕ is now of *fixed* form and the variation of the function is replaced by the variation of the parameters involved.

It is important to appreciate the nature of the step now taken. We have given up the consideration of any comprehensive variation in the function $x=\phi(t)$ and have limited ourselves to a more restricted variation described by parameters in a function of fixed form. If the parameters are few in number, the restriction is very severe. For example, using only three parameters, we may write

$$\phi(t)=\alpha t^2+\beta t+\gamma$$

so that our variable curve is limited to the class of parabolas with their axes vertical (parallel to Ox). But, by taking more and more parameters, we can make the field of variation of our function type more and more general. If a sufficiently large (but finite) number of parameters is selected, the restricted field of variation can be made to differ in few important respects from the complete field; we simply exclude from the latter the more unusual kinds of functions.

The problem to be solved is now greatly simplified. We seek the extreme values of $u=\int_{t_0}^{t_1} f(t)\,dt$ where $f(t)$ depends on the variable function $x=\phi(t)$. Limiting the latter in the way indicated, the expression to be integrated becomes $f(t\,;\ \alpha, \beta, \gamma, \ldots)$ where $\alpha, \beta, \gamma, \ldots$ are parameters and where the form of the function f is fixed. It follows that the value of the integral u depends only on the parameters $\alpha, \beta, \gamma, \ldots$ and we require its maximum or minimum value for all variations in the parameters allowed by the boundary conditions. The problem of the calculus of variations is thus reduced to a problem of extreme values of an ordinary function of several variables $\alpha, \beta, \gamma, \ldots$. Functionals are changed back into functions and we proceed on familiar lines.* The extent to which the simplified problem approximates to the original one depends on the number of parameters we care to take. The most important thing about the analysis which follows is that it is quite independent of how many parameters there are, provided only that their number is finite. The solution we obtain is not perfectly general but it provides an approximation sufficient for all practical purposes.

20.4 Solution of the simplest problem.

The function $f(t)$, which gives the variable u on integration, depends in some definite way upon the variable function $x=\phi(t)$. The dependence usually includes, not only ϕ itself, but also the various derivatives of ϕ. Of the many possible cases, only the simplest will be analysed here, the case where $f(t)$ depends on the variable t, on the function $\phi(t)$ and on the first derivative $\phi'(t)$:

$$f\left(t,\, x,\, \frac{dx}{dt}\right)=f\{t,\, \phi(t),\, \phi'(t)\}.$$

* The step from functions to functionals is reversed. A functional can be regarded as a function of an infinity of variables. In solving problems of functionals, we approximate by taking a function of a large number of variables (the parameters $\alpha, \beta, \gamma, \ldots$). This is in line with the methods always adopted for dealing with the infinite and the step made here is, after all, not unusual.

This is a function of t given in the function of functions form. But, since the function ϕ included is of variable form, the form of the function f to be integrated is also variable. We assume, further, that the boundary conditions are such that the variable curve $x=\phi(t)$ passes through two fixed points $A(t_0, x_0)$ and $B(t_1, x_1)$. The problem to be solved is thus :

The extreme values of the integral $u=\int_{t_0}^{t_1} f\left(t, x, \frac{dx}{dt}\right) dt$ are required for all possible variations in the function $x=\phi(t)$, such that $\phi(t_0)=x_0$ and $\phi(t_1)=x_1$ where $(t_0\ x_0)$ and (t_1, x_1) are fixed points.

The problems instanced in 20.2 above are both of this form. In the first problem of the surface of revolution, we have

$$f\left(t, x, \frac{dx}{dt}\right)=x\sqrt{1+\left(\frac{dx}{dt}\right)^2}$$

and in the problem of John Bernouilli

$$f\left(t, x, \frac{dx}{dt}\right)=\sqrt{\frac{1}{x}\left\{1+\left(\frac{dx}{dt}\right)^2\right\}}.$$

In solving the problem, we impose the limitations on the variation of $\phi(t)$ already described, taking this function in the form

$$x=\phi(t\,;\ \alpha, \beta, \gamma, \ldots)$$

where ϕ is a fixed function (with a continuous derivative) and where $\alpha, \beta, \gamma, \ldots$ are parameters. Allotting arbitrary differential increments $\delta\alpha$, $\delta\beta$, $\delta\gamma$, ... to the parameters, we derive first the corresponding variations δx and $\delta x'$ in the function x and its derivative $x'=\frac{dx}{dt}$:

$$\delta x=\frac{\partial x}{\partial\alpha}\delta\alpha+\frac{\partial x}{\partial\beta}\delta\beta+\frac{\partial x}{\partial\gamma}\delta\gamma+\ldots$$

and
$$\delta x'=\delta\left(\frac{dx}{dt}\right)=\frac{\partial}{\partial\alpha}\left(\frac{dx}{dt}\right)\delta\alpha+\frac{\partial}{\partial\beta}\left(\frac{dx}{dt}\right)\delta\beta+\frac{\partial}{\partial\gamma}\left(\frac{dx}{dt}\right)\delta\gamma+\ldots$$
$$=\frac{d}{dt}\left(\frac{\partial x}{\partial\alpha}\right)\delta\alpha+\frac{d}{dt}\left(\frac{\partial x}{\partial\beta}\right)\delta\beta+\frac{d}{dt}\left(\frac{\partial x}{\partial\gamma}\right)\delta\gamma+\ldots$$
$$=\frac{d}{dt}\left(\frac{\partial x}{\partial\alpha}\delta\alpha+\frac{\partial x}{\partial\beta}\delta\beta+\frac{\partial x}{\partial\gamma}\delta\gamma+\ldots\right)=\frac{d}{dt}(\delta x).$$

It should be remarked that all the variations here are ordinary differentials and subject to the usual rules of differentiation. They are denoted by the symbol " δ " in order to distinguish them from the other differentials, dx and dt, we obtain when we consider x as a function of t, the values of the parameters being given. Hence, throughout our analysis here, " d " refers to variation in the variable t and " δ " to variation in the parameters $\alpha, \beta, \gamma, \ldots$.

The function $f(t, x, x')$ and the integral u can now be considered as dependent on the parameters $\alpha, \beta, \gamma \ldots$ and the variations in their values are obtained as

$$\delta f = \frac{\partial f}{\partial x}\delta x + \frac{\partial f}{\partial x'}\delta x' = \frac{\partial f}{\partial x}\delta x + \frac{\partial f}{\partial x'}\frac{d}{dt}(\delta x)$$

and

$$\delta u = \delta\left\{\int_{t_0}^{t_1} f(t, x, x')\,dt\right\} = \int_{t_0}^{t_1}(\delta f)\,dt = \int_{t_0}^{t_1}\left(\frac{\partial f}{\partial x}\delta x\right)dt + \int_{t_0}^{t_1}\left\{\frac{\partial f}{\partial x'}\frac{d}{dt}(\delta x)\right\}dt.$$

Now, using the result of 15.3 above,

$$\frac{d}{dt}\left[\frac{\partial f}{\partial x'}\delta x - \int\left\{\frac{d}{dt}\left(\frac{\partial f}{\partial x'}\right)\delta x\right\}dt\right] = \frac{d}{dt}\left(\frac{\partial f}{\partial x'}\delta x\right) - \frac{d}{dt}\left(\frac{\partial f}{\partial x'}\right)\delta x$$

$$= \frac{d}{dt}\left(\frac{\partial f}{\partial x'}\right)\delta x + \frac{\partial f}{\partial x'}\frac{d}{dt}(\delta x) - \frac{d}{dt}\left(\frac{\partial f}{\partial x'}\right)\delta x = \frac{\partial f}{\partial x'}\frac{d}{dt}(\delta x).$$

Hence, except for the addition of an arbitrary constant,

$$\int\left\{\frac{\partial f}{\partial x'}\frac{d}{dt}(\delta x)\right\}dt = \frac{\partial f}{\partial x'}\delta x - \int\left\{\frac{d}{dt}\left(\frac{\partial f}{\partial x'}\right)\delta x\right\}dt.$$

So
$$\int_{t_0}^{t_1}\left\{\frac{\partial f}{\partial x'}\frac{d}{dt}(\delta x)\right\}dt = \left[\frac{\partial f}{\partial x'}\delta x\right]_{t_0}^{t_1} - \int_{t_0}^{t_1}\left\{\frac{d}{dt}\left(\frac{\partial f}{\partial x'}\right)\delta x\right\}dt.$$

The expression for the variation in u then becomes

$$\delta u = \left[\frac{\partial f}{\partial x'}\delta x\right]_{t_0}^{t_1} + \int_{t_0}^{t_1}\left\{\frac{\partial f}{\partial x} - \frac{d}{dt}\left(\frac{\partial f}{\partial x'}\right)\right\}\delta x\,dt.$$

Since, by the boundary conditions, the curve $x = \phi(t)$ always passes through two fixed points at $t = t_0$ and $t = t_1$, it follows that the variation of x is zero, i.e. $\delta x = 0$, at these points. Hence,

$$\left[\frac{\partial f}{\partial x}\delta x\right]_{t_0}^{t_1} = 0,$$

and so
$$\delta u = \int_{t_0}^{t_1}\left\{\frac{\partial f}{\partial x} - \frac{d}{dt}\left(\frac{\partial f}{\partial x'}\right)\right\}\delta x\,dt.$$

The integral u is to have an extreme value for variation in the function $x = \phi(t)$ as obtained by varying the parameters $\alpha, \beta, \gamma, \ldots$. The necessary condition for this is that $\delta u = 0$ for all values of

$\delta\alpha$, $\delta\beta$, $\delta\gamma$, ... , i.e. for all values of δx. From the above expression for δu, this is only true if

$$\frac{\partial f}{\partial x}=\frac{d}{dt}\left(\frac{\partial f}{\partial x'}\right).$$

This result, known as *Euler's equation*, determines that function $x=\phi(t)$ which maximises or minimises the value of u. Since $\frac{\partial f}{\partial x}$ and $\frac{\partial f}{\partial x'}$ are partial derivatives of $f(t, x, x')$ and so functions of t, x and $x'=\frac{dx}{dt}$, the equation is a relation in x as a function of t which involves the derivative of x with respect to t, i.e. it is a differential equation which must be solved to give the function $x=\phi(t)$ we seek. Our problem in the calculus of variations is thus reduced to the relatively simple problem of integrating a differential equation. The methods for completing the solution are set out in Chapter XVI above. Euler's equation, however, is only a necessary condition for extreme values of u. There remains the problem of distinguishing between the different extreme values. The number of variables (i.e. the parameters α, β, γ, ...) is here very large and, as we have seen in the previous chapter, a general criterion for separating maximum from minimum values is not readily obtainable. We must content ourselves with the fact that, in simple practical cases, we can usually tell from general reasoning whether we have a maximum or a minimum value of u.

20.5 Special cases of Euler's equation.

Two particularly simple cases of the problem of the previous section often arise in practice :

(1) To find extreme values of $u=\int_{t_0}^{t_1} f\left(t, \frac{dx}{dt}\right) dt$, where the function to be integrated does not involve x explicitly.

Writing $x'=\frac{dx}{dt}$ and noting that $\frac{\partial f}{\partial x}=0$, Euler's equation becomes

$$\frac{d}{dt}\left(\frac{\partial f}{\partial x'}\right)=0,$$

giving $$\frac{\partial}{\partial x'}f(t, x')=a,$$

where a is some constant. This is a differential equation involving only the first derivative x', i.e. a simple differential equation of the first order which can be solved, in most cases, without difficulty.

(2) To find extreme values of $u=\int_{t_0}^{t_1} f\left(x, \frac{dx}{dt}\right) dt$, where the function to be integrated does not contain t explicitly.

Writing $x'=\frac{dx}{dt}$, differentiating $f(x, x')$ as a function of two variables each depending on t, and using Euler's equation $\frac{\partial f}{\partial x}=\frac{d}{dt}\left(\frac{\partial f}{\partial x'}\right)$:

$$\frac{df}{dt}=\frac{\partial f}{\partial x}\frac{dx}{dt}+\frac{\partial f}{\partial x'}\frac{dx'}{dt}=x'\frac{d}{dt}\left(\frac{\partial f}{\partial x'}\right)+\frac{\partial f}{\partial x'}\frac{dx'}{dt}=\frac{d}{dt}\left(x'\frac{\partial f}{\partial x'}\right).$$

Hence, f and $x'\frac{\partial f}{\partial x'}$, having equal derivatives with respect to t, can only differ by an arbitrary constant, i.e.

$$f(x, x')=x'\frac{\partial}{\partial x'}f(x, x')+a$$

where a is some constant. Again we have a differential equation of the first order, involving only x and its derivative x'. The solution, giving the function $x=\phi(t)$ we require, proceeds with little difficulty. It is to be noticed that the two problems of 20.2 above are both of this special form.

20.6 Examples of solution by Euler's equation.

The methods of solution developed here can be illustrated by the following three examples:

Ex. 1. To find extreme values of $u=\int_{t_0}^{t_1} t\left(\frac{dx}{dt}\right)^2 dt$ subject to the boundary conditions that the curve $x=\phi(t)$ passes through two fixed points $A(t_0, x_0)$ and $B(t_1, x_1)$.

We have the special case (1) of 20.5 with

$$f(t, x')=tx'^2 \quad \text{and} \quad \frac{\partial}{\partial x'}f(t, x')=2tx'.$$

Hence, $$2tx'=\text{constant}, \quad \text{i.e.} \quad \frac{dx}{dt}=\frac{a}{t}$$

where a is some constant. The integral is obtained at once as

$$x=\int\frac{a}{t}dt+\text{constant}, \quad \text{i.e.} \quad x=a\log t+b$$

where b is a second constant. The extreme value of u is thus given by a logarithmic curve.

The appropriate values of the constants a and b are to be found by using the boundary conditions. Since the curve must pass through the fixed points A and B, we have

$$x_0 = a \log t_0 + b \quad \text{and} \quad x_1 = a \log t_1 + b.$$

On solving these linear equations, the constants are given as

$$a = \frac{x_1 - x_0}{\log t_1 - \log t_0} \quad \text{and} \quad b = \frac{x_0 \log t_1 - x_1 \log t_0}{\log t_1 - \log t_0}.$$

In this way we fix that particular logarithmic curve which passes through A and B. The use of the boundary conditions in determining the relevant values of the constants of the solution is to be noticed.

Finally, the extreme value of u determined is

$$u = \int_{t_0}^{t_1} t\left(\frac{dx}{dt}\right)^2 dt = a^2 \int_{t_0}^{t_1} \frac{dt}{t} = a^2(\log t_1 - \log t_0) = \frac{(x_1 - x_0)^2}{\log t_1 - \log t_0}.$$

It is not possible to indicate here whether this is a maximum or a minimum value.

Ex. 2. To find the minimum value of $u = \int_{t_0}^{t_1} x\sqrt{1 + \left(\frac{dx}{dt}\right)^2}\, dt$ where the curve $x = \phi(t)$ is subject to the usual boundary conditions.

We have here the special case (2) of 20.5 with

$$f(x, x') = x\sqrt{1 + x'^2} \quad \text{and} \quad \frac{\partial}{\partial x'} f(x, x') = \frac{xx'}{\sqrt{1 + x'^2}}$$

and Euler's equation is

$$x\sqrt{1 + x'^2} = \frac{xx'^2}{\sqrt{1 + x'^2}} + a.$$

On multiplying up, squaring and collecting terms, we find

$$x'^2 = \frac{x^2 - a^2}{a^2} \quad \text{i.e.} \quad \frac{dx}{dt} = \frac{\sqrt{x^2 - a^2}}{a}.$$

Hence,
$$\int \frac{dx}{\sqrt{x^2 - a^2}} = \int \frac{dt}{a} + \text{constant} = \frac{t - b}{a}$$

where both a and b are constants. It is easily verified that

$$\frac{d}{dx}\left\{\log \frac{x + \sqrt{x^2 - a^2}}{a}\right\} = \frac{1}{\sqrt{x^2 - a^2}}$$

and the solution of our differential equation becomes

$$\log \frac{x + \sqrt{x^2 - a^2}}{a} = \frac{t - b}{a}, \quad \text{i.e.} \quad x + \sqrt{x^2 - a^2} = ae^{\frac{t-b}{a}}.$$

Now,

$$x-\sqrt{x^2-a^2}=\frac{(x-\sqrt{x^2-a^2})(x+\sqrt{x^2-a^2})}{x+\sqrt{x^2-a^2}}=\frac{x^2-(x^2-a^2)}{x+\sqrt{x^2-a^2}}=\frac{a^2}{ae^{\frac{t-b}{a}}}=ae^{-\frac{t-b}{a}}.$$

Adding, $$x=\tfrac{1}{2}a\left(e^{\frac{t-b}{a}}+e^{-\frac{t-b}{a}}\right)$$

which is the function giving the minimum value of u. The boundary conditions provide the values of a and b in terms of (t_0, x_0) and (t_1, x_1):

$$x_0=\tfrac{1}{2}a\left(e^{\frac{t_0-b}{a}}+e^{-\frac{t_0-b}{a}}\right)\quad\text{and}\quad x_1=\tfrac{1}{2}a\left(e^{\frac{t_1-b}{a}}+e^{-\frac{t_1-b}{a}}\right).$$

The minimised value of u is then given by

$$u=\int_{t_0}^{t_1}x\sqrt{1+\left(\frac{dx}{dt}\right)^2}\,dt\quad\text{where}\quad\frac{dx}{dt}=\frac{\sqrt{x^2-a^2}}{a},$$

i.e. $$u=\int_{t_0}^{t_1}\frac{x^2}{a}\,dt=\frac{1}{4}\,a\int_{t_0}^{t_1}\left\{e^{\frac{2(t-b)}{a}}+e^{-\frac{2(t-b)}{a}}+2\right\}dt.$$

This integral can be evaluated fairly easily.

The curves corresponding to the function type $x=\frac{1}{2}a\left(e^{\frac{t-b}{a}}+e^{-\frac{t-b}{a}}\right)$ are called "catenaries". The curve which gives the minimum value of u is thus that catenary which passes through the two fixed points A and B given by the boundary conditions. The shape of this curve is shown in Fig. 105. The solution of the problem of 20.2, Ex. 1, is now obtained; the catenary is the curve which gives a surface of revolution of smallest area. It is to be noticed that the nature of this geometrical interpretation of the problem shows that the value of u we have found is a minimum and not a maximum.

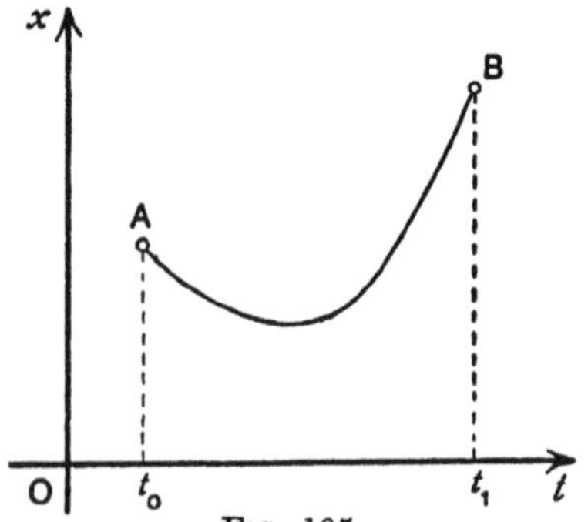

FIG. 105.

Ex. 3. To find the minimum value of $u=\int_0^{t_1}\sqrt{\frac{1}{x}\left\{1+\left(\frac{dx}{dt}\right)^2\right\}}\,dt$ subject to the usual boundary conditions.

This is again the special case (2) of 20.5 with

$$f(x, x')=\sqrt{\frac{1+x'^2}{x}}\quad\text{and}\quad\frac{\partial}{\partial x'}f(x, x')=\frac{x'}{\sqrt{x(1+x'^2)}}$$

and Euler's equation is

$$\sqrt{\frac{1+x'^2}{x}}=\frac{x'^2}{\sqrt{x(1+x'^2)}}+a.$$

We find, on multiplying up, squaring and collecting terms,

$$x'^2=\frac{1-a^2x}{a^2x} \quad \text{i.e.} \quad \frac{dx}{dt}=\frac{1}{a}\sqrt{\frac{1-a^2x}{x}}.$$

The integral of this differential equation is

$$\int\sqrt{\frac{x}{1-a^2x}}\,dx=\int\frac{dt}{a}+\text{constant}=\frac{t-b}{a}$$

where both a and b are arbitrary constants. It is now only a matter of evaluating the integral shown. This can be done only with the aid of "trigonometric" functions and we cannot proceed further here.* It can be stated, however, that the relation between x and t obtained on completing the integration is represented by a curve of a well-known class, the class of "cycloids". The solution of our problem, taking account as before of the boundary conditions, is given by that cycloid which passes through the two fixed points O and A. It is this curve which gives the minimum time of descent from O to A (see the brachistochrone problem of 20.2 above).

20.7 A dynamic problem of monopoly.

A monopolist sells a good X on a market consisting of a competitive group of consumers. If he produces an output x per unit of time, the cost of production is given by $\Pi(x)$ and remains unchanged throughout the period from $t=t_0$ to $t=t_1$ which we consider. It is assumed that the market price of X varies continuously over the period and is represented by some function $p(t)$. The demand of the competitive group of consumers is assumed to involve a "speculative" element and to be described by the demand law given in 16.7 above:

$$x=\phi\{p(t),\ p'(t)\}.$$

The market demand thus varies over the given period according to the course taken by the price of X. The monopolist's profit per unit of time when he produces an output x per unit of time and sells it at the price $p(t)$ is given by

$$xp(t)-\Pi(x).$$

This is the *rate of profit* at any time t and depends on the price and

* The solution of the problem can be written:

$$t=b+a'(\theta-\sin\theta),\quad x=a'(1-\cos\theta),$$

where a' is an arbitrary constant obtained from a. The required relation between x and t is obtained by eliminating the parameter θ. The expressions $\sin\theta$ and $\cos\theta$ are trigonometric ratios of the angle θ radians.

the rate of change of the price at this time. The *total profit* in the given period from $t=t_0$ to $t=t_1$ is obtained by the addition of the profits in successive units of time. Since the variation of price and output is assumed to be continuous, the addition can be represented by an integral, i.e. total profit is

$$u=\int_{t_0}^{t_1}\{xp-\Pi(x)\}\,dt.$$

For a given course of price over time, the demand x of the market is determined at each moment and there is a definite value of u corresponding. Different values of u are obtained for different courses of price. Hence, the monopolist's total profit depends on the form of the price function over time, on the course taken by price. It is assumed, finally, that the ruling price at the initial moment $t=t_0$ is given as p_0 and that the subsequent course of price is fixed by the monopolist so as to maximise his total profit in the whole period from $t=t_0$ to $t=t_1$. It is required to determine the course of price, i.e. the form of the price function $p(t)$, actually fixed by the monopolist. The corresponding output at successive moments is then derived from the given demand law.*

As a first problem, it is taken that the choice of the monopolist is restricted by the fact that the final price (p_1 at $t=t_1$) is given in addition to the initial price. (This artificial assumption will be dropped when it has served its purpose.) We have then a problem in the calculus of variations: to find the function $p(t)$ which maximises the value of $u=\int_{t_0}^{t_1}\{xp-\Pi(x)\}\,dt$, where $x=\phi(p,p')$ and where the boundary conditions give $p=p_0$ when $t=t_0$ and $p=p_1$ when $t=t_1$. This is the special case (2) of 20.5 with

$$f(p,p')=xp-\Pi(x)\quad\text{where}\quad x=\phi(p,p')$$

as the function to be integrated to give u.

* The problem is based on the work of Roos and Evans, see Evans, *Mathematical Introduction to Economics* (1930), pp. 143 *et seq.* Notice that, if prices do not change continuously but are fixed at definite intervals, then the monopolist's total profit is a function of a finite number of prices (one for each of the intervals in which prices are fixed). The problem of maximising total profit is then relatively simple, a problem of the maximum value of a function of several variables. It appears that here, as in many problems of capital theory, the discontinuous form is both more realistic and more simple from the mathematical point of view.

Hence,

$$\frac{\partial}{\partial p'}f(p, p')=p\frac{\partial x}{\partial p'}-\frac{d\Pi}{dx}\frac{\partial x}{\partial p'}=\left(p-\frac{d\Pi}{dx}\right)\frac{\partial x}{\partial p'}$$

and Euler's equation gives

$$xp-\Pi(x)=p'\left(p-\frac{d\Pi}{dx}\right)\frac{\partial x}{\partial p'}+\text{constant.}$$

This is a first-order differential equation in the function $p(t)$, remembering that $x=\phi(p, p')$ is a given function of p and $p'=\frac{dp}{dt}$. The solution of the equation gives the function $p(t)$ required except that it involves two arbitrary constants (the constant above and a constant of integration). The appropriate values of the constants are given by the two boundary conditions fixing the initial and final prices. The problem is thus theoretically determinate and the following particular case shows how an actual solution can be obtained when we know the forms of the demand and cost functions.

Suppose that the demand function is linear, $x=ap+b+cp'$, and the cost function quadratic, $\Pi=\alpha x^2+\beta x+\gamma$, where the six coefficients are all constants. The equation for $p(t)$ is now

$$p(ap+b+cp')-\Pi=cp'\left(p-\frac{d\Pi}{dx}\right)+\text{constant,}$$

i.e.

$$c\frac{d\Pi}{dx}\frac{dp}{dt}+ap^2+bp-\Pi=\text{constant.}$$

Differentiating with respect to t and collecting terms,

$$c^2\frac{d^2\Pi}{dx^2}\frac{d^2p}{dt^2}+ac\frac{d^2\Pi}{dx^2}\frac{dp}{dt}+\left(2ap+b-a\frac{d\Pi}{dx}\right)=0.$$

But, $\frac{d\Pi}{dx}=2\alpha x+\beta=2c\alpha\frac{dp}{dt}+2a\alpha p+(2b\alpha+\beta)$ and $\frac{d^2\Pi}{dx^2}=2\alpha$.

The differential equation for $p(t)$ then reduces to

$$2c^2\alpha\frac{d^2p}{dt^2}-2a(a\alpha-1)p+(b-2ab\alpha-a\beta)=0,$$

i.e.

$$\frac{d^2p}{dt^2}=\lambda^2(p-\bar{p}),$$

where

$$\bar{p}=\frac{b-2ab\alpha-a\beta}{2a(a\alpha-1)}\quad\text{and}\quad\lambda^2=\frac{a(a\alpha-1)}{c^2\alpha}$$

(λ^2 is positive if the constants have the signs appropriate to the normal case, i.e. $a<0$ and $\alpha>0$). The solution of this simple second-order differential equation is known to be

$$p=\bar{p}+Ae^{\lambda t}+Be^{-\lambda t}$$

where A and B are constants of integration. The solution can be checked by differentiation. The constants $\bar{p}$ and λ are given in terms of the coefficients of the demand and cost laws. The constants A and B are to be found in terms of the initial and final prices by substituting $p=p_0$ when $t=t_0$ and $p=p_1$ when $t=t_1$. The problem is solved and the course of prices over time, as determined by the monopolist, is given uniquely.

Returning to the general case, given the final price p_1, we have shown that the price function $p(t)$ and the maximised value of total profits u can be obtained. The total profits (as maximised) must be a function of p_1, say $u=F(p_1)$. To complete the problem, the monopolist has only to fix that final price (of all possible final prices) which maximises u as a function of p_1, i.e. which gives the largest total profits of all. For this, we put

$$\frac{du}{dp_1}=0 \quad \text{subject to} \quad \frac{d^2u}{dp_1^2}<0$$

and determine p_1 from the resulting equation.

20.8 Other problems in the calculus of variations.

The simple problem in the calculus of variations treated above is capable of generalisation in a number of directions.* For example, the function in the integral u may involve second and higher order derivatives of the variable function ϕ, or the variable function may depend, not on one variable t, but on several variables $t_1, t_2, t_3, \ldots$. Further, the extreme value of u may be required subject to a number of given side relations in the variable function, the problem of relative maximum and minimum values. But perhaps the most useful extension of the problem arises when the integral u depends on several variable functions instead of on only one such function. This extension can be illustrated by the case where there are two variable functions, $x=\phi(t)$ and $y=\psi(t)$, and where the integral to

* See Courant, *op. cit.*, pp. 507-20.

be maximised or minimised is of the simple form (analogous to that of 20.4 above):

$$u=\int_{t_0}^{t_1} f\left(t,\, x,\, y,\, \frac{dx}{dt},\, \frac{dy}{dt}\right) dt$$

subject to the boundary conditions that x and y both assume fixed values when $t=t_0$ and $t=t_1$.

It can be shown (following the argument of 20.4) that, when the parameters of the variable functions are changed so that δx and δy are the resulting increments in the functions themselves, then the variation of u is

$$\delta u=\int_{t_0}^{t_1}\left[\left\{\frac{\partial f}{\partial x}-\frac{d}{dt}\left(\frac{\partial f}{\partial x'}\right)\right\}\delta x+\left\{\frac{\partial f}{\partial y}-\frac{d}{dt}\left(\frac{\partial f}{\partial y'}\right)\right\}\delta y\right]dt$$

where $x'=\dfrac{dx}{dt}$ and $y'=\dfrac{dy}{dt}$. The necessary condition for an extreme value of u is that $\delta u=0$ for all variations δx and δy. Hence,

$$\frac{\partial f}{\partial x}=\frac{d}{dt}\left(\frac{\partial f}{\partial x'}\right) \quad \text{and} \quad \frac{\partial f}{\partial y}=\frac{d}{dt}\left(\frac{\partial f}{\partial y'}\right),$$

i.e. there are two differential equations (each similar to Euler's equation above) to be solved for the functions $x=\phi(t)$ and $y=\psi(t)$.

An economic example of the present problem can be taken from a theory of saving developed by Ramsey.* There are two variable factors of production in a community, i.e. labour B and capital C, each of which is taken as homogeneous in nature. If amounts b and c of the factors are used, the total product (or income) of the community is given by the production function: product $=f(b, c)$. The variations of labour and capital over time are assumed to be continuous and represented by the functions $b(t)$ and $c(t)$. The rate of growth of capital, or the saving of the community, is $\dfrac{dc}{dt}$ at any time. The consumption (or expenditure) of the community is then

$$a(t)=f(b, c)-\frac{dc}{dt}. \qquad \text{.........................(1)}$$

It is assumed that the utility of the community's consumption is measurable and dependent only on the amount of the consumption, $u=\phi(a)$. Similarly, the disutility (assumed measurable) of labour

* See Ramsey, *A Mathematical Theory of Saving*, Economic Journal, 1928. A modified version of this theory is given here.

is taken as $v = \psi(b)$. Over a period of time (t_0, t_1), the total net utility of the community is

$$U = \int_{t_0}^{t_1} \{\phi(a) - \psi(b)\}\, dt$$

where future utilities are not discounted. This quantity U is a functional of the variable functions $b(t)$ and $c(t)$, representing the changes in the employment of labour and capital over time. The function $a(t)$ is expressed in terms of $b(t)$ and $c(t)$ by (1) above. It is assumed that the amount of labour and capital employed at the initial time $(t = t_0)$ are given and that the " optimum " amounts to be aimed at $(t = t_1)$ are also known. Subject to these boundary conditions, the community is taken as fixing the amount of work it does and the amount of saving it makes over time to maximise total utility. We have, therefore, to find the functions $b(t)$ and $c(t)$ which maximise the value of U, subject to the usual boundary conditions.

Write $c' = \frac{dc}{dt}$ and

$$U = \int_{t_0}^{t_1} F(b, c, c')\, dt,$$

where $\quad F(b, c, c') = \phi(a) - \psi(b) \quad$ and $\quad a = f(b, c) - c'$.

The conditions for the maximum of U are

$$\frac{\partial F}{\partial b} = \frac{d}{dt}\left(\frac{\partial F}{\partial b'}\right) = 0 \quad \text{and} \quad \frac{\partial F}{\partial c} = \frac{d}{dt}\left(\frac{\partial F}{\partial c'}\right).$$

The first condition gives

$$\frac{\partial F}{\partial b} = \phi'(a)\frac{\partial f}{\partial b} - \psi'(b) = 0,$$

i.e.
$$\frac{\partial f}{\partial b} = \frac{\psi'(b)}{\phi'(a)}. \quad \text{.................................(2)}$$

The second condition gives

$$\phi'(a)\frac{\partial f}{\partial c} = \frac{\partial F}{\partial c} = \frac{d}{dt}\left(\frac{\partial F}{\partial c'}\right) = \frac{d}{dt}\{-\phi'(a)\},$$

i.e.
$$\frac{\partial f}{\partial c} = -\frac{1}{\phi'(a)}\frac{d}{dt}\{\phi'(a)\}. \quad \text{.........................(3)}$$

The functions $a(t)$, $b(t)$ and $c(t)$ are to be found from the differential equations (1), (2) and (3), the boundary conditions being used to evaluate the arbitrary constants that appear on integration. The solution of the problem is determinate in general.

The partial derivatives $\frac{\partial f}{\partial b}$ and $\frac{\partial f}{\partial c}$ of the production function measure the marginal products of labour and capital at any time and can thus be associated with wages and the rate of interest respectively. The rate of wages, by (2), equals the ratio of the marginal disutility of labour to the marginal utility of consumption. The rate of interest, by (3), is equal to the proportional rate of decrease of the marginal utility of consumption over time. Assuming that the rate of interest never becomes negative, the marginal utility of consumption falls over time either until it vanishes or until the rate of interest becomes zero. Further, if $\phi'(a)$ is a decreasing function of a (decreasing marginal utility), then consumption rises over time until its marginal utility or the rate of interest is zero.

An expression for the rate of saving can be found as follows. Using the results (1), (2) and (3), we have

$$\begin{aligned}
\frac{d}{dt}\{\phi'(a)f(b, c)\} &= \phi''(a)\frac{da}{dt}f(b, c) + \phi'(a)\left(\frac{\partial f}{\partial b}\frac{db}{dt} + \frac{\partial f}{\partial c}\frac{dc}{dt}\right) \\
&= \phi''(a)\frac{da}{dt}f(b, c) + \psi'(b)\frac{db}{dt} - \frac{d}{dt}\{\phi'(a)\}\{f(b, c) - a\} \\
&= \phi''(a)\frac{da}{dt}f(b, c) + \psi'(b)\frac{db}{dt} - \phi''(a)\frac{da}{dt}\{f(b, c) - a\} \\
&= a\frac{d}{dt}\{\phi'(a)\} + \psi'(b)\frac{db}{dt}.
\end{aligned}$$

So
$$\begin{aligned}
\phi'(a)f(b, c) &= \int a\frac{d}{dt}\{\phi'(a)\}\,dt + \int \psi'(b)\frac{db}{dt}\,dt + \text{constant} \\
&= a\phi'(a) - \int \phi'(a)\frac{da}{dt}\,dt + \int \psi'(b)\frac{db}{dt}\,dt + \text{constant *} \\
&= a\phi'(a) - \{\phi(a) - \psi(b)\} + \text{constant},
\end{aligned}$$

i.e.
$$\phi'(a)\{f(b, c) - a\} = \text{constant} - \{\phi(a) - \psi(b)\}.$$

* Since $\frac{d}{dt}\{a\phi'(a)\} = a\frac{d}{dt}\{\phi'(a)\} + \phi'(a)\frac{da}{dt}$, we have

$$a\phi'(a) = \int a\frac{d}{dt}\{\phi'(a)\}dt + \int \phi'(a)\frac{da}{dt}\,dt,$$

i.e.
$$\int a\frac{d}{dt}\{\phi'(a)\}dt = a\phi'(a) - \int \phi'(a)\frac{da}{dt}\,dt.$$

Hence, using (1), we find

$$\frac{dc}{dt}=\frac{A-\{\phi(a)-\psi(b)\}}{\phi'(a)}, \qquad \text{.....................(4)}$$

where A is a constant.

Since U is to be a maximum (as opposed to a minimum), the excess of utility over disutility, $\phi(a)-\psi(b)$, must increase over time. Further, it can be taken that capital increases at a decreasing rate until it becomes stationary at the "optimum" at $t=t_1$. So, from (4), the excess of utility over disutility increases and tends to its maximum value A at $t=t_1$. Then, from (4), the rate of saving multiplied by the marginal utility of consumption at any time equals the amount by which the excess of utility over disutility falls short of the optimum amount A. As time goes on, consumption increases, the marginal utility of consumption decreases and the rate of saving decreases until it becomes zero.

EXAMPLES XX

Problems in the calculus of variations

1. Find the curve $x=\phi(t)$ which corresponds to an extreme value of $u=\int_{t_0}^{t_1} t^2\left(\frac{dx}{dt}\right)^2 dt$ and passes through two fixed points at $t=t_0$ and $t=t_1$. Generalise by considering $u=\int_{t_0}^{t_1} t^n\left(\frac{dx}{dt}\right)^2 dt$, where n is a fixed number, on similar lines.

2. Show that $u=\int_{t_0}^{t_1} x^2\left(\frac{dx}{dt}\right)^2 dt$ has an extreme value when the function $x=\phi(t)$ is of the form $x=a\sqrt{t-b}$. How do the boundary conditions that the curve passes through fixed points at $t=t_0$ and $t=t_1$ determine the constants a and b?

3. Show that an exponential curve gives the maximum or minimum values of $u=\int_{t_0}^{t_1} \frac{1}{x^2}\left(\frac{dx}{dt}\right)^2 dt$, subject to the conditions that the curve passes through fixed points at $t=t_0$ and $t=t_1$.

4. If $u=\int_{t_0}^{t_1}\sqrt{x\left\{1+\left(\frac{dx}{dt}\right)^2\right\}}\,dt$, where fixed values of x correspond to $t=t_0$ and $t=t_1$, find the function $x=\phi(t)$ which gives u its extreme values. Show that the corresponding curve is an arc of a parabola.

5. Light, travelling in a plane Oxy, follows a path $y=\phi(x)$ which is determined (by Fermat's principle) so that the time of transit is a minimum. This is equivalent to finding the function ϕ which minimises $u=\int_{x_0}^{x_1}\frac{1}{v}\sqrt{1+\left(\frac{dy}{dx}\right)^2}\,dx$, where v is a function of x and y denoting the velocity of light at various points. Assuming that v is a constant, show that the path of light between two fixed points is a straight line.

6. The velocity of light, in the problem of the previous example, increases in proportion to the height above the zero horizontal line Ox. Put $v=ky$ and show that the path of light between two fixed points is an arc of a circle.

7. If $f(x)$ is a given function and $u=\int_{t_0}^{t_1} f(x)\sqrt{1+\left(\frac{dx}{dt}\right)^2}\,dt$, where x has fixed values at $t=t_0$ and $t=t_1$, show that an extreme value of u is given by the function $x=\phi(t)$ defined by

$$\int \frac{dx}{\sqrt{a\{f(x)\}^2-1}}=t-b$$

where a and b are constants to be determined by the boundary conditions. Show that the previous three examples are particular cases of this general result.

8. A radio manufacturer (see Examples V, 15, and Examples VIII, 33) produces x sets per week at a total cost of £Π, where $\Pi=\frac{1}{25}x^2+3x+100$. The demand of the market is $x=75-3p+275\frac{dp}{dt}$ sets per week at any time when the price is £p per set. The manufacturer is a monopolist fixing the course of price over time so that his total profit is a maximum. The initial price ($t=0$) is £16 per set and it is required to obtain a price of £20 per set after 100 weeks ($t=100$). Show that, at any time t weeks after the initial week, the price per set is

$$p=15{\cdot}179+0{\cdot}171e^{\cdot 0333t}+0{\cdot}650e^{-\cdot 0333t}.$$

Take all figures correct to three decimal places and $\log_{10}e=0{\cdot}4343$.

9. In the problem of the previous example, show that the price falls to a minimum after about 20 weeks. What is this minimum? Draw a graph to illustrate the course of price over time.

10. In the solution of the problem of 20.7, show that

$$A=\frac{(p_1-\bar{p})e^{-\lambda t_0}-(p_0-\bar{p})e^{-\lambda t_1}}{e^{\lambda(t_1-t_0)}-e^{-\lambda(t_1-t_0)}}, \quad B=\frac{(p_0-\bar{p})e^{\lambda t_1}-(p_1-\bar{p})e^{\lambda t_0}}{e^{\lambda(t_1-t_0)}-e^{-\lambda(t_1-t_0)}}.$$

11. Under the conditions of the previous example, show that the price p falls in the period (t_0, t_1) from p_0 to a minimum and then rises to p_1. Verify that the minimum price occurs when

$$t=\frac{1}{2\lambda}\log\frac{B}{A}.$$

12. In the problem in the theory of saving analysed in 20.8, the production function $f(b, c)$ is linear and homogeneous. Show that

$$b\psi'(b)=a\phi'(a)+\frac{d}{dt}\{c\phi'(a)\}$$

and interpret this result. Show, also, that

$$\frac{1}{c}\frac{dc}{dt}=\frac{\partial f}{\partial c}-\frac{a\phi'(a)-b\psi'(b)}{c\phi'(a)}.$$

What is the rate of interest at the optimum time ($t=t_1$) when capital ceases to accumulate?

INDEX

[Numbers refer to pages; numbers in italics refer to examples appended to chapters.]

MATHEMATICAL METHODS

ECONOMIC APPLICATIONS

AUTHORS

PRINTED IN GREAT BRITAIN BY ROBERT MACLEHOSE AND CO. LTD.
THE UNIVERSITY PRESS, GLASGOW